Methods in Molecular Biology

Series Editor
John M. Walker
School of Life and Medical Sciences
University of Hertfordshire
Hatfield, Hertfordshire, AL10 9AB, UK

For further volumes:
http://www.springer.com/series/7651

Gene Therapy for Neurological Disorders

Methods and Protocols

Edited by

Fredric P. Manfredsson

Translational Science and Molecular Medicine, Michigan State University, College of Human Science, Grand Rapids, Michigan, USA

Humana Press

Editor
Fredric P. Manfredsson
Translational Science and Molecular Medicine
Michigan State University, College of Human Science
Grand Rapids, Michigan, USA

ISSN 1064-3745 ISSN 1940-6029 (electronic)
Methods in Molecular Biology
ISBN 978-1-4939-3270-2 ISBN 978-1-4939-3271-9 (eBook)
DOI 10.1007/978-1-4939-3271-9

Library of Congress Control Number: 2015953099

Springer New York Heidelberg Dordrecht London

Printed on acid-free paper

Humana Press is a brand of Springer
Springer Science+Business Media LLC New York is part of Springer Science+Business Media (www.springer.com)

Preface

Gene therapy of the nervous system, a technique once utilized by a few select laboratories, is now a commonplace research tool used around the world. Not only is gene therapy a useful utility in treating and creating preclinical models, but this technology has also demonstrated success in the clinic, in terms of both safety and efficacy [1, 2].

Gene therapy is a valuable tool that is being increasingly utilized to model neurodegenerative disorders [3]. One reason for this is the inherent ability of gene therapy to control genetic expression in both a spatial and temporal manner. For example, using this precision of gene therapy to model neurodegenerative disorders enables researchers to overcome any developmental compensations that may occur with germ line manipulations [4, 5], to create lesions that are restricted to one hemisphere or specific circuits, and to easily titrate the genetic material of interest [6], among other benefits. Of course, these benefits of gene therapy also translate to the use of gene therapy for the delivery of therapeutic genes in preclinical models of neurological disorders [7–11]. That being said, after over 15 years of experience in gene therapy, it has become clear to me that a significant amount of crucial knowledge necessary to design and execute a *successful* gene therapy experiment often fails to be disseminated in a normal format (i.e., via scientific manuscripts). Rather, this esoteric, yet essential knowledge is either briefly mentioned or solely propagated via word of mouth. Therefore, it is all too common that studies involving gene therapy manipulations produce results that vary between investigators (e.g., Ref. 12). Although such discrepancies are not the result of any wrongdoing, their occurrence adds to the "mysticism" sometimes associated with gene therapy and could serve to reduce the enthusiasm for taking on similar projects in the future. Thus, one purpose of this book is to dispel any confusion and provide a clear and detailed road map of how to successfully design and execute a gene therapy experiment in order to obtain consistent results.

As science progresses and new discoveries are made, the boundaries of gene therapy are rapidly expanding: Gene therapy vehicles are continuously undergoing development and are becoming more readily available, delivery methods are continuously being developed, and transgene cassettes are becoming more and more refined. This leaves the researcher with a plethora of decisions that must be considered before undertaking a gene therapy experiment. In this volume I have invited experts from around the world to share their expertise in finite areas of neurological gene therapy. The compilation of protocols and instructive chapters in this book are intended to give researchers, clinicians, and students of all levels a foundation upon which future gene therapy experiments can be designed. When one designs experiments involving gene therapy of the nervous system, several aspects need to be considered before experiments are designed: What delivery vehicle do you use? Will you produce this vector? How will you ensure that your vector retains stability? What expression system best fits your needs? What route will you choose to deliver your gene therapy agent? How will you model the neurodegenerative disorder that you aim to investigate, and what are the proven methods to treat these disorders in preclinical models? This book is aimed to address all these important considerations as well as to disseminate the aforementioned bits of arcane information that are very important to consider during the course of experimentation.

Finally, the penultimate goal for many gene therapists is to see their product eventually end up in the clinic as a treatment for neurological disorders. Although gene therapy has progressed to the clinic, this is not a straightforward path as several variables such as age and disease status have to be considered. Several chapters in this volume will also discuss special considerations that need to be addressed when translating experimental approaches to the clinic.

1. Maguire AM et al. (2009) Age-dependent effects of RPE65 gene therapy for Leber's congenital amaurosis: a phase 1 dose-escalation trial. Lancet 374: 1597–1605
2. Marks WJ Jr. et al. (2010) Gene delivery of AAV2-neurturin for Parkinson's disease: a double-blind, randomised, controlled trial. Lancet. 9: 1164–1172
3. Kirik D et al. (2002) Parkinson-like neurodegeneration induced by targeted overexpression of alpha-synuclein in the nigrostriatal system. J Neurosci 22: 2780–2791
4. Gorbatyuk OS et al. (2010) In vivo RNAi-mediated alpha-synuclein silencing induces nigrostriatal degeneration. Mol Ther18: 1450–1457
5. Kanaan NM, Manfredsson FP (2012) Loss of functional alpha-synuclein: a toxic event in Parkinson's disease? J Parkinsons Dis 2:249–267
6. Manfredsson FP et al. (2009) Tight Long-term dynamic doxycycline responsive nigrostriatal GDNF using a single rAAV vector. Mol Ther 17: 1857–1867
7. McBride JL et al. (2003) Structural and functional neuroprotection in a rat model of Huntington's disease by viral gene transfer of GDNF. Exp Neurol 181: 213–223
8. Gombash SE et al. (2012) Striatal pleiotrophin overexpression provides functional and morphological neuroprotection in the 6-hydroxydopamine model. Mol Ther 20: 544–554
9. Carty NC et al. (2008) Adeno-associated viral (AAV) serotype 5 vector mediated gene delivery of endothelin-converting enzyme reduces Abeta deposits in APP + PS1 transgenic mice. Mol Ther 16: 1580–1586
10. Azzouz M et al. (2004) VEGF delivery with retrogradely transported lentivector prolongs survival in a mouse ALS model. Nature 429: 413–417
11. King GD et al. (2008) High-capacity adenovirus vector-mediated anti-glioma gene therapy in the presence of systemic antiadenovirus immunity. J Virol 82: 4680–4684
12. Klein RL et al. (2006) Efficient neuronal gene transfer with AAV8 leads to neurotoxic levels of tau or green fluorescent proteins. Mol Ther 13: 517–527

Grand Rapids, MI, USA *Fredric P. Manfredsson*

Contents

Contributors

Brenda Auffinger • *The University of Chicago Medicine, Chicago, IL, USA*

Mariela Moreno Ayala • *Instituto de Investigaciones Biomédicas (CONICET), Facultad de Medicina, Universidad de Buenos Aires, Buenos Aires, Argentina*

Gregory J. Baker • *Department of Neurosurgery, The University of Michigan School of Medicine, Ann Arbor, MI, USA; Department of Cell and Developmental Biology, The University of Michigan School of Medicine, Ann Arbor, MI, USA*

Krystof S. Bankiewicz • *Department of Neurological Surgery, University of California San Francisco, San Francisco, CA, USA*

Matthew J. Benskey • *Department of Translational Science & Molecular Medicine, Michigan State University, Grand Rapids, MI, USA*

Tomas Björklund • *Molecular Neuromodulation, Wallenberg Neuroscience Center, Lund University, Lund, Sweden*

Nicholas Boulis • *Department of Neurosurgery, Emory University, Atlanta, GA, USA*

Ludivine Breger • *Department of Experimental Medical Sciences, CNS Gene Therapy Unit, Wallenberg Neuroscience Center, Lund University, Lund, Sweden*

John Bringas • *Department of Neurological Surgery, University of California San Francisco, San Francisco, CA, USA*

Corinna Burger • *Department of Neurology, Medical Sciences Center, University of Wisconsin-Madison, Madison, WI, USA*

Marianela Candolfi • *Instituto de Investigaciones Biomédicas (CONICET), Facultad de Medicina, Universidad de Buenos Aires, Buenos Aires, Argentina*

Michael J. Castle • *Research Institute of the Children's Hospital of Philadelphia, Philadelphia, PA, USA; Department of Neurosciences, University of California-San Diego, La Jolla, CA, USA*

Maria G. Castro • *Department of Neurology, The University of Michigan School of Medicine, Ann Arbor, MI, USA; Department of Cell and Developmental Biology, The University of Michigan School of Medicine, Ann Arbor, MI, USA*

Randall L. Clark • *Division of Neuroscience, Oregon National Primate Research Center, Beaverton, OR, USA*

Benjamin Combs • *Department of Translational Science and Molecular Medicine, College of Human Medicine, Michigan State University, Grand Rapids, MI, USA*

Xue-Feng Ding • *Beijing Institute of Basic Medical Sciences, Beijing, China*

Anthony Donsante • *Department of Neurosurgery, Emory University, Atlanta, GA, USA*

Megan F. Duffy • *Department of Translational Science & Molecular Medicine, Michigan State University, Grand Rapids, MI, USA; Neuroscience Graduate Program, Michigan State University, Grand Rapids, MI, USA*

Marta Dzaman • *Department of Neurology, The University of Michigan School of Medicine, Ann Arbor, MI, USA; Department of Cell and Developmental Biology, The University of Michigan School of Medicine, Ann Arbor, MI, USA*

MING FAN • *Beijing Institute of Basic Medical Sciences, Beijing, China*
GREGORY FISCHER • *Department of Anesthesiology, Medical College of Wisconsin, Milwaukee, WI, USA*
D. LUKE FISCHER • *Department of Translational Science & Molecular Medicine, Michigan State University, Grand Rapids, MI, USA; MD/PhD Program, Michigan State University, Grand Rapids, MI, USA; Neuroscience Graduate Program, Michigan State University, Grand Rapids, MI, USA*
KEVIN D. FOUST • *Department of Neuroscience, Ohio State University, Columbus, OH, USA*
GUANGPING GAO • *University of Massachusetts Medical School, Worcester, MA, USA*
DOMINIC J. GESSLER • *University of Massachusetts Medical School, Worcester, MA, USA*
SARA E. GOMBASH • *Department of Translational Science & Molecular Medicine, Michigan State University, Grand Rapids, MI, USA; Neuroscience Graduate Program, University of Cincinnati, Cincinnati, OH, USA*
MARCIA N. GORDON • *Molecular Pharmacology and Physiology Department, Byrd Alzheimer Institute, University of South Florida, Tampa, FL, USA*
STACY D. GRUNKE • *Department of Neuroscience, Baylor College of Medicine, Houston, TX, USA*
BRADLEY HAMMOND • *Michigan State University, East Lansing, MI, USA*
JAMES G. HECKER • *Department of Anesthesiology and Pain Medicine, Harborview Medical Center, University of Washington School of Medicine, Seattle, WA, USA*
MATTHEW L. HIRSCH • *Gene Therapy Center, University of North Carolina, Chapel Hill, NC, USA; Department of Ophthalmology, University of North Carolina, Chapel Hill, NC, USA*
QUINN H. HOGAN • *Department of Anesthesiology, Medical College of Wisconsin, Milwaukee, WI, USA*
KEN-ICHI INOUE • *Systems Neuroscience Section, Primate Research Institute, Kyoto University, Inuyama, Japan*
JOANNA L. JANKOWSKY • *Department of Neuroscience, Baylor College of Medicine, Houston, TX, USA; Department of Neurology and Neurosurgery, Huffington Center on Aging, Baylor College of Medicine, Houston, TX, USA*
NEHA KAMRAN • *Department of Neurosurgery, The University of Michigan School of Medicine, Ann Arbor, MI, USA; Department of Cell and Developmental Biology, The University of Michigan School of Medicine, Ann Arbor, MI, USA*
NICHOLAS M. KANAAN • *Department of Translational Science and Molecular Medicine, College of Human Medicine, Michigan State University, Grand Rapids, MI, USA; Neuroscience Program, Michigan State University, Grand Rapids, MI, USA*
J. ROBERT KANE • *The University of Chicago Medicine, Chicago, IL, USA*
SHIGEKI KATO • *Department of Molecular Genetics, Institute of Biomedical Sciences, Fukushima Medical University School of Medicine, Fukushima, Japan*
CHRISTOPHER J. KEMP • *Department of Translational Science & Molecular Medicine, Michigan State University, Grand Rapids, MI, USA*
JULIUS W. KIM • *The University of Chicago Medicine, Chicago, IL, USA*
JI-YOEN KIM • *Department of Neuroscience, Baylor College of Medicine, Houston, TX, USA*
ANDREW KNEYNSBERG • *Department of Translational Science and Molecular Medicine, College of Human Medicine, Michigan State University, Grand Rapids, MI, USA; Neuroscience Program, Michigan State University, Grand Rapids, MI, USA*

KENTA KOBAYASHI • *Section of Viral Vector Development, National Institute of Physiological Sciences, Okazaki, Japan*
KAZUTO KOBAYASHI • *Department of Molecular Genetics, Institute of Biomedical Sciences, Fukushima Medical University School of Medicine, Fukushima, Japan*
JEFFREY H. KORDOWER • *Department of Neurological Sciences, Rush University Medical Center, Chicago, IL, USA*
DAVID L. KREULEN • *Department of Physiology, Michigan State University, East Lansing, MI, USA*
SEBASTIAN KÜGLER • *Department of Neurology, Center Nanoscale Microscopy and Physiology of the Brain (CNMPB), University Medicine Göttingen, Göttingen, Germany*
MACIEJ S. LESNIAK • *The University of Chicago Medicine, Chicago, IL, USA*
PEDRO R. LOWENSTEIN • *Department of Neurology, The University of Michigan School of Medicine, Ann Arbor, MI, USA; Department of Cell and Developmental Biology, The University of Michigan School of Medicine, Ann Arbor, MI, USA*
CECILIA LUNDBERG • *Department of Experimental Medical Sciences, CNS Gene Therapy Unit, Wallenberg Neuroscience Center, Lund University, Lund, Sweden*
FREDRIC P. MANFREDSSON • *Translational Science and Molecular Medicine, Michigan State University, College of Human Science, Grand Rapids, MI, USA*
DAMIEN MARSIC • *Department of Pediatrics, College of Medicine, University of Florida, Gainesville, FL, USA*
JODI L. MCBRIDE • *Division of Neuroscience, Oregon National Primate Research Center, Beaverton, OR, USA; Department of Behavioral Neuroscience, Oregon Health and Science University, Portland, OR, USA; Department of Neurology, Oregon Health and Science University, Portland, OR, USA*
ZACHARY T. MCEACHIN • *Department of Neurosurgery, Emory University, Atlanta, GA, USA*
RAMIN A. MORSHED • *The University of Chicago Medicine, Chicago, IL, USA*
CHRISTIAN MUELLER • *UMASS Medical School, Neurology & Gene Therapy Center, Worcester, MA, USA*
KEVIN R. NASH • *Molecular Pharmacology and Physiology Department, Byrd Alzheimer Institute, University of South Florida, Tampa, FL, USA*
NICOLE K. POLINSKI • *Department of Translational Science & Molecular Medicine, Michigan State University, Grand Rapids, MI, USA; Neuroscience Graduate Program, Michigan State University, Grand Rapids, MI, USA*
JIAN QIAO • *The University of Chicago Medicine, Chicago, IL, USA*
LUIS QUINTINO • *Department of Experimental Medical Sciences, CNS Gene Therapy Unit, Wallenberg Neuroscience Center, Lund University, Lund, Sweden*
ERNESTO A. SALEGIO • *Department of Neurological Surgery, University of California San Francisco, San Francisco, CA, USA; Laboratory for CNS Repair, Brain and Spinal Cord Injury Center, Department of Neurological Surgery, University of California San Francisco, San Francisco, CA, USA*
R.J. SAMULSKI • *Gene Therapy Center, University of North Carolina, Chapel Hill, NC, USA; Department of Pharmacology, University of North Carolina, Chapel Hill, NC, USA*
JASON G. SMITH • *Department of Microbiology, University of Washington, Seattle, WA, USA*
CARYL E. SORTWELL • *Department of Translational Science & Molecular Medicine, Michigan State University, Grand Rapids, MI, USA*

MASATOSHI SUZUKI • *Department of Comparative Biosciences, The Stem Cell and Regenerative Medicine Center, University of Wisconsin-Madison, Madison, WI, USA*
CLIVE N. SVENDSEN • *Board of Governors Regenerative Medicine Institute, Cedars-Sinai Medical Center, Los Angeles, CA, USA*
MASAHIKO TAKADA • *Systems Neuroscience Section, Primate Research Institute, Kyoto University, Inuyama, Japan*
GABRIELA TORO CABRERA • *UMASS Medical School, Neurology & Gene Therapy Center, Worcester, MA, USA*
HEIKKI T. TURUNEN • *Department of Ophthalmology, Schepens Eye Research Institute, Massachusetts Eye and Ear Hospital, Harvard Medical School, Boston, MA, USA*
LUK H. VANDENBERGHE • *Department of Ophthalmology, Schepens Eye Research Institute, Massachusetts Eye and Ear Hospital, Harvard Medical School, Boston, MA, USA*
ERIKA ELGSTRAND WETTERGREN • *Department of Experimental Medical Sciences, CNS Gene Therapy Unit, Wallenberg Neuroscience Center, Lund University, Lund, Sweden*
JOHN H. WOLFE • *Research Institute of the Children's Hospital of Philadelphia, Philadelphia, PA, USA; Department of Pediatrics, Perelman School of Medicine, University of Pennsylvania, Philadelphia, PA, USA; W.F. Goodman Center for Comparative Medical Genetics, School of Veterinary Medicine, University of Pennsylvania, Philadelphia, PA, USA*
SONYA J. WOLF • *Gene Therapy Center, University of North Carolina, Chapel Hill, NC, USA; Department of Ophthalmology, University of North Carolina, Chapel Hill, NC, USA*
DEBORAH YOUNG • *Department of Pharmacology & Clinical Pharmacology, Centre for Brain Research, School of Medical Sciences, University of Auckland, Auckland, New Zealand*
HONGWEI YU • *Department of Anesthesiology, Medical College of Wisconsin, Milwaukee, WI, USA*
SERGEI ZOLOTUKHIN • *Department of Pediatrics, College of Medicine, University of Florida, Gainesville, FL, USA*

Part I

Introduction

Chapter 1

Introduction to Viral Vectors and Other Delivery Methods for Gene Therapy of the Nervous System

Fredric P. Manfredsson

Abstract

The use of gene therapy in neuroscience research has become common place in many laboratories across the world. However, contrary to common belief, the practical application of viral or non-viral gene therapy is not as straightforward as it may seem. All too often investigators see their experiments fail due to low-quality third-party vectors or due to a lack of knowledge regarding the proper use of these tools. For example, researchers often find themselves performing experiments using the wrong methodology (e.g., using the wrong type of vector or mishandling the vector to the point where the efficacy is significantly reduced) resulting in experiments that potentially fail to accurately answer a hypothesis, or the generation of irreproducible data. Thus, it is important for investigators that seek to utilize gene therapy approaches to gain a basic understanding of how to apply this technology. This includes understanding how to appropriately design and execute an experiment, understanding various delivery vehicles (e.g., what virus to use), delivery methods (e.g., systemic versus intracranial injections), what expression system to use, and the time course involved with a particular expression system. This chapter is intended to present an overview of this fundamental knowledge, providing the researcher with a decision tree upon which to build their gene therapy experiment.

Key words Gene therapy, AAV, Adeno-associated virus, Lentivirus, Adenovirus, Electroporation, Plasmid, Recombination, Vector production

1 Delivery Vehicle

The first consideration for a researcher will be the type of vehicle used to deliver the genetic material to the animal. There is a rather extensive selection of vectors available to researchers today; thus, this decision will be based upon a multitude of factors: (1) How big (i.e., how many nucleotide bases) does this genetic material encompass? (2) What is the target structure and what is the target cell (i.e., phenotype)? (3) What is the experiment trying to accomplish? For instance, are you aiming to deliver a therapeutic gene, or target tumor cells for destruction? (4) What laboratory resources

Fredric P. Manfredsson (ed.), *Gene Therapy for Neurological Disorders: Methods and Protocols*, Methods in Molecular Biology, vol. 1382, DOI 10.1007/978-1-4939-3271-9_1,

are available? Do you produce your own vector (viral or non-viral) or do you receive it from a collaborator or vector core? Table 1 summarizes the main characteristics of the vectors described in this chapter.

Table 1
Typical vectors and their main characteristics

Vector	Capacity	Flexibility	Considerations
rAAV			
	4.7-5 kb	Tropism can be changed by pseudotyping and directed evolution.	The capsid may illicit an immune response when delivered peripherally, or with readministration centrally.
		Limited to neurons, astrocytes, and oligodendrocytes.	Genome does not integrate, thus, infection of dividing cells will result in genome dilution.
		Capacity can be enhanced by split genome approaches.	
Retrovirus/lentivirus			
	9 kb	Tropism can be changed by inserting various proteins in viral envelope.	Very easy to produce.
		Virtually all cell types in the nervous system can be transduced.	Integrates into host cell genome, thus, viral genome dilution is not a concern.
			Safety concerns remain.
Adenovirus			
	35 kb	Tropism can be altered by inserting various proteins in capsid.	Concerns with longevity of expression.
		Transduces largely neurons and astrocytes.	Some concerns with natural cytotoxicity.
			Very time consuming to produce.
DNA/RNA			
	Unlimited	Uptake of nucleic acids is not sensitive to a particular cellular phenotype.	Does not utilize viral infection pathways and has to be delivered to target cells (e.g., electroporation or lipid-based transfection).
			Is sensitive to dilution by cell division.

1.1 Virus

Recombinant viral vectors are commonly used for the delivery of genetic material to the nervous system. Viruses have evolved over eons to bypass a host's immune system and deliver its genetic material in to a host cell. Recombinant viral genomes are engineered to take advantage of the virus' ability to infect a host cell while removing the ability of the virus to replicate itself, or to produce damage to the host cell [1]. This book covers the three most commonly used viral vectors: adeno-associated virus (AAV), lentiviruses (LV)/retroviruses, and adenovirus (Ad). Each of these vectors has its distinct advantages and disadvantages that need to be considered when planning an experiment.

1.1.1 AAV

AAV is by far the most commonly utilized gene therapy vector in the nervous system due to its strong transduction profile and its established safety profile both in animals and humans. Recombinant AAV (rAAV) retains only two small genetic elements from the wildtype genome: the two inverted terminal repeats (ITRs), which are required for the packaging of the viral genome in to the viral capsid. Since the discovery of AAV, a wealth of various AAV serotypes have been identified, each differing in the capsid surface. Viral entry to a host cell is mediated by interactions of the viral capsid with cell surface receptors, thus differences in capsid structure between serotypes result in viruses with unique cellular tropism [2, 3]. Therefore, today it is commonplace to package the AAV2 genome (ITRs) in the capsid that best suits the experiment. This is called pseudotyping, and the resultant vector is typically denoted rAAV 2/X, where 2 denotes the genomic origin of the ITRs and the X denotes the genomic origin of the capsid. AAV transduces neurons, astrocytes, and to a lesser extent, microglia. Based on the promoter chosen to express the transgene, expression can be guided to either of these cellular populations. One of the drawbacks of rAAV is the limited transgene capacity of the viral vector. Due to its small size, the AAV capsid can contain roughly ~4.7–5 kb of genetic material, and attempts to package genomes larger than this often result in lower packaging efficiency and reduced titers. Therefore, gene therapy approaches that require transfer of multiple genomes usually utilize viral vectors with greater transgene capacity such as lentivirus. One consideration with the use of rAAV is that the recombinant genome does not integrate. This means that if one aims to target dividing cells of the nervous system, genome dilution due to cell division will occur. In contrast, expression is maintained in nondividing cells (i.e., neurons) for the lifetime of the cell. Another limitation to the use of AAV is that production, or packaging, is quite labor intensive and often requires specialized equipment (*see* Chapter 7).

1.1.2 Lentiviruses and Retroviruses

Retroviruses and lentiviruses (a subclass of the retrovirus family) are characterized by the ability to integrate in to the host-cell genome, and thus these vectors are not subject to genomic dilution with cell division. Although these two viruses are very similar, non-lentiviral retroviruses cannot traverse the nuclear membrane, thus their use is limited to only actively dividing cells. Lentiviruses on the other hand are a specialized subgroup of retroviruses which encode additional genes that facilitate nuclear import, and thus infect dividing and nondividing cells alike. Although the safety profile of the latest generation of retroviruses has been vastly improved, some safety concerns still remain: the generation of replication-competent lentivirus through recombination, mobilization (rescue and replication of the recombinant genome by the superinfection of another lentivirus such as HIV [4]), or insertional mutagenesis (such as the disruption of an essential gene or a tumor suppression gene by the insertion of the retroviral genome [5]). To combat the latter, integration-deficient LV can maintain stable expression in neurons in vivo [6]. However, remaining concerns has not precluded the use of these vectors in clinical trials. Similar to rAAV, modifications can be made to the wild-type virus in order to change its tropism (i.e., pseudotyping). This is accomplished by inserting various surface glycoproteins into the viral envelope. The archetypical vector contains the vesicular stomatitis virus (VSV) G-protein (VSV-G) in the envelope. But more recent studies have demonstrated an improved, and altered, tropism in the nervous system using surface proteins from other viral families such as *Lyssavirus* (rabies) and *Filovirus* (Ebola) [7–10]. The larger size of this family of viruses allows for the inclusion of roughly 9 kb of genetic material, nearly twice the transgene capacity of rAAV. Another distinct advantage to the use of LV and retroviruses in the laboratory is the ease of production, which consists of filtering and centrifugation of the culture media (*see* Chapter 8).

1.1.3 Adenovirus

Early-generation adenoviruses were plagued by cytotoxicity and immunogenicity, limiting the utility of this virus as a gene therapy vector. The latest generation of "gutless" vectors has had much of the viral coding sequence removed, providing for a safer and more efficient gene therapy vehicle. This high-capacity recombinant adenovirus (rAd) has the capacity to carry roughly 35 kb of genetic material making it an ideal vehicle for the delivery of large genes. However, for optimal packaging efficiency this virus requires a certain minimal amount of genetic material in the viral genome. Thus, if the transgene cassette is too small, specific "stuffer" DNA needs to be included in the viral genome [11]. Despite the fact that acute inflammation can be detected following stereotaxic delivery to the brain, transgene expression with gutless Ad persists for over 1 year [12, 13]. In contrast, the use of earlier generation Ad has been favored in the treatment of CNS neoplasms, such as glioblastoma multiforme, due to its natural cytotoxicity (*see* Chapter 31). Similarly, the natural immunogenicity of Ad has made it an ideal vector for vaccination studies. Comparatively, Ad production is relatively time consuming (*see* Chapter 9), restricting the use of this virus to studies and trials where the specific attributes of this vector are needed. Similar to rAAV and LV, recombinant adenoviruses can be pseudotyped by incorporating various attachment proteins on the viral surface resulting in enhanced neuronal tropism [14].

1.2 Non-viral

In contrast to the use of viral vectors, non-viral vectors do not rely upon the evolved capabilities of viruses to insert genetic material in to the target cell. Instead, non-viral gene-therapy relies upon the use of a physical force, or DNA conjugated to branched chemicals, in order to force the cell to take up the exogenous DNA, typically in the form of linear DNA, plasmid DNA, or siRNA. A variety of techniques have been devised to achieve this, including the gene gun, electroporation, or the use of branched molecules such as polyethylenimine [15]. Although this approach is beneficial in that it does not require the complexities involved in viral production, the efficacy and longevity of non-viral gene therapy are often inferior to viral-mediated gene therapy.

1.3 Produce or Purchase

Once a gene delivery vehicle has been selected, the second decision is whether to produce the vector in-house, or to utilize a core service (either academic or industry) to produce the vector. Moreover, if the choice to outsource vector production is made, what type of service is needed: Will you design, clone, grow, and send the viral genome to a core for packaging or will you utilize a "boutique" service where the complete process (genomic design, cloning, and packaging of the vector) is outsourced? This may depend on how frequently viral vectors will be utilized, as setting up production in-house can be both costly and time-consuming. However, once a laboratory has a working production protocol, outsourcing production is often more expensive and time consuming.

1.4 Growing of Genome

If one chooses to clone and grow the viral genome, there is a certain degree of esoteric knowledge that must be understood, as the cloning and handling of each type of viral vector genome carries with it certain nuances that are unique to that virus. For instance, the AAV and lentiviral genomes are contained within standard plasmids, and during the production of the recombinant virus these plasmids are inserted in to the production cell (e.g., HEK 293 cells) together with plasmids encoding helper functions and essential viral components (*see* Chapters 7 and 8). The AAV ITRs are genetic structural elements that are relatively unstable and readily recombine. However, this poses a problem as the ITRs are absolutely required for packaging of viable particles, and the functional titer is directly related to the ratio of plasmids with intact versus recombined ITRs. In order to grow these AAV genome plasmids, one must therefore utilize recombination deficient bacterial cells (e.g., SURE cells from Agilent). Moreover, reducing the growing temperature (from 37 °C to 30 °C) and the time of plasmid growth (from roughly 16 to 12–14 h) further improves the integrity of the population of ITRs. Finally, it is imperative that the integrity of the ITRs is checked prior to packaging using a restriction enzyme that gives a unique patterning following recombination (e.g., sma1 for AAV2 ITRs). The ITRs also create a structural barrier for certain manipulations of the viral genome. For instance, any type of mutagenesis that involves a PCR amplification of the entire bacterial plasmid is unlikely to succeed as the complimentary terminal repeats may cause PCR "skipping." Thus, sub-cloning of the area of interest into a holding plasmid should be done prior to such manipulations. Similarly, the lentiviral LTRs (contained in the transfer plasmid) are also prone to recombination, thus, care should be taken while growing this plasmid in bacteria as well. In addition to the genomic component, when producing lentivirus, one must pay attention to what components must be provided in *trans* during packaging as the *cis* acting elements change with each generation of lentiviruses. For instance, the third generation of lentivirus requires the concomitant transfection of four plasmids during production: one transfer plasmid, two packaging plasmids, and one envelope plasmid.

Many plasmids containing viral genomes, helper and transfer plasmids, envelope plasmids, and various expression cassettes can be found at Addgene (http://www.addgene.org) or the National Gene Vector Repository (https://www.ngvbcc.org).

1.5 Handling and Storage

One of the greatest sources of discrepancies between seemingly identical gene therapy studies lies in how the virus is handled once it is produced. Again, this involves esoteric knowledge that is rarely published, if ever. For instance, both AAV and LV are sensitive to freeze-thawing, where the titer drops with each subsequent freeze cycle. In fact, in our hands we never freeze AAV, as this virus is stable at 4 °C for years. Similarly, following production, LV is immediately aliquoted and stored at −80 °C, but each aliquot is never re-frozen.

Outside a host cell, a viral vector can be defined as a complex protein/molecule. The protein structures that envelope these viruses are incredibly "sticky." Accordingly, viral particles will aggregate in solutions with low ionic strength, or via the adherence of capsids to any contact surfaces. In order to prevent the loss of viral particles (and to avoid re-titering of the vector preparation) we have implemented certain precautions outlined in Chapter 14. Briefly, every surface that comes into contact with the virus is siliconized (e.g., by treating the surface with Sigmacote®

(Sigma-Aldrich)). This reduces the surface charges which otherwise would facilitate virus binding. Alternatively, one can use an identical virus to "coat" surfaces that will come in contact with the vector. Regardless, despite taking all these measures, it is safe to assume that if a virus is transferred (e.g., aliquoted), that some virus will be lost to the procedure itself.

2 Target Cell Population

In designing your gene therapy experiment another important consideration is what population of cells you are aiming to transduce: Do I need to transduce only neurons, or astrocytes, or both? Is off-target transduction acceptable? How much volume of the brain do I need to cover? These questions highlight several variables that must be considered when designing your experiment in order to successfully implement a rational approach that best fits your particular situation: (1) the type and pseudotype of viral vector injected, (2) the expression cassette utilized to express your genetic payload, and (3) delivery method.

2.1 Pseudotype Variation

rAAV exhibits a very strong tropism for neurons, and to a lesser extent astrocytes. However, achieving microglial transduction using rAAV has proven difficult, and LV is a better viral candidate for this purpose. Regardless of whether you are using rAAV, LV, or Ad there are additional strategic variations that may be employed to further modulate the tropism of the viral vectors in order to target different cellular populations with various efficiencies. For instance, in the case of lentivirus, incorporating glycoproteins from lymphocytic choriomeningitis virus or Mokola virus in the envelope will efficiently guide expression specifically to neurons and astrocytes, respectively [7]. Similarly, by incorporating protein domains from canine Ad serotype 2 in to Ad5, the tropism of Ad shifts to neurons [14].

2.2 Expression Cassette

In many instances, ubiquitous expression is not a problem and the choice of expression cassette/promoter is largely dictated by the level of expression desired. However, in other instances researchers may need to guide expression to a phenotypically distinct cell type in the nervous system. In this case, specialized expression cassettes must be utilized.

Alternatives are possible, you to a specific subset of cells in the nervous system is in the choice of a specific promoter. A variety of both neuronal and non-neuronal promoters have been designed and validated in vivo (*see* Chapter 6). Moreover, the Pleiades Promoter Project (http://www.caneucre.org) [16] utilized a bioinformatics approach to design human mini-promoters (less than 4 kb) for a large number of proteins expressed in the brain. Although only a handful of these promoters have been validated in vivo, these promoter constructs are available in plasmid form through Addgene.

An alternate way of controlling expression is through the incorporation of micro RNA (MIR)-binding sites in an untranslated region of your transcript. The recognition of these cleavage sites results in the ablation of the transcript in cells expressing the specific MIR. For instance, the incorporation of MIR-9 (expressed in neurons but not microglia) binding sites effectively silences transcription in neurons [17].

Finally, FLEX vectors contain a loxP-flanked transgene coding sequence that is inverted with respect to the promoter (ON), or in line with the promoter (OFF). When used in combination with animals expressing CRE recombinase under the control of a phenotype-specific promoter, CRE-mediated recombination will “flip” the coding sequence either activating (ON) or disabling (OFF) transcription specifically within cells expressing CRE [18].

2.3 Delivery Method

Delivery methods are covered in more detail in Chapters 14–24. However, to some extent the delivery method can also dictate the type of cell that expresses the genetic payload. Direct stereotaxic injections of either Ad, LV, or rAAV have the benefit of containing the viral vector within a confined volume (with some spread along white matter tracts). Ventricular delivery of rAAV in the neonate transduces large portions of the CNS, but the cell specificity depends on the serotype and the age of the animal during the injection [19]. Similarly, systemic delivery of rAAV transduces either neurons or astrocytes, or both, again depending on the age of the animal and the serotype used. However, systemic delivery of rAAV also transduces a wide range of organ systems outside of the nervous system [20, 21], and care must be taken as this may confound the study.

2.4 Enriching Vector Targeting via Directed Evolution

In some instances, the approaches outlined in this chapter do not result in the effective transduction of certain populations of cells within the nervous system. In an effort to effectively target hard-to-infect cells, a recent approach has been to use standard molecular evolution techniques to generate viral capsid libraries based on the capsids of the various serotypes of AAV [22]. Following the injection of these highly diverse viral populations in to animals, scientists can isolate the cell/or tissue of interest, and from those infected cells/tissue, isolate and sequence the specific capsid mutations that facilitated the transduction. Perhaps one of the best examples of this methodology has been the identification of capsids that effectively transverses the seizure-affected mature blood–brain barrier [23].

3 Fundamental Modes of Gene Therapy

In a most basic sense, gene therapy for neurological disorders can be divided in to different basic theoretical approaches: (1) gene overexpression, (2) gene silencing, and (3) gene editing. Each of these methodologies brings forth certain nuances that are important to consider when engineering your genetic payload.

3.1 Overexpression

Transgene overexpression is a common approach in CNS gene therapy where the intended effect is to (ectopically) overproduce a certain protein. In specific cases this protein may be intended to elicit a therapeutic or symptomatic benefit whereas in other experiments the gene is intended to model a disease by producing a toxic effect. If you aim to deliver a neuroprotective protein you need to consider whether the protective mode of action is intracellular or transcellular. For instance, in Parkinson's disease the overexpression of members of the glial cell line-derived neurotrophic factor (GDNF) family of ligands, such as GDNF and neurturin, are currently being tested clinically for their efficacy to halt neurodegeneration. Preclinical testing using both rAAV and LV has established that these proteins exhibit a strong neurotrophic potential [24]; however, the effect is mediated via transmembrane signaling. Thus, in order to achieve this effect the protein has to be released in to the surrounding area from the transduced cell. This is accomplished by including a signal peptide targeting the protein for release. In this approach, the phenotype of cells directly targeted with the genetic material is not important as long as the correct anatomical location is targeted [25]. In contrast, many approaches require that the transgene is active within specific target cells, this may include both therapeutic and disease modeling experiments. In PD, mutations of the E3 ligase parkin has been attributed to neurodegeneration of cells in the substantia nigra (SN) in familial forms of the disease [26]. Overexpression of this protein, specifically in nigral neurons, produces both a neuroprotective effect and symptomatic benefit [27, 28]. Similarly, the protein alpha-synuclein (α-syn) has been implicated to play a central role in the neurodegeneration of SNc neurons and gene therapy-mediated overexpression of this protein specifically in nigral neurons is a commonly used model of the disease (*see* Chapter 26) [29, 30].

3.2 Silencing

In many cases, neurodegenerative disorders are caused by mutations resulting in a toxic gain of function. For instance, a CAG expansion in the gene encoding the protein huntingtin ultimately gives rise to Huntington's disease. One preclinical gene therapy approach aimed at treating this devastating disorder has been the use of RNA silencing techniques such as short-hairpin RNA (shRNA) or MIR. These approaches have both advantages and disadvantages; however, preclinical studies show beneficial effects on disease progression after reducing the expression of mutant huntingtin [31–33]. An important note of caution with the use of shRNA or MIR-mediated silencing is that of off-target toxicity. This has been observed with the use of both shRNAs and MIRs, in part due to the inhibition of nuclear export due to the overwhelming of the nuclear transporter, exportin-5 [34]. It is thus imperative that any experiment involving shRNAs or MIRs be controlled properly. Common practice is to include a group of subjects receiving a titer-matched scrambled (no cellular mRNA target) shRNA or MIR control. However, this type of control does not provide any target engagement (i.e., full assembly of the RNA-induced silencing complex (RISC) complex) and thus, an ideal control would be to include a shRNA/MIR targeted against an expressed cellular mRNA that when silenced does not produce a phenotype.

Another important consideration when thinking about taking a knockdown approach is the target itself. Proteins such as huntingtin and α-syn may normally provide roles that are crucial to cellular function and survival. Thus, significantly reducing the mRNA, although it may ultimately rescue the disease phenotype, may produce other deleterious effects. For example, high-level knockdown of the protein α-syn, although implicated as a toxic component of Parkinson's disease, results in severe neurodegeneration [35], which has been hypothesized to be due to a disruption in the putative role of α-syn in dopamine handling [30]. One way such concerns may be circumvented is to utilize what is referred to as a "kill and replace" approach. In this instance, the shRNA or MIR is accompanied by a "hardened" copy of the target wild-type cDNA. Hardened refers to the inclusion of several silent mutations which renders the replacement copy insensitive to the shRNA or the MIR.

3.3 Editing

More recently, another means by which to remove (or replace) a toxic gene has been to apply vector based genome editing. Recent advances have yielded engineered nucleases which bind to a specific sequence in the genome and induce a double-stranded break. This activates the cell's DNA repair machinery. Repair by the nonhomologous end joining pathway will ligate the break with the likelihood of a frameshift and the formation of a premature termination codon. If a template is included in the gene therapy, the gene can be repaired by homology-directed repair, although this is a low-efficiency event likely to only occur in cycling cells such as astrocytes and microglia [36]. Of these new nucleases, zinc-finger nucleases (ZFNs) and the clustered regularly interspaced short palindromic repeats (CRISPR) systems are best suited for gene therapy due to the ability to fit these systems in to standard viral vectors [37, 38]. However, transcription activator-like effector nucleases (TALENs) can be utilized using non-viral delivery [39] or Ad.

Although these systems are in their infancy, and specificity and potential off-target effects still needs to be evaluated, genome editing holds great promise, both as research tools and as potential therapeutic modalities.

4 Expression Cassettes

In some instances, standard expression cassettes are not sufficient to produce the desired effect, perhaps the transgene is too large to be contained within a single viral genome, multiple genes are required for a desired effect, or cyclic expression (e.g., regulated expression) is needed.

4.1 Split Genomes

The carrying capacity of rAAV and LV is ~4.7 and 9 kb, respectively, an absolute size limitation based on the intraviral volume. If a larger gene is needed few options are available. Certainly, a viral vector with higher capacity such as adenovirus can be utilized. However, a split-genome approach using rAAV can be utilized (Chapter 2). This approach allows you to split the viral genome in to multiple viral vectors, and upon superinfection, the genome (containing the full coding sequence) gets reconstructed via homologous recombination or nonhomologous end joining.

4.2 Multicistronic Cassettes

In some cases multiple coding sequences are required in order to produce a desired effect. In this case, and with space permitting, multiple cistrons can be contained within the same genome (Chapter 3). Multicistronic cassettes can be engineered in several ways; using separate promoters encoding each sequence, utilizing a single promoter and an internal ribosome entry site (IRES) within the mRNA, or using a coding sequence where the polypeptide sequence contains proteolytic cleavage site, such as the foot-and-mouth 2A peptide [40]. The latter is advantageous since all the proteins are produced at equal ratios, whereas the expression from multiple promoters or IRES sequences does not follow simple stoichiometry.

4.3 Regulated Vectors

A desirable approach is the use of regulatable expression cassettes [41]. These promoters can be based on the classic tet-operon [42], the rapamycin dimerization switch [43], or the inclusion of a destabilizing domain in the peptide amongst others [44]. In all these examples the standard premise is the same: A systemically delivered agent (e.g., dietary doxycycline) interacts with either vector-encoded transcriptional elements (tet-system or rapamycin system) or the peptide itself (destabilized peptide system) and controls transgene expression or protein stability, respectively. Moreover, the dose of this agent dictates the amount of gene product produced [25]. A similar approach is the use of conditional promoters. For instance, pathophysiological conditions in neurological disorders often include a hypoxia or oxidative stress. Researchers have taken advantage of this situation by incorporating hypoxia-response elements [45] or antioxidant response elements [46] as promoter elements in viral vectors, producing expression only during the presence of these pathological events.

5 Special Considerations in Experimental Design

In the simplest gene therapy experiment, the only control that is required is a vector that contains the same expression cassette, with a reporter such as green fluorescent protein (GFP) replacing the gene of interest. In this case, you are controlling for every aspect of the gene therapy (such as the utilization of cellular machinery to produce high levels of a transgene) and sham surgery, or vehicle, controls are not necessary. However, some caution is urged, especially for long-term experiments involving high levels of expression, for the use of GFP and similar proteins, as toxicity has been reported in some instances [47].

There are, however, decisions that must be made *a priori* when designing certain experiments involving gene therapy. For instance, how will you assess transduction, and how much transduction do you want? Or if you are knocking down a protein (i.e., silencing expression), how will you determine how much you are reducing levels of this protein? The answers to these questions may seem straightforward, but they are not. For instance, if you hypothesize that the gene therapy will result in neurodegeneration, in order to assess transduction you should include experimental groups that show the level of overexpression or knockdown at some time point before the cells die. Similarly, if you are superimposing a therapeutic during the course of this neurodegeneration, at the very least, groups should be included that are sacrificed at the onset of the therapy in order to get a snapshot of transduction at that time. Moreover, many experiments call for the mixing of multiple vectors. This requires the inclusion of several groups that control for the dilution of each individual vector in the mix, as well as for any competition for receptors that may occur between the increased level of viral particles. Table 2 outlines standard experimental groups to consider in gene therapy studies.

Assessing the level of transduction can in many cases be a straightforward process using standard histological (e.g., immunohistochemistry) and/or biochemical methodologies (e.g., quantitative real-time PCR). There are however, situations that warrant extra planning. For instance, if your gene therapy (overexpression or silencing) involves a protein that has a high level of endogenous

Table 2
Examples of standard experimental designs in gene therapy experiments

Experiment	Control groups
Overexpression	
	• Titer-matched vector with reporter gene
Toxic transgene	
	• Titer-matched vector with reporter gene
	• Subset of animals sacrificed prior to toxicity in order to assess transduction
Toxic transgene with therapeutic overlay	
	• Titer-matched vector with reporter gene
	• Subset of animals (both treatment and control) sacrificed at the time of therapeutic onset in order to assess transduction at this time
Concomitant delivery of toxic and therapeutic vectors	
	• Mix of therapeutic and toxic vectors (treatment group)
	• Mix of therapeutic and control vector (control for volume and receptor competition)
	• Mix of toxic vector and control vector (control for volume and receptor competition)
	• Equal volume and total titer control vector (negative control)

expression, how will you assess transduction? In the case of overexpression you may need to utilize the coding sequence from another species (e.g., human) to facilitate transgene detection in the rodent. Alternatively, you can include epitopes (e.g., an HA-tag) to your transgene in order to be able to demonstrate that your transgene is present. If none of these alternatives are possible, you can utilize in situ hybridization probing either the recombinant genome or the specific message [48]. Silencing experiments also pose unique challenges when the targeted gene is ubiquitously expressed, including in non-transduced cells (e.g., microglia). For instance, standard micro-dissections of the injected area will in this case thus include non-transduced cells expressing the targeted gene. In such cases, the incorporation of a reporter cassette (such as GFP) in your vector is crucial. This will facilitate the isolation of transduced cells in order to accurately determine the level of knockdown in vivo.

6 Conclusions

The use of gene therapy to study, model, and treat neurological disorders has become increasingly popular in the last decade. What was previously envisioned as a futuristic and esoteric tool used by only a few research groups around the world, is now one of the more prevalent methods to perform genetic manipulations and target validation in vivo. However, it is often the case that data generated using gene therapy can be difficult to reproduce between groups. This is not because the results are not true. Rather, discrepancies occur because the arcane, but crucial, knowledge utilized by staunch gene therapists is disseminated by the water cooler, and often not published. For instance, the failure to propagate the knowledge about the inherent variation in production of viral vectors between laboratories, the fact that handling the vector the wrong way will have a significant effect on titer, the fact that the utilized pseudotype dictates specificity and efficacy, all play a role in the variation seen, and may sometime serve as to influencing researchers to take a different approach. This volume, and this chapter, is intended to disseminate some of this obscure knowledge, as well as to give the reader insight into some of the more advanced methodology used in gene therapy today.

References

1. Manfredsson FP, Mandel RJ (2010) Development of gene therapy for neurological disorders. Discov Med 9:204–211
2. Benskey MJ et al (2015) Targeted gene delivery to the enteric nervous system using AAV: a comparison across serotypes and capsid mutants. Mol Ther 23:488–500
3. Burger C et al (2004) Recombinant AAV viral vectors pseudotyped with viral capsids from serotypes 1, 2, and 5 display differential efficiency and cell tropism after delivery to different regions of the central nervous system. Mol Ther 10:302–317
4. Evans JT, Garcia JV (2000) Lentivirus vector mobilization and spread by human immunodeficiency virus. Hum Gene Ther 11:2331–2339
5. Hacein-Bey-Abina S et al (2003) LMO2-associated clonal T cell proliferation in two patients after gene therapy for SCID-X1. Science 302:415–419
6. Yanez-Munoz RJ et al (2006) Effective gene therapy with nonintegrating lentiviral vectors. Nat Med 12:348–353
7. Cannon JR, Sew T, Montero L, Burton EA, Greenamyre JT (2011) Pseudotype-dependent lentiviral transduction of astrocytes or neurons in the rat substantia nigra. Exp Neurol 228:41–52
8. Liehl B et al (2007) Simian immunodeficiency virus vector pseudotypes differ in transduction efficiency and target cell specificity in brain. Gene Ther 14:1330–1343
9. Watson DJ, Kobinger GP, Passini MA, Wilson JM, Wolfe JH (2002) Targeted transduction patterns in the mouse brain by lentivirus vectors pseudotyped with VSV, Ebola, Mokola, LCMV, or MuLV envelope proteins. Mol Ther 5:528–537
10. Manfredsson FP, Mandel RJ (2011) The development of flexible lentiviral vectors for gene transfer in the CNS. Exp Neurol 229:201–206
11. Alba R, Bosch A, Chillon M (2005) Gutless adenovirus: last-generation adenovirus for gene therapy. Gene Ther 12 Suppl 1: S18–S27
12. Semkova I et al (2002) Autologous transplantation of genetically modified iris pigment epithelial cells: a promising concept for the treatment of age-related macular degeneration and other disorders of the eye. Proc Natl Acad Sci U S A 99:13090–13095
13. Soudais C, Skander N, Kremer EJ (2004) Long-term in vivo transduction of neurons throughout the rat CNS using novel helper-dependent CAV-2 vectors. FASEB J 18:391–393

14. Lewis TB, Glasgow JN, Harms AS, Standaert DG, Curiel DT (2014) Fiber-modified adenovirus for central nervous system Parkinson's disease gene therapy. Viruses 6:3293–3310
15. Sakae M et al (2008) Highly efficient in vivo gene transfection by plasmid/PEI complexes coated by anionic PEG derivatives bearing carboxyl groups and RGD peptide. Biomed Pharmacother 62:448–453
16. Portales-Casamar E et al (2010) A regulatory toolbox of MiniPromoters to drive selective expression in the brain. Proc Natl Acad Sci U S A 107:16589–16594
17. Akerblom M et al (2013) Visualization and genetic modification of resident brain microglia using lentiviral vectors regulated by microRNA-9. Nat Commun 4:1770
18. Saunders A, Johnson CA, Sabatini BL (2012) Novel recombinant adeno-associated viruses for Cre activated and inactivated transgene expression in neurons. Front Neural Circuits 6:47
19. Chakrabarty P et al (2013) Capsid serotype and timing of injection determines AAV transduction in the neonatal mice brain. PLoS One 8, e67680
20. Towne C, Raoul C, Schneider BL, Aebischer P (2008) Systemic AAV6 delivery mediating RNA interference against SOD1: neuromuscular transduction does not alter disease progression in fALS mice. Mol Ther 16:1018–1025
21. Bevan AK et al (2011) Systemic gene delivery in large species for targeting spinal cord, brain, and peripheral tissues for pediatric disorders. Mol Ther 19:1971–1980
22. Maheshri N, Koerber JT, Kaspar BK, Schaffer DV (2006) Directed evolution of adeno-associated virus yields enhanced gene delivery vectors. Nat Biotechnol 24:198–204
23. Gray SJ et al (2010) Directed evolution of a novel adeno-associated virus (AAV) vector that crosses the seizure-compromised blood–brain barrier (BBB). Mol Ther 18:570–578
24. Manfredsson FP, Okun MS, Mandel RJ (2009) Gene therapy for neurological disorders: challenges and future prospects for the use of growth factors for the treatment of Parkinson's disease. Curr Gene Ther 9:375–388
25. Manfredsson FP et al (2009) Tight Long-term dynamic doxycycline responsive nigrostriatal GDNF using a single rAAV vector. Mol Ther 17:1857–1867
26. Kitada T et al (1998) Mutations in the parkin gene cause autosomal recessive juvenile parkinsonism. Nature 392:605–608
27. Manfredsson FP et al (2007) rAAV-mediated nigral human parkin over-expression partially ameliorates motor deficits via enhanced dopamine neurotransmission in a rat model of Parkinson's disease. Exp Neurol 207:289–301
28. Vercammen L et al (2006) Parkin protects against neurotoxicity in the 6-hydroxydopamine rat model for Parkinson's disease. Mol Ther 14:716–723
29. Gombash SE et al (2013) Morphological and behavioral impact of AAV2/5-mediated overexpression of human wildtype alpha-synuclein in the rat nigrostriatal system. PLoS One 8, e81426
30. Kanaan NM, Manfredsson FP (2012) Loss of functional alpha-synuclein: a toxic event in Parkinson's disease? Journal of Parkinson's disease 2:249–267
31. Ramaswamy S, Kordower JH (2012) Gene therapy for Huntington's disease. Neurobiol Dis 48:243–254
32. Boudreau RL et al (2009) Nonallele-specific silencing of mutant and wild-type huntingtin demonstrates therapeutic efficacy in Huntington's disease mice. Mol Ther 17:1053–1063
33. McBride JL et al (2011) Preclinical safety of RNAi-mediated HTT suppression in the rhesus macaque as a potential therapy for Huntington's disease. Mol Ther 19:2152–2162
34. Grimm D et al (2006) Fatality in mice due to oversaturation of cellular microRNA/short hairpin RNA pathways. Nature 441:537–541
35. Gorbatyuk OS et al (2010) In vivo RNAi-mediated alpha-synuclein silencing induces nigrostriatal degeneration. Mol Ther 18:1450–1457
36. Chan F, Hauswirth WW, Wensel TG, Wilson JH (2011) Efficient mutagenesis of the rhodopsin gene in rod photoreceptor neurons in mice. Nucleic Acids Res 39:5955–5966
37. Doudna JA, Charpentier E (2014) Genome editing. The new frontier of genome engineering with CRISPR-Cas9. Science 346:1258096
38. Urnov FD, Rebar EJ, Holmes MC, Zhang HS, Gregory PD (2010) Genome editing with engineered zinc finger nucleases. Nat Rev Genet 11:636–646
39. Wright DA, Li T, Yang B, Spalding MH (2014) TALEN-mediated genome editing: prospects and perspectives. Biochem J 462:15–24
40. Szymczak AL et al (2004) Correction of multi-gene deficiency in vivo using a single 'self-cleaving' 2A peptide-based retroviral vector. Nat Biotechnol 22:589–594
41. Manfredsson FP, Bloom DC, Mandel RJ (2012) Regulated protein expression for in vivo gene therapy for neurological disorders: progress, strategies, and issues. Neurobiol Dis 48:212–221

42. Gossen M, Bujard H (1992) Tight control of gene expression in mammalian cells by tetracycline-responsive promoters. Proc Natl Acad Sci U S A 89:5547–5551
43. Wachs FP et al (2006) Transforming growth factor-beta1 is a negative modulator of adult neurogenesis. J Neuropathol Exp Neurol 65:358–370
44. Quintino L et al (2013) Functional neuroprotection and efficient regulation of GDNF using destabilizing domains in a rodent model of Parkinson's disease. Mol Ther 21:2169–2180
45. Tang Y, Jackson M, Qian K, Phillips MI (2002) Hypoxia inducible double plasmid system for myocardial ischemia gene therapy. Hypertension 39:695–698
46. Hurttila H et al (2008) Oxidative stress-inducible lentiviral vectors for gene therapy. Gene Ther 15:1271–1279
47. Klein RL et al (2006) Efficient neuronal gene transfer with AAV8 leads to neurotoxic levels of tau or green fluorescent proteins. Mol Ther 13:517–527
48. Grabinski TM, Kneynsberg A, Manfredsson FP, Kanaan NM (2015) A Method for Combining RNAscope In Situ Hybridization with Immunohistochemistry in Thick Free-Floating Brain Sections and Primary Neuronal Cultures. PLoS One 10, e0120120

Part II

Expression Cassettes

Chapter 2

Delivering Transgenic DNA Exceeding the Carrying Capacity of AAV Vectors

Matthew L. Hirsch, Sonya J. Wolf, and R.J. Samulski

Abstract

Gene delivery using recombinant adeno-associated virus (rAAV) has emerged to the forefront demonstrating safe and effective phenotypic correction of diverse diseases including hemophilia B and Leber's congenital amaurosis. In addition to rAAV's high efficiency of transduction and the capacity for long-term transgene expression, the safety profile of rAAV remains unsoiled in humans with no deleterious vector-related consequences observed thus far. Despite these favorable attributes, rAAV vectors have a major disadvantage preventing widespread therapeutic applications; as the AAV capsid is the smallest described to date, it cannot package "large" genomes. Currently, the packaging capacity of rAAV has yet to be definitively defined but is approximately 5 kb, which has served as a limitation for large gene transfer. There are two main approaches that have been developed to overcome this limitation, split AAV vectors, and fragment AAV (fAAV) genome reassembly (Hirsch et al., Mol Ther 18(1):6–8, 2010). Split rAAV vector applications were developed based upon the finding that rAAV genomes naturally concatemerize in the cell post-transduction and are substrates for enhanced homologous recombination (HR) (Hirsch et al., Mol Ther 18(1):6–8, 2010; Duan et al., J Virol 73(1):161–169, 1999; Duan et al., J Virol 72(11):8568–8577, 1998; Duan et al., Mol Ther 4(4):383–391, 2001; Halbert et al., Nat Biotechnol 20(7):697–701, 2002). This method involves "splitting" the large transgene into two separate vectors and upon co-transduction, intracellular large gene reconstruction via vector genome concatemerization occurs via HR or nonhomologous end joining (NHEJ). Within the split rAAV approaches there currently exist three strategies: overlapping, trans-splicing, and hybrid trans-splicing (Duan et al., Mol Ther 4(4):383–391, 2001; Halbert et al., Nat Biotechnol 20(7):697–701, 2002; Ghosh et al., Mol Ther 16(1):124–130, 2008; Ghosh et al., Mol Ther 15(4):750–755, 2007). The other major strategy for AAV-mediated large gene delivery is the use of fragment AAV (fAAV) (Dong et al., Mol Ther 18(1):87–92, 2010; Hirsch et al., Mol Ther 21(12):2205–2216, 2013; Lai et al., Mol Ther 18(1):75–79, 2010; Wu et al., Mol Ther 18(1):80–86, 2010). This strategy developed following the observation that the attempted encapsidation of transgenic cassettes exceeding the packaging capacity of the AAV capsid results in the packaging of heterogeneous single-strand genome fragments (<5 kb) of both polarities (Dong et al., Mol Ther 18(1):87–92, 2010; Hirsch et al., Mol Ther 21(12):2205–2216, 2013; Lai et al., Mol Ther 18(1):75–79, 2010; Wu et al., Mol Ther 18(1):80–86, 2010). After transduction by multiple fAAV particles, the genome fragments can undergo opposite strand annealing, followed by host-mediated DNA synthesis to reconstruct the intended oversized genome within the cell. Although, there appears to be growing debate as to the most efficient method of rAAV-mediated large gene delivery, it remains possible that additional factors including the target tissue and the transgenomic sequence factor into the selection of a particular approach for a specific application (Duan et al., Mol Ther 4(4):383–391, 2001; Ghosh et al., Mol Ther 16(1):124–130, 2008;

Fredric P. Manfredsson (ed.), *Gene Therapy for Neurological Disorders: Methods and Protocols*, Methods in Molecular Biology, vol. 1382, DOI 10.1007/978-1-4939-3271-9_2,

Hirsch et al., Mol Ther 21(12):2205–2216, 2013; Trapani et al., EMBO Mol Med 6(2):194–211, 2014; Ghosh et al., Hum Gene Ther 22(1):77–83, 2011). Herein we discuss the design, production, and verification of the leading rAAV large gene delivery strategies.

Key words Large gene delivery, Split AAV method, Overlapping, Trans-splicing, Hybrid, Fragment, Adeno-associated virus, Concatemer, Vectorcapacity

1 Introduction

1.1 AAV Vectorology

Adeno-associated virus (AAV) is a non-autonomous parvovirus that requires a helper virus, such as adenovirus, to complete its life cycle. AAV remains the smallest known virus and is comprised of a protein capsid (20–24 nm) and a single-strand DNA genome of approximately 4.7 kb. The AAV genome contains at least three open reading frames encoding replication, capsid, and capsid assembly proteins [14]. An important feature of the AAV genome is that it is flanked on both sides by 145-nucleotide inverted terminal repeats (ITRs) which are necessary for viral replication, packaging, and site-specific integration [14]. In the 1980s, the AAV genome was cloned from nature and the controlled production of WT AAV was demonstrated in cell culture in the presence of adenovirus [15]. Soon thereafter, it was discovered that the AAV genome could be substituted for a sequence of choice. Thus, recombinant AAV (rAAV) particles could be produced having transgenic genomes [15, 16]. The only viral DNA elements necessary for replication and packaging during rAAV production are the ITRs flanking the transgenic DNA in *cis*, as the *rep* and *cap* genes can be supplied in *trans* using a separate plasmid [16]. Shortly after these seminal observations it was demonstrated that adenovirus could be substituted by its partial genome in plasmid form, which allowed the production of rAAV in the absence of contaminating adenovirus [17]. Despite over 25 years of rAAV optimizations for diverse applications, this method of rAAV production predominantly remains unchanged.

Regarding the transduction efficiency, rAAV has proven the most efficient and safe method of gene delivery for sustained mammalian cell transduction. The favorable attributes of rAAV include (1) its non-pathogenicity, (2) ability to transduce nondividing and dividing cells, (3) its broad tissue tropism conferred by various natural and mutant serotypes, (4) the persistence of rAAV genomes as primarily episomes with very low levels of integration into the host chromosome, and (5) the ability to confer long-term transgene expression following a single injection [14]. Given these favorable attributes, well over 100 rAAV clinical trials have been performed to date for diverse diseases with notable successes for the treatment of hemophilia B and Leber's congenital amaurosis [18, 19].

Despite rAAV's popularity and clinical success, it does have a major limitation in that the AAVcapsid cannot package sequences greater than about 5 kb [1]. This packaging limitation is an obstacle for treatment of genetic diseases requiring larger transgenes such as Duchenne muscular dystrophy, hemophilia A, and cystic fibrosis. However, to overcome the packaging limitations, creative intracellular large gene reconstruction strategies have been developed including the split vector approaches (overlapping, trans-splicing, and hybrid vectors) and fAAV vector transduction [1].

1.2 Split rAAV Large Gene Delivery

Split vector large gene delivery approaches take advantage of the observation that rAAV genomes undergo concatemerization via non-homologous end joining (NHEJ) and intermolecular HR. Previous reports have shown that rAAV genomes form concatemer in liver, muscle, and cell lines [9, 20–23], which likely aids in the persistence and long-term expression of the transgenic cassette. Following uncoating, rAAV genomes readily circularize through intramolecular ITR linkages and over time dimer and multimeric concatemers are generated that persist as episomes [2, 3, 9, 21, 22]. This inherent tendency for intermolecular genome associations serves as a requirement for all rAAV large gene delivery approaches herein. Split rAAV vectors rely on the division of a large transgene (and expression requirements) into two rAAV vectors (termed vector A and vector B herein) that are distinct and produced independently [4, 5]. Within the split vector approaches there is a primary division of overlapping and trans-splicing rAAV vectors [4–6].

1.3 Overlapping Vector Approach

In the overlapping vector approach, vectors A and B display a region of sequence homology to promote intermolecular HR, thus generating the large transgene by recombining vectors A and B as depicted (Fig. 1). The efficiency of homologous recombination correlates with the degree of sequence overlap, and the actual sequence, of the vector A and vector B genomes. Increased size of the overlapping sequence can lead to an increase in homologous recombination [4]. This increased homologous recombination can also lead to a slight hindrance in the overall objective of rAAV large gene therapy, as increasing the overlapping sequence effectively decreases the overall transgene capacity [23]. Therefore the limitation for the overlapping vector approach is largely dependent on the recombination efficiency of the overlapping sequence and the propensity for episomal homologous recombination in the transduced cell.

1.4 Trans-splicing Technique

The trans-splicing technique also relies on a co-transduction approach in which the transgene is split such that one vector (vector A) contains a promoter, a 5′ portion of the gene, followed by a splice donor sequence (Fig. 2). The downstream vector (vector B) contains a splice acceptor sequence the remaining 3′ cDNA (or

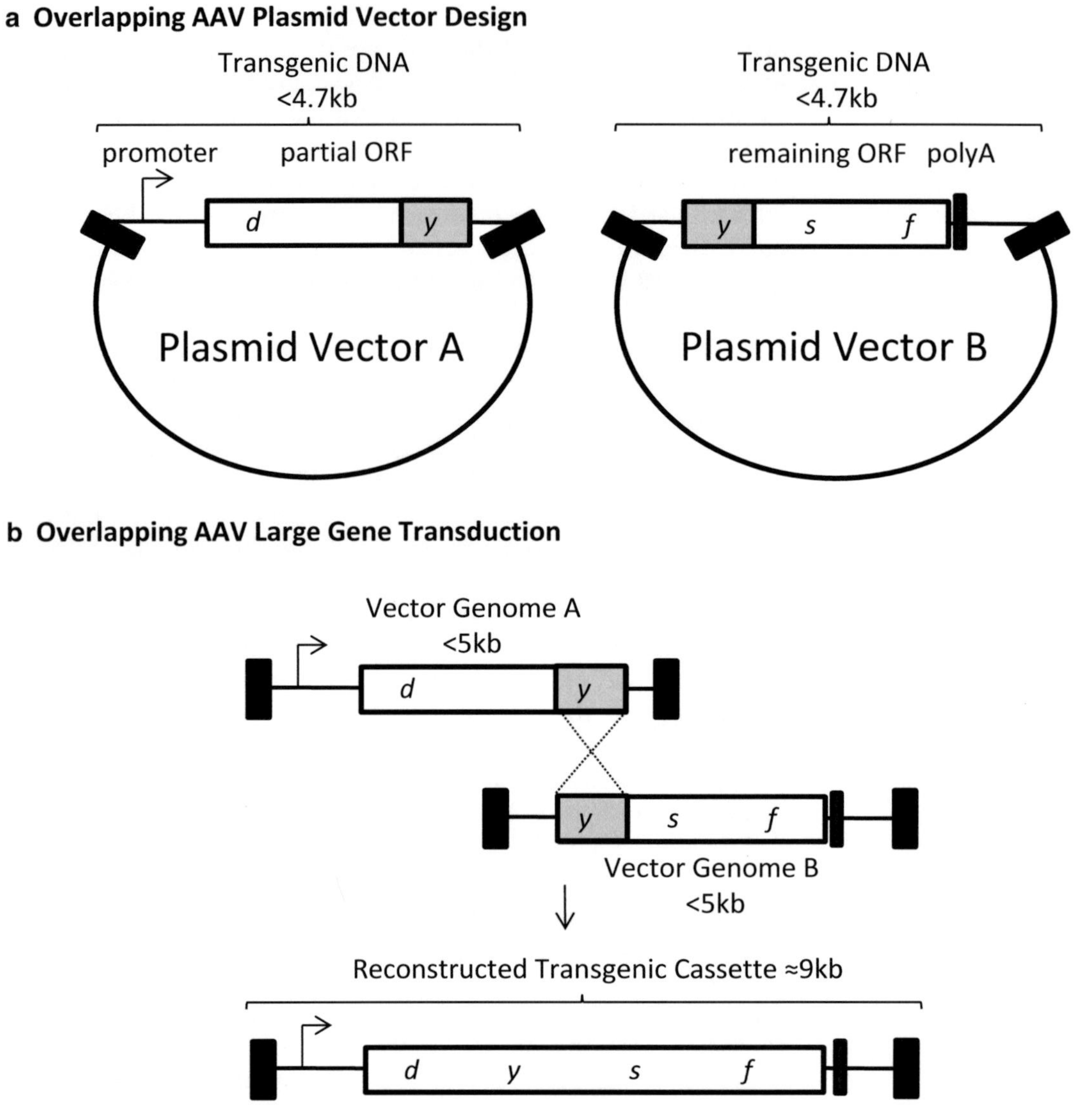

Fig. 1 Overlapping AAV genome design and transduction. (**a**) The transgenic cassette, containing the depicted elements, is split into two separate vectors such that there is a homologous sequence overlap (*gray*). The total size of the transgenic DNA must be less than 4.7 kb. (**b**) Following co-transduction of vector genomes A and B, homologous recombination at the overlapping sequence occurs to reconstruct the depicted large transgenic DNA cassette of approximately 9 kb. *Black box* AAV inverted terminal repeat

gene), and a poly A tail [4]. A concatemerization event in the correct orientation (theoretically about a 16 % chance) generates a single DNA molecule containing the intended large transgene expression cassette (Fig. 2) [4, 24]. Upon production of the pre-mRNA the intron is spliced out along with the ITR recombination junction, establishing an intact open reading frame large than can be packaged in a single vector [23, 25]. The trans-splicing method has been used in several Duchenne muscular dystrophy mouse

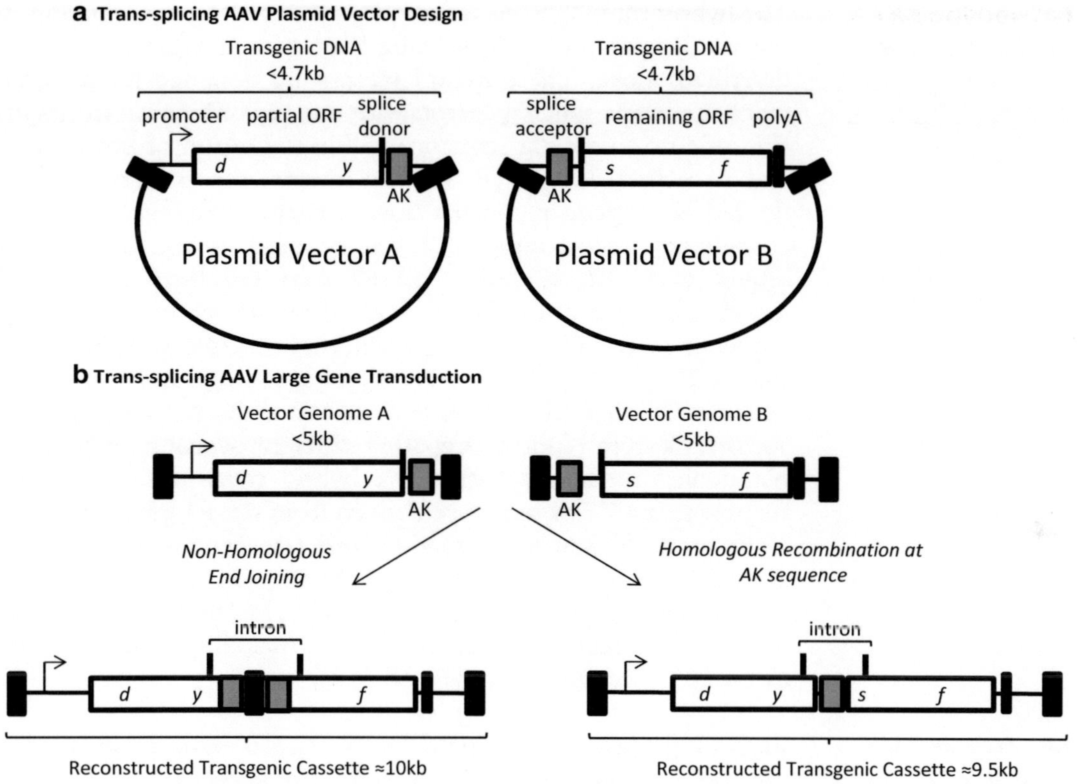

Fig. 2 Trans-splicingAAV vector design and transduction. (**a**) The genetic elements of the two required AAV plasmids are depicted. Plasmid A contains a promoter, a partial open reading frame terminated at a splice donor site, followed by a recombinogenic sequence (hybrid vectors only, AK). The vector B plasmid contains a recombinogenic sequence (hybrid vectors only, AK), a splice acceptor site, the remaining ORF followed by a poly A sequence. The transgenic DNA in each plasmid is less than 4 kb. (**b**) Following co-transduction vector genomes A and B can undergo end joining (*left*) or homologous recombination at the AK sequence (*right*) to reconstruct a large transgenic cassette containing the depicted intron. *Black box* AAV inverted terminal repeat

models and has shown detectable restoration levels of dystrophin post-transduction [26–28]. There are also studies showing efficient whole-body and retina transduction that may also be promising for efficiently restoring large transgene expression [7, 29]. Even though efficient transduction is noticed in these experiments the transduction efficiency of these trans-splicing vectors is decreased compared to that of a single intact vector in skeletal muscle, eye, and liver [9]. The reason for this decreased transduction efficiency is due to several factors including (1) the efficiency of multiple vector transduction of a single cell, (2) the efficiency of the cellular machinery to reconstruct the large transgene in the desired orientation, (3) the decreased accumulation of mRNA due to the instability of the pre-mRNA, and (4) the gene splitting site [30]. The use of synthetic introns, including exonic splicing enhancers, can be used to at least partially counteract these rate-limiting steps [26, 31, 32].

1.5 Hybrid Trans-splicing Technique

The hybrid trans-splicing technique is essentially a combination of the overlapping and trans-splicing large gene delivery approaches described above. These hybrid vectors are designed in the same manner as trans-splicing vectors with one key difference: the inclusion of an overlapping sequence within the intron of both the 5′ and 3′ vectors (Fig. 2) [6, 13]. To increase the likelihood of the desired large gene reconstruction, particular sequences that are considered "recombinogenic" can be used as the overlapping sequence [6, 12, 13, 33]. Following co-transduction, the full-length transgene can either be formed by HR at the overlapping sequence or the ITR homology. Alternatively, the ITRs may join via NHEJ [9, 20]. Then, transcription, splicing and translation generate the desired protein as described above for trans-splicing vectors. Current reports comparing different overlapping "recombinogenic" sequences within the hybrid trans-splicing strategy suggest that a 77 nt sequence derived from the F1 phage genome is the most efficient in promoting large transgene reconstruction, at least in the tested cells [12]. This hybrid AK vector context, as well as the normal trans-splicing approach, has demonstrated relevance for the treatment of retinal diseases including Stargardt's disease and Usher syndrome type 1B [12].

1.6 Fragment AAV Large Gene Delivery

Regarding rAAV large gene delivery, reports have demonstrated the ability of the AAVcapsid to package genomes over 5 kb [34–37]. For instance, Allocca et al. surveyed the packaging capacity of different capsid serotypes [34]. In the eye, it was demonstrated that AAV serotype 5 capsid, in particular, could package an expression cassette containing the adenosine triphosphate-binding cassette (ABCA4) at an unprecedented size of 8.9 kb, which was verified by alkaline gel electrophoresis [34]. Importantly, this AAV5-ABCA4 vector mediated effective transduction and resulted in the correction of disease phenotypes observed in the retina of a Stargardt's disease mouse model [34]. The excitement for this groundbreaking report was later tempered by the reports of several groups that concluded that rAAV genomes over 5 kb were not packaged in their entirety, but instead DNA "fragments" of the intended oversized cassette were encapsidated (Fig. 3). Characterization of this process, demonstrated that oversized genome packaging initiates from the 3′ end and the external 5′ end is truncated when the capacity of the capsid is reached [8, 10, 11]. However, this process remains largely uncharacterized as heterogeneous DNA species (all <5 kb) are packaged in the apparently intact AAV capsids [8–11, 38]. Interestingly, these fAAV preparations retained their ability to mediate large gene transduction, albeit at a reduced efficiency compared to intact AAV in a tissue- and administration-dependent manner. A mechanistic eval-

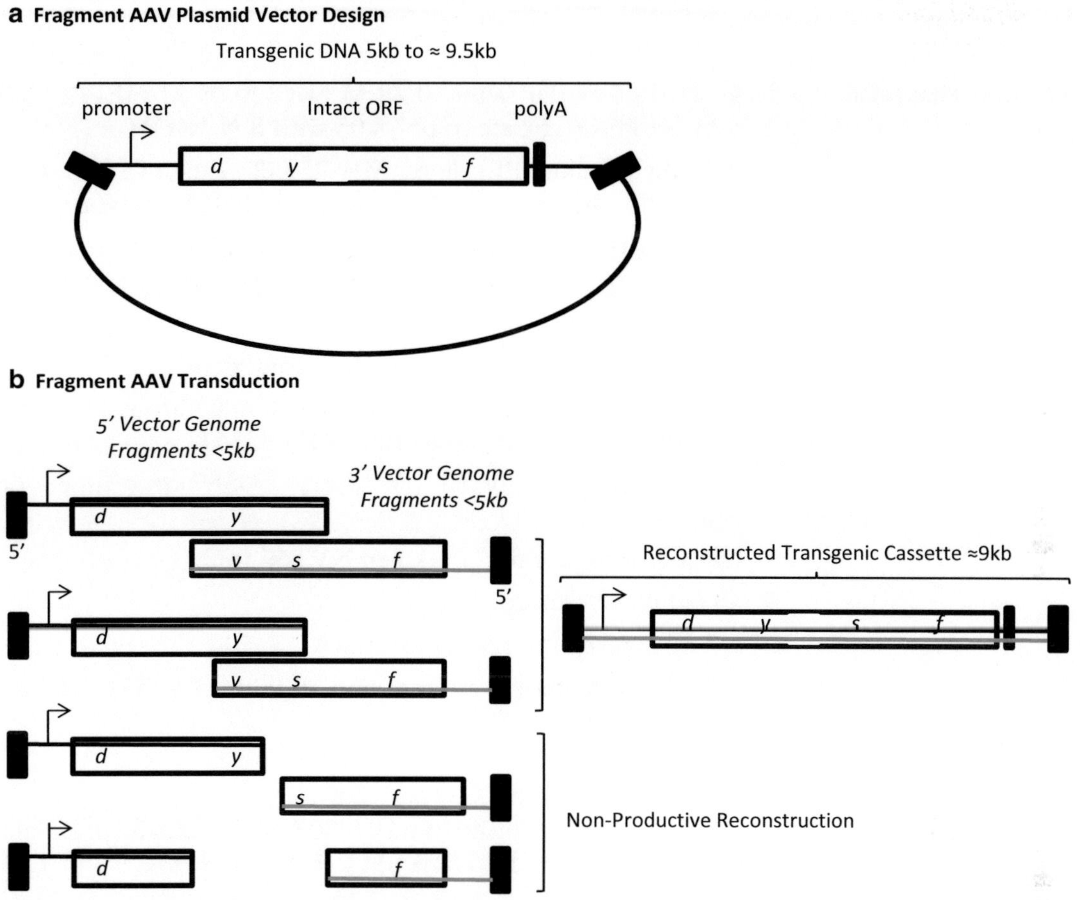

Fig. 3 Fragment AAV vector design and transduction. (**a**) A single plasmid with an intact transgenic cassette is depicted. (**b**) During AAV vector production different sized genome fragments of the over-sized transgenic cassette are packaged. Fragments of both polarities are packaged (*gray and black lines*) starting at the 3′ end. Opposite polarity fragments containing regions of complementarity can anneal followed by strand synthesis to generate a large transgenic cassette. Contaminating fragment genome species below half of the intended transgenic cassette do not contain strand complementarity and are non-productive for transduction. *Black box* AAV inverted terminal repeat

uation of fAAV transduction in cell culture suggested that a large gene reconstruction event relies on HR, as the repair proteins Rad51C and XRCC3 were necessary while the NHEJ mediator DNA-PKcs was dispensable [9]. Despite the lack of an explanation for the earlier report demonstrating AAV5 large gene packaging [34], what is clearly evident is that a new rAAV large gene strategy emerged (fAAV) which, as in the other strategies, relies on the intracellular reconstruction of the oversized genome fragments, perhaps by canonical homology directed repair, at least in vitro [1, 9].

2 Materials

2.1 rAAV Production

1. 2× HEPES-buffer saline: 0.28 M NaCl, 0.05 M HEPES, 1.5 mM Na_2HpO_4, titrate to pH 7.05 with 5 N NaOH.
2. Polyethylenimine (PEI) linear MW 25,000: 1 mg of PEI in 1 ml of 1× PBS, Adjust pH to 4 or 5 with 12 N HCl (*see* **Note 1**).
3. DMEM: 10 % FBS, 1× penicillin/streptomycin.
4. DNaseI digestion buffer: 10 mM Tris–HCl pH 7.5, 10 mM $MgCl_2$, 50 U/ml DNase I.
5. Sonicator (Branson Sonifier 250, or equivalent).
6. Beckman Quick-Seal polypropylene (16 mm × 76 mm) centrifuge tubes (Beckman, Indianapolis IN, USA).
7. Ultracentrifuge (Sorvall Discovery 90SE centrifuge, or equivalent).
8. Ultracentrifuge rotor (Beckman NVT65 rotor, or equivalent).
9. 15 cm tissue culture plates.
10. 1.7 ml polystyrene microfuge tubes.
11. Slide-A-Lyzer dialysis cassette with 20,000 (MWCO) and 0.5–3.0 ml capacity.
12. HEK 293 cells.

2.2 Southern Dot Blot

1. DNaseI digestion buffer (6 ml): 5.808 ml ddH_2O, 60 μl DNase (10 μg/μl), 60 μl 1 M Tris pH 7.5, 60 μl $MgCl_2$, 12 μl 1 M $CaCl_2$.
2. Proteinase K solution (10 ml solution): 6.9 ml ddH_2O, 2 ml 5 M NaCl, 1 ml 10 % Sarkosyl, 100 μl proteinase K (10μg/μl).
3. 0.5 M NaOH (10 ml): 1 ml 5 M NaOH, 9 ml ddH_2O.
4. 0.4 M Tris pH 7.5 (100 ml): 40 ml 1 M Tris pH 7.5, 60 ml ddH_2O.
5. Hybond-XL nylon membrane (GE Healthcare Life Sciences, Pittsburgh, PA, USA).
6. Filter paper.
7. 96-well dot blot manifold (or equivalent).
8. Church buffer.
9. Radioactive probe labeling kit (*see* **Note 2**).
10. High-salt wash (500 ml solution): 445 ml ddH_2O, 50 ml 20 % SSC, 5 ml 10 % SDS.
11. Low-salt wash (500 ml system): 492.5 ml ddH_2O, 2.5 ml 20× SSC, 5 ml 10 % SDS.

2.3 Alkaline Gel Southern Blot

1. Loading dye (6×): 80 μl 5 M NaOH, 10 μl 0.5 M EDTA, 0.18 g Ficoll, 730 μl H_2O, xylene cyanol for color.
2. Alkaline gel: Add 0.5 g agarose to 50 ml water (1 %), microwave to dissolve, let cool then add 500 μl 5 M NaOH and 100 μl 0.5 M EDTA (*see* **Note 3**).
3. Gel running buffer: 10 ml 5 M NaOH, 2 ml 0.5 M EDTA, 988 ml H_2O.
4. Transfer buffer: 80 ml of 5 M NaOH in 920 ml of deionized water.
5. Hybond-XL nylon membrane the size of the gel.
6. Two sheets of filter paper slightly bigger than the size of the gel.
7. One sheet of filter paper the width of the gel and about 1½ ft long for the transfer bridge.
8. A pan that can hold the transfer buffer.
9. A stack of paper towels.

2.4 Quantitative Polymerase Chain Reaction

1. DNaseI mixture (listed above in Subheading 2.2, Southern Blot materials).
2. Proteinase K solution (listed above).
3. 10 mM Tris (pH 8).
4. 2× SYBR mix (Life Technologies, Grand Island, NY, USA).
5. Forward primer (20 μM) (*see* **Note 4**).
6. Reverse primer (20 μM) (*see* **Note 4**).
7. Sterile H_2O.

3 Methods

3.1 Split rAAV Vectors: Construct Design

1. Overlapping technique: To design your two split overlapping vectors, first construct plasmid vector A by placing a promoter followed by the transgene such that the total transgenic sequence is <4.7 kb (excluding the 2 ITR sequences which are each about 150 nucleotides) (Fig. 1). Next, place the remaining transgene sequence on plasmid vector B followed by a poly-adenylation sequence (Fig. 1). Again, the total size of the transgenic DNA needs to be <4.7 kb. Realistically, the overall size of the transgene cassette will dictate the maximum region of vector genome overlap (which is a requirement), with the basic understanding that increased region of homology will be more likely to stimulate intermolecular recombination, and thus the reconstructed large transgene cassette [23].
2. Trans-splicing technique: Divide the transgene approximately near the middle, ideally at a conserved splicing junction.

Then, add a canonical intron donor sequence immediately following the vector A ORF (Fig. 2). Any sequence between this donor signal and the vector A 3′ ITR will be intronic (Fig. 2). To generate vector B, an intronic sequence is positioned downstream of the 5′ ITR and terminated by an intron acceptor sequence (Fig. 2). Then, the remainder of the ORF (prematurely truncated in vector A is positioned upstream of a poly-A sequence prior to the 3′ ITR of vector B (Fig. 2)). As with the overlapping AAV vectors, and the overall rationale of this methods chapter, the total size of intact vectors (including the ITRs) needs to be <5 kb. Designs of trans-splicing constructs in the literature provide examples of the genetic arrangements and intronic sequences used previously [4, 6, 7, 12, 13, 24, 27].

3. Hybrid trans-splicing technique: For this large gene AAV vector format, follow the design of trans-splicing vectors with a single exception, place the same overlapping sequence within the intron on the 3′ end of vector A and the 5′ end of vector B as depicted in Fig. 2. One consideration is the maintenance of a valid intronic branch point to facilitate splicing. Currently, there are only a few reports on overlapping sequences employed in hybrid AAV vectors and evaluation of the sequences used therein should be further investigated [6, 12, 13, 27].
4. Fragment AAV vectors: construct design: A key distinction between fAAV large gene transduction and the split AAV vectors described above is that for the later approaches two distinct plasmid vectors (A and B) need to synthesized and employed for independent vector preparations whereas, a single plasmid containing the ITR sequences is used for fAAV (Fig. 3). Since DNA strands of both polarity are truncated and packaged during the viral vector production, the single plasmid design format is the following (5′ to 3′): (1) ITR, (2) promoter, (3) the oversized transgene, (4) a poly-A sequence, and (5) the flanking ITR sequence (Fig. 3). As genomic fragments will be packaged in both strand orientations, and need to anneal at a region of sequence complementarity, the total fAAV plasmid size between the ITR sequences should not exceed 9 kb.

3.2 rAAV Production

1. 24 h prior to transfection, split a confluent 15 cm plate of 293 cells at a 1:3 ratio into 4–6, 15 cm plates [final volume of culture medium (DMEM 10 % FBS 1× Penn/Strep) is approximately 20 ml].
2. At approximately 70 % confluency, perform the triple transfection using polyethylenimine (PEI) using the plasmids and amounts described in Table 1 below and elsewhere [39].

Table 1
Transfection mixture for rAAV production

Transfection reagents	Amount per 15 cm plate
pXX680 (Ad helper functions)	12 μg
pRepCap (replication and capsid genes)	10 μg
pITR-transgene (packaged sequence)	6 μg
Serum-free DMEM	500 ml
PEI	100 μl

3. Add the three plasmids to a tube followed by the serum free DMEM and then the PEI. Immediately vortex the mixture and let sit at RT for 5 min. Add the transfection mixture dropwise to a single 15 cm plate of HEK 293 cells and gently swirl the plate to achieve uniform distribution.
4. 48–72 h post-transfection harvest the transfected 293 cells for intracellular vector purification (*see* **Note 5**).
5. Collect the HEK-293 cells from all four plates and place in a centrifuge tube (*see* **Note 6**).
6. Pellet cells by centrifuging at 2500 rpm for 5 min.
7. Remove the supernatant, wash the cells with 10 ml of 1×PBS, and spin again.
8. Resuspend the washed cell pellet in 10 ml of 1×PBS and keep on ice.
9. Sonicate each sample on ice for 25 pulses with a Branson Sonifier set at cycle 50 and output 5 (or equivalent).

3.3 rAAV CsCl Gradient Purification

1. Following sonication, add 100 μl of the DNaseI solution to the samples and incubate at 37 °C for 1 h.
2. Bring samples to a final volume of 11.7 ml using 1×PBS.
3. Add 6.6 g of CsCl to each sample, vortex immediately to dissolve, and place on ice.
4. Repeat the sonication (**step** 7) and proceed to the next step.
5. Load samples into a Beckman Quick-Seal polyallomer (16 mm×76 mm) centrifuge tubes (about 11.5 ml will fit with a little leftover which can be discarded) and seal the tube with heat (*see* **Note** 7).
6. Centrifuge at 65,000 rpm in a Beckman NVT65 rotor using a Sorvall Discovery 90SE centrifuge (or equivalent) for at least 12 h.

7. After the centrifugation, collect gradient fractions by placing a 21-gauge needle in the bottom of the tube and an 18-gauge needle in the top to let liquid flow (*see* **Note 8**). Collect 1 ml fractions (about 20 drops).
8. Determine your peak fractions containing the most packaged viral genomes using the Southern dot blot hybridization method described below. In the case of fAAV, CsCl gradient fractions should be analyzed on an alkaline gel to determine packaged genome size as reporter [9] (*see* **Note 9**).
9. Combine peak fractions (*see* **Note 10**) and dialyze in 1× PBS using a Slide-A-Lyzer dialysis cassette (or equivalent) at 4 °C with constant stirring for at least 12 h (approximately 1 L of PBS to 1 ml of vector fraction is sufficient).
10. Remove the samples from the dialysis cassettes, aliquot into micro-centrifuge tubes and store at −80 °C. All analyses of vector characterizations should be performed on thawed aliquots (*see* **Note 11**).

3.4 Split rAAV Vector Production

1. Vectors A and B are produced independently using this protocol (i.e., 2 separate transfections/vector preparations; Figs. 1 and 2). Final titer after dialysis can be determined by Q-PCR (*see* **Note 12**).

3.5 fAAV Vector Production

1. One ITR plasmid vector is used for production as described (single preparation). It is important to keep fractions one to nine for characterization of packaged genome size by alkaline gel Southern blotting (*see* **Note 13**) as the genome size of the transducing species must be over ½ of the total sequence to be packaged including the ITRs and sequence between them (Fig. 3) [9]. For fAAV preparations, the final titer should be performed by Q-PCR using multiple primers sets, as not all regions are packaged at similar efficiencies. An effective way to do this is to perform Q-PCR using a primer set designed to amplify the conserved ITRs, with the assumption that as only the 3′ ITR is packaged while the 5′ ITR was removed during production.

3.6 AAV Final Titer

1. Perform a DNAseI digestion to remove rAAV genomes and plasmidDNA that are not encapsidated. 10 μl of virus + 90 μl of DNaseI solution and vortex. Incubate for 1 h at 37 °C.
2. To inactivate the DNAseI, add 6 μl of 0.5 M EDTA and thoroughly vortex.
3. To degrade the AAVcapsid, add 120 μl of Proteinase K solution and vortex.
4. Incubate at 55 °C for ≥2 h (sample can be left at 55 °C overnight).

5. Heat the sample for 10 min at 95 °C to inactivate proteinase K.
6. Dilute the sample at least 100-fold in clean water or 10 mM Tris-HCl (pH 8) and use this solution as a template for Q-PCR.
7. Make plasmidDNA standards for the Q-PCR reaction. Dilute plasmid DNA (ITR plasmid used in the initial transfection) to 10 ng/μl in 10 mM Tris-HCl (pH 8) in 1.5 ml siliconized tubes. Vortex and spin briefly (to collect contents) in every step. Make standard serial 1:10 dilutions from 50 pg to 0.05 fg.
8. Prepare the Q-PCR SYBR master reaction (5 μl 2× SYBR mix, 0.25 μl forward primer (20 μM), 0.25 μl reverse primer (20 μM), 2.5 μl H_2O). The final volume of each reaction is 10 μl. Set up an 8 μl master mix per reaction as follows and use a twofold dilution series for the viral genome preparations (*see* **Note 14**). Although total amplified double-strand DNA as measured by SYBR stain is described herein, probe-specific methods for detection can also be used.
9. Pipette 8 μl of master mix into each well.
10. Add 2 μl of DNA (vector sample or plasmid standard**) to a well of a 96-well plate.
11. Add 2 μl of dH_2O or 10 mM Tris-HCl solution as a no template control.
12. Seal the plate and spin down @ 1800 rpm for 10 s.
13. Perform the Q-PCR in a thermocycler using the manufacturer's recommended conditions specific to the primer annealing temperature.

3.7 Southern Dot Blot for Vector Genome: Sample Preparation

1. In an appropriate tube or well add 10 μl of the CsCl vector fraction to 100 μl of DNaseI digestion buffer (solution 1). For final titer determination by this method a twofold dilution series can be used (in triplicate) starting with a max of 10 μl of the dialyzed vector preparation.
2. Vortex and incubate at 37 °C for 1 h to remove unpackaged DNA.
3. Stop the digestion by adding 6 μl of 0.5 M EDTA and vortex to mix thoroughly.
4. To digest the vector capsid, add 120 μl of the proteinase k solution (solution 2).
5. Vortex and incubate at 55 °C for a minimum of 2 h (up to 24 h is ok)
6. Add 144 μl of 0.5 M NaOH.
7. Vortex and incubate at RT for 30 min to denature the DNA.

3.8 Southern Dot Blot for Vector Genome: Plasmid Standard Preparation

1. Using the AAV ITR plasmid that contains the transgenic sequence, prepare a plasmid standard from 50 ng to 5 ng to generate a standard curve (*see* **Note 15**).
2. Denature the plasmid standard by add 300 μl of 0.5 M NaOH, and 1 M NaCl (solution 4) to each tube.
3. Incubate at RT for 30 min.

3.9 Southern Dot Blot for Vector Genome: Preparation of Dot Blot Apparatus

1. Cut a HyBond XL nylon membrane and two sheets of filter paper to fit the size of the dot blot apparatus such that all 96 wells are covered.
2. Wet membrane and the two sheet of filter paper in solution 5.
3. Place the two sheets of filter paper down first followed by the membrane such that sample loading will be directly on the membrane. Then place the top of the dot blot apparatus on tightly to prevent sample leakage.
4. Attach a vacuum to the apparatus for 5 min.
5. Stop the vacuum.
6. Pipette each sample and standard into an individual well.
7. Wait for 5 min.
8. Apply vacuum for 10 min.
9. Wash each well with 300 μl of H_2O.
10. Vacuum for an additional 10–15 min.
11. Remove the membrane from the apparatus and UV cross-link the DNA to the membrane.
12. Place the membrane in a glass hybridization tube containing Church buffer to cover the membrane and pre-hybridize for 30 min at 60 °C using a rotisserie to provide constant rotation.
13. Add a radio-labeled probe and let incubate with membrane overnight in the rotisserie at 60 °C (*see* **Note 16**).
14. Wash the membrane two times in the high-salt wash for 5 min.
15. Wash the membrane in the low-salt solution for 30 min.
16. Wrap membrane in saran wrap or a sheet protect.
17. Expose to film or a phospho-imaging screen for 2–3 h.
18. Develop film or scan the screen and quantify the dot intensity (*see* **Note 17**).
19. Prepare a curve using the dot intensities of the standards and determine the number of viral genomes/μl as previously described [39].

3.10 Packaged Genome Characterization: Alkaline Gel Southern Blot

1. Remove unpackaged DNA by adding 10 μl of the DNaseI solution (solution 1) to 10 μl of virus (≈1–5e^9 viral genomes; *see* **Note 19**).
2. Incubate for 30 min at 37 °C.

3. Add 4 μl of 0.5 M EDTA and vortex to inactivate the DNaseI.
4. Add the gel loading dye to make a 1× solution.
5. Add 2 μl of 10 % SDS to disrupt the AAVcapsid exposing the viral genome (*see* **Note 18**).
6. Vortex for 10 s and let sit at RT for 20 min.
7. Spin at 12,000 rpm for 20 min.
8. Load the alkaline gel along with a proper ladder/marker (*see* **Note 19**).
9. Normally two to three fractions are combined for dialysis. In general, fractions corresponding to the intensity of the 25 ng plasmid standard or greater are worth moving forward with.
10. Run the gel at 15 V overnight in alkaline gel electrophoresis buffer (*see* **Note 20**).
11. When gel has finished running cut the top wells off and any other part of the gel that may not be needed.
12. Soak the gel in transfer buffer for 10 min with mild agitation.
13. Measure the size of the gel, so you can cut your filter paper and HyBond-XL membrane accordingly.
14. Perform Southern transfer using standard procedures [39, 40].
15. Once the transfer is complete UV cross-link the membrane.
16. Hybridize the DNA on the membrane to a radio-labeled probe in Church buffer as described for the Southern dot blot method (*see* **Notes 16** and **21**). The wash and detection conditions are also the same as described above.
17. Intact AAV genomes (<5 kb) package primary a single DNA species the size of the transgenic DNA cassette and the ITRs, whereas fAAV packages heterogeneous DNA species of different sizes which are separated in the gradient based on density. For transduction competent fAAV preparations, the fractions containing the largest DNA species are the most desirable to pool, dialyze, and prior to determining the final titer (*see* **Note 9**). We have previously reported explicit examples of both intact and fAAV genome species found in the different fractions of the CsCl gradient and correlated them to both transduction efficiency and the refractive index of the CsCl fraction [9].

3.11 AAV Large Gene Transduction Verification

1. Following production and characterization of your AAV vector preparations large transgene synthesis will, of course, require verification. To do this follow standard procedures, such as Western and northern blots, but be sure to always included the proper controls. For instance, all split AAV vector approaches rely on co-transduction of AAV vectors. Therefore, it will be necessary to perform transduction with each vector independently, and in concert, to be certain that large transgene pro-

duction only occurs if both vector types are used. In the case of fAAV preparations, functional transduction can be performed on all fractions prior to dialysis to determine which fractions are competent for large transgene production [9]. Then, the size of the packaged genome fragments can be correlated to functional large transgene production and guide the decision of which fractions to dialyze (only packaged genome fragments greater than half the size of the transgenic cassette, theoretically, will mediate transduction).

4 Notes

1. PEI is a potent transfection reagent that is relatively inexpensive. A large batch can be made and stored in aliquots at –20 °C to provide consistency. We routinely store small aliquots at 4 °C for up to 1 month.
2. For most applications we use a random primed labeling kit and P^{32} CTP due to the strength of the signal. However, end labeling kits and other types of non-radioactive detection can be used as well.
3. It is important to add the 5 M NaOH and 0.5 M EDTA to the cooled alkaline gel solution prior to polymerization. If added when the solution is too hot it will turn a yellowish color and should be discarded.
4. Primer design for Q-PCR follows normal considerations in regards to length, nucleotide composition, etc.
5. At the time of harvest the cells should appear smaller and rounder in morphology as they approach the pXX680 induced cytopathic effect. If the cells appear completely healthy then it is likely that the transfection is likely poor. At this point the vector preparation should be terminated and repeated with further optimized transfection conditions.
6. Although not described herein, AAV vectors can also be purified from the culture medium at the time of cell harvest [41].
7. It is important to make sure that all tubes are precisely balanced in the centrifuge.
8. It is a good idea to use one needle to punch a hole in the bottom of the tube and then place a new in the created hole to ensure a constant flow.
9. Only genomes greater than ½ of the total cassette size retain the ability for single strand annealing and therefore productive transduction.

10. If the Proteinase K solution is added without the EDTA being added first it will lead to the degradation of the viral DNA.
11. Repeated freeze/thawing of viral vector aliquots should be avoided.
12. Determination of the exact AAV vector titer is difficult and therefore a Southern dot blot should be used to confirm the PCR data.
13. SYBR Gold (Life Technologies) nucleic acid staining of the alkaline gel can be used for AAV genome detection instead of transferring the DNA to a membrane and detecting via probe hybridization.
14. Primer design follows standard protocols and it is wise to target at least two positions within the transgenic vector cassette. If the signal is too strong, then you can let the membrane sit until the signal reduces in intensity.
15. Digestion of the pAAV-ITR plasmid to recover a plasmid fragment representing the exact sequence to be packaged is the preferred species to use as the standard.
16. Both end-labeled defined probes and those generated via random-primed labeling can be used and prepared using kits following the manufacturer's recommendation. If using a random-primed labeling kit a digested fragment representing the packaged transgene cassette is ideal as the reaction template. Nonradioactive probe detection is also acceptable.
17. Dot intensity for the Southern dot blot can be quantified using the Image J program.
18. The Proteinase K solution can be used in lieu of the 10 % SDS.
19. Linear fragments of the AAV ITR plasmid at several different sizes work nicely for this, although probe homology must be accommodated.
20. It is important to remove air bubbles between the membrane and the alkaline gel to not disrupt the transfer.
21. To decrease the nonspecific background, probes for the Southern blot applications can be column purified prior to hybridization.

Acknowledgments

This work is supported by the Jain Foundation and the NIH (RO1AR064369-01A1, R01AI072176-06A1).

References

1. Hirsch ML, Agbandje-McKenna M, Samulski RJ (2010) Little vector, big gene transduction: fragmented genome reassembly of adeno-associated virus. Mol Ther 18(1):6–8
2. Duan D et al (1999) Formation of adeno-associated virus circular genomes is differentially regulated by adenovirus E4 ORF6 and E2a gene expression. J Virol 73(1):161–169
3. Duan D et al (1998) Circular intermediates of recombinant adeno-associated virus have defined structural characteristics responsible for long-term episomal persistence in muscle tissue. J Virol 72(11):8568–8577
4. Duan D, Yue Y, Engelhardt JF (2001) Expanding AAV packaging capacity with trans-splicing or overlapping vectors: a quantitative comparison. Mol Ther 4(4):383–391
5. Halbert CL, Allen JM, Miller AD (2002) Efficient mouse airway transduction following recombination between AAV vectors carrying parts of a larger gene. Nat Biotechnol 20(7):697–701
6. Ghosh A et al (2008) A hybrid vector system expands adeno-associated viral vector packaging capacity in a transgene-independent manner. Mol Ther 16(1):124–130
7. Ghosh A et al (2007) Efficient whole-body transduction with trans-splicing adeno-associated viral vectors. Mol Ther 15(4):750–755
8. Dong B, Nakai H, Xiao W (2010) Characterization of genome integrity for oversized recombinant AAV vector. Mol Ther 18(1):87–92
9. Hirsch ML et al (2013) Oversized AAV transduction is mediated via a DNA-PKcs independent, Rad51C-dependent repair pathway. Mol Ther 21(12):2205–2216
10. Lai Y, Yue Y, Duan D (2010) Evidence for the failure of adeno-associated virus serotype 5 to package a viral genome > or = 8.2 kb. Mol Ther 18(1):75–79
11. Wu Z, Yang H, Colosi P (2010) Effect of genome size on AAV vector packaging. Mol Ther 18(1):80–86
12. Trapani I et al (2014) Effective delivery of large genes to the retina by dual AAV vectors. EMBO Mol Med 6(2):194–211
13. Ghosh A, Yue Y, Duan D (2011) Efficient transgene reconstitution with hybrid dual AAV vectors carrying the minimized bridging sequences. Hum Gene Ther 22(1):77–83
14. Mitchell AM et al (2010) AAV's anatomy: roadmap for optimizing vectors for translational success. Curr Gene Ther 10(5): 319–340
15. Samulski RJ et al (1982) Cloning of adeno-associated virus into pBR322: rescue of intact virus from the recombinant plasmid in human cells. Proc Natl Acad Sci U S A 79(6): 2077–2081
16. Samulski RJ et al (1983) Rescue of adeno-associated virus from recombinant plasmids: gene correction within the terminal repeats of AAV. Cell 33(1):135–143
17. Samulski RJ, Chang LS, Shenk T (1989) Helper-free stocks of recombinant adeno-associated viruses: normal integration does not require viral gene expression. J Virol 63(9): 3822–3828
18. Manno CS et al (2006) Successful transduction of liver in hemophilia by AAV-factor IX and limitations imposed by the host immune response. Nat Med 12(3):342–347
19. Testa F et al (2013) Three-year follow-up after unilateral subretinal delivery of adeno-associated virus in patients with Leber congenital Amaurosis type 2. Ophthalmology 120(6): 1283–1291
20. Inagaki K et al (2007) The role of DNA-PKcs and artemis in opening viral DNA hairpin termini in various tissues in mice. J Virol 81(20):11304–11321
21. Choi VW, McCarty DM, Samulski RJ (2006) Host cell DNA repair pathways in adeno-associated viral genome processing. J Virol 80(21):10346–10356
22. Choi VW, Samulski RJ, McCarty DM (2005) Effects of adeno-associated virus DNA hairpin structure on recombination. J Virol 79(11): 6801–6807
23. Sun L, Li J, Xiao X (2000) Overcoming adeno-associated virus vector size limitation through viral DNA heterodimerization. Nat Med 6(5): 599–602
24. Hirsch ML et al (2009) AAV recombineering with single strand oligonucleotides. PLoS One 4(11), e7705
25. Nakai H, Storm TA, Kay MA (2000) Increasing the size of rAAV-mediated expression cassettes in vivo by intermolecular joining of two complementary vectors. Nat Biotechnol 18(5): 527–532
26. Koo T et al (2014) Triple trans-splicing adeno-associated virus vectors capable of transferring the coding sequence for full-length dystrophin protein into dystrophic mice. Hum Gene Ther 25(2):98–108
27. Ghosh A et al (2009) Systemic trans-splicing adeno-associated viral delivery efficiently transduces the heart of adult mdx mouse, a model

for Duchenne muscular dystrophy. Hum Gene Ther 20(11):1319–1328

28. Lai Y et al (2008) Design of trans-splicing adeno-associated viral vectors for Duchenne muscular dystrophy gene therapy. Methods Mol Biol 433:259–275
29. Reich SJ et al (2003) Efficient trans-splicing in the retina expands the utility of adeno-associated virus as a vector for gene therapy. Hum Gene Ther 14(1):37–44
30. Xu Z et al (2004) Trans-splicing adeno-associated viral vector-mediated gene therapy is limited by the accumulation of spliced mRNA but not by dual vector coinfection efficiency. Hum Gene Ther 15(9):896–905
31. Lai Y et al (2006) Synthetic intron improves transduction efficiency of trans-splicing adeno-associated viral vectors. Hum Gene Ther 17(10):1036–1042
32. Yang J et al (1999) Concatamerization of adeno-associated virus circular genomes occurs through intermolecular recombination. J Virol 73(11):9468–9477
33. Zhang Y, Duan D (2012) Novel mini-dystrophin gene dual adeno-associated virus vectors restore neuronal nitric oxide synthase expression at the sarcolemma. Hum Gene Ther 23(1):98–103
34. Allocca M et al (2008) Serotype-dependent packaging of large genes in adeno-associated viral vectors results in effective gene delivery in mice. J Clin Invest 118(5):1955–1964
35. Hermonat PL et al (1997) The packaging capacity of adeno-associated virus (AAV) and the potential for wild-type-plus AAV gene therapy vectors. FEBS Lett 407(1):78–84
36. Grieger JC, Samulski RJ (2005) Adeno-associated virus as a gene therapy vector: vector development, production and clinical applications. Adv Biochem Eng Biotechnol 99:119–145
37. Wu J et al (2007) Self-complementary recombinant adeno-associated viral vectors: packaging capacity and the role of rep proteins in vector purity. Hum Gene Ther 18(2):171–182
38. Grose WE et al (2012) Homologous recombination mediates functional recovery of Dysferlin deficiency following AAV5 gene transfer. PLoS One 7(6), e39233
39. Grieger JC, Choi VW, Samulski RJ (2006) Production and characterization of adeno-associated viral vectors. Nat Protoc 1(3):1412–1428
40. Hirsch ML, Samulski RJ (2014) AAV-mediated gene editing via double-strand break repair. Methods Mol Biol 1114:291–307
41. Lock M et al (2010) Rapid, simple, and versatile manufacturing of recombinant adeno-associated viral vectors at scale. Hum Gene Ther 21(10):1259–1271

Chapter 3

Expression of Multiple Functional RNAs or Proteins from One Viral Vector

Tomas Björklund

Abstract

In this chapter, we will cover the available design choices for enabling expression of two functional protein or RNA sequences from a single viral vector. Such vectors are very useful in the neuroscience-related field of neuronal control and modulation, e.g., using optogenetics or DREADDs, but are also desirable in applications of CRISPR/Cas9 in situ genome editing and more refined therapeutic approaches. Each approach to achieving this combined expression has its own strengths and limitations, which makes them more or less suitable for different applications. In this chapter, we describe the available alternatives and provide tips on how they can be implemented.

Key words Bicistronic, Adeno-associated viral vector, Lentiviral vector, Transgene expression, In vivo gene transfer, Gene therapy, Genome editing, Bidirectional promoter, Ribosome skipping, IRES, Cloning

1 Introduction

Studies utilizing viral vector technology and in vivo gene transfer, using HIV-1-derived lentiviral and adeno-associated viral (AAV) vectors, have seen a major increase in the popularity over the last 10 years. This is due to the enormous flexibility these viral vectors provide, the relative ease of production of lentiviral vectors and the growing availability of viral vector cores that can produce high-quality AAV vectors. There have also emerged a number of broadly applicable applications such as induction and differentiation of induced pluripotent stem cChells [1], optogenetics [2], chemogenetics [3], and in situ genome editing [4] that all greatly benefit from the use of viral vector technology. Another reason is that a number of very promising viral vector therapies have now reached the clinic [5–7] and gene therapy has finally attracted large investments from big pharmaceutical companies, furthermore stressing that its applications will have a big impact on the future of clinical medicine. In clinical gene therapy as well as in

Fredric P. Manfredsson (ed.), *Gene Therapy for Neurological Disorders: Methods and Protocols*, Methods in Molecular Biology, vol. 1382, DOI 10.1007/978-1-4939-3271-9_3, © Springer Science+Business Media New York 2016

many of the advanced experimental designs pursued using viral vectors, there is often a need to be able to deliver more than one gene to the host. These may be multiple enzymes required to synthesize one therapeutic molecule [8], multiple transcription factors to revert a somatic cell to an IPS cell [9] or both a template RNA molecule and a nuclease for genome editing [10]. While mixing two viral vector preparations may be perfectly acceptable for some experimental settings, it is often beneficial to be able to have two or more functional sequences expressed from the same viral vector particle. The benefits from such an approach are multiple; the expression ratio between the two genes will then not be dependent on production titers and are thus more predictable, each target cell has then a higher chance of expressing both genes as a double-hit is guaranteed and the clinical translation will be significantly easier as the therapeutic agent is better defined.

The genome of viruses has been fine-tuned and optimized with evolution to result in sequences that are very compact and often reused (one promoter driving the expression of multiple genes and the same coding sequence giving rise to multiple proteins through splicing or leading and lagging strand translation). When we replace this genome with our promoters and genes of interest, we therefore have a very small space to work within. That size restriction often means that we are limited in our choice of promoters and other sequences and that they all will interact in sometimes unpredictable ways. This limitation becomes even more apparent when you aim to express more than one functional sequence in one viral vector genome.

In this chapter, we will present multiple strategies to achieve this and present the pros and cons of each approach (*see* Table 1).

2 Materials

Descriptions of the components in this chapter are mainly theoretical and described as a basis for in silico cloning and vector design. For the application in a specific vector design and generation of final plasmids, we refer to other descriptions of molecular cloning techniques, e.g., Molecular Cloning: A Laboratory Manual by Green and Sambrook. However, of note is that multiple approaches described in this chapter are very sensitive to frame-shifts and thus require verbatim cloning. Therefore, it is highly recommended to base the generation of fusion protein and ribosome-skipping vectors (described below) on gene synthesis products or Gibson assembly [11] and not on "traditional" restriction enzyme-based cloning. As gene synthesis is rapidly becoming cheaper and quicker, this is also an attractive alternative for design of other approaches described here as well.

Table 1
Comparison of alternative approaches for dual gene expression in viral vectors

Strategy	Products	Stoichiometry	Advantages	Disadvantages	Potential applications	References
Fusion protein (Fig. 1a)	Protein	1:1 predictable	+ Intracellular co-localization of the two proteins is guaranteed + Very space-efficient thus good for large genes + Inference of gene expression can be conducted both "upstream" and "downstream"	– One protein has to be compatible with N-terminal fusion and the other to C-terminal fusion to a second protein – Protein stability, function, and immunoreactivity may change compared to the two separate proteins	• Optogenetics/ DREADD • Multi-gene-dependent enzyme replacement	[14] [15]
Ribosome-skipping sequences (Fig. 1b)	Protein	1:1 total amount but unpredictable amounts of fusion protein	+2A sequences are very short + Inference of gene expression can be conducted "upstream"	– The biological principles behind the ribosome skipping are not entirely elucidated, making sequence design a "trial-and-error" approach	• IPS cell generation • Correction of multi-gene deficiency • Optogenetics/ DREADD	[19] [20] [32]

(continued)

Table 1 (continued)

Strategy	Products	Stoichiometry	Advantages	Disadvantages	Potential applications	References
Internal ribosome entry site (Fig. 1c)	Protein	Cell type-dependent, normally varying between 1:0.5 and 1:0.01	+ Inference of gene expression can be conducted "upstream"	–Stoichiometric ratio is very cell type-dependent making in vitro prediction of in vivo function difficult	• Optogenetics/ DREADD • Multi-gene-dependent enzyme replacement	[7] [33]
Bidirectional promoters (Fig. 1d)	Any Pol II expressed RNA (protein or miRNA product)	Promoter and UTR dependent	+ Expression strength of each RNA can be tailored slightly independently through modification of UTR sequences	– No inference of gene expression can be conducted – Can be difficult to implement in RNA base vectors as anti-sense RNA is produced during production	• Tet/Dox-regulated vectors • Marker gene for quantitative in vitro applications	[22] [24] [25] [34]
Dual promoters (Fig. 1e–g)	Any coding or non-coding RNA (Pol II or Pol III promoter driven)	Promoter and UTR dependent	+ Each expression cassette can be tailored individually + Pol II and Pol III promoters can be mixed	– Risk for promoter interference – No inference of gene expression can be conducted	• CRISPR/Cas9 in situ genome editing • Tet/Dox-regulated vectors • Enzyme replacement	[8] [23] [25] [26] [27]

3 Methods

3.1 Polycistronic Vectors

Polycistronic vector designs are the most commonly used approaches to enable the expression of two different proteins from one viral vector genome. The advantage with these approaches is that they are relatively easy to design. However, one limitation is that they are limited to expressing sequences coding for proteins. Therefore, they will not fulfill all needs in vector design. In addition, they also have other limitations, specific to each polycistronic approach, that we will discuss further on in the chapter. The three approaches covered under this section are fusion protein approach, ribosome-skipping sequences, and internal ribosome entry sites.

3.1.1 Fusion Proteins

Fusion proteins (Fig. 1a) are not technically derived from a polycistronic operon as the transcribed RNA codes only for a single open reading frame, albeit translated into a protein with (hopefully) all the functions of the two proteins fused together. However, it is the simplest way to achieve nearly the same thing as the two polycistronic approaches described below and is therefore a natural place to start.

The design of a fusion protein requires some understanding of the proteins involved with regards to their structure, signaling/trafficking peptides, and requirements to interact with other cellular components such as cell membrane or the mitochondria. The first criterion for a successful fusion protein is that they can reside in a location in the cell compatible with the function of both proteins, i.e., a cell-surface receptor with trans-membrane domains could be compatible with a cytosolic protein if they are fused in the intracellular end of the protein while it could not be compatible with a mitochondria-associated protein.

The second step in the design of a fusion protein is to understand the optimal order between the two polypeptides. For this step there is no universal recommendation. As noted above, there may be situations where a fusion must be, say at the C-terminal of a Type I or III trans-membrane protein for the second protein to be located in the cytosol and not on the cell surface (if this is desirable). If both proteins are soluble and can reside in the same cellular compartment, e.g., two enzymes in the production cascade of a neurotransmitter, then either of the two orders may work. However, one combination may provide superior function of the complex. In some of these cases, in vitro or in vivo comparisons are required before settling on the final design.

The final component of the fusion protein is the insertion of a polypeptide linker. While a linker is not required in every design of a fusion protein, it increases the chances of success, as it is providing flexibility in the protein structure and thus will allow both proteins fused to fold in as natural way as possible.

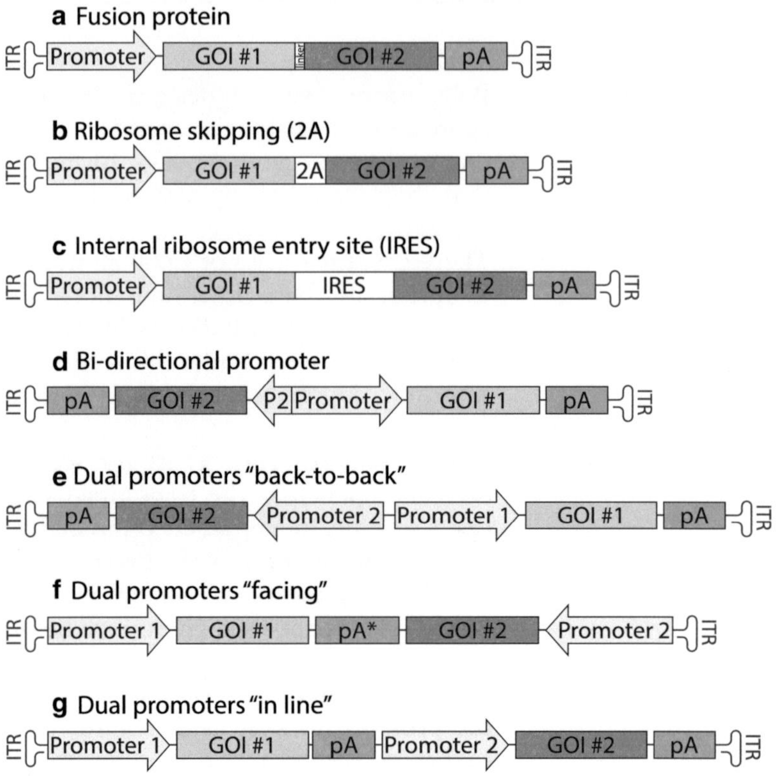

Fig. 1 Alternative designs for the expression of two proteins from one AAV vector. Schematic drawings showing the seven different design options applicable when the goal is to express two proteins (GOI #1 and #2) from the same viral vector. In this case the examples show an adeno-associated viral (AAV) vector design. The expression cassette is flanked by two inverted terminal repeats (ITR) enabling the packaging and complementary strand synthesis of the single-stranded DNA in the virus. Details on each of the seven constructs are found in the main text. Abbreviations: *GOI* gene of interest, *pA* poly-adenylation sequence, *pA** a pA sequence with bidirectional activity, e.g., the full-length sv40 pA, *2A* ribosome-skipping sequence, *IRES* internal ribosome entry site, *P2* short promoter fragment, e.g., from the CMV

Polypeptide linkers can be divided into three functional subgroups: flexible, ridged, and cleavable linkers. For more information *see* reference [12].

Applications of fusion proteins

The simplest example of a fusion protein is the addition of a short polypeptide "Tag" to either the N-terminal or the C-terminal of the expressed protein. Such tags are beneficial for a number of different functions. They are designed and evolved in close conjunction with the development of high affinity and specificity antibodies targeted toward that specific polypeptide sequence. This allows the tags to be used for recombinant protein purification or identification and visualization of the expressed protein in situ in the tissue even in cases where, either there are no available antibodies raised toward that specific protein of interest, or when you need to distinguish the ectopic expression from the endogenous, identical one.

Another interesting use of fusion proteins in viral vectors is the fusion of camelid nanobodies to RNA binding proteins or transcription factors for interaction with other proteins [13]. The development of the nanobody has risen from the unique forms of homodimeric heavy-chain antibodies (not requiring any light-chain components), to date found only in camels and related species. The variable region of these antibodies is intrinsically stable and can be truncated to form a nanobody with very high affinity to the antigen. In the paper of Ekstrand et al. 2014, they fused a nanobody (raised toward the GFP protein) to the ribosomal protein L10a and expressed it using a Cre-dependent AAV vector [14]. In vivo this was combined with a retrogradely transported canine adenovirus (CAV) to allow for a three-factor identification of dopamine cells that project to the nucleus accumbens of DAT-cre mice. Post-mortem, the nanobody-L10a fusion enabled GFP-dependent pull-down of transcribed RNA, enriched for the targeted cell population.

Fusion proteins are also commonly used in Optogenetic and DREADD studies where the channel rhodopsin or G-protein-coupled receptor is expressed using viral vectors (most commonly AAV vectors). This allows for both the identification of transduced cells in situ for live cell recording and for confirmation that the channels are correctly transported intracellularly to the site of stimulation with the optrode/ligand. The latter turned out to be a significant issue initially for the use of halorhodopsins in optogenetics [15, 16].

3.1.2 Controlling the Ribosome

In the evolutionary pressure of viral vectors to become smaller and more efficient, many viruses have developed related but slightly different approaches to achieve translation of all required virus-derived proteins from as short DNA/RNA sequence as possible. Such feats have been sometimes achieved through alternative initiation of translation of one sequence such as the VP1, 2, and 3 capsid proteins of AAV or to allow for some proteins to be expressed in *trans* from the same sequence that expresses a completely different protein in *cis*. Such approaches, while very elegant are highly sequence-dependent and are thus not universally applicable for viral vector design.

Two other approaches that viruses have developed are however much more broadly applicable, ribosome skipping and internal ribosome entry sites.

Ribosome-skipping sequences

Viruses in the *Picornaviridae* virus family, such as foot-and-mouth disease virus (and other related viruses) have developed a short consensus nucleotide sequence that is very unfavorable for the translation into a polypeptide sequence [17]. This sequence causes the ribosome to fail with addition of an amino acid at the C-terminal end during translation [18]. However, it does not prevent from the continued insertion of t-RNA molecule 3′ of the site. Thus, the translation continues but the new polypeptide generated is not linked to the polypeptide generated upstream of the virally derived sequence. The end result is two proteins generated from the same mRNA molecule but that are handled independently inside the cell. Such a sequence is called a ribosome-skipping sequence.

Commonly used ribosome-skipping sequences include a conserved 2A-like motif and will here collectively be referred to as just 2A sequences (Fig. 1b). The greatest advantage of 2A sequences in the application of viral vectors is that they are all very short. With a length between 21 and 30 nucleotides, they add very little to the size of the construct, something especially crucial for AAV vectors, but also when you aim to overexpress more than two proteins from the same lentiviral vector.

Secondly, the 2A sequences provide the most predictable stoichiometric ratio between the two proteins in the vector, around 1:1. However, this should not be taken as that the ribosome-skipping event is totally predictable or that all proteins generated are split into two. Neither of these is the case. The cleavage with a 2A sequence will ever be 100 % and there will always be a fraction of the protein residing as an uncleaved, fusion protein. The fraction of all protein that will be uncleaved is dependent both on the sequence the 2A is inserted into and the combination of cell type and 2A sequence utilized. Not enough is known about the 2A sequences to make this universally predictable and thus validation of each new construct is required inside the correct cell type.

1. Applications of ribosome skipping: One of the most popular current applications of 2A sequences is the utilization in single vector-mediated reprogramming of somatic cells into IPS cells using the four Yamanaka factors, Klf4, Oct4, Sox2, and c-Myc [1]. When fused in reading frame, this construct harboring three different 2A sequences E2A, P2A, and T2A inserted into a lentiviral vector expressing all genes under a single CMV promoter, this was capable of generating IPS cells from all three germ layers [19]. Interestingly, the majority of IPS cells generated contained only a single integration, showing that an approach of mixing four separate lentiviral vectors would in such a case prove potentially very inefficient.

2. Potential issues and recommendations: While the basic concept of the 2A sequence is straightforward, some extra attention is required in the design of the constructs. The first thing to be aware of is that the 2A sequence leaves traces in the final polypeptide. The 2A consensus sequence in itself codes for Asp-Val/Ile-Glu-X-Asn-Pro-Gly-Pro. Normally, the cleavage occurs just before the last proline (Pro) leaving only a single amino acid on the N-terminal of the downstream protein while the C-terminal of the upstream protein will have seven extra amino acids [20]. Thus, for a successful outcome, the upstream protein must handle a larger C-terminal peptide without disruption of the function. In addition, there will always be a fraction of the expressed protein that will remain as a fusion protein with an 8aa linker. Such species should preferably be functional as well, but at least not detrimental or toxic for the system studied.

   A second design consideration that the 2A sequences share with the fusion proteins above is that all three sequences (first and second protein and the 2A sequence) need to be in the same reading frame. Thus, it is strongly recommended to utilize either modern cloning techniques such as Gibson assembly [11] or ordering of the sequence from gene synthesis service providers.

3. Internal ribosome entry site: The internal ribosome entry site, IRES is probably the most well-known approach for the generation of polycistronic vectors and very widely utilized (Fig. 1c). As with the 2A sequences, the IRES was first discovered in viral sequences [21]. The sequence attracts the ribosome to assemble in the middle of an mRNA molecule and initiates translation at a new start codon. It is attractive for a number of reasons. The first is that the IRES is easy to clone into a vector and to get it to function satisfactory. It does not have to be aligned to any specific reading frame. All three com ponents (the upstream and downstream proteins and the IRES) can all be in different reading frames. The second reason is that the resulting proteins are always two separate proteins and no remnants of the IRES are found on either protein. Lastly, through the expression of the protein downstream of the IRES, the presence of the protein upstream can be reasonably well inferred. However, the opposite is not guaranteed and inferring anything about the expression level of the first protein based on the second one is not advisable. The reason for this is that the activity of the IRES is very cell type- and context-dependent. Thus the stoichiometric ratio between the two proteins expressed will vary significantly. The one thing that is certain though is that the second protein will never be more expressed than the first protein (if correct Kozak sequences are used for both reading frames).

Multiple promoters

All three approaches described above have been relying on a single promoter to generate one operon that is then translated into one or more proteins with multiple functions. Another strategy, however is to design viral vectors that would produce two or more species of mRNA from each singe vector-derived DNA molecule. Such vectors rely on multiple sequences that would promote transcription.

4. Bidirectional promoters: Most RNA Polymerase type II (Pol II)-based promoters have the ability to initiate transcription of DNA into mRNA not only in the forward direction (*cis*) but also in the reverse direction (*trans*). However, the efficiency of transcription is much lower in *trans*. To increase the efficiency of transcription in this direction, the 3′ fraction of the CMV promoter can be fused in reverse to the 5′ end of a preferred Pol II promoter, e.g., PGK [22]. This results in a bidirectional promoter that inherits most of the expression profile of the main promoter (in this example PGK) and will produce two transcripts with reasonably good efficiency (Fig. 1d). As this construct has a single entry site for the polymerase complex, promoter interference is considered less of an issue than when using dual promoters (described below). However, there are also aspects of the bidirectional promoter that are important to be aware of. The first is that the bidirectional promoters are all Pol II. Thus if one of the transcripts you aim to express is a short hairpin RNA or guide RNA, you will have to use two separate promoters (*see* below). Another potential area of concern is the utilization of a bidirectional promoter in RNA-based viral vectors such as lentiviruses. In this case, transcription in the reverse direction will generate an RNA during production that is complimentary to the RNA produced to be packaged during the production phase and may reduce the final titer. Thus if very high titer LV vectors are required, e.g., for intra-parenchymal injection into the CNS, one of the other approaches presented in this chapter may be preferable.
5. Dual promoters: Dual promoters in viral vectors have recently gained significant interest, especially in the context of the in situ genome editing using the Crispr/Cas9-based approaches (*see* below). The combination of two functionally independent promoters adds significant advantages over the strategies proposed above. First, it provides the option to express RNAs of different functional categories, as one or both of the functional promoters can be of the RNA Polymerase III (Pol III) family. Such promoters enable exact, consistent expression of short, non-coding RNAs such as small interfering RNAs (siRNA) or guide RNA (gRNA) utilized in gene silencing or genome editing respectively. Such RNA species cannot be expressed using

the Pol II family of promoters required for coding RNAs and the type of polycistronic constructs described above. However, in many cases, it is advantageous to be able to express both one coding and one non-coding RNA from a single AAV vector and this has therefore become the most utilized application of a dual promoter-based AAV construct.

6. Combining Pol II and Pol III promoters: The first use of combined Pol II/III AAV vectors was most likely the application in the field of gene silencing. While full-length pre-microRNAs can be expressed by Pol II promoters, the simpler and more broadly utilized short-hairpin RNAs (shRNA) require a precise transcription start and thus have to depend on a Pol III promoter. Visualization of the AAV-derived shRNA is non-trivial and due to the self-amplifying nature of the siRNA generated, the expression strength cannot be unequivocally determined. Therefore, many studies have included simultaneous expression of a marker gene from the same construct. Such approach is facilitated due to the small size of both commonly used Pol III promoters (U6 and H1) and the shRNA's short length.

   Similarly, the expression of the short guide synthetic RNA (sgRNA), used as a template and hybridization domain for the CRISPR (Clustered Regularly Interspaced Short Palindromic Repeats)/Cas9 genome editing, requires the use of a Pol III promoter. As the gene coding for the Cas9 nuclease from Streptococcus pyogenes is in itself over 4 kb in size, most AAV vector systems have had to deliver the Cas9 gene in a separate vector from the sgRNA (Cas9 from other species are now being characterizes as they are sometimes significantly smaller). In lentiviral vectors, however, the sgRNA and Cas9 gene can fit well in a single construct [23] often resulting in a LV vector of around 8 kb in size.

7. Dual Pol II promoters: There are a number of scenarios when you wish to express two genomic sequences that can be separately translated into protein. The most obvious application is when you wish to apply a regulated gene therapy using a single viral vector construct such as the tetracycline/doxycycline-regulated system (tet-on/-off). In these systems, it is potentially very beneficial to express both the trans-activator and the gene of interest (under the control of the trans-activator-regulated promoter) from the same viral vector. However, neither fusion proteins, internal ribosome-skipping sequences, IRES sequences or bidirectional promoters would suffice in this setting as none of these approaches allow for independent promoter control of the two operons as is required for the regulation to work (the trans-activator expression is normally driven by a constitutive promoter such as CMV). In AAV, three alternatives have been utilized; two separate promoters

placed "back-to-back" (Fig. 1e, similar to the bidirectional promoter but containing two full, functional promoters) [24], two "facing" promoters (Fig. 1f) at each end of the vector construct just next to the ITR [25] and two promoters placed in the same orientation (Fig. 1g), here called "in line" [8, 26]. The back-to-back approach is complicated by the fact that most promoters also have some, albeit lower, promoter activity also in the reverse orientation. In many cases this may not have severe consequences, but in a tet-on/off system, this reverse activity leads to significant leakage of the system [25]. It is possible that the leakage can be reduced through the use of a potent insulator sequence between the promoters, but such design comes at the expense of construct size.

The facing promoters have been least utilized in the literature. The reason for this is unclear, but when utilized in the tet-on system, the leakage of the system in the absence of doxycycline was as low as with the in-line vector design described below, but the activation was only two-thirds as strong, leading to a smaller dynamic range (fold difference between the inactive and active state) [25]. In-line promoters can be a very potent alternative. This design keeps the promoters furthest away from each other (recombinant AAV genomes concatemerize into episomal plasmids where promoters from "facing" vectors would come in close proximity to each other over the ITR), and allows for the expression strength and pattern of each gene to be controlled independently. In both the tet-on and the tet-off system, this has resulted in the largest dynamic range of the regulation with a very strong "on" state [25, 27].

The dual Pol II promoter approach also has the advantage over the other approaches that the expression ratio of each of the genes can be tightly controlled through the independent utilization of expression enhancing sequences such as cell-specific promoters, WPRE, introns and poly-A sequences. They can all here be applied independently to the two genes [8, 28]. However, in the design process, it is also important to take into account effects such as homologous recombination (discussed in Subheading 4) and read-through during transcription. The latter can be avoided using strong poly-A sequences and insulator sequences.

4 Notes

1. General design notes: When you aim to design a viral vector construct consisting of multiple functional domains, it is easy to gravitate toward using the same element multiple times, e.g., two copies of the same promoter. Such a design has the

advantage of achieving as similar expression pattern as possible of the two genes of interest. However, is does come with its own set of complications. First and foremost is the increased risk of homologous recombination. Especially in single-stranded DNA viruses such as the AAV, this recombination can be quite efficient and result in a mixed population vector batch where a fraction of the vector contains only the second gene (when they are placed in the same orientation). The efficiency of such recombination is dependent on both the length and the sequence of the repeated domain (with recombination increasing with the increased length). If the design of the vector is done such that the stoichiometric ratio increase with removal of the first gene is not detrimental (i.e., the second gene is the rate-limiting gene), such recombination may not necessarily be a showstopper, but it is important to take this phenomenon into account already in the design phase.

Other common areas of repeated sequence utilization is when more than two genes are expressed using a single promoter with the ribosome-skipping sequences or in cases where multiple poly-A sequences are used. Fortunately, in both these cases, multiple sequence alternatives exist for both of these classes, and therefore homology can be minimized without deteriorating the function.

2. Special considerations for RNA-based viruses, e.g., lentiviral vectors: Designing the expression cassette to be utilized in RNA vectors such as HIV-derived lentiviral or retroviral vectors needs a special attention. This is due to the fact that transcription of plasmid-derived DNA into the viral RNA is a key component of the viral vector production. This is achieved in mammalian cells (often HEK293T or Hela cells) so any ubiquitous promoter utilized inside the viral vector construct (to drive the transgene of interest) will also be expressed during production. While this is also true for DNA viruses, the noncoding aspects of RNA viruses can have detrimental effects on the titer of the RNA virus. The following vector components require extra attention when utilized in the design:
   (a) Poly-adenylation (pA) sequences
   (b) The WPRE sequence
   (c) Promoters placed in reverse (*trans*) direction
   (d) Micro-RNA target sequences
   (e) Introns

   The poly-adenylation (poly-A) sequence has a number of biological functions that all work together to make the amount of protein produced per copy DNA to be maximized. The two major functions are to terminate transcription from DNA to RNA and to induce a repeat sequence of adenine nucleotides at the 3′ end of the RNA that delays degradation and improves trafficking from the nucleus to the ribozomes in the cytosol. When a poly-A sequence is inserted inside an RNA virus expression plasmid, between the LTR sequences in *cis*, this will result in that the transcription into RNA is efficiently terminated there during production and any down-stream sequence, including the second LTR will not be transcribed. This will in turn be very detrimental for successful vector assembly and functional titer. To make matters more complicated, many poly-A sequences, especially those derived from viral sequences (very efficient and compact and thus very popular in vector design), such as the sv40 poly-A contain efficient poly-A function in both directions (terminating early and late genes respectively in the wild-type sv40 virus). Thus, if a poly-A sequence is desired in *trans*, in combination with a bidirectional promoter or a single promoter placed in *trans*, then care must be taken to choose a poly-A sequence with only activity in one direction (such as the synthetic poly A by Levitt et al., or a 5′ truncated sv40 poly-A [29]).

The Woodchuck hepatitis virus Post-transcriptional Regulatory Element (WPRE) is a virally derived sequence containing RNA hairpin structures and three functional domains. When inserted as the 3′ untranslated region (UTR) of the transgene of interest, is promotes a significantly increased level of protein per genome copy. This appears to be achieved through multiple mechanisms including stabilization of mRNA and maybe also trafficking of mRNA from the nucleus out to the cytosol. The inclusion of the WPRE is very common in lentiviral vectors, as a poly-A sequence cannot be utilized. However, the direction of the WPRE sequence is very important, as placement of the sequence in reverse significantly reduces the transgene expression compared to a vector without any WPRE sequence [30]. Unpublished data from our lab show that this is due to early RNA termination, resulting in suboptimal lenti production. Thus, lentiviral designs where a WPRE sequence is utilized in *trans* in combination with, e.g., a bidirectional promoter will most likely result in very low production titers and are therefore discouraged. An alternative in this case is then to utilize a CTE domain, which also has a hairpin structure and promotes trafficking of the RNA from the nucleus into the cytosol. However, little data exist to show that it has any added benefit over the use of a poly-A sequence placed in *trans* (with no *cis* acting poly-A activity).

There are a number of scenarios where the placement of the transgene expression cassette in reverse orientation (*trans*) can be very beneficial, e.g., if you aim to express sequences that are modified on the mRNA level, e.g., sequences containing introns.

If placed in *cis*, such intronic sequence would be spliced out during lenti production and resulting in a vector inserted with only exonic sequences. However, when such sequences are placed in *trans*, they should be retained also in the lentiviral vector.

Similarly, micro-RNA targets when designed to include perfect match for micro-RNAs expressed in the production cell line may also result in a decreased production titer when placed in *cis* orientation. Such vectors may also benefit from placing the expression cassette in *trans*.

When utilizing such a design you need to be aware that it also can come at a cost of reduction in viral titers. This is due to the fact that the RNA expressed from the internal promoter, placed in *trans*, will be complementary to the full lentiviral RNA containing the LTRs. Such double-stranded RNA activates the antiviral defense system in the production cell line through the activation of the dsRNA-activated inhibitor (DAI) enzyme. The activation of DAI is then phosphorylating eIF-2, halting the protein production in the cell. Fortunately, there are ways to reduce this antiviral defense system, e.g., using the adenoviral virus-associated RNA I [31]. This RNA, expressed by a Pol III promoter in a separate helper plasmid during production (e.g., pAdVantage from Promega) inhibits the DAI enzyme and thereby significantly improving the production titer. In our hands, the addition of this plasmid in the production increases the titers of any lentiviral vector and can thus be included without detriment even when a promoter in *trans* is not included.

References

1. Takahashi K, Yamanaka S (2006) Induction of pluripotent stem cells from mouse embryonic and adult fibroblast cultures by defined factors. Cell 126:663–676
2. Deisseroth K, Schnitzer MJ (2013) Engineering approaches to illuminating brain structure and dynamics. Neuron 80:568–577
3. Giguere PM, Kroeze WK, Roth BL (2014) Tuning up the right signal: chemical and genetic approaches to study GPCR functions. Curr Opin Cell Biol 27:51–55
4. Hsu PD, Lander ES, Zhang F (2014) Development and applications of CRISPR-Cas9 for genome engineering. Cell 157:1262–1278
5. Maude SL, Frey N, Shaw PA et al (2014) Chimeric antigen receptor T cells for sustained remissions in leukemia. N Engl J Med 371: 1507–1517
6. Maguire AM, High KA, Auricchio A et al (2009) Age-dependent effects of RPE65 gene therapy for Leber's congenital amaurosis: a phase 1 dose-escalation trial. Lancet 374: 1597–1605
7. Palfi S, Gurruchaga J-M, Ralph GS et al (2014) Long-term safety and tolerability of ProSavin, a lentiviral vector-based gene therapy for Parkinson's disease: a dose escalation, open-label, phase 1/2 trial. Lancet 383:1138–1146
8. Cederfjäll E, Sahin G, Kirik D et al (2012) Design of a single AAV vector for coexpression of TH and GCH1 to establish continuous DOPA synthesis in a rat model of Parkinson's disease. Mol Ther 20:1315–1326
9. Sommer CA, Stadtfeld M, Murphy GJ et al (2009) Induced pluripotent stem cell generation using a single lentiviral stem cell cassette. Stem Cells 27:543–549

10. Swiech L, Heidenreich M, Banerjee A et al (2014) In vivo interrogation of gene function in the mammalian brain using CRISPR-Cas9. Nat Biotechnol 33(1):102–106
11. Gibson DG, Young L, Chuang R-Y et al (2009) Enzymatic assembly of DNA molecules up to several hundred kilobases. Nat Methods 6: 343–345
12. Chen X, Zaro JL, Shen W-C (2013) Fusion protein linkers: property, design and functionality. Adv Drug Deliv Rev 65:1357–1369
13. Muyldermans S (2013) Nanobodies: natural single-domain antibodies. Annu Rev Biochem 82:775–797
14. Ekstrand MI, Nectow AR, Knight ZA et al (2014) Molecular profiling of neurons based on connectivity. Cell 157:1230–1242
15. Gradinaru V, Thompson KR, Deisseroth K (2008) eNpHR: a Natronomonas halorhodopsin enhanced for optogenetic applications. Brain Cell Biol 36:129–139
16. Gradinaru V, Zhang F, Ramakrishnan C et al (2010) Molecular and cellular approaches for diversifying and extending optogenetics. Cell 141:154–165
17. Luke GA, de Felipe P, Lukashev A et al (2008) Occurrence, function and evolutionary origins of "2A-like" sequences in virus genomes. J Gen Virol 89:1036–1042
18. Donnelly ML, Luke G, Mehrotra A et al (2001) Analysis of the aphthovirus 2A/2B polyprotein 'cleavage' mechanism indicates not a proteolytic reaction, but a novel translational effect: a putative ribosomal "skip". J Gen Virol 82: 1013–1025
19. Shao L, Feng W, Sun Y et al (2009) Generation of iPS cells using defined factors linked via the self-cleaving 2A sequences in a single open reading frame. Cell Res 19:296–306
20. Szymczak AL, Workman CJ, Wang Y et al (2004) Correction of multi-gene deficiency in vivo using a single "self-cleaving" 2A peptide-based retroviral vector. Nat Biotechnol 22:589–594
21. Pelletier J, Sonenberg N (1988) Internal initiation of translation of eukaryotic mRNA directed by a sequence derived from poliovirus RNA. Nature 334:320–325
22. Amendola M, Venneri MA, Biffi A et al (2005) Coordinate dual-gene transgenesis by lentiviral vectors carrying synthetic bidirectional promoters. Nat Biotechnol 23:108–116
23. Sanjana NE, Shalem O, Zhang F (2014) Improved vectors and genome-wide libraries for CRISPR screening. Nat Methods 11:783–784
24. Fagoe ND, Eggers R, Verhaagen J et al (2014) A compact dual promoter adeno-associated viral vector for efficient delivery of two genes to dorsal root ganglion neurons. Gene Ther 21:242–252
25. Chenuaud P, Larcher T, Rabinowitz JE et al (2004) Optimal design of a single recombinant adeno-associated virus derived from serotypes 1 and 2 to achieve more tightly regulated transgene expression from nonhuman primate muscle. Mol Ther 9:410–418
26. Kügler S, Lingor P, Schöll U et al (2003) Differential transgene expression in brain cells in vivo and in vitro from AAV-2 vectors with small transcriptional control units. Virology 311:89–95
27. Manfredsson FP, Burger C, Rising AC et al (2009) Tight long-term dynamic doxycycline responsive nigrostriatal GDNF using a single rAAV vector. Mol Ther 17:1857–1867
28. Björklund T, Hall H, Breysse N et al (2009) Optimization of continuous in vivo DOPA production and studies on ectopic DA synthesis using rAAV5 vectors in Parkinsonian rats. J Neurochem 111:355–367
29. Levitt N, Briggs D, Gil A et al (1989) Definition of an efficient synthetic poly(A) site. Genes Dev 3:1019–1025
30. Zufferey R, Donello JE, Trono D et al (1999) Woodchuck hepatitis virus posttranscriptional regulatory element enhances expression of transgenes delivered by retroviral vectors. J Virol 73:2886–2892
31. O'Malley RP, Mariano TM, Siekierka J et al (1986) A mechanism for the control of protein synthesis by adenovirus VA RNAI. Cell 44: 391–400
32. Stachniak TJ, Ghosh A, Sternson SM (2014) Chemogenetic synaptic silencing of neural circuits localizes a hypothalamus→midbrain pathway for feeding behavior. Neuron 82:797–808
33. Parnaudeau S, O'Neill P-K, Bolkan SS et al (2013) Inhibition of mediodorsal thalamus disrupts thalamofrontal connectivity and cognition. Neuron 77:1151–1162
34. Amendola M, Giustacchini A, Gentner B et al (2013) A double-switch vector system positively regulates transgene expression by endogenous microRNA expression (miR-ON vector). Mol Ther 21:934–946

Chapter 4

Regulated Gene Therapy

Ludivine Breger, Erika Elgstrand Wettergren, Luis Quintino, and Cecilia Lundberg

Abstract

Gene therapy represents a promising approach for the treatment of monogenic and multifactorial neurological disorders. It can be used to replace a missing gene and mutated gene or downregulate a causal gene. Despite the versatility of gene therapy, one of the main limitations lies in the irreversibility of the process: once delivered to target cells, the gene of interest is constitutively expressed and cannot be removed. Therefore, efficient, safe and long-term gene modification requires a system allowing fine control of transgene expression.

Different systems have been developed over the past decades to regulate transgene expression after in vivo delivery, either at transcriptional or post-translational levels. The purpose of this chapter is to give an overview on current regulatory system used in the context of gene therapy for neurological disorders. Systems using external regulation of transgenes using antibiotics are commonly used to control either gene expression using tetracycline-controlled transcription or protein levels using destabilizing domain technology. Alternatively, specific promoters of genes that are regulated by disease mechanisms, increasing expression as the disease progresses or decreasing expression as disease regresses, are also examined. Overall, this chapter discusses advantages and drawbacks of current molecular methods for regulated gene therapy in the central nervous system.

Key words Tet-responsive, Doxycycline, Promoter, Zinc finger-based transcription factor, Destabilizing domain, Trimethoprim

1 Introduction

The possibility to regulate transgene expression has been discussed in the gene therapy field for a long time (*see*, e.g., [1, 2]). In clinical settings, regulated transgene expression would allow for increased or decreased transgene levels in response to clinical need. Regulating transgene expression would ideally provide a means to avoid adverse effects due to continuous overexpression of therapeutic genes. Furthermore, the ability to turn transgene expression off and on offers experimental advantages when studying causal effects of gene transfer in disease models.

Fredric P. Manfredsson (ed.), *Gene Therapy for Neurological Disorders: Methods and Protocols*, Methods in Molecular Biology, vol. 1382, DOI 10.1007/978-1-4939-3271-9_4, © Springer Science+Business Media New York 2016

Many different regulated gene expression systems have been developed and most operate at transcriptional levels. In this chapter we discuss three different approaches to achieve regulation of genes by gene therapy. Two are active at the transcriptional level and target either transgenic or endogenous genes. The third example regulates protein stability rather than transcriptional activity and represents a novel approach to transgene regulation that may be utilized for gene therapy to the brain.

1.1 Inducible Promoters: Tetracycline-Controlled Transgene Expression

Various drug-dependent induced technologies have been developed to control gene expression in mammalian cells. However, the most common and widely used remains the tetracycline-controlled promoter activity developed by Gossen and colleagues more than 20 years ago [3, 4]. The tetracycline systems take advantage of the tetracycline-resistance operon derived from the Tn10-resistant *E. coli* strain. In these bacteria, tetracycline-resistant mediated promoters are repressed by the binding of the tetracycline-dependent repressor (TetR) on the tetracycline operator (TetO). In the presence of the antibiotic tetracycline, the TetR is prevented from binding its operator, thus allowing transcription of the genes. Two main variants of controlled expression were developed based on this mechanism: the Tet-Off and Tet-On system. The first one uses a fusion of DNA-binding domain of the TetR, obtained from the Tn10 *E. coli*, and the C-terminal transcription activation domain of virion protein 16 of herpes simplex virus (VP16). The resulting DNA is placed under the control of a tissue/cell-specific promoter, therefore allowing expression of the tetracycline-controlled transactivator (tTA) in desired cell types. Controllable expression of a gene of interest is obtained by placing the target gene under control of a minimal promoter sequence of the cytomegalovirus promoter (CMV) fused with TetO. In the absence of the antibiotics tetracycline, the tTA, expressed in a cell-specific manner, will bind to the TetO, thus initiating the transcription of the target gene (Fig. 1). Administration of tetracycline switches off the system. Indeed, by binding the tTA, tetracycline induces conformational changes, preventing tTA from binding and activating the TetO, therefore blocking the transcription of the downstream target gene. In opposition, the Tet-On system required the presence of tetracycline to allow transcription of the target gene. Indeed, the reverse tetracycline-controlled transcriptional activator system (Tet-On), although based on the same principle, has the complete opposite effect. In more details, a mutant Tet repressor was fused to VP16, altogether coding for the rtTA, which can bind to TetO only after conformational changes occurring while binding tetracycline.

Although both systems are commonly used in neuroscience research, it is considered preferable to use a Tet-On approach for the development of gene therapy for the treatment of neurologi-

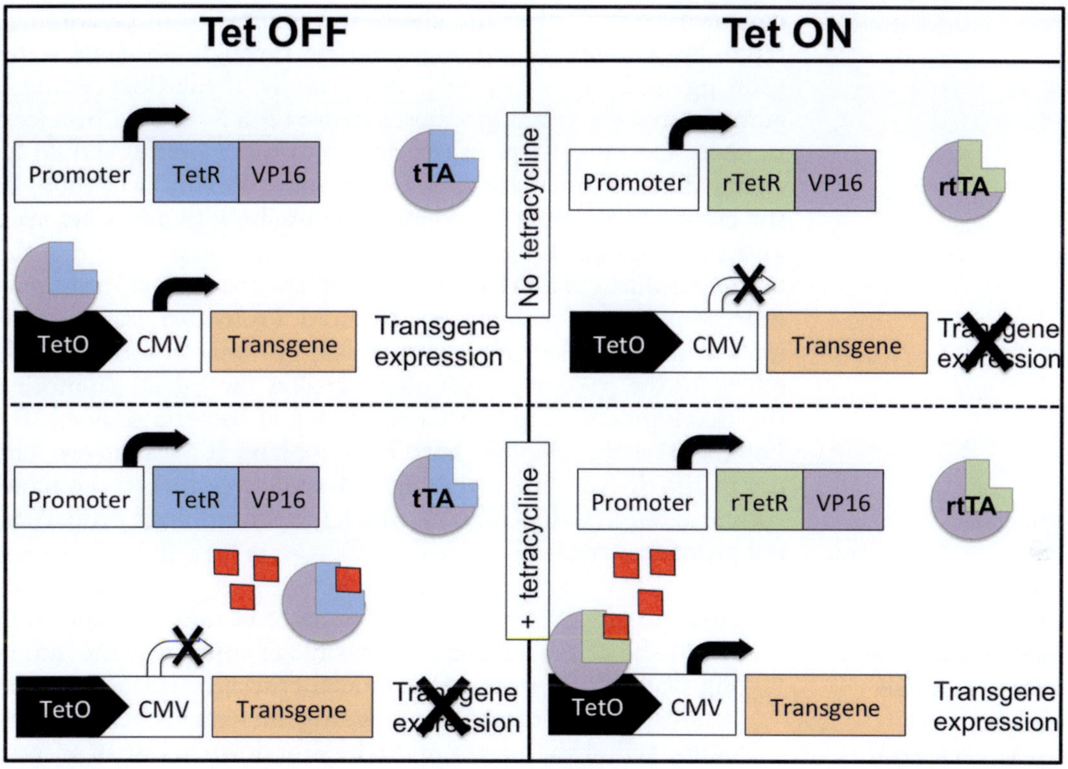

Fig. 1 Schematic representation of the Tet-Off (*left*) and Tet-On (*right*) system, in the presence (*higher panel*) or absence (*lower panel*) of tetracycline (*red squares*). Tet-Off: the transactivator (tTA) is expressed under the control of a specific promoter. In the absence of tetracycline, it binds the operator (TetO) and activates the CMV promoter, leading to transgene expression, while in the presence of tetracycline, the tTA undergoes conformation changes and is no longer able to bind the TetO which ends the transcription of the transgene. Tet-On: in the absence of tetracycline, the reverse transactivator (rtTA) is unable to bind the TetO and no transcription can occur, when in the presence of tetracycline it changes conformation and is enabled to bind the TetO, allowing transcription of the transgene

cal disorders. Indeed, an approach where transgene expression is normally repressed and will only occur when patients are submitted to treatment with the inducer is considered safer. The original inducing drug used to activate the Tet-On system was a tetracycline, but other derivatives have been used. Among theme, doxycycline, another antibiotic, is currently the most widely used as it has a low cost and a long half-life and crosses the blood-brain barrier easily [5]. However, it has been shown that the half-life of doxycycline can be reduced by 50 % when co-administered with other neurological treatments [6]. Patients suffering from neurological disorders are usually treated with various cocktails of drugs. It is therefore important to bear in mind that inducers remain active drugs, which could interact with other treatments that the patients might be on. Doxycycline has very limited side effect and has been safely used in the clinic. In rodents, it can be administered by gavage (20–50 mg/day) or through drinking water (200 μg to 2 mg/ml) [7–9]. Importantly, if the doxycycline is administered via drinking water, sucrose (2–5 %) should be added to cover the bitter taste of the drug and water bottles should be changed every other day as the drug loses stability over time. In the clinic, doxycycline is administered orally, with doses between 100 and 200 mg/day for adults. It is of course important that the dose of inducer required to reach therapeutic level of transgene expression remains below that threshold. Prolonged use of antibiotics as inducer rises important issues, not only in terms of side effect for the patients, but it also increases the risk of promoting the development of antibiotic resistance in bacteria strains [10]. For that reason, scientists have been looking at alternatives. The use of the doxycycline metabolite (4-epidoxycyclin) or the tetracycline agonist (GR333076X), which has no antibiotic properties, is a promising option.

1.2 The Five Golden Rules for a Clinically Relevant Inducible Transgene Expression System

1. Absence of basal expression: In order to be safe, the expression should be as close to zero as possible in absence of the inducing drug ("off" state). It is indeed crucial to ensure that the residual level of expression of transgene remains below therapeutic action so the system can be shut down in case of adverse effects.
2. Rapid, dose-dependent induction: The expression of the transgene should occur rapidly after administration of the inducing drug. The level of expression of the transgene should be dependent of the dose dependency inducer administered. The inducer should be have a long half-life and be able to cross the blood-brain barrier. Finally, protein levels should be within the therapeutical range.
3. Quick shut down: In order to better manage potential side effects, the expression of the transgene should stop rapidly after discontinuation of the drug treatment. However, the stability of the therapeutic molecule will also influence the duration of ongoing adverse events.
4. Specificity: The transgene should be expressed in a discrete area and/or cells types in the brain. This will ensure maximal and localized effect while reducing the risk of adverse effect.
5. Limited immune response: The delivery of viral vectors into the brain requires surgical intervention, thus compromising the blood-brain barrier. The rupture of the wall, normally isolating the brain from circulating white blood cells, might trigger an immune response against the exogenous protein. To minimize immunogenicity, it is important to use human genes and to avoid contamination with animal products (e.g., serum in culture medium).

1.3 In a Nutshell

Main advantages: The Tet system allows fine control of level of expression of the transgene and can be shut down if necessary. Doxycycline is commonly used in the clinic to treat infections; it is a potent, low-cost, cheap, and safe inducer. Finally, the Tet system has been extensively characterized and tested in animal models of neurological disorder (for review, *see* ref. 11).

Main drawbacks: One of the main challenges associated with inducible expression approaches for gene therapy is the leakiness of the system. The existence of basal level of expression of the transgene in "off state" raises serious concern about the controllability of the system. Research is currently ongoing to improve tightness of the system, including reduction of nonspecific transactivator-TetO binding. The second issue concerns the triggering of an immune response and inflammation by the Tet transactivator. As the majority of the population have been in contact with the herpes simplex virus from which the VP16 part of the rTA has been derived, this system can be particularly immunogenic [12].

Important things to consider: The Tet systems comprise two elements, the rtTA and the transgene cassettes (Fig. 1) that can be delivered either separately or by the same vector. However, it is possible that altogether, these constructs exceed the cloning capacity of certain viral vectors (e.g. adeno associated virus ≈4.5 kb). Although the use of a dual-vector approach is possible, it results in reduced expression of the transgene, as each cell has to be transduced by the two vectors to allow gene expression. The single vector approach is therefore highly recommended. Different configurations of the Tet system have been developed and adapted for single vector approaches: either using right-facing cistrons, where the transgene and the rtTA are placed one after the other, or using a bidirectional promoter to drive the expression of the transgene, on one side, and the rtTA on the other side. Finally, the areas of the brain and the cell types to be targeted should be carefully chosen as ectopic expression can influence efficacy and safety of the treatment.

1.4 Autoregulated Promoters

An autoregulated system could be an alternative to using a drug-regulated system, such as the tetracycline and rapamycin systems. These have the advantage that no proteins of nonmammalian origin have to be over expressed and no exogenous administration of a regulating drug is required. Instead, autoregulated systems are based on a promoter or regulatory elements from an endogenous gene. Examples of promoters that have been used in autoregulated vectors for CNS gene therapy are the glial fibrillary acidic protein (GFAP) promoter and the enkephalin (ENK) promoter [13]. GFAP is expressed in astrocytes and is upregulated in the gliotic reaction following a lesion. ENK has been shown to be upregulated in striatal neurons of the indirect pathway following dopamine depletion in Parkinson's disease (PD). The authors showed that vectors containing these promoters have a similar expression pattern in rat striatum as the endogenous proteins in animals subjected to lesions or dopamine depletion, respectively. The transgene expression was also responsive to decreases in inflammation and restoration of dopamine levels. The hypoxia-responsive element (HRE) from the erythropoietin gene has also been used to achieve autoregulated transgene expression in the CNS [14]. In this system, nine copies of the HRE sequence were coupled to a SV40 minimal promoter. No transgene expression was detected in healthy mouse brain, but transgene expression could be detected following transient middle cerebral artery occlusion. These results show that promoter elements or promoters of genes regulated by disease or changes in the cells environment could be used to create autoregulated vectors.

Zinc finger-based transcription factors (ZFTFs) can be used to regulate the transcription of endogenous genes. ZFTF consists of several connected zinc fingers, which determine binding specificity, a nuclear localisation signal and an activating or repressing domain. A ZFTF consisting of six zinc fingers recognizes an 18-base pair sequence and is regarded as specific for one site in the human genome. Most studies using ZFTF in the CNS have used ZFTF designed to target and upregulate the endogenous vascular endothelial factor (VEGF) gene. VEGF plays a role in angiogenesis, but has also been shown to have neuroprotective and neurotrophic effects [15]. Beneficial effect on cell survival and motor behavior has so far been reported in rat models of stroke, spinal cord injury, and traumatic brain injury following injection of viral vectors carrying the VEGF ZFTF gene [16–18]. ZFTF has also been used in studies on the neurodegenerative disorders Huntington's disease (HD) and PD. In the HD study, the authors used a ZFTF designed to target extended CAG repeats [19]. By this approach, they were able to specifically knockdown the mutant huntingtin allele and improve motor behavior in an HD mouse model. In the PD study, the authors used a ZFTF designed to upregulate the endogenous glial cell line-derived neurotrophic factor (GDNF) gene [20]. This potent neurotrophic factor has been shown to promote the survival of dopaminergic neurons. Expression of the GDNF ZFTF in a rat model of PD reduced the loss of dopaminergic neurons and improved motor behavior.

Transcription activator-like effector based transcription factors (TALE-TFs) can also be used to regulate the transcription of an endogenous gene. TALE-TFs consist of several connected DNA-binding repeats derived from natural TALEs found in *Xanthomonas*, a nuclear localization signal and an activating or repressing domain. The DNA-binding domain of TALE-TFs is more modular than the domain found in ZFTF. While each finger in a ZFTF recognize three to four base pairs and neighboring fingers affect each other, each DNA-binding repeat in a TALE-TF recognise only one base pair without any influence from neighboring repeats. The use of TALE-TFs to regulate endogenous genes is still a fairly new technology and studies using TALEs in the CNS has therefore been few. However, a recent study combined light-inducible transcriptional effector technology with a customised TALE DNA-binding domain to create an optically controlled TALE-TF [21]. The authors used a two-component system where the first component contained the TALE DNA-binding domain coupled to CIB1 and the second component contained a light-sensitive cryptochrome 2 protein coupled to an activator. Upon exposure to blue light, CIB1 and cryotochrome 2 combine to create a functional TALE-TF. The study showed that this technology could be used to upregulate the endogenous metabotropic glutamate receptor 2 gene (Grm2) in the mouse prefrontal cortex. Upregulation of Grm2 was also possible in a more traditional TALE-TF setting using a TALE DNA-binding domain coupled directly to the activator.

Both ZFTF and TALE-TFs have the advantage that all splice variants of the gene is produced since both technologies function at the level of transcription. This is essential when overexpressing certain genes. For instance, overexpression of VEGF using cDNA for only one splice variant leads to the formation of leaky vessels. By contrast, overexpression of all the splice variants using a ZFTF designed to target the endogenous gene leads to new fully functional vessels [22].

1.5 Destabilizing Domains

It is also possible to regulate gene expression at the protein level using destabilizing domains (DD). These are protein domains that have been mutated to be readily ubiquitinated and consequently targeted for destruction to the ubiquitin-proteasome system. Depending on the protein used to create the DD, it is possible to have small-molecule ligands that shield the DD from degradation, thereby stabilizing it. Creating a fusion protein containing the DD and a protein of interest will target the whole fusion protein to degradation that can be rescued in the present of the shielding small molecule ligands. Therefore, the DD system can be used to regulate gene expression by regulating protein stability of proteins fused to DD.

To date, three different proteins have been used to engineer DD: FK506- and rapamycin-binding protein (FKBP) using the synthetic ligand shield-1 as the stabilizing ligand [23], *E. coli* dihydrofolate reductase (DHFR) using trimethoprim (TMP) as the stabilizing ligand [24] and estrogen receptor ligand binding domain (ERLBD) using hydroxytamoxifen (4OHT) as the ligand [25]. From the three DD, DHFR and ERLBD can be used to regulate gene expression in the brain as TMP [26] and 4OHT [27] can cross the blood-brain barrier.

Initial characterization of YFP fused to DHFR-based DD (YFP-DD) [24] showed that YFP-DD was efficiently regulated in the brain of rats. In a following study using YFP-DD [28], it was shown that YFP-DD expression could be reversibly regulated with peak expression 3 weeks after TMP treatment was initiated and returned to background levels 3 weeks after TMP treatment ceased.

Furthermore, placing the DD in the C or N terminus of the fusion protein influenced the stability and induction of YFP-DD. C-terminal placement of DD led to more consistent expression after induction and lower background expression when the system was turned off. Moreover, the C-terminal YFP-DD could be regulated in a dose-response manner between 0.01 and 0.2 mg/ml TMP.

The DHFR-DD system has also been used to regulate GDNF. The first-generation GDNF DD fusion proteins (GDNF-DD) resulted in limited induction of C and N terminal GDNF-DD [28]. Subsequent analysis [29] indicated that the C-terminal GDNF-DD was not efficiently processed and the N-terminal GDNF-DD had impaired secretion. To address these issues, second generation of GDNF-DD were created where the N-terminal placement of DD was optimized and an additional furin-cleavage site was added to the DD placed on the C-terminal DD.

In vitro validation assays indicated that while second-generation N-terminal GDNF-DD was efficiently secreted, it had a high background expression when the system was not induced. On the other hand, the second-generation C-terminal GDNF-DD was efficiently secreted and had a negligible background when the system was off.

Second-generation C-terminal GDNF-DD was validated in vivo in the striatum of rats. Three weeks of TMP induction was sufficient to elicit a robust GDNF-DD expression. When compared to wild-type GDNF, GDNF-DD secreted 4–6 times less protein [29]. However, the amount of secreted GDNF-DD was functional as it was sufficient to activate signaling pathways in target cells. The group of GDNF-DD animals that did not receive TMP had only minimal expression of GDNF-DD that was not functional.

GDNF-DD was also tested in a 6-hydroxydopamine induced model of PD and when induced showed neuroprotective effects comparable to wild-type GDNF. Animals where GDNF-DD was not induced showed low levels of GDNF-DD that was not functional. Moreover, the GDNF-DD animals not given TMP were comparable to YFP-DD control animals. This suggested that second-generation GDNF-DD could be regulated to therapeutic levels in vivo and exhibited a very tight regulation in vivo, event in neurodegenerative disease models.

The DD system has several advantages as the system needs only the fusion protein and has negligible expression when the system is not induced. There are also considerations for the use of the DD system. The system has a lower dynamic range of induction when compared to tetracycline-based inducible systems; therefore it is suitable for secreted proteins or proteins that do not require very high levels of expression. Moreover, the design of the fusion protein is empirical and the ideal placement of the DD needs to be validated. Due to the posttranslational nature of the regulation, the regulated proteins may be detected at low levels and need to be validated using functional assays to ensure that any residual expression inert. Although, the use of DD system in the brain is still in its infancy, preliminary studies show great promise and the system seems especially suited for gene regulation in the brain.

1.6 Guidelines for Creation and Validation of DD-Regulated Proteins

Designing DD fusion proteins: N and C-terminal DD placements need to be designed and validated for every protein of interest to assess optimal DD placement. Also glycine linkers need to be added [24] ensure minimal steric hindrance from the DD. For secreted proteins an extra furin cleavage site can be added [29].

In vitro validation: After cloning, production of viral vectors and transduction of cells, the DD can be induced using a concentration of 10 μM TMP for 24 h [23, 24, 28, 29]. A functional assay should be designed to ensure that the fusion between the protein of interest and DD is functional and that there is no leakage of the DD when it is not induced. This is of special importance for proteins that go through the secretory pathway as a reservoir of DD will be present at the endoplasmic reticulum [29, 30]. Although, this DD reservoir is not functional, it needs to be considered when validating candidates.

In vivo TMP treatments: For in vivo induction, 0.01–0.5 mg/ml TMP should be given in the drinking water of animals, continuously for at least 3 weeks to ensure a robust induction [28, 29]. YFP-DD studies indicate the animals need at least 3 weeks without TMP that to ensure that expression of DD reverts to basal levels. Similarly, to the in vitro situation, there will be a low nonfunctional expression of proteins fused to DD, particularly in the case of secreted proteins [29]. Although this background expression is inert, it needs to be accounted and assessed using functional assays in vivo.

2 Summary

The possibility to regulate gene expression by gene therapy is indeed a promising future avenue for gene therapy to the brain. It will, however, need further development and characterization to become a viable clinical option. The work includes analysis of immunological responses, regulation of repeated cycles and over long periods of time, and of course many efficacy parameters in relevant in vivo models.

References

1. Cress D. E. (2008) The need for regulatable vectors for gene therapy for Parkinson's disease. Exp Neurol 209:30–3
2. Kordower J. H. and Olanow C. W. (2008) Regulatable promoters and gene therapy for Parkinson's disease: is the only thing to fear, fear itself? Exp Neurol 209:34–40
3. Gossen M, Bujard H (1992) Tight control of gene expression in mammalian cells by tetracycline-responsive promoters. Proc Natl Acad Sci U S A 89:5547–5551
4. Gossen M, Freundlieb S, Bender G et al (1995) Transcriptional activation by tetracyclines in mammalian cells. Science 268:1766–1769

5. Cunha BA (2000) Minocycline versus doxycycline in the treatment of Lyme neuroborreliosis. Clin Infect Dis 30:237–238
6. Penttila O, Neuvonen PJ, Aho K et al (1974) Interaction between doxycycline and some antiepileptic drugs. Br Med J 2:470–472
7. Okoye G, Zimmer J, Sung J et al (2003) Increased expression of brain-derived neurotrophic factor preserves retinal function and slows cell death from rhodopsin mutation or oxidative damage. J Neurosci 23:4164–4172
8. Kafri T, van Praag H, Gage FH et al (2000) Lentiviral vectors: regulated gene expression. Mol Ther 1:516–521
9. Corti O, Sanchez-Capelo A, Colin P et al (1999) Long-term doxycycline-controlled expression of human tyrosine hydroxylase after direct adenovirus-mediated gene transfer to a rat model of Parkinson's disease. Proc Natl Acad Sci U S A 96:12120–12125
10. Gonzalez-Zorn B, Escudero JA (2012) Ecology of antimicrobial resistance: humans, animals, food and environment. Int Microbiol 15:101–109
11. Naidoo J, Young D (2012) Gene regulation systems for gene therapy applications in the central nervous system. Neurol Res Int 2012: 595410
12. Xu F, Sternberg MR, Kottiri BJ et al (2006) Trends in herpes simplex virus type 1 and type 2 seroprevalence in the United States. JAMA 296:964–973
13. Jakobsson J, Rosenqvist N, Marild K et al (2006) Evidence for disease-regulated transgene expression in the brain with use of lentiviral vectors. J Neurosci Res 84:58–67
14. Shen F, Fan Y, Su H et al (2008) Adeno-associated viral vector-mediated hypoxia-regulated VEGF gene transfer promotes angiogenesis following focal cerebral ischemia in mice. Gene Ther 15:30–39
15. Greenberg DA, Jin K (2005) From angiogenesis to neuropathology. Nature 438:954–959
16. Liu Y, Figley S, Spratt SK et al (2010) An engineered transcription factor which activates VEGF-A enhances recovery after spinal cord injury. Neurobiol Dis 37:384–393
17. Siddiq I, Park E, Liu E et al (2012) Treatment of traumatic brain injury using zinc-finger protein gene therapy targeting VEGF-A. J Neurotrauma 29:2647–2659
18. D'Onofrio PM, Thayapararajah M, Lysko MD et al (2011) Gene therapy for traumatic central nervous system injury and stroke using an engineered zinc finger protein that upregulates VEGF-A. J Neurotrauma 28:1863–1879
19. Garriga-Canut M, Agustin-Pavon C, Herrmann F et al (2012) Synthetic zinc finger repressors reduce mutant huntingtin expression in the brain of R6/2 mice. Proc Natl Acad Sci U S A 109:E3136–E3145
20. Laganiere J, Kells AP, Lai JT et al (2010) An engineered zinc finger protein activator of the endogenous glial cell line-derived neurotrophic factor gene provides functional neuroprotection in a rat model of Parkinson's disease. J Neurosci 30:16469–16474
21. Konermann S, Brigham MD, Trevino AE et al (2013) Optical control of mammalian endogenous transcription and epigenetic states. Nature 500:472–476
22. Rebar EJ, Huang Y, Hickey R et al (2002) Induction of angiogenesis in a mouse model using engineered transcription factors. Nat Med 8:1427–1432
23. Banaszynski LA, Chen LC, Maynard-Smith LA et al (2006) A rapid, reversible, and tunable method to regulate protein function in living cells using synthetic small molecules. Cell 126:995–1004
24. Iwamoto M, Bjorklund T, Lundberg C et al (2010) A general chemical method to regulate protein stability in the mammalian central nervous system. Chem Biol 17:981–988
25. Miyazaki Y, Imoto H, Chen LC et al (2012) Destabilizing domains derived from the human estrogen receptor. J Am Chem Soc 134: 3942–3945
26. Tu YH, Allen LV Jr, Fiorica VM et al (1989) Pharmacokinetics of trimethoprim in the rat. J Pharm Sci 78:556–560
27. Hayashi S, McMahon AP (2002) Efficient recombination in diverse tissues by a tamoxifen-inducible form of Cre: a tool for temporally regulated gene activation/inactivation in the mouse. Dev Biol 244:305–318
28. Tai K, Quintino L, Isaksson C et al (2012) Destabilizing domains mediate reversible transgene expression in the brain. PLoS One 7:e46269
29. Quintino L, Manfre G, Wettergren EE et al (2013) Functional neuroprotection and efficient regulation of GDNF using destabilizing domains in a rodent model of Parkinson's disease. Mol Ther 21:2169–2180
30. Sellmyer MA, Chen LC, Egeler EL et al (2012) Intracellular context affects levels of a chemically dependent destabilizing domain. PLoS One 7:e43297

Chapter 5

Design of shRNA and miRNA for Delivery to the CNS

Gabriela Toro Cabrera and Christian Mueller

Abstract

Neurologic diseases tend to target various areas of the central nervous system (CNS) and can therefore result in paralysis, dementia, and death. Neurodegenerative diseases distinguish themselves from other diseases by affecting nerve cells, which unlike many other cells in our body cannot regenerate when severely injured. The discovery of RNA interference (RNAi) has enabled scientist to design new therapeutic approaches based on specific gene silencing rather than the canonical gene therapy through gene augmentation. Two types of molecules can be used for viral vector-mediated gene silencing: short hairpin RNAs (shRNAs) and artificial microRNAs (miRNAs) that have the ability to enter the RNAi pathway. Although both shRNAs and miRNAs can be used to silence genes, they enter the RNAi pathway at different points. Unlike shRNAs, miRNAs require an additional cleavage step inside the nucleus before being exported to the cytoplasm. These molecules can then be incorporated into the RNA-induced silencing complex (RISC) which utilizes sequence complementarity to recognize target mRNAs and activate either translational repression, in the case of partial complementarity, or induce mRNA cleavage in the case of complete complementarity. Elevated amounts of shRNAs, which are commonly driven by strong polymerase III promoters, can cause saturation of the endogenous RNAi machinery due to competition between endogenous and artificial molecules. Switching to a DNA polymerase II promoter is an alternative to reduce shRNA production, thereby reducing toxicity. Even though the molecules are designed to target specific mRNAs there may be off-target effects due to nonspecific binding that must be accounted for during the design process. In this chapter we discuss the design and in vitro screening of shRNAs and artificial miRNAs.

Key words miRNA, shRNA, RNAi, AAV, Knockdown

1 Introduction

There are many well-known human neurodegenerative diseases, such as Parkinson's, Alzheimer's, and Huntington's, that specifically affect neurons [1]. Neuronal cells, unlike other cells in the body are non-regenerating; therefore if they suffer damage they may deteriorate and die without being replaced [2]. Neuronal loss can lead to many symptoms of neurologic disease such as ataxia (uncoordinated movements) and/or dementia (loss of mental health and higher thought processing) [1–6]. Every year the number of patients diagnosed with one of these illnesses increases [7, 8]. Currently there is

Fredric P. Manfredsson (ed.), *Gene Therapy for Neurological Disorders: Methods and Protocols*, Methods in Molecular Biology, vol. 1382, DOI 10.1007/978-1-4939-3271-9_5, © Springer Science+Business Media New York 2016

an increased interest in neuroscience research a lot of whcih is focused in pursuit of a treatment that will slow the progression and/or cure these diseases. To this end many approaches are constantly being developed. The discovery of RNAinterference (RNAi) as well as the use of viral vectors as mediators for delivery have given gene therapy the possibility to treat diseases caused by a toxic gain of function, such as Huntington's and ALS [9].

The RNAi mechanism entails the transcription of sequence-specific small RNAs that bind in a complementary manner to their target messenger RNA (mRNA) forming double-strand RNA (dsRNA). The cell's innate defense response to dsRNA is to induce its degradation [10]. This response evolved to regulate gene expression or as protection against exogenous, single-stranded DNA seen, for example, from viruses [11, 12].

RNAi was first observed in plants in 1990, followed by its description in the nematode *Caenorhabditis elegans* in 1998 by Craig Mello and Andrew Fire [13]. Small RNAs such as short hairpin RNAs (shRNAs) and microRNAs (miRNAs) enter the RNAi pathway and act as post-transcriptional gene regulators [9, 10, 13]. Researchers have designed artificial short hairpin RNAs (shRNAs) and micro RNAs (miRNAs) to target specific gene sequences.

Engineered miRNAs are designed based on the endogenous miRNA biogenesis. They are commonly transcribed from genes by RNA polymerase II (pol II) or less frequently by RNA polymerase III (pol III). Pol II promoters transcribe precursors for small RNAs, mRNA, snRNA and microRNAs. Pol III are strong promoters that not only synthesize small RNAs but are also transcribing housekeeping genes that are expressed abundantly throughout different cell types [14].

After being transcribed, miRNA transcripts are capped at the 5′ end, and polyadenylated at the 3′ end. This structure is termed primary miRNA (pri-miRNA). Pri-miRNAs will then be recognized by the protein/enzyme complex DGCR8/Drosha and further cleaved to precursor miRNAs (pre-miRNAs) [15]. Pre-miRNA hairpins are then structurally recognized by the nuclear receptor Exportin-5, which will guide it to the cytoplasm [15]. In the cytoplasm the RNase III enzyme Dicer serves as a molecular ruler that will recognize dsRNA and cleave a specified distance from the 3′ end overhang [16], or 5′ end phosphate group [17] resulting in a 22-nucleotide miRNA:miRNA duplex [16]. This duplex is made up of a guide strand, which will enter the RNA-induced silencing complex (RISC), and a passenger strand, which will be degraded. Once in the RISC complex the guide strand will bind to its target mRNA. If the guide strand, the guide is perfectly complementary to the target sequence the result will be cleavage by the catalytic component of RISC, argonaut 2 (AGO2) [9, 15, 18]. However if there is only partial complementarity at the seed sequence (2–8 nts) with the target sequence the result will be translational repression.

Short hairpin RNAs are commonly transcribed from pol III promoters (e.g., U6 or H1), pol II and cell-specific promoters can be alternative option, although scarcely used [19]. Unlike miRNAs they enter the RNAi pathway as hairpins skipping cleavage by Drosha and being shuttled through Exportin-5 to the cytoplasm where they follow the same miRNA Dicer-RISC route. A diagram of the RNAi pathway can be found in ref. 9. Although these molecules are designed to only bind to their target, there may be instances where there are off-target effects. Given that miRNAs can regulate translation by just binding at the seed sequence it is difficult to repress all off-target effects [20]. However, the engagement of the slicer activity form AGO2 relies on more extensive base pairing as mentioned above, this the off-target effects for this can be more easily controlled. Several bioinformatics tools and websites exist that will allow the user to query their sequence to find theoretical targets to minimize the off-target effects.

A debatable issue regarding shRNAs driven by strong pol III promoters has been cell toxicity caused by overload of the RNAi machinery. Saturation occurs when artificial molecules and endogenous microRNAs compete to enter this pathway and exit it through Exportin-5 [9, 21]. Researchers have compared toxicity levels between miRNAs driven by pol III promoters and shRNAs; they found that although miRNAs have an additional saturable step in the nucleus, they cause less toxicity than shRNAs. Results by Maczuga et al. also show that although using a pol II (CMV) promoter leads to reduced toxicity compared to pol III-driven shRNAs, the overall effectiveness of miRNAs was more consistent than that of shRNAs [22].

Additionally, vector-derived genes can transcribe artificial miRNAs by either pol II or pol III; allowing researchers to use pol II tissue specific promoters. These factors have encouraged investigators to opt for the use of miRNAs for long-term expression and silencing rather than using shRNAs for therapeutical purposes.

This protocol will describe how to design a miRNA and an shRNA, and how to design a proper in vitro screen prior to packaging these constructs in either recombinant adeno-associated viruses (rAAV) or lentivirus for an in vivo delivery. The ultimate goal is to design molecules that will potentially be delivered to the CNS for therapeutic purposes.

2 Materials

2.1 Selecting RNA Target Sites for Designing shRNA or miRNA (Requires Internet Access)

1. Acquire the target mRNA sequence from a database, suggested site: National Center for Biotechnology information (NCBI, http://www.ncbi.nlm.nih.gov/).
2. Bioinformatics software: Work suite that supports molecular biology data and in silico pre-analysis, Clone Manager (Sci-Ed Software, Morrisville, NC, USA) or equivalent.

3. Selection of target sites using a structure prediction software (e.g., RNA Fold, http://rna.tbi.univie.ac.at/cgi-bin/RNAfold.cgi).
4. Oligonucleotides: From Integrated DNA Technologies (Integrated DNA Technologies, Coralville, IA, USA) or similar.

2.2 Cloning the Artificial shRNA, miRNA

1. Cloningplasmid (*see* **Note 1**).
2. Corresponding restriction enzymes to digest cloningplasmid (*see* **Note 2**).
3. Running buffer: Tris-acetate-EDTA (Sigma-Aldrich, St. Louis, MO, USA).
4. Agarose (Sigma-Aldrich, St. Louis, MO, USA).
5. Gel chamber and power supply.
6. Ethidium bromide.
7. 1 kb DNA ladder.
8. Annealing buffer (Integrated DNA Technologies, Coralville, IA, USA).
9. Restriction enzyme required to linearize plasmid (*see* **Note 2**).
10. Gel extraction kit.
11. T4 DNA ligase.
12. Competent cells (e.g., sure2, neb-5alpha).
13. Terrific broth (TB).
14. Ampicillin.
15. Incubator.
16. DNA extraction kit: Qiagen or equivalent.
17. DNA sequencing service to confirm shRNA or miRNA by sequencing: Integrated DNA Technologies or equivalent.

2.3 Validate shRNA, miRNA In Vitro in a Cell Line

1. Large DNA extraction kit: Qiagen Plasmid Midi Kit (Qiagen, Germantown, MD, USA) or equivalent.
2. Cell line selection for knockdown screening: HEK 293T or similar.
3. Growth media for cells: DMEM, 10 % fetal bovine serum, 1 % penicillin-streptomycin antibiotic.
4. Transfection reagent: JetPrime (VWR International, Radnor, PA, USA), Lipofectamine 2000 (Life Technologies, Carlsbad, CA, USA), or equivalent.
5. RNA extraction solution: Trizol (Life Technologies, Carlsbad, CA, USA).

2.4 RT-qPCR to Quantify Knockdown of Targets

1. cDNA kit.
2. PCR thermal cycler.

3. TaqMan probes for specific gene targets and controls: Life Technologies Custom assay (Life Technologies, Carlsbad, CA, USA).
4. Endogenous control: GAPDH probe (Life Technologies, Carlsbad, CA, USA).
5. TaqMan master mix (Life Technologies, Carlsbad, CA, USA).
6. Micro Amp Fast optical 96 well reaction plate with Barcode, 0.1 ml (Life Technologies, Carlsbad, CA, USA).
7. MicroAmp Optical adhesive film (Life Technologies, Carlsbad, CA, USA).
8. Real-time qPCR machine.

2.5 Preparing DNA for Packaging in Viral Vectors or Lentivirus

1. Mega Prep kit: Qiagen or equivalent.

3 Methods

3.1 Selecting RNA Target Sites and Designing shRNA or miRNA

1. Determine the target sequence to be silenced. Both miRNAs and shRNAs can be designed to be species specific, or to a conserved target sequence across various species. They can also be designed to target all possible variants produced by a gene or a single variant (*see* **Note 3**). The sequences can be obtained at the National Center for Biotechnology Information (NCBI) http://www.ncbi.nlm.nih.gov/refseq/.
2. Analyze the folding structure of the target sequence. The software will compute an algorithm for basepair probabilities, thermodynamics and circular RNA folding (*see* **Note 4**). Input the target sequence to the software using the predetermined software settings. The prediction software will allow the visualization of the secondary structure of the target sequence. Find loops with areas of single-stranded RNA that are at least 22 nucleotides long to serve as target sites.

3.2 Designing a microRNA (Fig. 1)

1. Copy the whole sequence from the loop in Subheading 3.1, **step 2**, and paste it into a word document.
2. Within this sequence look at the 5′ end for an A or a U to be the first base of your microRNA, this will improve thermodynamic stability, and allow for better AGO2 loading [23].
3. Then select the following 19–21 nucleotides (*see* **Notes 5** and **6**).
4. Check that the sequence selected is unique to your target by searching the BLAST database (http://blast.be-md.ncbi.nlm.nih.gov/Blast.cgi).

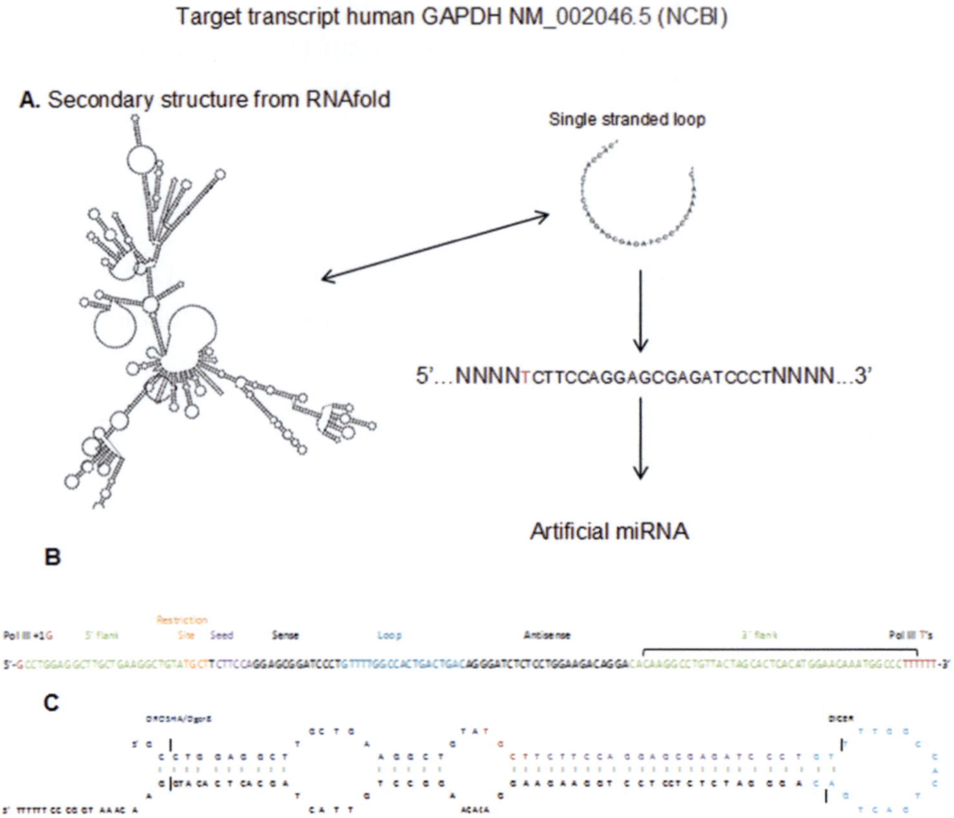

Fig. 1 An example of a microRNA designed against the human housekeeping gene GAPDH NM_002046.5 (NCBI), both flanks and loop have been modified and adapted from human miR155. (**a**) Indicates the folding structure of GAPDH obtained from the software, zoomed into the loop to be targeted. (**b**) Starting from the 5′ end the microRNA contains flank sequences, which will form part of the structure and cleavage sites recognized by Drosha. The 5′ flank is followed by the 22-nucleotide sequence that contains a seed sequences from nucleotides 2–8, and a mismatched loop. The antisense sequence is the reverse compliment of the sense sequence with nucleotides from positions 10–11 deleted to form an unpaired bulge. The sequence has a +G modification that should be added if cloning into a plasmid with a U6 promoter, and a tail of 6 Ts as the terminator. Panel (**c**) is an image of the folding structure of the designed microRNA

5. Check the GC content of the selected site; a 20–70 % range is preferable. The sequence selected for example shown in Fig. 1a has a GC content of 57.1 %. You can use a calculator available online or count the number of Gs and Cs and divide by the total number of nucleotides in the sequence.
6. A basal UG motif immediately after the stem-loop at position 14 shows an enhancement in Dicer binding [23]; adding this to the design is left to the researchers' discretion.
7. Select at least five possible target regions (*see* **Note 7**).

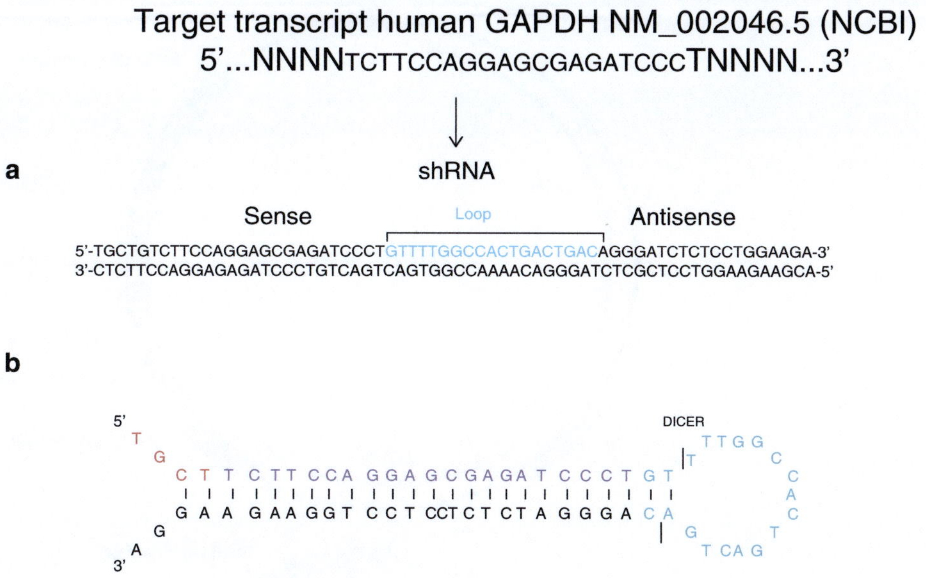

Fig. 2 An example of an shRNA designed to target the human housekeeping gene GAPDH NM_002046.5 (NCBI). The loop has been modified and adapted from hsa-miR155. In panel (**a**) starting from the 5′ end is a restriction site for cloning; followed by the sense 22-nucleotide sequence, then a 19-nucleotide loop that contains the sites for Dicer cleavage, and finally the antisense 22-nucleotide sequences with bases in positions 10–11 deleted to form a bulge. The bottom strand is the reverse complementarity of the top strand. In panel (**b**) is an image of the folding structure of the shRNA

3.3 Backbone Design

1. Either endogenous miR155 or miR30 (*see* **Note 11**) are commonly used as backbones to design the artificial microRNA and shRNA. For the microRNA design both the 3′ and 5′ flanks, as well as the loop region from the endogenous miRNA of choice, will be adapted as a backbone (Fig. 1). For the shRNA design only the loop region is adapted as the backbone (Fig. 2).
2. Find 5′ and 3′ flank sequences of 50–100 nts, from an endogenous microRNA (*see* **Note 11**).
3. The 5′ and 3′ flanks contain the structural cleavage areas for the Drosha-DGCR8 machinery to sit; they are 40 nt upstream, and downstream, of the pre-miRNA hairpin; these flanking sequences complement each other and leave single-stranded areas at the base of the stem loop that the microprocessor will recognize as substrate. Refer to refs.[18, 24–26].
4. In addition to the flanks also identify a loop from a naturally occurring microRNA. Dicer will recognize the hairpin shaped pre-mRNA. Refer to ref. [27] on specific details for loop design and refs. [28, 29] for problems from improper Dicer cleavage.
5. Copy and paste the flank sequences into a word document.

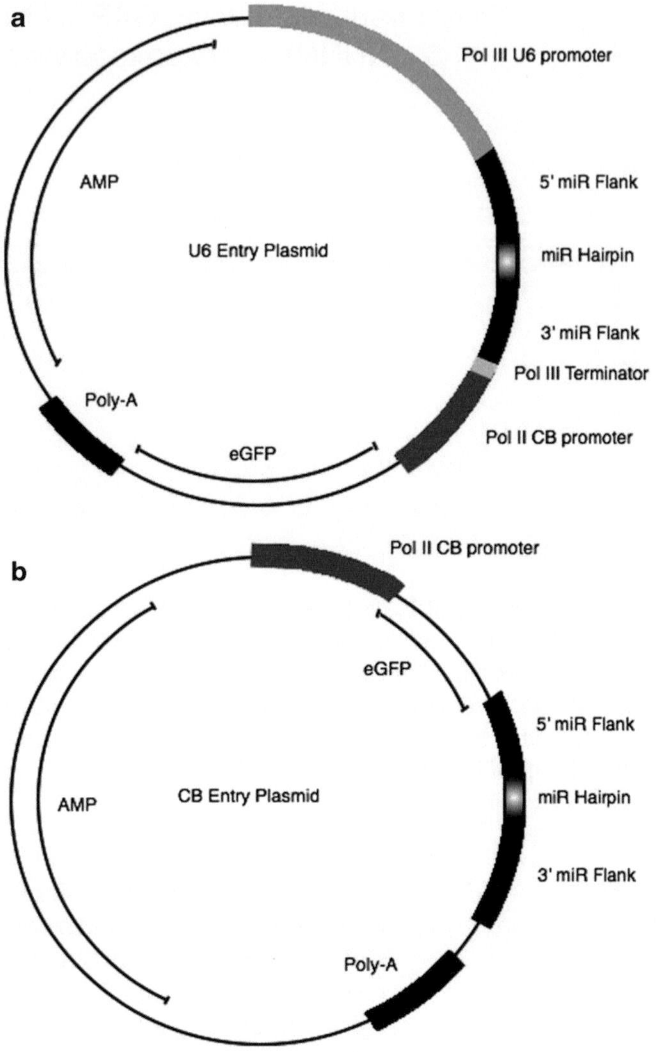

Fig. 3 An example of the entry plasmid used for cloning a microRNA/ shRNA. (**a**) U6 entry plasmid contains a U6 promoter followed by either the miR155 endogenous flanks or the shRNA directly. An additional Chicken beta actin (CB-A) promoter driving the expression of a reporter gene (GFP) can be added. (**b**) CB entry plasmid contains a CB promoter driving GFP, followed by the miR155 endogenous flanks or the shRNA directly

3.4 Introduce the 20–22-Nucleotide Sequence into the Backbone

1. Immediately after 5′ flank sequence, paste the 20–22-nucleotide sequence that is perfectly complementary to the target sequence, this stretch of nucleotides will also contain the seed located in positions 2–8.
2. After the 22 nucleotides, paste the loop sequence adapted from the naturally occurring microRNA.
3. After the loop, paste the reverse complement of the 22 nucleotides modified by having a deletion in positions 10–11 creating

a bulge. The presence of the bulge will allow for the sense strand to be preferentially loaded into AGO2.

4. Select the whole sequence described above and reverse complement it. This will be the bottom strand 3′–5′.
5. If a pol III promoter such as U6 is being used add an additional +1G at the 3′ end of the U6 promoter sequence, before the 5′ flank sequence of the miRNA [30].
6. After the 3′ flank sequence add the pol III terminator formed by 4–6 Ts (TTTTTT) and the restriction site of choice for cloning. This stretch of Ts will create an overhang at the 3′ end necessary for Dicer recognition [30].
7. Proper controls should include a microRNA that targets an unrelated sequence to control for saturation of the RNAi machinery, promoter, or reporter toxicity.

3.5 shRNA Design (Fig. 2)

1. In a word document paste the 22-nucleotide sequence complementary to the target selected.
2. This sequence will then be followed by a 19-nucleotide loop adapted from an endogenous microRNA (human miR155 in this case).
3. After the loop add the reverse complement of the 22 nucleotides.
4. The end result will be a hairpin, where the conserved loop will contain the cleavage sites needed for Dicer to remove the loop and leave the dsRNA duplex.
5. Select a restriction enzyme to use for cloning the shRNA to an expression vector.
6. Decide whether to use a U6 or H1 promoter. When using a U6 promoter a +1-G must be added at the 3′ end of the promoter.

3.6 Cloning of the miRNA or shRNA

1. Reconstitute the oligonucleotides with nuclease-free water to 100 μm concentration.
2. Anneal the oligonucleotides: In a beaker bring water to boil at 95 °C, add the tubes containing the annealing reaction (Table 1). Keep at 95 °C for 5 min, then shut off the heat, and allow to cool to room temperature. Store at −20 °C.
3. Check plasmid and oligonucleotides: The plasmid containing the promoter of choice, reporter, etc. must be digested with the enzyme selected for cloning to confirm the presence of the expected cut site and size of the band. Make a 1 % agarose gel to run the annealed oligonucleotides and the vector side by side to ensure a 1:1 ratio for an adequate ligation.
4. Prepare the ligation mixture as described in Table 2 and ligate at 16 °C overnight.

Table 1
Annealing reaction

Component	Volume
5′–3′ oligo	2 μl
3′–5′ oligo	2 μl
Annealing buffer	196 μl

Table 2
Ligation reaction

Vector	1 μl
Insert	1 μl
Quick T4 DNA ligase	1 μl
Buffer	2 μl
Nuclease-free water	15 μl
Volume reaction	20 μl

5. Follow transformation protocol included with your selected super-competent cells. Plate the reactions and incubate overnight. The next day pick multiple colonies and grow individually in 5 ml of terrific broth media containing the antibiotic used for selection at 37 °C, shaking. Do not grow for more than 13 h.
6. Extract your plasmidDNA from 2 ml of the bacterial growth media using a miniprep kit.

   Select a restriction site that will allow you to clearly see the vector and the insert. Digest and run in a 1 % agarose gel.
7. Additionally take positive samples and send them for sequencing to confirm the insert (miRNA/shRNA) is properly cloned into the plasmid (*see* **Note 8**). Once the sequencing results are received the sequences can be compared and aligned to an expected cloning map. Plasmid cloning maps can be designed in programs such as clone manager that facilitate cloning projects and analysis.

3.7 Screening of the miRNA or shRNA

1. Select a cell line that highly expresses your gene of interest. For screening purposes ensure that is has high transfection efficiency and is easy to culture. A common choice is human embryonic kidney cells (HEK293) (*see* **Note 12**).
2. Perform a transfection optimization prior to the experiment by testing different reagents, concentrations, and incubation times. Suggested reagents include Lipofectamine or Jet Prime. Reporter genes are commonly used to look at transfection effi-

Table 3
cDNA reaction

	Component volume/reaction (μl)	
Component	**+ RT reaction**	**– RT control**
2× RT buffer	10	10
20× Enzyme mix	1	–
RNA sample	2 μg- up to 9 μl	2 μg- up to 9 μl
Nuclease-free H_2O	Bring to 20 μl	Bring to 20 μl
Total per reaction	20	20

ciency. They can either be cloned into the same construct with the microRNA or can be co-transfected at a 1:1 ratio. After transfection collect the cells at 24–72 h depending on the protein turnover for your gene of interest.

3. Ensure that proper experimental controls are included (*see* **Note 9**): (1) A microRNA control that targets an unrelated sequence. It will control for saturation of the RNAi machinery. Scrambled miRNA sequences are not recommended due to the possibility of mistargeting. (2) A control plasmid with only the selected promoter and reporter (i.e., CMV-GFP) to control for toxicity. (3) A sham transfection, where all transfection components are added except for DNA. (4) An un-transfected well, to control for cell growth, viability, and contamination.
4. To collect cells, remove cell media and wash 1× with PBS solution. Follow Trizol reagent protocol to extract RNA and/or protein.

3.8 Analyze Extent of Knockdown by RT-qPCR or Western Blot

1. After extraction measure RNA concentration and make cDNA using the reaction mixture outlined in Table 3.
2. Incubate the reaction for 37 °C for 60 min. Stop the reaction by heating to 95 °C for 5 min and hold at 4 °C.
3. Order TaqMan primer-probes to the gene of interest. Dilute cDNA 1:10 ratio with sterile water and follow TaqMan assay protocol. Include endogenous controls; common house keeping genes are GAPDH and HPRT.
4. Analyze knockdown by the delta delta CT method [31].
5. If antibodies against the gene of interest are available western blot may be used to assess the level of knockdown.
6. Quantify amount of protein from Subheading 3.7, **step 4**
7. Run a western blot using standard methodology (*see* **Note 10**).

4 Notes

1. A plethora of plasmids are available for the purpose of cloning the shRNA or microRNA. Researchers are urged to use a plasmid that fits their purpose (e.g., contains a promoter of choice, specific antibiotic resistance, and convenient restriction sites for cloning). Entry plasmids can be acquired at http://www.addgene.org/vector-database. It is left to the researcher's discretion the addition of a CB promoter driving a reporter (GFP).
2. Required restriction enzymes depend on the choice of cloningplasmid.
3. To test the selected constructs in vivo, the artificial miRNAs/shRNAs can be packaged in either recombinant adeno-associated virus (rAAV) or lentivirus. Something to be considered for the in vivo work is the nature of the constructs, if they are designed to be species specific to humans, a transgenic model may be needed. However the microRNAs/shRNAs designed against humans might cross-react with other species if the target sequence is conserved.
4. A suggested site is RNAfold: http://rna.tbi.univie.ac.at/cgi-bin/RNAfold.cgi. The software shows the secondary structures as well as loops of unpaired single strands. These loops are the regions of interest. Keeping in mind that miRNA and shRNA design are based on statistical folding predictions multiple sequences should be tested.
5. Avoid having restriction sites within the target regions, to avoid unwanted cleavage.
6. Avoid having stretches of AAAA or TTTTTs, for these are common terminator sites for pol III promoters.
7. One common problem is not finding a 21–22-nucleotide single-stranded sequence. In that case the seed sequence should be sufficient, although the specificity and efficiency might vary. Other prediction programs might result in different secondary structures.
8. A suggested site is Eurofins genomic: http://www.operon.com. This site will request the primers needed to sequence the sample. They can be sent or synthesized upon request. The primers should be designed against flanking regions outside of the microRNA/shRNA.
9. The in vitro screening should be done in biological replicates to be able to perform statistical analysis.
10. If efficient knockdown is achieved, the results of the RT-qPCR and Western blot should correlate. Less mRNA will result in less protein.

11. The most common miRNAs used as backbones are miR 155 and miR30, the sequences for the flank regions can be found online at the NCBI website. You can also use graphic imager to visualize the pri-miRNA structure vs. the mature miRNA.
12. If there are no cell lines that either (a) express sufficient levels of the gene of interest or (b) are not easy to culture/transfect. You can choose to co-transfect the target, in this case you should have a reporter linked to the target sequence, and this will allow you to determine if the assay is working. Make sure to include proper controls (e.g., WT untransfected, target plasmid alone, miR/shRNA against target alone).

References

1. Sathasivam S (2010) Motor neurone disease: clinical features, diagnosis, diagnostic pitfalls and prognostic markers. Singapore Med J 51(5):367–372, quiz 373
2. Nowakowski RS (2006) Stable neuron numbers from cradle to grave. Proc Natl Acad Sci U S A 103(33):12219–12220
3. Harvey RJ, Skelton-Robinson M, Rossor MN (2003) The prevalence and causes of dementia in people under the age of 65 years. J Neurol Neurosurg Psychiatry 74(9):1206–1209
4. Mercy L et al (2008) Incidence of early-onset dementias in Cambridgeshire, United Kingdom. Neurology 71(19):1496–1499
5. Ratnavalli E et al (2002) The prevalence of frontotemporal dementia. Neurology 58(11):1615–1621
6. Williams DB, Floate DA, Leicester J (1988) Familial motor neuron disease: differing penetrance in large pedigrees. J Neurol Sci 86(2-3):215–230
7. Ferri CP et al (2005) Global prevalence of dementia: a Delphi consensus study. Lancet 366(9503):2112–2117
8. de Lau LM, Breteler MM (2006) Epidemiology of Parkinson's disease. Lancet Neurol 5(6):525–535
9. Borel F, Kay MA, Mueller C (2013) Recombinant AAV as a platform for translating the therapeutic potential of RNA interference. Mol Ther 22(4):692–701
10. Davidson BL, McCray PB Jr (2011) Current prospects for RNA interference-based therapies. Nat Rev Genet 12(5):329–340
11. Napoli C, Lemieux C, Jorgensen R (1990) Introduction of a chimeric chalcone synthase gene into petunia results in reversible co-suppression of homologous genes in trans. Plant Cell 2(4):279–289
12. van der Krol AR et al (1990) Flavonoid genes in petunia: addition of a limited number of gene copies may lead to a suppression of gene expression. Plant Cell 2(4):291–299
13. Fire A et al (1998) Potent and specific genetic interference by double-stranded RNA in Caenorhabditis elegans. Nature 391(6669):806–811
14. Vannini A, Cramer P (2012) Conservation between the RNA polymerase I, II, and III transcription initiation machineries. Mol Cell 45(4):439–446
15. Zamore SLAPD (2013) Diversifying microRNA sequence and function. Nat Rev Mol Cell Biol 14:475–488
16. Macrae IJ et al (2006) Structural basis for double-stranded RNA processing by Dicer. Science 311(5758):195–198
17. Park JE et al (2011) Dicer recognizes the 5′ end of RNA for efficient and accurate processing. Nature 475(7355):201–205
18. Zeng Y, Yi R, Cullen BR (2005) Recognition and cleavage of primary microRNA precursors by the nuclear processing enzyme Drosha. EMBO J 24(1):138–148
19. Pan Q et al (2012) A dynamic perspective of RNAi library development. Trends Biotechnol 30(4):206–215
20. Birmingham A et al (2006) 3′ UTR seed matches, but not overall identity, are associated with RNAi off-targets. Nat Methods 3(3):199–204
21. Grimm D (2011) The dose can make the poison: lessons learned from adverse in vivo toxicities caused by RNAi overexpression. Silence 2:8
22. Maczuga P et al (2012) Optimization and comparison of knockdown efficacy between polymerase II expressed shRNA and artificial

miRNA targeting luciferase and Apolipoprotein B100. BMC Biotechnol 12:42

23. Auyeung VC et al (2013) Beyond secondary structure: primary-sequence determinants license pri-miRNA hairpins for processing. Cell 152(4):844–858
24. Han J et al (2004) The Drosha-DGCR8 complex in primary microRNA processing. Genes Dev 18(24):3016–3027
25. Zeng Y, Cullen BR (2004) Structural requirements for pre-microRNA binding and nuclear export by Exportin 5. Nucleic Acids Res 32(16):4776–4785
26. Chen CZ et al (2004) MicroRNAs modulate hematopoietic lineage differentiation. Science 303(5654):83–86
27. Gu S et al (2012) The loop position of shRNAs and pre-miRNAs is critical for the accuracy of dicer processing in vivo. Cell 151(4):900–911
28. Lewis BP et al (2003) Prediction of mammalian microRNA targets. Cell 115(7):787–798
29. Siolas D et al (2005) Synthetic shRNAs as potent RNAi triggers. Nat Biotechnol 23(2):227–231
30. Boudreau RL, Monteys AM, Davidson BL (2008) Minimizing variables among hairpin-based RNAi vectors reveals the potency of shRNAs. RNA 14(9):1834–1844
31. Livak KJ, Schmittgen TD (2001) Analysis of relative gene expression data using real-time quantitative PCR and the 2(-Delta Delta C(T)) method. Methods 25(4):402–408

Chapter 6

Tissue-Specific Promoters in the CNS

Sebastian Kügler

Abstract

This chapter outlines some general principles of transcriptional targeting approaches using viral vectors in the central nervous system. Transcriptional targeting is first discussed in the context of vector tropism and appropriate delivery. Then, some of our own attempts to restrict expression of therapeutic factors to distinct brain cell populations are discussed, followed by a detailed description of the setscrews that are available for these experiments. A critical discussion of current stumbling blocks and necessary developments to achieve clinical applicability of advanced targeted vector systems is provided.

Key words Transcriptional targeting, AAV, Promoter, Transcription blocker, Enhancer, Mir-binding site, Vector tropism

1 Transcriptional Control in the Context of Other Targeting Strategies

Viral vectors can be targeted to specific brain cell populations by different means: first, and probably the most important when discussing gene therapy in the brain, is appropriate stereotaxic delivery of the viral suspension into its target area. This may be relatively easy if small nuclei like the substantia nigra are to be transduced, and can be facilitated in larger brain areas like the caudate/putamen if secreted transgenes (e.g., neurotrophic factors) are the potentially curative molecule of choice. When thinking about the human cortex, for example in paradigms like Alzheimer's disease, the brain volume to target is tremendously larger than corresponding structures of rodent and nonhuman primate's brain, and delivery of viral particles to sufficient neurons will be a major concern. Alternatives to the invasive stereotaxic application route are under development, but are not likely to be clinically applicable in the near future. For example, localized opening of the blood-brain barrier by MRI-guided focused ultrasound [1] may be a forthcoming option after systemic application of the recombinant virus. However, current AAV vectors are for the most part captured in liver and muscle after intravenous injection [2], allowing only small

Fredric P. Manfredsson (ed.), *Gene Therapy for Neurological Disorders: Methods and Protocols*, Methods in Molecular Biology, vol. 1382, DOI 10.1007/978-1-4939-3271-9_6,

portion of such virus to reach the brain. De-targeting the capsid away from peripheral receptors needs to be achieved to make such technology safe and efficient.

The second level of targeting is the viral capsid: capsid components binding to high-affinity receptors in the brain will restrict transduction to small brain areas close to the needle injection tract, while capsids binding to low-affinity receptors may diffuse over much larger brain areas. The simple capsid structure of the various serotypes of the adeno-associated virus (AAV) is very much facilitating the development of mutants with desired transduction characteristics [3, 4]. Still, despite the availability of hundreds of natural serotypes and efficient screening methods for useful capsid mutations (as described elsewhere in this volume) the generation of a recombinant AAV specifically transducing distinct cell types of the brain (neurons, astrocytes, microglia, oligodendroglia, vascular epithelia) has not been fully achieved yet. The prototype AAV, AAV-2, which for more than a decade was the only AAV under investigation, fortunately demonstrated a highly neuronal tropism in the brain irrespective of transcriptional control elements exploited, and thus is still the vehicle of choice for neuron-directed gene therapy of the brain [5–7]. Many other AAV serotypes, however, although being favorable in terms of production up-scaling, transduction efficacy, diffusion, and escape from the human immune system, are not that much restricted to a single brain cell type, and a simple example is shown in Fig. 1: the hybrid serotype AAV-1/2, and the natural serotypes AAV-5 and AAV-6 are capable of transgene expression in both neurons and astroglia with equally high efficacy, simply depending on transcriptional control elements incorporated into the viral genome.

Transcriptional control elements are the third level of targeting therapeutic transgene expression to specific brain cell populations, and in terms of efficacy and feasibility appear to be the easiest to achieve. However, as illustrated in Fig. 1, targeting transgene expression by means of cell-type specific promoters or binding sites for micro-RNAs by no means alters the transduction properties of the viral capsid itself. Thus, detection of transgene expression only in neurons does not mean that the recombinant virus would not bind to glial cell surface receptors and enter astrocytes with equal efficacy, but only that in these cells the neuron-specific promoter is expressing with negligible activity. Although with the neuron- and astrocyte-specific promoters described in this chapter we never detected transgene expression in microglia, this does not prove absence of microglial transduction and potential antigen-processing and presentation. Indeed, at least some microglial transgene expression was suggested to occur after injection of AAV-2 and AAV-5 vectors into the brain which drove expression from the macrophage-specific F4/80 promoter [8].

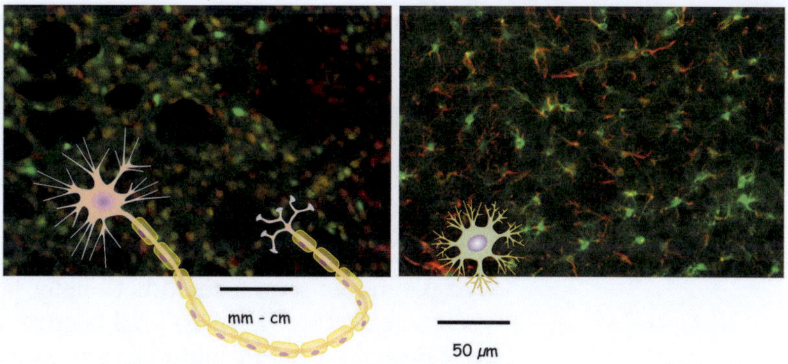

Fig. 1 Transcriptional targeting of AAV vectors in the rat striatum. Exclusively neuronal transgene expression is shown after AAV-5-hSYN-EGFP transduction (*left panel*; shown is an overlay of green = EGFP fluorescence with red = NeuN immunoreactivity); exclusively astrocytic transgene expression is shown after AAV-5-GFAP-EGFP transduction (*right panel*; shown is an overlay of green = EGFP fluorescence with red = GFAP immunoreactivity). Both vectors were injected into the brain at $6 \times 10e^9$ vg. Animals were sacrificed at 1 month after vector application and tissue sections were stained with anti-NeuN (neuronal marker) or anti-GFAP (astrocytic marker). The sketches show the transduced cell types and shall serve as reminders that striatal neurons project to far remote brain structures over several mm or even cm, making it likely that a secreted transgene like a neurotrophic factor is delivered to these potentially off-target sites, while astrocytes have a much more restricted area of influence. Identical results as shown here for AAV-5 were obtained with AAV vectors of the hybrid serotype 1/2 and serotype 6

2 A History of Our Attempts to Achieve Neuron- and Astrocyte-Specific Transgene Expression in the Brain

The chapters in this series are thought to be "protocols" for design of optimized viral vectors, with potential use in gene therapeutic applications. However, there is no simplified protocol, which will allow the reader to generate transcriptionally targeted viral vectors specifically for certain neuronal or glial brain cell populations, simply because the predictive powers of algorithms suggesting binding of cell-type-specific transcription factors to DNA sequences are quite limited. In addition, promoter elements defining cell specificity may reside far upstream from transcription initiation, making it necessary to use relatively bulky promoter constructs. While this is not a major problem for high-capacity (up to 30 kbp) adenoviral vectors, and may work with the cloning capacity of lentiviral vectors (up to 8 kbp), it represents a major hurdle for the AAV vectors with maximum capacities around 5 kbp. Thus, the only protocol to propose here is to simply go for trial and error: clone the respective promoter sequence into your vector of choice, and evaluate cell-type specificity over a wider range of titers, and, in case of AAVs, serotypes with different transduction and tissue spreading efficacies, and in different brain areas. Once suitable specificity has been achieved with a respective DNA sequence and AAV serotype, one can serially shorten the promoter sequence to make it more applicable especially for vectors with limited genome capacity, or combine the promoter with de-targeting miR-binding sites to further increase cell-type specificity (*see* description of transcriptional control elements in Subheading 3). However, how little one can trust even promoter sequences functionally proven in transgenic mice when transferring them to viral vectors will be illustrated by our work described below, aiming to transcriptionally target AAVs to astrocytes. Although the text will be somehow anecdotal, the author hopes that along the line of our experiments several important principles of transcriptionally targeting will become evident.

We started our attempts to target viral vectors specifically to neurons more than a decade ago, when still using first-generation adenoviral vectors of the human serotype 5 (Ad5). The capsid of these vectors had a relatively strong glial tropism, i.e., use of ubiquitous promoters such as the popular CMV promoter resulted in mainly astroglial transgene expression. For certain experiments we could still use this type of vector to achieve specific transduction of distinct neuronal populations, by making use of the retrograde transport of Ad5 capsids in nerve fibers: application of Ad5 with CMV promoter at the transected optic nerve resulted in transgene expression specifically in retinal ganglion neurons, which project into the optic nerve [9] (*see* Subheading 1). However, this vector was not suitable for neuron-specific transgene expression elsewhere in the brain.

Thus, we tested a variety of promoters in Ad5 vectors for their neuron specificity against the human CMV promoter: the human synapsin 1 gene promoter (hSYN; 470 bp), the rat neuron-specific enolase promoter (NSE; 1800 bp), the rat tubulin alpha-1 gene promoter (Ta1; 450 bp), and the human U1 snRNA gene promoter (U1; 400 bp) [10]. For a first assessment of any promoter fragment a good in vitro test system is important, allowing to quickly proceed through multiple constructs. We used primary neuron/glia co-cultures for this purpose, and found that the hSYN promoter offered the best overall performance in terms of promoter size, neuronal expression level, and rapid onset of transgene expression. The favorable properties of the hSYN promoter were then confirmed in the rat brain, where we for the first time achieved long-lasting and virtually 100 % neuron-specific transgene expression with a first-generation adenoviral vector [11]. These results were extended by Glover et al. [12] who showed that the hSYN promoter remained absolutely neuron specific even at largely increased expression levels, with high-titer vectors and incorporation of the mRNA stabilizing woodchuck hepatitis virus posttranscriptional regulatory element (WPRE [13]), resulting in greatly increased protein expression rate.

However, first-generation Ad5 vectors provoked inflammatory responses in the rodent brain when used at higher titers. Thus, we moved on to adeno-associated viral (AAV) vectors [14, 15]. Here again, we aimed to confirm that the hSYN promoter is an adequate pan-neuronal promoter. As tissue-specific promoters are frequently claimed to have only weak expression levels, we compared the hSYN promoter to the very strong mouse CMV (mCMV) promoter, under the assumption that the proposed neuronal tropism of the AAV-2 capsid would allow expression from both promoters in neurons only. In cultured primary neuron/glia co-cultures we found that the mCMV promoter resulted in exclusively astrocyte-specific expression, while the hSYN promoter restricted expression exclusively to neurons. Thus, a promoter with high activity in glia was able to "de-target" a so-called neuronotrophic AAV quite efficiently. Secondly, in terms of transgene expression levels, the mCMV promoter was a bit ahead of the hSYN promoter, but the latter kept up to equal expression level after only a few days in culture. Thus, a cell-type-specific promoter is by no means necessarily a weak promoter, although this may depend on enhancing elements as the WPRE. When testing both hSYN and mCMV promoter in AAV-2 vectors in different brain areas we elucidated another important principle: results obtained under primary cell culture conditions are not inevitably the same in the adult brain. In the retina, hSYN expressed exclusively in neurons, and mCMV very weakly in Müller glia; in the cortex hSYN expressed exclusively in neurons, while mCMV expressed very weakly in neurons but moderately in astrocytes; in the thalamus, both hSYN and mCMV expressed strongly and exclusively in neurons. Lessons to be learned from such experimentation was that (1) the hSYN promoter under all conditions demonstrated exclusively neuron-specific expression, but that (2) a promoter like the mCMV may have very different expression characteristics depending on use in cultured brain cells or in vivo, and even depending on brain area under investigation.

Since astroglia-specific expression was only partially achievable so far, we next exploited astrocyte-specific promoters proven to work as such in transgenic animals. Both the full-length 2.2 kbp glial fibrillary acidic protein (GFAP) promoter and its much shorter variant, the GfaABC1D promoter, drove transgene expression exclusively in astrocytes in transgenic mice [16]. With the full-length GFAP promoter we indeed achieved a virtually exclusive astroglial expression both in neuron/glia co-cultures and in the rodent brain [17]. These experiments were conducted with several AAV serotypes (hybrid-1/2, 5, and 6) and for all these vectors resulted in the same outcome: transgene expression was exclusively neuronal under control of the hSYN promoter, and exclusively astrocytic under the control of the GFAP promoter (*see* Fig. 1). Thus, for AAV serotypes with quite different tissue transduction properties (AAV-6: high-level transduction, restricted area;

AAV-1/2: high-level transduction, widespread; AAV-5: low-level transduction, widespread) a highly efficient and mutually exclusive transcriptional targeting was proven. An important lesson to learn from these experiments is: if a viral vector is described to result in transduction of a certain cellular population in a given tissue, it may very well be that a different promoter would result in a completely different outcome. It is important to remember that transcriptional targeting and targeting by modification of capsid tropism are two completely independent approaches.

After having achieved a strict astrocytic transgene expression, which was exploited to restrict the delivery of neurotrophic factors to defined brain areas by avoiding off-target delivery through long-range neuronal projections, we next sought to gain temporal control over transgene expression. We thus put together a regulated gene transfer system, which is based on the mifepristone-regulated GeneSwitch (GS) [18, 19]. Such regulated transgene expression systems (as described in detail elsewhere in this issue) require a second expression cassette to be expressed together with the therapeutic transgene of choice, the latter under control of the regulated promoter. Thus, space restrictions are likely to encounter when using small viruses as the AAVs, and thus we aimed to replace the bulky full-length GFAP promoter with a variant already cut down to less than 700 bp, the so-called GfaABC1D promoter [16]. By this means it would be possible to construct a single vector genome for both astrocyte-specific GS expression and regulated GDNF expression. Unfortunately, however, despite the fact that the GfaABC1D promoter restricted transgene expression exclusively to astrocytes in transgenic mice, and despite promising results in neuron/glia co-culture, this promoter was not able to restrict transgene expression to astrocytes in AAV vectors (expression was about 70 % neuronal and 30 % astroglial [19]). Thus, the predictive power of both transgenic animals and cell culture experiments was not sufficient to foresee the characteristics of this promoter in the AAV-5 and AAV-6 gene transfer vectors investigated. Addition of neuron-specific mir124-binding sites into the 3′-untranslated region of an expression cassette driven by the GfaABC1D promoter drastically diminished neuronal expression, but also impacted on expression levels in astrocytes (Taschenberger and Kügler, unpublished), rendering this de-targeting approach almost useless under the conditions applied, although it worked out very well in the context of lentiviral vectors [20].

3 Setscrews for Transcriptional Control

To summarize the transcriptional targeting approaches described above in a more general format the principal "screws" that can be important for cell-type-specific transgene expression are summarized in Fig. 2. Shown is the prototype genome of a recombinant AAV

Fig. 2 A prototype AAV vector genome containing the elements useful for transcriptional targeting

vector, but the basic principles are the same for lentiviral vectors. It should be taken into account, however, that AAVs mainly persist as epigenomic concatamers, while LVs integrate into the host cell genome, often at sites of active transcription. Thus, influences from neighboring genomic sequences cannot be ruled out, and results obtained for LVs must be experimentally confirmed with AAVs and vice versa.

The potential impact of each vector genome element on transcriptional targeting is described below.

"ITR": the inverted terminal repeats flanking the recombinant vector genome are necessary for packaging the genome into the viral capsid. These sequences demonstrate relatively weak but readily detectable promoter activity on their own, allowing to construct "promoter-less" genomes for incorporation of very large transgenes.

"TB": transcription blockers are artificial transcription stop sites, which can be used to shield a cell type-specific promoter from the influence of the ITR. We have found such TBs not to be necessary for promoters with high transcriptional activity, such as the neuron-specific hSyn promoter or the astrocyte-specific full-length GFAP promoter. However, up to three consecutive TB sequences were necessary to isolate an inducible promoter from the ITRs [19], and these elements may become quite important if weak cell-type-specific promoters are used.

"Promoter": The promoter sequence is the single most important element in transcriptional targeting approaches. While cell-type specificity is the overriding criterion, another important issue is size of the promoter sequence, especially for vectors with limited capacity as AAV. The pan-neuronal hSyn promoter is of only 470 bp in length, occupying less than 10 % of available genome size. The full-length GFAP promoter, however, is of 2200 bp in length, occupying almost 40 % of available genome size. Considering that enhancers such as the WPRE plus the polyadenylation site add another roughly 1000 bp to the transcriptional control elements, the residual transgene capacity is only 2000 bp. None the less, this will still allow expression of proteins of nominal molecular weight up to 70–75 kDa.

"Transgene": Proteins or regulatory RNAs expressed from a cell type-specific promoter are efficiently directed to their target cell by this simple approach. However, several proteins of therapeutic interest, such as lysosomal proteins or neurotrophic factors, are secreted from their primary target cell and can act on remote cell types or brain areas, which can be seen both as an advantage or disadvantage. For example, the neurotrophic factor GDNF is secreted from un-myelinated axons of the medial forebrain bundle after AAV-mediated expression in substantia nigra neurons, and could thereby trigger release of corticotrophin-releasing hormone in the thalamus [21]. Such off-target effects can be minimized or prevented by the more localized expression of GDNF in astrocytes [17] (and *see* Fig. 1), but this restriction in spatial distribution comes at the cost of limited tissue penetration of the therapeutic molecule.

"Enhancers": Incorporation of splice sites/introns [22] or the WPRE [13] to increase mRNA stability is routinely used to enhance protein expression levels in recombinant vectors. In our hands the WPRE increases protein expression by roughly 5- to 15-fold, depending on the transgene under investigation. Both the hSYN promoter for strict neuronal transgene expression and the full-length GFAP promoter for strict astrocytic transgene expression were cell-type specific if expression levels were enhanced by the WPRE. This holds true even if relatively high vector titers (up to $6 \times 10e^9$ vg per mouse striatum) were used, demonstrating their robust targeting properties. However, it has to be envisaged that an "upper limit" of specificity exists for every promoter, and removing enhancer elements from the expression cassette may substantially add to restriction of transgene expression to distinct brain cell types. This consideration is especially true for vectors with a broad transduction spectrum (*see* Fig. 1; and comments above). Likewise, in our inducible vectors we found it absolutely necessary to remove enhancer sequences in order to maintain lowest background expression in the non-induced state and the high neuronal specificity of the pTK promoter (almost 99 % neuron specific) driving expression of the GS regulatory protein [19].

"mirB": The de-targeting strategy by incorporation of cell type-specific micro-RNA-binding sites has been exploited extremely efficiently to target transgene expression of LV vectors to astrocytes [20]. The prototype VSV-G pseudotyped LV shows a highly neuronal tropism in the brain, making it necessary to replace the VSV-G coat with that of Mokola virus in order to allow at least partial astrocyte-specific transgene expression. Residual neuronal transgene expression was still significant, but was finally avoided by incorporation of several binding sites for the neuron-specific miR124 into the 3′-untranslated region of the expressed mRNA, resulting in degradation in neurons and exclusively astrocytic transgene expression. Care has to be taken to make sure that species-specific differences in the sequences of miR-binding sites are recognized.

"pA": According to our experiences with AAV vectors the polyadenylation site has only little (if any) influence on transcriptional targeting. In our hands the eukaryotic bovine growth hormone polyadenylation site (bGH) worked well under all conditions, but vectors terminating transcription with a viral SV40 polyadenylation site were equally effective.

4 Targeting of Neuronal Subpopulations?

At least for the many serotypes of the AAV vectors it can be anticipated that an almost exclusive targeting of transgene expression to either neurons or astroglia can be achieved by exploiting the hSyn or the full-length GFAP promoter under most conditions. However, the pan-neuronal hSyn promoter is unable to discriminate between different types of neurons (excitatory vs. inhibitory, glutamatergic vs. cholinergic vs. gabaergic vs. dopaminergic, etc.) and thus much of the "targeting" depends on appropriate stereotaxic delivery of the recombinant virus. Evidently, both preclinical research and therapeutic options would be greatly enhanced by availability of neuronal subtype-specific vectors. There are comprehensive resources available facilitating the search for such specific promoters, such as the Allen Brain Atlas [23], which maps expression of about 20,000 genes in the mouse brain, or the Pleiades Promoter Project ([24], and www.pleiades.org), which has functionally tested a large number of bioinformatics-based promoter constructs in the mouse brain. However, despite the efficient bioinformatics tools applied in the Pleiades project, and the proof of principle that certain promoters indeed restricted transgene expression to some sub-populations of neurons (but spatially rather than discriminating different neurotransmitter types) care has to be taken when trying to translate these findings to the situation of viral vectors. Pleiades' "mini promoters" were quite bulky (usually 3–4 kbp) and tested only as single-copy insertions into the *Hprt-locus* in transgenic mice. As it is an unrealistic assumption that viral vector mediated gene transfer results in single-copy delivery per cell it is by no means proven that any of Pleiades' promoters will work appropriately in AAV or LV or other types of viral gene transfer tools. The author of these lines would probably be more optimistic in this case if he would not have had the failure of a "proven transgenic promoter" in his own laboratory (*see* Subheading 2).

A certain specificity of several neuronal cell type-specific promoters has been achieved with lentiviral vectors [25], but this work also demonstrates that much is still to learn about the design of promoters which are exclusively active in distinct types of neurons or under distinct disease situations. Another quite successful approach at least for pre-clinical research is combination of viral vectors carrying floxed expression cassettes with the ever growing number of murine transgenic Cre driver lines [26], which "activate" the viral genome in a cell type-specific manner. As the transgenic mice accept much larger promoter regions than any viral vector construct, their Cre expression pattern is more likely to reflect the "endogenous" activity of a given gene. However, in terms of gene therapy, this combination of classical mouse genetics and viral gene transfer is of course not helpful.

Native transcriptional targeting of AAV vectors has been described for example for nigral dopaminergic neurons by exploiting a 2.5 kbp fragment of the tyrosine-hydroxylase (TH) promoter [27]. However, careful reading of such literature often shows that the respective virus was used at low titer in a spatially very restricted area, with only short-term analysis, and not tested in other brain areas. Thus, appropriate stereotaxic delivery would have demonstrated the same level of targeting. This criticism by no means disqualifies the work done, but is aimed to sensitize the reader of this chapter for the manifold problems associated with appropriate targeting of viral vectors.

5 Concluding Remarks

Clearly, transcriptional targeting offers great opportunities for preclinical research: brain diseases are by no means only affecting neurons, and astrocytes not only outnumber neurons in the brain but also serve essential supporting features and contribute directly to brain diseases. Studying the contribution of each of these brain cell populations independently will allow to open completely new venues to find cures for devastating disorders like AD and PD. If oligodendrocyte-specific vectors become available prevention of demyelination and stimulation of re-myelinating process may be beneficial for MS patients. Vectors allowing microglia-specific transgene expression may prove important in immunomodulatory studies, although transgene expression in immune cells may provoke serious safety concerns in clinical applications.

However, it has to be envisaged that clinical applicability of specifically targeted vectors is not to become reality within the next few years. Recent clinical trials in the brain relied on the relative neuro-tropism of AAV-2 and VSV-G pseudotyped LV vectors, allowing the use of ubiquitous promoters. Even the first genetic medication approved in the Western world, Glybera to treat lipoprotein lipase deficiency by intramuscular application, is an AAV-1 vector driving transgene expression from a simple CMV promoter. Thus, the advent of cell- and tissue-specific gene therapy vectors is yet to come.

References

1. Thevenot E, Jordao JF, O'Reilly MA et al (2012) Targeted delivery of self-complementary adeno-associated virus serotype 9 to the brain, using magnetic resonance imaging-guided focused ultrasound. Hum Gene Ther 23:1144–1155

2. Zincarelli C, Soltys S, Rengo G, Rabinowitz JE (2008) Analysis of AAV serotypes 1-9 mediated gene expression and tropism in mice after systemic injection. Mol Ther 16: 1073–1080

3. Mitchell AM, Nicolson SC, Warischalk JK, Samulski RJ (2010) AAV's anatomy: roadmap for optimizing vectors for translational success. Curr Gene Ther 10:319–340
4. Van Vliet KM, Blouin V, Brument N et al (2008) The role of the adeno-associated virus capsid in gene transfer. Methods Mol Biol 437:51–91
5. Hickey P, Stacy M (2013) AAV2-neurturin (CERE-120) for Parkinson's disease. Expert Opin Biol Ther 13:137–145
6. Kaplitt MG, Feigin A, Tang C et al (2007) Safety and tolerability of gene therapy with an adeno-associated virus (AAV) borne GAD gene for Parkinson's disease: an open label, phase I trial. Lancet 369:2097–2105
7. Christine CW, Starr PA, Larson PS et al (2009) Safety and tolerability of putaminal AADC gene therapy for Parkinson disease. Neurology 73:1662–1669
8. Cucchiarini M, Ren XL, Perides G, Terwilliger EF (2003) Selective gene expression in brain microglia mediated via adeno-associated virus type 2 and type 5 vectors. Gene Ther 10: 657–667
9. Kügler S, Klocker N, Kermer P et al (1999) Transduction of axotomized retinal ganglion cells by adenoviral vector administration at the optic nerve stump: an in vivo model system for the inhibition of neuronal apoptotic cell death. Gene Ther 6:1759–1767
10. Kügler S, Meyn L, Holzmuller H et al (2001) Neuron-specific expression of therapeutic proteins: evaluation of different cellular promoters in recombinant adenoviral vectors. Mol Cell Neurosci 17:78–96
11. Kügler S, Kilic E, Bahr M (2003) Human synapsin 1 gene promoter confers highly neuron-specific long term transgene expression from an adenoviral vector in the adult rat brain depending on the transduced area. Gene Ther 10:337–347
12. Glover CP, Bienemann AS, Heywood DJ et al (2002) Adenoviral-mediated, high-level, cell-specific transgene expression: a SYN1-WPRE cassette mediates increased transgene expression with no loss of neuron specificity. Mol Ther 5:509–516
13. Klein R, Ruttkowski B, Knapp E et al (2006) WPRE-mediated enhancement of gene expression is promoter and cell line specific. Gene 372:153–161
14. Kügler S, Lingor P, Scholl U et al (2003) Differential transgene expression in brain cells in vivo and in vitro from AAV-2 vectors with small transcriptional control units. Virology 311:89–95
15. Shevtsova Z, Malik JM, Michel U et al (2005) Promoters and serotypes: targeting of adeno-associated virus vectors for gene transfer in the rat central nervous system in vitro and in vivo. Exp Physiol 90:53–59
16. Lee Y, Messing A, Su M, Brenner M (2008) GFAP promoter elements required for region-specific and astrocyte-specific expression. Glia 56:481–493
17. Drinkut A, Tereshchenko Y, Schulz JB et al (2012) Efficient gene therapy for Parkinson's disease using astrocytes as hosts for localized neurotrophic factor delivery. Mol Ther 20: 534–543
18. Tereshchenko J, Maddalena A, Bähr M, Kügler S (2014) Pharmacologically controlled, discontinuous GDNF gene therapy restores motor function in a rat model of Parkinson's disease. Neurobiol Dis 65C:35–42
19. Maddalena A, Tereshchenko J, Bähr M, Kügler S (2013) Adeno-associated virus-mediated, mifepristone-regulated transgene expression in the brain. Mol Ther Nucleic Acids 2, e106
20. Colin A, Faideau M, Dufour N et al (2009) Engineered lentiviral vector targeting astrocytes in vivo. Glia 57:667–679
21. Manfredsson FP, Tumer N, Erdos B et al (2009) Nigrostriatal rAAV-mediated GDNF overexpression induces robust weight loss in a rat model of age-related obesity. Mol Ther 17:980–991
22. Hermening S, Kügler S, Bähr M, Isenmann S (2004) Increased protein expression from adenoviral shuttle plasmids and vectors by insertion of a small chimeric intron sequence. J Virol Methods 122:73–77
23. Sunkin SM, Ng L, Lau C et al (2013) Allen Brain Atlas: an integrated spatio-temporal portal for exploring the central nervous system. Nucleic Acids Res 41:D996–D1008
24. Portales-Casamar E, Swanson DJ, Liu L et al (2010) A regulatory toolbox of MiniPromoters to drive selective expression in the brain. Proc Natl Acad Sci U S A 107:16589–16594
25. Delzor A, Dufour N, Petit F et al (2012) Restricted transgene expression in the brain with cell-type specific neuronal promoters. Hum Gene Ther Methods 23(4):242–54
26. Tolu S, Avale ME, Nakatani H et al (2010) A versatile system for the neuronal subtype specific expression of lentiviral vectors. FASEB J 24:723–730
27. Oh MS, Hong SJ, Huh Y, Kim KS (2009) Expression of transgenes in midbrain dopamine neurons using the tyrosine hydroxylase promoter. Gene Ther 16:437–440

Part III

Viral Vector Production

Chapter 7

Small-Scale Recombinant Adeno-Associated Virus Purification

Corinna Burger and Kevin R. Nash

Abstract

Recombinant adeno-associated virus (rAAV) vectors have become increasingly popular in research and clinical trials due to their efficient gene transfer and long-term expression in tissues including brain. In addition, rAAV has demonstrated an impressive safety profile in gene therapy trials. The emergence of rAAV serotypes with different cell tropisms and distribution properties has allowed scientists to tailor serotypes to specific experimental needs. AAV does not have a cytopathic effect; therefore, purification methods require extraction of the viral vector from the cell. This involves gradient ultracentrifugation of the cellular extract sometimes followed by chromatography. This chapter describes a small-scale production method for rAAV purification from ten to twenty 15 cm plates of human embryonic kidney-derived 293B cells (HEK 293) cells that can yield approximately 300 μl of a 5×10^{12} to 1×10^{13} genome copies/ml viral preparation final concentration.

Key words Adeno-associated virus, Iodixanol gradient purification, Tissue culture

1 Introduction

Over the last 20 years scientists have developed and optimized methods for increasing viral vector titers and improving the purity of preparations of viral vectors (for a review *see* ref. 1). Large-scale production has also been developed for human gene therapy applications, using bioreactors, or insect cells [2]. The first step in production of rAAV involves the transfection of three plasmid constructs into HEK293 cells: (1) adenoviral helper genes necessary for replication of the rAAV DNA, (2) AAV rep and cap genes, necessary for capsid formation and replication into the cell, and (3) rAAV containing the gene of interest.

AAV is a single-stranded DNA (ssDNA) virus belonging to the Dependoparvovirus subfamily of the Parvoviridae family. The members of this subfamily are aptly named because they are dependent on a coinfection with a helper virus such as adenovirus or herpesvirus for efficient DNA replication and viral propagation;

Fredric P. Manfredsson (ed.), *Gene Therapy for Neurological Disorders: Methods and Protocols*, Methods in Molecular Biology, vol. 1382, DOI 10.1007/978-1-4939-3271-9_7,

in the absence of a helper virus virtually no DNA replication occurs. Therefore, in initial rAAV production methods a helper virus such as adenovirus (Ad) was used, which unfortunately leads to helper virus contamination in the rAAV preparations. More recently, the active helper proteins have been identified and incorporated into helper plasmids to eliminate the use of an active helper virus. The Ad genes E1a, E1b, E4, VA, and E2a are required for AAV helper function. With the exception of the E2a, which encodes the Ad DNA-binding protein (DBP), none of these genes code for enzymes that are directly involved in DNA replication, and deletion of DBP produces only a modest (fivefold) reduction on AAV DNA replication in vivo [3] and in vitro [4]. However, Ad DBP is required for efficient Rep gene transcription [5, 6]; it is possible that the only effect of Ad DBP is an indirect effect on gene expression of both Rep and the AAV capsid genes. Thus, in an Ad coinfection, AAV relies primarily on cellular enzymes for DNA replication. For AAV DNA replication please *see* [7]. The most common helper plasmid is pXX6 which contains the Ad helper functions, E2a, E4, and VA genes. The E1a and E1b are expressed in the HEK293 cells and necessitate the use of this cell line for rAAV production.

AAV encodes for four nonstructural proteins Rep 78, 68, 50, and 40 and the capsid proteins VP1, VP2, and VP3. Rep78 and 68 are involved in AAV DNA replication and both contain ATP-dependent DNA helicase and site-specific endonuclease activities. Rep 50 and 40 contain helicase activity and are believed to be involved in AAV DNA packaging. These required AAV proteins are typically incorporated into a second helper plasmid but may also be incorporated in a single plasmid with the Ad helper proteins. The lack the AAV terminal repeats on the Rep-Cap plasmid eliminates the possibility of generating wild-type virus.

The third plasmid required is one that contains your gene of interest flanked by AAV inverted terminal repeats (ITR), often referred to as just terminal repeats or TRs. The TRs are required for viral DNA replication as AAV replicates by a strand displacement method using a hairpin TR as a primer [7]. The TRs are also required for packaging of the ssDNA into the viral particles. The total size of the DNA from TR to TR is typically <5 kb as larger sizes cannot be fully packaged (AAV DNA 4.7 kb). This plasmid typically consists of a promoter driving the gene of interest with a poly-A termination signal such as SV40 Poly-A. The AAV TRs used for rAAV production are typically from AAV serotype 2. This necessitates the use of AAV2 Rep protein for efficient DNA replication and packaging. However, it has been found that the capsid serotype may be varied to generate a pseudotyped virus. These viruses although containing AAV2 rep can be packaged into a number of different capsid serotypes (most publications have utilized serotypes 1–10; consisting of rep2 and cap1-10).

Transfection of these three plasmids enables the activation of Rep-mediated rAAV DNA replication and packaging of the viral DNA into capsid particles. The rAAV is not released from the cell during the replication and packaging and therefore requires lysis of the cells and purification of the viral preparation from the cell extract. The most commonly used purification methods include density gradient purification with cesium chloride or iodixanol, or chromatography purification such as affinity chromatography [8] or ion exchange chromatography [9, 10] or a combination of density gradient followed by chromatography [11].

Here we provide a simple method for purifying small quantities of high-titer viral vectors. This protocol is designed for most AAV capsids using AAV2 terminal repeats (AAV2 TR).

2 Materials

Prepare all solutions using ultrapure water and analytical grade reagents. All solutions are sterilized before use. Here we report two different transfection methods:

2.1 Reagents for Calcium Phosphate Transfection (Transfection Method 1)

1. Transfection of the rAAV vector and helper DNA are carried out at a 1:1 molar ratio (vector containing adenoviral helper functions plus AAV rep and cap genes: rAAV containing the gene of interest) (*see* **Note 1**). E.g. 300 μg of rAAV DNA carrying gene of interest: 1 mg of Helper DNA (*see* **Note 2**).
2. $CaCl_2$ (2.5 M): Weigh 1.8 g of $CaCl_2$ *dihydrate* to a final volume of 5 ml in H_2O. Filter sterilize through a 0.22 μm filter attached to a 10 ml syringe. Make fresh each time.
3. Hepes-buffered saline, pH 7.05 (HBS; 2×): Weigh 16 g NaCl, 0.74 g KCl, 0.27 g $Na_2HPO_4{\cdot}H_2O$, 2 g dextrose (d-glucose), 10 g HEPES. Add water to 1 L, pH to 7.05 with 0.5 N NaOH (*see* **Note 3**). Filter sterilize. Dispense in 25 ml aliquots and store at −20 °C. Solution has 6-month shelf life.
4. EDTA/PBS (50 mM): Add 5 ml of 500 mM EDTA pH 8.0 to 495 ml PBS.

2.2 Reagents for PEI Transfection (Transfection Method 2)

1. Polyethylamine stock solution (0.323 g/L; PEI): Dissolve 25 kD linear PEI (Polysciences, Warrington, PA, USA) in endotoxin-free dH_2O that has been heated to ~80 °C and let dissolve. Cool to room temperature. Neutralize to pH 7.0, filter sterilize (0.22 μm), aliquot and store at −20 °C; a working stock can be kept at 4 °C. If taking an aliquot from the freezer, prior to use thaw PEI at 37 °C and then heat in a 55 °C water bath until PEI is clear again.

Table 1
Iodixanol solutions

	Optiprep	5 M NaCl	5×TD	Water	Final volume
15 %	45 ml	36 ml	36 ml	63 ml	180 ml
30 %	60 ml	–	24 ml	36 ml	120 ml
40 %	68 ml	–	20 ml	12 ml	100 ml
60 %	100 ml	–	–	–	100 ml

2.3 Viral Vector Purification

1. Cell lysis buffer: 150 mM NaCl, 50 mM Tris pH 8.5. Add 8.8 g of NaCl, 6.1 g Tris base. pH to 8.5. Add water to 1 L. Filter sterilize.
2. TD buffer (5×): 5× PBS, 5 mM $MgCl_2$, 12.5 mM KCl. Add 250 ml of 20× PBS, 5 ml of 1 M $MgCl_2$ and 0.93 g KCl. Filter sterilize.
3. Iodixanol gradient solutions: Iodixanol (ThermoFisher Scientific, Waltham, MA, USA). Prepare iodixanol solutions as outlined in Table 1 and filter sterilize (*see* **Note 4**).
4. Benzonase nuclease (Novagen, San Diego, CA, USA; or Sigma-Aldrich, St Louis, MO, USA).
5. Ultracentrifuge and rotor.
6. Ultracentrifuge tubes: Sorvall 35 ml, ultracrimp tubes for rotor T-865 (ThermoFisher Scientific, Waltham, MA, USA) or Beckman Optiseal tubes for rotor Ti70 (Beckman Coulter, Pasadena, CA, USA).
7. Apollo 20 ml centrifugal quantitative concentrators with 150 kDa Molecular cutoff (Orbital sciences, Topsfield, MA, USA).
8. 1.5 ml Eppendorf *Siliconized* tubes (ThermoFisher Scientific, Waltham, MA, USA) (*see* **Note 5**).

2.4 Viral Vector Preparation for Titering

1. DNase I 2 U/µl (Ambion/Life technologies, Grand Island, NY, USA).
2. 10× DNase buffer: 100 mM Tris pH 7.5, 25 mM $MgCl_2$, 5 mM $CaCl_2$.
3. Proteinase K (Ambion/Life technologies, Grand Island, NY, USA).
4. Proteinase K buffer: 100 mM Tris pH 8100 mM EDTA, 10 % SDS.
5. Lactated Ringers solution or PBS.
6. Glycoblue (Life Technologies, GrandIsland, NY, USA).

2.5 Viral Vector Quantification (Two Different Methods: Real-Time PCR and Dot Blot): Real-Time PCR

1. RT PCR primers: Designed to specific construct or promoter following standard molecular procedures.
2. DNA standards (preferably from the plasmid used to make your virus) of 500, 100, 20, 5, 1, and 0.2 pg (*see* **Note 6**).
3. Viral DNA samples. Run your viral samples in triplicate.
4. Sybr green master mix (2×) (Takara, Mountain View, CA, USA).

2.6 Viral Vector Quantification (Two Different Methods: Real-Time PCR and Dot Blot): Dot Blot

1. 1× Alkaline buffer: 0.4 M NaOH, 10 mM EDTA.
2. Hybridization buffer: 7 % SDS, 1 mM EDTA, 0.25 M $NaHPO_4$ pH 7.2.
3. Wash buffer: 1 % SDS, 1 mM EDTA, 40 mM $NaHPO_4$ pH 7.2.
4. DNA probe: 1 mM each dGTP, dCTP, dTTP, plus 0.84 mM dATP and 0.16 mM biotinylated dATP (Enzo Life Sciences, Farmingdale, NY, USA) (*see* **Note 7**).
5. Hybridization oven and hybridization tubes.
6. Hybond-N membrane (GE Life Sciences, Pittsburgh, PA, USA).
7. Whatman filter paper.
8. UV cross-linker.
9. Streptavidin-conjugated antibody: Streptavidin IRDye-800CW (LI-COR Biotechnology, Lincoln, NE, USA), or equivalent.
10. LI-COR Odyssey scanner (LI-COR Biotechnology, Lincoln, NE, USA), or equivalent.

3 Methods

Carry out all procedures under sterile conditions in a biosafety cabinet.

3.1 Tissue Culture

1. Human embryonic kidney-derived 293B cells (HEK 293) will be grown on 15 cm tissue culture dishes. Cells are cultured in Dulbecco's modified Eagle's medium (DMEM) supplemented with 10 % fetal bovine serum (FBS), penicillin (100 units/ml), and streptomycin (100 μg/ml) at 37 ° C in a 5 % CO_2 incubator. First thaw cells and add to a 10 cm tissue culture dish. Grow until they are confluent. Then pass them to a 15 cm dish and let them grow until confluent. The volume of added medium for a 15 cm plate is 20 ml.
2. 10 plate prep: Grow cells to confluence in a 15 cm dish and then passage the cells 1/10 to generate 10 × 15 cm plates from each original 15 cm plate. If cells are healthy the plates should be ~70 % confluent and ready for transfection after 3 days.

3. 20 plate prep: Keep passing 15 cm dish until you have 12 dishes. Pass cells in the 12 dishes 1:2 once they are confluent, so they will be ready for transfection the following day (*see* **Note 8**).

3.2 Calcium Phosphate Transfection (Method 1)

1. Check plates. Cells should be 70 % confluent the day of transfection.
2. Thaw 2× HBS and keep at 37 °C until ready to use.
3. Make $CaCl_2$ solution.
4. Calculate the amount of DNA needed for transfection: Helper DNA containing adenoviral and rep and cap AAV functions are transfected at a 1:1 molar ratio. For 20 dishes you need 900 μg of helper (Ad and rep and cap helper function plasmid) and 300 μg of rAAV. Add water to a 50 ml conical tube followed by 2.5 ml $CaCl_2$ (final $CaCl_2$ concentration = 0.25 M), and then add plasmid DNAs. The total volume should be 25 ml (*see* **Note 9**). If performing a triple transfection, e.g., using pXX6 (Ad helper DNA) with a pAAV (rep and cap helper DNA), the following amounts of DNA are used. For 20 dishes rAAV vector (300 μg), pAAV serotype plasmid (300 μg), and pXX6 Ad helper plasmid (700 μg) (*see* **Note 10**).
5. Add $CaCl_2$ and DNA mixture to the conical tube containing 25 ml of 2× HBS dropwise and bubble air through the 2×HEPES solution with a 2 ml serological pipette hooked to a pipet aid while adding the DNA/$CaCl_2$ solution dropwise with a P1000 pipetman. Do not vortex transfection cocktail.
6. Add transfection mix to cells with media dropwise, 2.5 ml per 15 cm plate.
7. Incubate at 37 °C for 48–65 h.

3.3 PEI Transfection (Method 2)

1. Prior to transfection heat PI 55 °C to dissolve clear.
2. The amounts of DNA outlined above in **step 4** in Subheading 3.2 (calcium phosphate transfection) can be used here and added to a sterile 50 ml tube.
3. To the DNA also add 2.5 ml of 1.5 M NaCl and sterile water to a final volume of 25 ml.
4. In a second tube mix by vortexing briefly the following: 8 ml PEI (0.323 g/L), 2.5 ml 1.5 M NaCl, and 14.5 ml water.
5. Add PEI mix to DNA mix and briefly vortex for ~20 s.
6. Incubate at room temperature for 15 min.
7. Add DNA/PEI mixture to cells: 2.5 ml/15 cm dish.
8. Harvest transfected cells 48–65 h post-transfection (*see* **Note 11**).

3.4 Harvesting and Lysis of Transfected Cells

Do this procedure in a biosafety cabinet.

1. Aspirate media from cells.
2. Add 6 ml of cold EDTA/PBS to each plate (50 mM EDTA in PBS) to detach cells (Alternatively, the cells can be scraped into the cell culture media and then centrifuged, **step 5** below).
3. Collect cells in a sterile 200–250 ml plastic centrifuge bottle (size depending on rotor).
4. Add a few extra ml of PBS to each plate to get the remaining cells.
5. Centrifuge at 5000 × *g*, 4 °C, for 10 min to pellet cells.
6. Discard supernatant.
7. Store cells at −80 °C, or continue to the next step (if not storing cell pellets, then do an extra freeze step (**step 10** below for a total of three freeze-thaws).
8. Thaw cells from harvest (in 250 ml centrifuge bottle).
9. Resuspend in 9 ml lysis buffer containing protease inhibitors. Once resuspended transfer suspension to a 50 ml conical tube and top to 18 ml with lysis buffer. Vortex.
10. Freeze for 10 min in an ethanol dry ice bath. Thaw for 15 min at 37 °C, and vortex. Do a total of 2 freeze/thaws (note, be careful to label the sides and tops of the 50 ml tubes in case the ethanol removes your marker from the side of the tube).
11. Benzonase treatment: Add 15 μl of 1 M $MgCl_2$ to the cell suspension and vortex. Add 3 μl of Benzonase and vortex. Incubate at 37 °C for 30 min.
12. Centrifuge for 20 min at 4000 × *g* (4800 rpm in Sorvall 14 × 50 F13 rotor).
13. Decant lysate (supernatant) into 50 ml conical tube. Lysate can also be stored at −80 °C until further use.
14. Save a 5 μl aliquot to run on SDS-PAGE gel.

3.5 Iodixanol Gradient Purification

1. Divide the ~18 ml of lysate into two ultracentrifuge tubes, 9 ml per tube. After pipetting lysate into the ultracentrifuge tubes, underlay lysate in each tube with the following concentrations of iodixanol in this order: 9 ml of 15 % iodixanol, 6 ml of 25 % iodixanol, 5 ml of 40 % iodixanol, and 5 ml of 60 % iodixanol. Basically the gradient is laid from the top layer finishing with the bottom layer so as not to disturb the layers by pouring on top of each layer (*see* Fig. 1). The samples have to be pipetted using a sterile glass Pasteur pipette. Since a Pasteur pipette cannot accurately measure the volume, a graduated pipette can be used to deliver the required amount of iodixanol into a sterile 15 ml tube prior to loading the gradient. Alternatively, layer the iodixanol step gradients from 15 % to

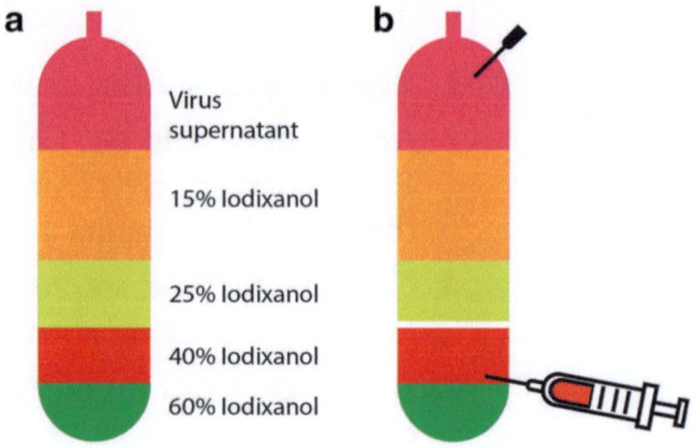

Fig. 1 Diagram showing the layout of the iodixanol gradient (**a**) before, and (**b**) after ultracentrifugation. Care needs to be taken when aspirating the AAV fraction to avoid the cellular extract (white band in **b**)

the 60 % (top to bottom), and then carefully drop-wise layer the virus preparation on top. This reduces the exposure to the virus preparation while layering the iodixanol.

2. Cap and seal tubes as per manufacturer's instructions. Confirm that tubes are balanced by weighing them. Tubes are then placed into the rotor. Ensure that caps are placed on top of tubes; otherwise the tubes will crush due to the amount of g-force.
3. Centrifuge at 65k rpm (350,000 × *g*) at 18 °C for 1 h or 60k rpm for 2 h (*see* **Note 12**). Set for slow deceleration.
4. Set up a stand with a clamp large enough to hold the centrifuge tube, and place tube in clamp. You should see the 40–60 % interphase. The 25–40 % interphase should have a white band.
5. Swab top and bottom with alcohol.
6. Vent top with needle, 18G recommended (Fig. 1).
7. Pull the 40 % iodixanol band and interphase between the 60 % and 40 % bands with a 5–10 ml syringe with an 18G needle (larger gauge needle reduces sheer effects on the virus). Collect approximately 4 ml of 40 % and 2 ml of the 60 % phase (*see* **Note 13** and Fig. 1). At this point it is recommended to run a quick western blot to confirm the presence of capsid protein in your 40 % fraction (10–20 μL sample). The 5 μL of crude can be run at the same time.

3.6 Buffer Exchange and Concentration

After confirmation that viral proteins are present in the preparation (as indicated by Western analysis), samples need to have iodixanol removed and the virus concentrated.

1. Wash concentrator according to the manufacturer's instructions.
2. Add ~14 ml of the appropriate buffer (lactated ringers or PBS) to the concentrator, and mix with iodixanol virus preparation (~6 ml). For Apollo concentrators a volume of buffer needs to

be placed in the lower half of the tube to give the desired final volume of concentrate (*see* the manufacturer's instructions).

3. Place the concentrator in the centrifuge and spin according to the manufacturer's recommendations.
4. The flow through is discarded and the dilution of virus and the spin are repeated as above. This will effectively exchange ~95 % of the iodixanol buffer with your buffer of choice.
5. These steps are repeated to ensure that you have removed all the iodixanol (typically five spins).
6. The volume remaining in the concentrator following the last sample spin should be ~200–300 μl. Carefully remove the sample and transfer to a silicone-treated tube.
7. Wash the concentrator by placing 15 ml of the exchange buffer into the concentrator, mix and leave overnight at 4 °C. Repeat the above concentrating procedure until the remaining volume is ~200 μl. Add this to the first 300 μl for a final volume of ~500 μl.
8. Mix the sample and remove whatever aliquots are necessary for titering.

3.7 Virus Titering Using RT-PCR

1. Take 4 μl of the concentrated virus and add 20 μl of 10× DNase buffer, 2 μl DNase, 174 μl of ddH_2O for a total volume 200 μl. Incubate at 37 °C for 1 h.
2. To the 200 μl DNase sample add: 22 μl of 10× proteinase buffer, 2 μl proteinase K (10 mg/ml stock) for a total volume 224 μl. Incubate at 37 °C for 1 h.
3. To proteinase sample add an equal volume (224 μl) of phenol/chloroform. Vortex 3 min. Microfuge for 5 min at max speed. Save aqueous layer in new 1.5 ml Eppendorf tube. Repeat extraction.
4. Add 1/10 volume of 3 M NaOAc, vortex. Add 1 μl glycoblue, and vortex. Add 2.5 volumes of 100 % EtOH, and vortex. Precipitate O/N at −80 °C. Microfuge for 20 min at max speed.
5. Discard supernatant.
6. Add 100 μl of 70 % EtOH.
7. Microfuge for 5 min at max speed.
8. Discard supernatant and air-dry for 5 min.
9. Resuspend DNA in 40 μl H_2O.
10. Set up a standard curve: Use a 5 log spanning serial dilution of rAAV DNA spanning from 100 pg to 0.01 pg of the vector plasmid. Do each real time PCR reaction in triplicate. Assume that one particle of AAV contains 1 copy of single-stranded DNA, and that 1 pg of DNA of a 7 kb plasmid corresponds to 1.3×10^5 plasmid molecules.

11. Set up the real-time PCR reactions are as follows: 10 μL 2× Sybr Green master mix, 4 μL DNA (standard or from viral preparation), 0.4 μL forward primer (10 pmol/μl), 0.4 μL reverse primer (10 pmol/μl), and 5.2 μL dH_2O for a total of 20 μL reaction volume per sample (*see* **Note 14**).

3.8 Virus Titering Using Dot Blot Protocol

1. Set the hybridization oven to 60–65 °C.
2. Set up standards: 1:2 serial dilution of a plasmid at known concentration. Dilute plasmid DNA into alkaline buffer (0.4 M NaOH, 10 mM EDTA) such that you have a final concentration of 6 ng/μL in the starting tube with a total volume of 500 μL. Serial dilute 1:2, placing 250 μL from the first tube into a second tube with 250 μL of alkaline buffer. Vortex and repeat. Make at least eight standard dilutions.
3. Virus: Start with diluting 25 μL of virus preparation into 225 μL of alkaline buffer. Serial dilute 1:2 to make at least five samples. Dilute 125 μL of tube one into 125 μL of alkaline buffer in a second tube, vortex, and repeat.
4. Wet the Hybond-N membrane and Whatman filter paper for the dot blot apparatus (use of water is ok). Clamp together.
5. Load 100 μL of the samples and standard into the wells (best to load two different standards) (This will mean that the standard plasmid amount starts at 600 ng and is diluted serially to at least 4.7 ng). Apply vacuum. Once all the wells have been vacuumed dry, remove vacuum. Remove top of dot blot apparatus and mark membrane for orientation (pencil is the best marker that will not be washed away).
6. Place membrane on clingwrap and cross-link DNA to membrane using UV crosslinker (optimal or autocross link function on machine).
7. Place membrane in hybridization tube (with DNA side up in the tube) with 10 ml of Hybridization buffer and set into the hybridization oven (do not forget to add counter balance tube). Incubate for 1–2 h.
8. Denature probe 95–100 °C for 5 min, and then place probe on ice for 5 min. Then add directly to the hybridization tube with the membrane. Use ~300 ng of probe in the hybridization tube. Incubate overnight (for probe preparation by PCR *see* **Note 15**).
9. Wash membrane in hybridization oven 3 × 15-min washes with wash buffer (1 % SDS, 40 mM $NaHPO_4$, pH 7.2, 1 mM EDTA).
10. Place membrane in 5–10 ml of wash buffer in western blotting dish. Add streptavidin IRDye-800CW antibody at a concentration of 1:5000–10,000. Incubate at room temperature for 2 h.

11. Wash membrane 3 × 5 min PBS-tween.
12. Measure blot on Li-cor Odyssey scanner.
13. Signal intensity of the virus is compared to the plasmid standard curve. Avogadro's number is used to calculate vector genomes from the ng/ml of DNA.

4 Notes

1. Vectors containing Ad/rep/cap functions can be obtained from a number of companies or university vector cores.
2. We strongly recommend that the DNA is purified by CsCl ultracentrifugation to obtain a high titer preparation. Methods like Qiagen endofree purification are not optimal and result in decreased vector titers.
3. The pH of this solution is critical for transfection efficiency. The optimal range for calcium phosphate transfection is between 6.9 and 7.1.
4. Protect solutions from light. Cover with aluminum foil or store in the dark.
5. It is important to store rAAV in siliconized tubes to prevent the virus to stick to plastic. The titer will drop dramatically.
6. Serial dilutions of standard DNA can be made ahead of time and store frozen to avoid having to make standards each time.
7. The biotinylated dATP may be substituted with a ^{32}P-dATP if a radioactive probe is preferred.
8. It is a good idea to use a brand new bottle of DMEM + 10 % FCS for the passage the night before transfection. This will ensure a good buffering capacity of the media during the critical calcium phosphate transfection step.
9. Calculate volume of DNA to get 300 μg of rAAV and 1 mg of helper DNA. To that volume add 2.5 ml (for the $CaCl_2$). Subtract this total volume from 25. This is the volume of water you need to add to this mixture.
10. The amount of rAAV plasmid can be reduced at least 50 % if the gene of interest is toxic to the HEK293 cells. In our experience the titer may be on the low side (~10^{12}) but can be a better titer than when using the higher concentration of rAAV where a significant number of cells are dying due to the transgene expression.
11. This transfection can be performed with only 10 × 15 cm dishes to yield virus quantities for 10^{12} to 10^{13} vg/ml final. This reduces the amount of plasmid DNA required.
12. Do not spin at a lower temperature or the iodixanol might precipitate.

13. Avoid the interphase between the 40 % and 25 % bands. It is important not to extract any of the white material at the 25–40 % interphase as this would contaminate your viral preparation with junk proteins. Leaving a small amount of the 40 % layer is recommended so as to avoid this contaminant.
14. Keep samples on ice until ready to run the reactions. Cycle conditions are primer specific. Primer selection, annealing temperatures, and plate read temps need to be optimized ahead of time.
15. Use PCR methodology to generate a probe for blot. Plasmid and primers used depend on vector that is to be titered. Primers for the transgene or promoter can be used and rAAV DNA plasmid as template. Use biotinylated dNTPs in a 50 μL reaction. A PCR cleanup kit can remove unincorporated biotin nucleotides (e.g., Zymo Research, Irvine, CA). Probe can be made prior to use and stored at −20 °C prior to denaturation and use. Immediately before use the probe must be denatured and placed on ice as described above.

References

1. Ayuso E, Mingozzi F, Bosch F (2010) Production, purification and characterization of adeno-associated vectors. Curr Gene Ther 10(6):423–436
2. Virag T, Cecchini S, Kotin RM (2009) Producing recombinant adeno-associated virus in foster cells: overcoming production limitations using a baculovirus-insect cell expression strategy. Hum Gene Ther 20(8):807–817
3. Carter BJ, Antoni BA, Klessig DF (1992) Adenovirus containing a deletion of the early region 2A gene allows growth of adeno-associated virus with decreased efficiency. Virology 191:473–476
4. Ward P et al (1998) Role of the adenovirus DNA-binding protein in in vitro adeno-associated virus DNA replication. J Virol 72(1):420–427
5. Janik JE et al (1989) Efficient synthesis of adeno-associated virus structural proteins requires both adenovirus DNA binding protein and VA I RNA. Virology 168:320–329
6. Chang LS, Shenk T (1990) The adenovirus DNA-binding protein stimulates the rate of transcription directed by adenovirus and adeno-associated virus promoters. J Virol 64: 2103–2109
7. Nash K, Chen W, Muzyczka N (2008) Complete in vitro reconstitution of adeno-associated virus DNA replication requires the minichromosome maintenance complex proteins. J Virol 82(3):1458–1464
8. Zolotukhin S et al (1999) Recombinant adeno-associated virus purification using novel methods improves infectious titer and yield. Gene Ther 6(6):973–985
9. Binny CJ, Nathwani AC (2012) Vector systems for prenatal gene therapy: principles of adeno-associated virus vector design and production. Methods Mol Biol 891:109–131
10. Chahal PS et al (2014) Production of adeno-associated virus (AAV) serotypes by transient transfection of HEK293 cell suspension cultures for gene delivery. J Virol Methods 196: 163–173
11. Zolotukhin S et al (2002) Production and purification of serotype 1, 2, and 5 recombinant adeno-associated viral vectors. Methods 28(2): 158–167

Chapter 8

Lentivirus Production and Purification

Matthew J. Benskey and Fredric P. Manfredsson

Abstract

Lentiviral (LV) vectors offer unique advantages over other gene delivery systems, namely the ability to integrate transgenes into the genome of both dividing and nondividing cells. Detailed herein is a simple protocol for the production LV vectors, describing the triple transfection of an LV transfer vector and LV helper plasmids into HEK-293 cells, and the subsequent purification of virions from the cellular media. The current protocol is versatile, and can be easily modified to fit the specific needs of the researcher in order to produce relatively high-titer LV vectors which can be used to transduce a wide variety of cells both in vitro and in vivo.

Key words Lentivirus, VSV-G, Viral vector, Gene delivery, Gene integration

1 Introduction

Lentivirus (LV) is an enveloped, plus-strand RNAvirus belonging to the family of retroviridae [1]. LV vectors, derived from the human immunodeficiency virus (HIV-1), are powerful genetic tools that offer several advantages over other viral vector systems. Principle amongst the advantages of LV vectors is the ability to integrate genetic material into the genome of dividing and nondividing host cells, both in vitro and in vivo [2, 3]. LV vectors can provide efficient, stable, and long-term gene expression in terminally differentiated cells such as immune cells or neurons [4–6]. Due to the ability of LV vectors to integrate genetic material into the host genome, progeny produced from the infected cell will harbor the same transgene, an attribute that is exploited in applications such as stem cell therapy [7, 8]. Further, LV vectors offer a relatively large carrying capacity (8–9 Kb) and exhibit low immunogenicity, making these vectors extremely useful gene delivery tools.

Progress in the production of LV vectors has led to the development of replication incompetent LV vectors, which can be produced easily with minimal effort. Replication incompetent LV vectors are produced by replacing all viral genes from the LV

Fredric P. Manfredsson (ed.), *Gene Therapy for Neurological Disorders: Methods and Protocols*, Methods in Molecular Biology, vol. 1382, DOI 10.1007/978-1-4939-3271-9_8, © Springer Science+Business Media New York 2016

genome with a transgene of interest, leaving only the *cis*-acting elements essential for viral packaging (i.e., long terminal repeats (LTR), tat activation region, Ψ packaging signal, Rev-responsive element, and polypurine tracts). The removed components of the viral genome are supplied in *trans* by separate plasmids encoding the viral genes necessary for packaging and envelope pseudotyping. The transfer vector containing the transgene of interest and the separate helper plasmids are co-transfected into a mammalian cell line, and LV vectors can be harvested and purified from the media. Providing the necessary viral genes in separate plasmids greatly reduces the possibility of mobilization and the production of a replication competent virus. Further, this system also affords the ability to easily mix and match glycoproteins from alternative enveloped viruses in order to optimize the expression levels and tropism of LV vectors specifically to the desired application [9, 10].

Here we describe a simple and efficient protocol for packaging a vesicular stomatitis virus glycoprotein (VSV-G) pseudotyped LV vector. The current protocol describes the triple transfection of a transfer vector and two LV helper plasmids, and the subsequent purification and concentration of LV vectors from cellular media by ultracentrifugation [11]. This protocol is flexible and can be easily scaled up or down in order to meet the needs of the researcher. Utilizing the current protocol we have generated LV-vectors with titers in the range of 2×10^{12} viral genomes (vg)/mL, that are capable of transducing neurons within an adult animal in vivo, as well as difficult-to-transfect cells in vitro (Fig. 1).

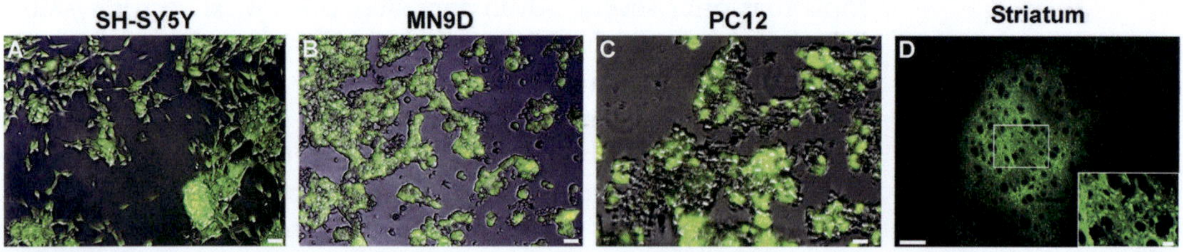

Fig. 1 GFP expression following transduction with a VSV-G pseudotyped LV vector. LV vectors produced using the current protocol efficiently transduces cells in vitro (**a–c**) and in vivo (**d**). SH-SY5Y (**a**), MN9D (**b**), or PC12 (**c**) cells were plated and grown to approximately 50 % confluency. Cells were transduced with the VSV-G pseudotyped LV vector expressing GFP under the control of the cytomegalovirus/chicken β-actin hybrid promoter at a multiplicity of infection (MOI) of 100. GFP expression was visualized 48 h post-transduction. To demonstrate LV vector transduction in vivo, adult male rats were injected with 2 μL of the VSV-G pseudotyped LV vector expressing GFP under the control of the cytomegalovirus/chicken β-actin hybrid promoter (2.07×10^{12} vg/mL) into the striatum. Animals were sacrificed 4 weeks post-transduction, and native GFP fluorescence was visualized in the striatum (**d**). The *inset in panel* (**d**) shows a higher magnification image of the area within the *box* in panel (**d**). Scale bar in panel (**a**)–(**c**) represents 50 μm. Scale bar in panel (**d**) and inset represent 200 μm and 50 μm, respectively

2 Materials

2.1 Plasmids and Cells

1. Human Embryonic Kidney 293 T (HEK-293 T) Cells (ATCC, Manassas, VA, USA).
2. pNHP packaging vector (Addgene, Cambridge, MA, USA; *see* **Notes 1** and **2**).
3. pHEF-VSVG (Addgene, Cambridge, MA, USA; *see* **Note 3**).
4. pFIN transfer vector (Addgene, Cambridge, MA, USA; *see* **Note 4**).

2.2 Reagents and Supplies

1. Penicillin/Streptomycin (Pen/Strep; 10,000 units/mL).
2. Fetal bovine serum (FBS).
3. TrypLE Express 1× dissociation reagent (Thermo Fisher Scientific, Grand Island, NY, USA).
4. 25 kDa linear Polyethylenimine (PEI) (Polysciences, Warrington, PA, USA).
5. Dulbecco's modified Eagle media (DMEM) containing high glucose, glutamine, and sodium pyruvate.
6. 1× Dulbecco's phosphate-buffered saline (DPBS) containing calcium and magnesium.
7. 0.22 μm Stericup filter unit (EMD Millipore, Billerica, MA, USA).
8. 1.5 M NaCl: Add 87.66 g of NaCl to 800 mL of dH_2O, stir until completely dissolved. Adjust to 1 L with dH_2O and filter sterilize with a 0.22 μm filter unit.
9. 0.45 μm Stericup-HV filter units (EMD Millipore, Billerica, MA, USA).
10. Centricon Plus-70 Centrifugal filter units (EMD Millipore, Billerica, MA, USA).
11. 38.5 mL conical thinwall polyallomer ultracentrifuge tubes or equivalent (Beckman Coulter, Indianapolis, IN, USA).
12. T175 Culture flasks.
13. SW32Ti rotor (Beckman Coulter, Indianapolis, IN, USA) or equivalent.
14. Beckman Coulter L-100XP Ultracentrifuge (Beckman Coulter, Indianapolis, IN, USA) or equivalent.
15. Sorvall RC 6 superspeed centrifuge (Thermo Fisher Scientific, Grand Island, NY, USA) or equivalent.
16. Real-time PCR system: Applied biosystems 7500 (Thermo Fisher Scientific, Grand Island, NY, USA) or equivalent.

17. Lenti-X-qRT-PCR kit (Clonetech, Mountain View, CA, USA).
18. Standard tissue culture equipment.

3 Methods

3.1 Prepare Reagents

1. Prepare complete HEK-293 T media (DMEM with 10 % FBS and 1 % Pen/strep). Add 100 mL of FBS and 10 mL of Pen/strep to 890 mL of DMEM (*see* **Note 5**).
2. Prepare PEI transfection reagent. Heat 200 mL of dH_2O to 80 °C in a beaker on a stirring hotplate. Add 80.75 mg of PEI to the beaker and stir with a stir bar. Stir at 80 °C until the PEI is completely dissolved. Adjust pH to 8.0 with hydrochloric acid. Adjust to 250 mL with dH_2O. Filter sterilize with a 0.22 μm filter unit (*see* **Note 6**).
3. Prepare the required amount of plasmidDNA (for each respective plasmid) using a standard plasmid DNA preparation method. DNA must be endotoxin free.

3.2 Transfection and Viral Purification

1. Seed the T175 flask with HEK-293 T cells. Grow HEK-293 T cells in a T75 starter flask prior to seeding the T175. Once the T75 starter flask is at 90–95 % confluency aspirate media from the flask and gently rinse with sterile PBS. Incubate cells with 4 mL of dissociation reagent (e.g., TrypLE Express) at 37 °C for 5 min. Add 6 mL of media containing serum to deactivate the dissociation reagent, and triturate with 25–30 full strokes of a serological pipette to create a single cell suspension.
2. Determine cell number per mL using a hemocytometer.
3. Calculate the amount of the single cell suspension needed to seed the T175 flask with 3×10^7 HEK-293 T cells. Add cells to flask and bring the total volume of the flask to 28 mL with warm (37 °C), complete HEK-293 T media.
4. Incubate T175 Flask at 37 °C with 5 % CO_2 overnight or until 80–90 % confluency is reached.
5. Prepare transfection mixture. To create the complete transfection mixture, two separate solutions are first prepared, a DNA solution and a PEI solution.
6. Prepare the DNA solution. In a 50 mL conical add 21.3 μg pNHP, 8 μg pHEF-VSVG, 7.2×10^{11} copies of the pFIN transfer vector (*see* **Note 7**), and 148.75 μL of 1.5 M NaCl. Bring total volume of DNA solution to 1.44 mL with dH_2O and mix well.
7. Prepare PEI solution. In a separate 50 mL conical add 460.59 μL PEI and 148.75 μL 1.5 M NaCl to 1.04 mL dH_2O (*see* **Note 8**). Mix well.

8. Prepare the complete transfection solution by adding the PEI solution dropwise to the DNA solution. Vortex vigorously for 1–2 min (*see* **Note 9**).
9. Incubate the complete transfection solution at room temperature for 20 min. The transfection solution should turn cloudy.
10. Remove all media from the 80–90 % confluent T175 flask. Mix the complete transfection solution with 28 mL of complete HEK-293 T media in a graduated cylinder. Gently add the transfection-media mixture to the T175 flask and be sure all cells are covered.
11. Incubate the T175 flask at 37 °C and 5 % CO_2.
12. Collect the supernatant containing virus (*see* **Note 10**). The supernatant is collected at 24 h post-transfection and again at 40 h post-transfection. At the 24-h time point, collect the viral media in a 50 mL polypropylene conical. Replace the media with 28 mL of warm (37 °C) complete HEK-293 T media.
13. Spin the collected viral media at 783 × *g* in the RC 6 (or equivalent) centrifuge for 5 min to pellet cell debris. Collect the supernatant and store in a fresh tube at 4 °C.
14. At the 40 h time point, again collect the media and remove debris by centrifugation as in **steps 12** and **13**.
15. Pool all viral media from the 24- and 40-h time points. Filter the viral media through the 0.45 μm Stericup-HV filter.
16. Split the viral media into two thinwall polyallomer ultracentrifuge tubes (*see* **Notes 11** and **12**).
17. Carefully load the filled ultracentrifugation tubes into the SW32Ti rotor and spin at 80,000 × *g* for 2 h at 4 °C.
18. Remove the supernatant from the viral pellet by carefully aspirating the media from the ultracentrifugation tube (*see* **Note 13**).
19. After removing the supernatant resuspend the viral pellet by gently overlaying the pellet with 10 μL of sterile DPBS in the conical (*see* **Note 14**).
20. Seal the conical tube with parafilm and store at 4 °C overnight.
21. After overnight incubation, place the sealed conical on ice. Gently shake the conical on an orbital rotating shaker for 2 h [11].
22. Aliquot the resuspended virus into working aliquots and store at −80 °C (*see* **Notes 15**).
23. Utilize one of your aliquots from **step 22** and titer your viral prep using the Lenti-X titering kit following the manufacturer's instructions.

4 Notes

1. The current protocol utilizes the second-generation LV packaging system in which all necessary genes are separated into three distinct plasmids (the transfer vector, envelope plasmid and packaging plasmid). A third-generation packaging system is also available, in which the necessary genes are further separated into four distinct plasmids (transfer vector, envelope, and two packaging plasmids). Although the third generation system offers increased biosafety, it is also more cumbersome due to the increased number of plasmids. However, the current protocol is amenable to third generation packaging systems.
2. The pNHP packaging helper plasmid encodes the following genes: Gag (encodes the capsid, nucleocapsid and matrix structural proteins), Pol (encodes reverse transcriptase, protease and integrase proteins), Tat (encodes protein for transactivation of transcription from viral LTR), Rev (encodes protein that mediates export of viral RNAs from nucleus), as well as vpr (mediates cell cycle arrest and assists in nuclear import) and vpu (enhances virion production and degradation of CD4 receptor in the host cell).
3. pHEF-VSVG encodes the envelope glycoproteins derived from the VSV. LV vectors pseudotyped with VSV-G exhibit a broad host cell range; however, the tropism, titers, and expression of levels LV vectors can be altered by pseudotyping vectors with glycoproteins from alternative enveloped viruses [9, 10].
4. The pFIN LV transfer vector contains a green fluorescent protein (GFP) gene followed by the woodchuck posttranscriptional response element driven by the EF1α promoter. This expression cassette is flanked by HIV-1 LTRs. The pFIN vector also contains components from the LV genome absolutely necessary for viral packaging. If desired, the promoter and transgene can be easily replaced with a promoter or transgene of interest using standard cloning methods.
5. All work should be done in a culture hood using sterile technique.
6. Very little HCl is required to bring the PEI solution to a pH of 8.0. If too much HCl is added, use NaOH to bring the pH back to 8.0. Aliquot the PEI in 12 mL aliquots and store at −80 °C for future use.
7. The size (in nucleotide bases) of the genetic material to be packaged in the lentivirus vector will alter the size of the pFIN plasmid and the corresponding weight of DNA needed for

transfection. To maintain accuracy, determine the molecular weight of the pFIN plasmid containing your genetic material of interest and then calculate the total amount (in nanograms) necessary to transfect 7.2×10^{11} copies of the pFIN plasmid. This can be done easily using an online DNA molecular weight calculator.

8. PEI must be completely homogenous prior to use. If using a previously frozen aliquot, completely thaw at room temperature. Once thawed, heat solution to 55 °C to ensure all PEI is in solution. Let cool to room temperature before use.
9. Adding the PEI dropwise to the DNA mixture prevents the rapid precipitation of DNA and ensures a homogenous DNA/PEI solution. It is helpful to place the DNA mix on a stir plate and gently stir the DNA mixture while adding the PEI mixture dropwise.
10. The quality of the transfection can be determined by viewing the cells under a fluorescent microscope. Alternatively if the GFP transgene was removed, transfection efficiency can be determined by observing the VSV-G mediated fusion of HEK-293 T cells into multinucleated cell syncytia [11].
11. If making larger preparations of lentivirus (using more flasks or larger flasks) it is necessary to concentrate the viral media prior to ultracentrifugation. This can be done using a column concentrator and centrifugation (e.g., Centricon-70 ultra centrifugation column [11]) or using tangential flow filtration [12].
12. Be sure to fill the ultracentrifugation tubes to the required level according to the manufacturer's instructions or the tube will collapse during centrifugation.
13. To remove the final few mL of media from the tube, tilt the tube at a 45° angle and continue to aspirate the media as it flows down the side of the tube away from the pellet. Do not disturb the pellet at the bottom of the tube.
14. Due to the inherently "sticky" nature of the viral envelope and exposed glycoproteins, it is necessary to use siliconized pipette tips and tubes when handling the virus. For a detailed protocol on siliconizing *see* Chapter 14.
15. LV cannot be refrozen after thawing; as such it is best practice to aliquot the virus into small working aliquots.

Acknowledgement

This work was supported by the Saint Mary's Foundation.

References

1. Knipe DM, Howley PM, Griffin DE (2001) Fundamental virology. Lippincott Williams & Wilkins, Philadelphia. p84
2. Naldini L et al (1996) In vivo gene delivery and stable transduction of nondividing cells by a lentiviral vector. Science 272:263–267
3. Naldini L (1998) Lentiviruses as gene transfer agents for delivery to non-dividing cells. Curr Opin Biotechnol 9:457–463
4. Blomer U et al (1997) Highly efficient and sustained gene transfer in adult neurons with a lentivirus vector. J Virol 71:6641–6649
5. Akkina RK et al (1996) High-efficiency gene transfer into CD34+ cells with a human immunodeficiency virus type 1-based retroviral vector pseudotyped with vesicular stomatitis virus envelope glycoprotein G. J Virol 70:2581–2585
6. Kafri T, Blomer U, Peterson DA, Gage FH, Verma IM (1997) Sustained expression of genes delivered directly into liver and muscle by lentiviral vectors. Nat Genet 17:314–317
7. Case SS et al (1999) Stable transduction of quiescent CD34(+)CD38(−) human hematopoietic cells by HIV-1-based lentiviral vectors. Proc Natl Acad Sci U S A 96:2988–2993
8. Sutton RE, Wu HT, Rigg R, Böhnlein E, Brown PO (1998) Human immunodeficiency virus type 1 vectors efficiently transduce human hematopoietic stem cells. J Virol 72:5781–5788
9. Cannon JR, Sew T, Montero L, Burton EA, Greenamyre JT (2011) Pseudotype-dependent lentiviral transduction of astrocytes or neurons in the rat substantia nigra. Exp Neurol 228:41–52
10. Cronin J, Zhang X-Y, Reiser J (2005) Altering the tropism of lentiviral vectors through pseudotyping. Curr Gene Ther 5:387–398
11. Semple-Rowland SL, Berry J (2014) Use of lentiviral vectors to deliver and express bicistronic transgenes in developing chicken embryos. Methods 66:466–473. doi:10.1016/j.ymeth.2013.06.026
12. Geraerts M, Michiels M, Baekelandt V, Debyser Z, Gijsbers R (2005) Upscaling of lentiviral vector production by tangential flow filtration. J Gene Med 7:1299–1310

Chapter 9

Viral Vector Production: Adenovirus

Julius W. Kim, Ramin A. Morshed, J. Robert Kane, Brenda Auffinger, Jian Qiao, and Maciej S. Lesniak

Abstract

Adenoviral vectors have proven to be valuable resources in the development of novel therapies aimed at targeting pathological conditions of the central nervous system, including Alzheimer's disease and neoplastic brain lesions. Not only can some genetically engineered adenoviral vectors achieve remarkably efficient and specific gene delivery to target cells, but they also may act as anticancer agents by selectively replicating within cancer cells.

Due to the great interest in using adenoviral vectors for various purposes, the need for a comprehensive protocol for viral vector production is especially apparent. Here, we describe the process of generating an adenoviral vector in its entirety, including the more complex process of adenoviral fiber modification to restrict viral tropism in order to achieve more efficient and specific gene delivery.

Key words Adenovirus, Viral vector production, Adenoviral gene therapy, Adenovirus fiber modification, Adenovirus tropism

1 Introduction

Adenoviral vector production is the process by which replication-defective viral vectors or replication-competent viruses are generated and amplified for use in laboratory experimentation [1, 2]. In general, the incorporation of a specific gene of interest into the viral vector is made possible through means of homologous recombination [3]. The common generation of a recombinant adenoviral vector is done by introducing a gene of interest into the E1 region of an adenoviral vector backbone. With that as a model, we hereby provide a detailed protocol describing how to construct and use a shuttle vector carrying a gene of interest to generate desired adenoviral vectors, with or without viral fiber modification, through homologous recombination with BJ5183 cells. The processes of rescuing and up-scaling a virus including culturing of virus producing cells as well as viral purification are also described in detail.

Fredric P. Manfredsson (ed.), *Gene Therapy for Neurological Disorders: Methods and Protocols*, Methods in Molecular Biology, vol. 1382, DOI 10.1007/978-1-4939-3271-9_9,

2 Materials

Each solution must be prepared within a sterile hood except for the bacterial culture materials. Cell-involved materials must be stored in 4 °C and bacterial culture related materials can be stored at room temperature except for antibiotics, which should be stored at –20 °C.

2.1 Human Embryonic Kidney 293 Cell Line Culture

1. Cell growth media: 5 mL of 100× stock of penicillin sodium (100 units/mL) and streptomycin (100 μg/mL), 5 mL of 100× stock of glutamine (200 mM, light sensitive), and 50 mL of fetal bovine serum (FBS) added to Dulbecco's modified Eagle's medium (DMEM, 500 mL).
2. Virus infection media: 10 mL of FBS, 5 mL of penicillin/streptomycin stock, and 5 mL of glutamine added to 500 mL of DMEM.
3. Cell-freezing media: 10 % Dimethyl sulfoxide (DMSO) in FBS.
4. Cell-detaching reagent: 1.5 mL (T75 flask) or 4 mL (T175 flask) of 0.25 % Trypsin-EDTA solution.

2.2 Bacterial Culture and Cloning

1. Luria-Bertani (LB) Broth.
2. LB Agar.
3. 50 mg/mL (1000×) Kanamycin sulfate.
4. 50 mg/mL (1000×) Ampicillin, sodium salt.

2.3 Cell Transfection

1. Attractine (Qiagen, Germantown, MD, USA).
2. Opti-MEM® (Life Technologies, Grand Island, NY, USA).

2.4 Virus Purification

1. 5 mM HEPES buffer, pH 7.8: Dilute 2.5 mL of 1 M HEPES (*see* **Note 1**) buffer pH 7.8 in 497.5 mL of water.
2. 1.33 g/mL CsCl: 113.550 mg CsCl in 250 mL of 5 mM HEPES buffer.
3. 1.45 g/mL CsCl: 152.250 mg CsCl in 250 mL of 5 mM HEPES buffer.
4. 21.5-gauge needles.
5. 5 mL syringes.
6. Beckman SW41 rotor with buckets, or equivalent.
7. 13.2 mL Beckman SW41 ultraclear tubes or equivalent.

2.5 Dialysis

1. Slide-A-Lyzer 10 K dialysis cassette (Thermo Scientific, Waltham, MA, USA).
2. Float buoys (Thermo Scientific, Waltham, MA, USA).
3. Dialysis buffer: 10 % glycerol in 1× phosphate-buffered saline (PBS).

4. 1 L beaker.
5. Magnetic stir bar.
6. Stir plate.

2.6 Virus Titration

1. Virus lysis buffer: 1 % SDS in 1× PBS.

3 Methods

3.1 Cell Culture of HEK 293 Cell Line

1. In order to generate an adenoviral vector, a stable human embryonic kidney (HEK) 293 cell line culture is needed for transfection purposes. When passaging each serial HEK 293 cell line culture, a determined quantity of cells for a given flask size is needed as noted in Table 1.
2. At confluency, the passaged cells should have the documented quantity as listed. The number of cells may be quantified through a trypan blue exclusion method using a hemocytometer.

3.2 Propagating the Initial Culture from a Frozen Stock

1. A cryotube of stock HEK 293 cells is noted to contain approximately $2.0–3.0 \times 10^6$ cells [4–6] (*see* **Notes 2** and **3**).
2. Add 14 mL of growth media to a T75 flask.
3. Thoroughly mix the growth media throughout the flask to ensure that it is evenly distributed.
4. Thaw the frozen stock of HEK 293 cells in a water bath set to a temperature of 37 °C until it becomes completely thawed.
5. Transfer the previously frozen, now thawed, cell line stock from the cryotube into the T75 flask containing the 14 mL of growth media.
6. Incubate the HEK 293 cells with growth media in 37 °C, 5 % CO_2, humidified incubator.
7. After 20–24 h, aspirate and discard the growth media from the T75 flask. Immediately following, apply 14 mL of fresh and

Table 1
Approximate cell numbers and amount of media in different flasks

		Cell number	Amount of media for culture	Amount of media for a virus infection
1	6 well	1.1×10^6 in one well	2 mL	1 mL
2	T25	3×10^6	4 mL	3 mL
3	T75	$1–2 \times 10^7$	14 mL	10 mL
4	T175	$3–5 \times 10^7$	24 mL	20 mL

pre-warmed (in a water bath set to 37 °C) growth media. Please advise that it will take about 2–3 days for the HEK 293 cells to recover from the propagation and, thus, fully grow in a T75 flask.

3.3 Propagation (Subculture) of HEK 293 Cell Culture

1. Aspirate and discard the existing growth media from the T75 flask and wash the attached cells with approximately 5 mL of PBS for the purpose of removing any excess of dead cells that were not removed through the initial growth media aspiration (*see* **Note 4**).
2. Add 1.5 mL of 0.25 % trypsin-EDTA solution to the T75 flask containing the attached HEK 293 cells, thoroughly spreading it entirely across the flask, and incubate the flask until there is evident cell detachment present (placing the incubating flask into a 37 °C, 5 % CO_2 incubator may accelerate cell detachment via trypsin activation).
3. Add 5 mL of fresh 10 % FBS DMEM to the T75 flask, washing the cells on the flask's side to the bottom of the flask.
4. Homogenize the HEK 293 cell line culture in the growth media through repeated pipetting (the total volume of the T75 flask should be consistent with 6.5 mL).
5. From one T75 flask, you can make 3× T75 flask (outlined in **steps 6** and 7) or 1× T175 flask and 1× T75 flask (outlined in **steps 8** and **9**).
6. From a single T75 flask, three separate T75 flasks can be processed. Aliquot the volume of the original T75 flask, which should be 6.5 mL in volume, to each of the three new flasks in an even manner.
7. Add 12 mL of a fresh 10 % FBS, DMEM into each of the three T75 flask. The new, total volume should now be 14–15 mL.
8. If using 1× T175 and 1× T75 add 5.5 mL from #4 into a T175 flask. Keep 1.5 mL in the existing T75 flask.
9. Add 18 mL of growth media to T175 (total 24–25 mL) and 14 mL of media into T75 (total 14–15 mL) (of note, cells in the flask will grow a bit slower since the initial amount of cells will be lower).
10. Spread the cells gently, yet thoroughly, over the bottom surface of the flask and return it to a 37 °C, 5 % CO_2, incubator.

3.4 Generation of Frozen Stocks of HEK 293 Cells

1. It is recommended to have at least $2.0–3.0 \times 10^6$ cells per 1 mL of freezing medium in a cryotube [4, 7]. As described in the Table 1, T75 flask can contain approximately 1.0×10^7 cells at confluency, ergo, three to four cryotubes of frozen HEK 293 cell stock can be processed. Please note that if the confluency

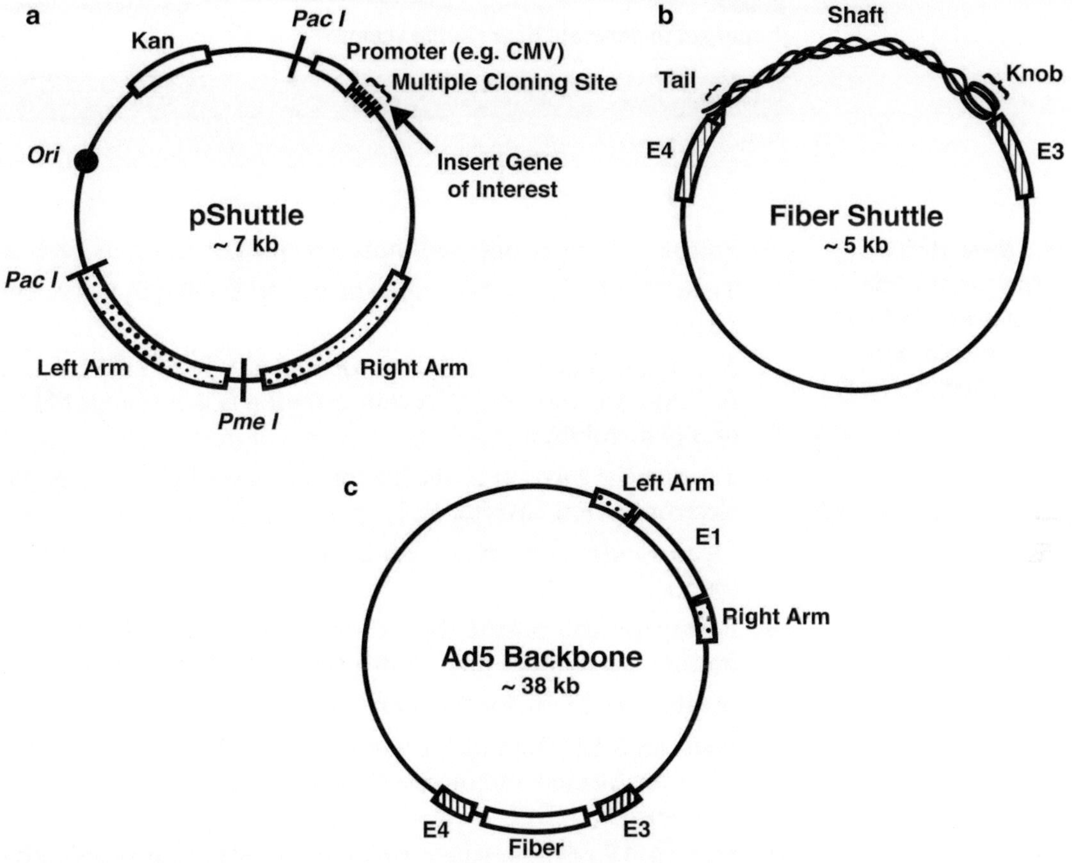

Fig. 1 Shuttle vectors and adenoviral backbone. (**a**) pShuttle vector(s) can be used to replace the E1 gene with the chosen gene of interest, thereby modifying the virus in a way that it is now replication incompetent. (**b**) A fiber shuttle vector can be used to modify the wild-type fiber, thus changing the viral tropism. (**c**) Commercial AdEasy or wild-type adenovirus type 5 may be used as a backbone

is ill defined, it may be of particular interest to quantify the number of cells in order to achieve the desired $2.0–3.0 \times 10^6$ cells/mL [5, 8] (*see* **Note 5**).

3.5 Generation of a Shuttle Vector (Fig. 1)

1. Insertion of a gene of interest into the E1 region of the plasmid (*see* **Notes 6–8**). There are commercially available shuttle vectors that insert the gene of interest into E1 region. Through the usage of a multi-cloning site (MCS), a gene of interest can be inserted per the instructions of a standard molecular cloning method [3, 9–12].
2. Modification of viral fiber(s) to alter viral tropism: Using the primers shown in Table 2 with adenoviral DNA as a template, the region of E3 (left arm)-fiber-E4 (right arm) can be obtained and inserted into any preferable common vector [3, 10–13].

Table 2
Primer set to generate fiber shuttle vector

Forward (SalI)	GCAGTCGACTCTAGAAATGGACGGAATTATTA
Reverse (NotI)	ACGCGGCCGCACCGGGAGGTGGTGAATTACAA

3.6 Generation of Adenoviral Vector with a Gene of Interest Through Homologous Recombination (Fig. 2)

1. Prepare the PmeI-digested shuttle vector generated in Table 2.
2. Thaw 20 μl of electro-competent BJ 5813 cells [3, 9–12, 14–16] on ice.
3. Add PmeI-digested shuttle vector and Ad5 backbone (such as AdEasy): the ratio of DNA concentration is 3:1 in a maximum of 3 μl in volume.
4. Transfer the mixture to the bottom of a pre-chilled, 1 mm gap electroporation cuvette.
5. Electro-pulse the cuvette containing the mixture at 1800 V for 5 s.
6. Resuspend the pulsed mixture in 1 mL of LB medium and transfer the mixture into 14 mL bacterial culture tube.
7. Incubate at 37 °C for 40–50 min.
8. Plate on 3 LB/Kan plates with 500 μl, 300 μl and 100 μl of the transformed mixture and incubate overnight at 37 °C (*see* **Note 9**).
9. Pick 10–15 of the smallest, single (isolated) colonies and grow each in 2 mL of LB/Kan broth overnight at 37 °C with vigorous shaking (*see* **Note 10**).
10. Complete mini-prep as follows [11] (*see* **Note 11**).
11. Transfer the bacteria grown LB into 1.5 mL tubes.
12. Centrifuge at 14,000 rpm for 10 min.
13. Aspirate all excess LB (a bacteria pellet should formed at the bottom).
14. Add 300 μl of homogenizing buffer (commercially called P1 buffer) and homogenize well by thorough vortexing.
15. Add 300 μl of alkaline buffer (commercially called P2 buffer) and mix through inverted shaking.
16. Add 300 μl of acidic (neutralizing) buffer (commercially called P3 buffer) and mix through inverted shaking.
17. Centrifuge at 14,000 rpm for 15 min.
18. Transfer only the clear solution to a 1.5 mL tube (be cautious to remove any white debris). (However, if any white debris is collected, an additional centrifugation is necessary for its removal.)

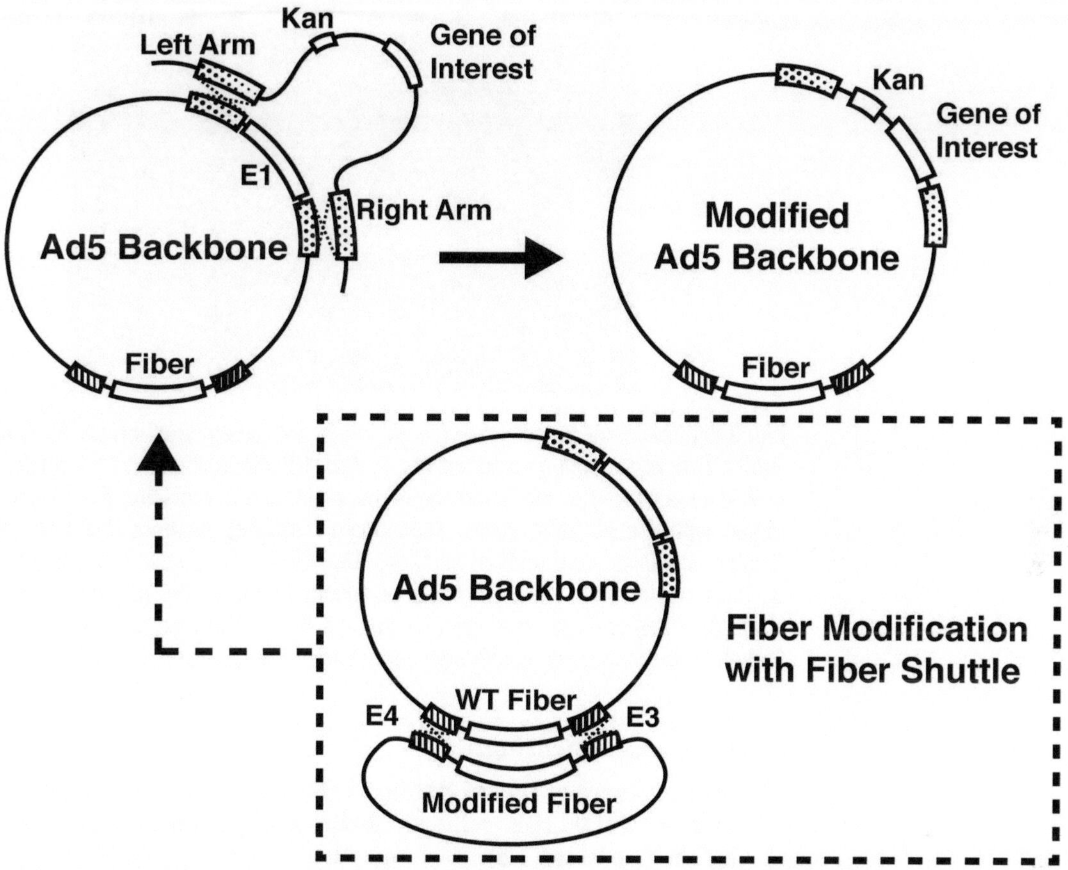

Fig. 2 Generation of plasmid DNAs of modified adenovirus by way of homologous recombination. Plasmid DNA of the adenoviral backbone and that of pShuttle (or pFiber shuttle) are co-transformed into BJ5183 cells. The resultants are selected by negative antibiotic resistance and the relative size of resultants' DNAs (Fig. 3). Sequencing verified recombined DNA from BJ5183 cells should be retransformed into DH5α cells to achieve a large amount of stable DNAs

19. Add 560 μl of isopropanol and vortex.
20. Centrifuge at 14,000 rpm for 15 min.
21. Aspirate the supernatant (the DNA should be near the bottom of the tube).
22. Add 400 μl of 70 % ethanol.
23. Centrifuge at 14,000 rpm for 10 min.
24. Aspirate out the supernatant (the DNA should be near the bottom of the tube).
25. Dry completely.
26. Add 20–30 μl of dH_2O.
27. Run on 1 % agarose gel (electrophoresis).
28. Check the size of the plasmid DNAs (Fig. 3).

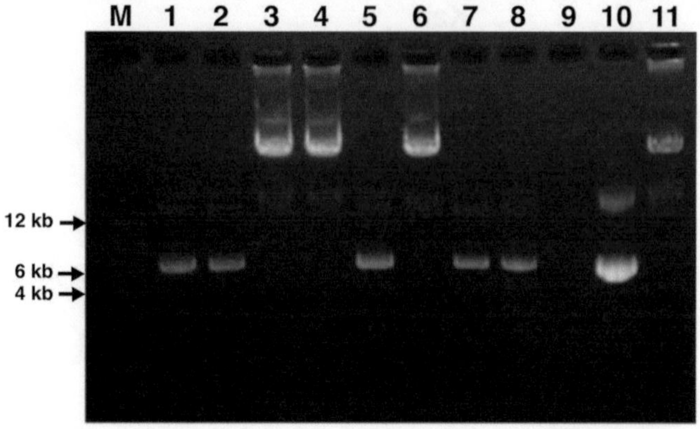

Fig. 3 Verification of homologous recombinants per gel electrophoresis. M: DNA ladder. 1–8: Homologous recombinants. 9: Void. 10: pShuttle vector. 11: Viral DNA of wild-type Ad5. After the homologous recombination of replacing the E1 gene region with the pShuttle vector (kanamycin resistant), plasmid DNAs of the resulting colonies were purified and separated on a 1 % agarose gel. Since the BJ cells can contain either homologous recombinants or the pShuttle vector, alone, and may grow in kanamycin LB, lanes 3, 4, and 6 are the possibly correct homologous recombinants, although verification by sequencing is required

29. Verify via sequencing.
30. Re-transform sequence verified DNAs into electro-competent DH5α or DH10B cells to obtain a large amount of stable DNAs (*see* **Note 12**).

3.7 Transfection of Adenoviral Vector in HEK 293 Cells

1. Plate HEK 293 cells (or 911 cells) in a 6-well plate approximately 12–20 h prior to transfection. The confluency should be approximately 50–70 % at the time of transfection. Transfection of one well may be sufficient for virus generation and for further amplification. However, it is highly recommended that multiple wells be used for transfection as they will yield higher initial titers and quicker amplifications [7, 12, 17].
2. Digest 10 μg of recombinant adenoviral plasmids with *Pac*I (*see* **Note 13**).
3. Precipitate PacI digested plasmid DNA as described in **steps 4–10**.
4. Add 400 μL of ethyl alcohol (ethanol, *see* **Note 14**).
5. Centrifuge at maximum speed for approximately 10–15 min.
6. Aspirate and discard the isopropyl alcohol (isopropanol) supernatant (a white precipitate of DNA should be visible at the bottom of the tube).
7. Add 400 μL of isopropyl alcohol (isopropanol).
8. Centrifuge at maximum speed for approximately 10–15 min.

9. Aspirate and discard the isopropyl alcohol (isopropanol) supernatant and allow the DNA to dehydrate.
10. Add 20 μL of dH_2O (*see* **Note 15**).
11. Perform a transfection with a lipid-based transfection reagent according to the manufacturer's instructions (e.g., with Attractine from Qiagen as described in **steps 12–17**).
12. Mix 5 μg of PacI-digested DNA and 15 μL of Attractine in 150 μL of Opti-MEM.
13. Vortex thoroughly and incubate for 15 min at room temperature.
14. Add 1 mL of Opti-MEM into each well.
15. Add 1 mL of Opti-MEM into the mixture.
16. After thorough pipetting, add the mixture to the HEK 293 cell plate in a dropwise manner.
17. Remove the growth media containing transfection mixture approximately 4–6 h later, and add 2 mL of fresh 2 % FBS of DMEM (it is advisable to check the cell quality 2 h into the incubation period, changing growth media if necessary).
18. After 72 h, aspirate and discard all media and add fresh 2 % FBS of DMEM growth media.
19. Incubate the transfected HEK 293 cells in a 37 °C, 5 % CO_2 incubator for 10–15 days. Typically, it takes at least 14–20 days to rescue a virus from transfected cells. During this period, the media should not be changed unless there is an observable change in growth media color from red to yellow. If observed, add 1 mL of fresh 2 % growth media.
20. Harvest all the cells when 80 % of the cells in a well show cytopathic effect (CPE) (Fig. 4). Place the collected mixture in a 50 mL tube.

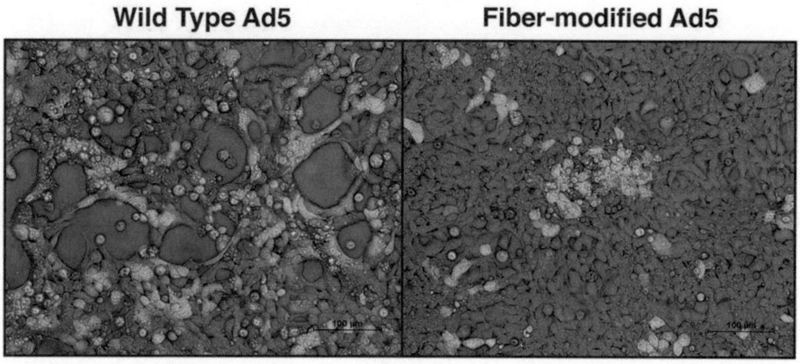

Fig. 4 Cytopathic effect (CPE) of adenovirus on cells: (**a**) Observed CPE of plaque-forming wild-type virus. (**b**) No apparent CPE of fiber-modified adenovirus (plaque-less virus)

Table 3
Different up-scaling steps

1	Type	Typical fast growing wild-type-like virus from purified virus stock (100 MOI)
	Propagation step	1× T75 → 3× T175 → 20× T175
2	Type	Typical fast-growing wild-type-like virus from transfection on 6 well
	Propagation step	One well from 6 well → 1× T75 → 3× T75 → 6× T175 → 20× T175
3	Type	Structure-modified slow-growing virus
	Propagation step	One well from 6 well → 1× T25 → 1× T75 → 1× T175 → 5× T175 → 20× T175

21. Freeze the mixture completely at −80 °C and thaw it in a 37 °C water bath.
22. Repeat **step 7** three times in preparation for the following propagation procedures.

3.8 Propagation of Adenovirus

1. The growth rate of a virus varies depending on its structural characteristics and the characteristics of the inserted gene of interest. Appropriated up-scaling steps are required to acquire maximum amount of adenovirus (~10^{12} vp/mL) in a stock [7, 12, 18] (*see* **Note 16**). Different up-scaling steps are depicted in Table 3.

3.9 1st Up-Scaling of Virus

1. Prepare HEK 293 cells in a T75 flask (the appropriated cellular density should be consistent with approximately 10^7 cells in T75 flask).
2. Prepare virus containing media.
3. Infection with purified virus stock. Calculate and make ten plaque-forming units (PFU) or 100 viral particles of multiple of infections (MOI) per cell in 10 mL of 2 % FBS infection media.
4. Infection from a rescued virus in a 6-well. After three times of "freeze and thaw," centrifuge the sample at 4000 rpm for 30 min, aspirate the supernatant, and add 2 % FBS infection media to make the final volume equivalent to 10 mL.
5. Remove the growth media from the T75 flask and add the virus-containing media to the opposite side of the flask where cells are growing.
6. Incubate in a 37 °C, 5 % CO_2, incubator for approximately 2–5 days until there is an appreciable cytopathic effect (CPE) evidenced at a consistency of approximately 80–100 %.

7. Harvest the mixture of virus-infected cells and infection media in a 50 mL tube.
8. Centrifuge at 1200 rpm for 5 min.
9. Aspirate and discard the infection media, retaining approximately 5 mL.
10. Perform three cycles of "freeze and thaw."
11. Keep at −80 °C until the second up-scaling.

3.10 Second Up-Scaling of a Virus

1. Thaw the tube containing the virus from **step 11** in Subheading 3.9 and centrifuge at 4000 rpm for 30 min.
2. Prepare cultures of HEK 293 cells in 3× T175 flasks.
3. Aspirate the supernatant from **step 1** and add 2 % FBS adenovirus infection media to allow for a total volume of 60 mL (each of the three T175 flasks requires 20 mL of infection media).
4. Add 20 mL of the mixture of adenovirus in 2 % infection media to each T175 flask to the opposite side of the flask where cells are growing.
5. Follow **steps 6–11** described in first up-scaling.

3.11 Third Up-Scaling of Virus

1. Over the course of serial passaging, prepare HEK 293 cell cultures in 20× T175 flasks.
2. Thaw the tube containing the virus from the second up-scaling (**step 5** in Subheading 3.10) and centrifuge at 4000 rpm for 30 min.
3. Aspirate the supernatant from **step 2** and add 2 % FBS adenovirus infection media to allow for a total volume of 400 mL (a T175 flask needs 20 mL: 20 × T175 = 400 mL).
4. Add the mixture of adenovirus and infection media to each flask to the opposite side of the flask where cells are growing.
5. Follow **steps 6–11** described in 1st up-scaling.
6. Aspirate and discard all infection media from the eight tubes.
7. Combine all centrifuged cells in one 50 mL tube with a fresh 4 mL of 2 % FBS infection media.
8. Freeze and thaw the tube three times and store at −80 °C until purification.

3.12 Cesium Chloride (CsCl) Purification and Dialysis

1. This method of adenovirus purification involves separation of viral particles based on the known density of intact adenovirus particle (1.42). To maximize the purity of adenovirus, two rounds of CsCl gradient-based purification are recommended using a CsCl gradient of 1.33–1.45 g/mL [19].

3.13 Adenovirus Purification with CsCl

1. Thaw the virus containing cell lysates from **step 8**, Subheading 3.11 and centrifuge at 4000 rpm for 30 min at 4 °C.
2. Prepare the CsCl gradient as follows (first CsCl gradient centrifugation).
3. Add 4 mL of 1.33 g/mL CsCl gradient in SW41 ultraclear tube.
4. Add 4 mL of 1.45 g/mL CsCl gradient at the bottom of the tube.
5. Add 4 mL of supernatant from the viral cell lysate (**step 1**) to the top of the tube.
6. Fill up the tube with 2 % FBS infection media, forming a concave meniscus.
7. Prepare a balance tube.
8. Add 4 mL of 1.33 g/mL CsCl gradient in SW41 ultraclear tube.
9. Add 4 mL of 1.45 g/mL CsCl gradient at the bottom of the tube.
10. Fill up the tube with 2 % FBS infection media, forming a concave meniscus.
11. Centrifuge at 18,000 rpm for 4 h at 4 °C.
12. Collect the lowest observed band with a 21.5-gauge needle connected to a 5 mL syringe (Fig. 5a) (*see* **Note 17**).
13. Prepare the second concentration gradient for centrifugation.
14. Add 4 mL of 1.33 g/mL CsCl gradient.
15. Add 4 mL of 1.45 g/mL CsCl gradient.
16. Add the collected viruses.

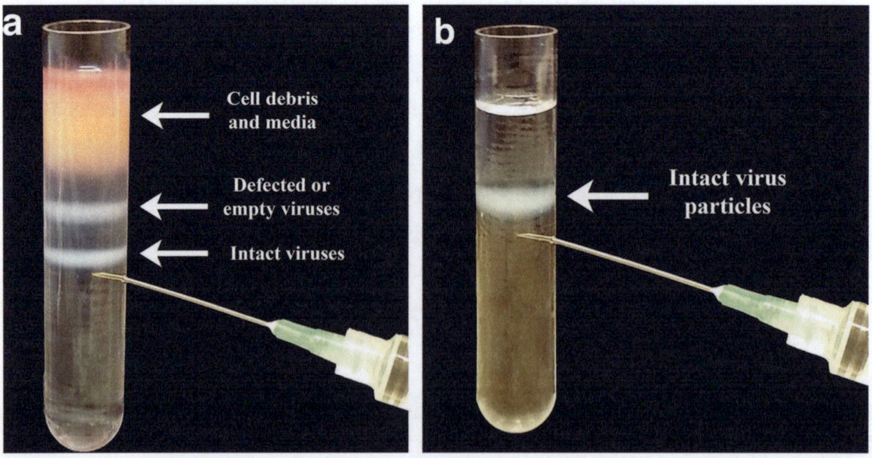

Fig. 5 CsCl gradient-based virus purification. (**a**) Gradient-based layers after initial 4 h of centrifugation. Among the layers, only the intact virus layer (gradient of 1.42 g/mL) should be collected with a 21.5-gauge needle connected to a 5 mL syringe, and used for the second centrifugation. (**b**) A single, pure viral band after overnight centrifugation

17. Fill up the tube with HEPES buffer.
18. Prepare a balance with the same composition as in the **step 6** except for the incorporation of the virus.
19. Centrifuge at 35,000 rpm at 4 °C for 12–16 h.
20. After 12–16 h of centrifugation, there should be only one band observed. Collect the observed band with a 21.5-gauge needle connected to a 5 mL syringe (Fig. 5b). If there is an obvious second band observed, perform the third centrifugation with the collected low-appearing band.

3.14 Adenovirus Dialysis

1. Dialysis is required to remove any remaining CsCl, which is known to be cytotoxic and to collect pure virus particles in 10 % Glycerol in 1× PBS.
2. Prepare a standard dialysis buffer with a total volume of 4 L (stored at 4 °C).
3. Thoroughly wet a dialysis cassette in the prepared dialysis buffer.
4. Insert the purified viral lysate into the dialysis cassette (a total volume may not exceed 3 mL).
5. In a 1 L beaker containing 1 L of dialysis buffer and a magnetic stir rod, place the dialysis cassette containing the viral lysate with the buoyant (injection site) side in the supine position (i.e., facing upward).
6. Place the beaker on a magnetic stirring plate in a cold room (4 °C).
7. Perform this dialysis treatment 3–4 times with 1 L of a fresh dialysis buffer on a time consistency of approximately 2–3 h for each treatment (keep 1 mL of the dialysis buffer from the last treatment for the virus titration).
8. After 3–4 times of full dialysis treatments, remove the virus from the buffer with 21.5-gauge needle connected to a 5 mL syringe.
9. These purified virus particles can be aliquoted in desirable amounts. Commonly, aliquots of 10 μL in 20–40 tubes are made. The remaining virus is then aliquoted into 1.5 mL tubes with 100 μL per tube.
10. Store the aliquots of the purified virus at −80 °C.

3.15 Virus Titration: Virus Particle Analysis in O.D. 260 nm (VP/mL) (See Note 18)

1. Prepare blank solution: 100 μL of dialysis buffer and 10 μL of virus lysis buffer.
2. Prepare virus sample solution: 100 μL of purified virus + 10 μL of virus lysis buffer (if preferred, serial diluted virus samples can be tested).
3. Incubate samples for 30 min at 56 °C with vigorous shaking.

4. Read your samples on O.D. 260 nm.
5. Calculate the virus particle (VP titer using the formula below [20]):

$$VP / mL = 260 \text{ nm} \times \text{dilution factor} \times 1.1 \times 10^{12}$$

4 Notes

1. 1 M HEPES buffer can be purchased or prepared in the laboratory. To prepare 1 L of 1 M HEPES buffer dissolve 238.3 g HEPES in 800 mL of water. Mix and adjust pH to 7.8 with NaOH. Adjust volume to 1 L with water. Filter sterilize with a 0.45 μm filter.
2. Take extra protective care in culturing HEK 293 cells since it takes approximately 2–3 weeks for the virus to be rescued from the HEK 293 cells. During this period, cells will grow without any antibiotics. Any contaminants which are inert during normal cell culture will be detrimental to the following virus rescue period.
3. Do *not* wipe out 70 % ethanol after spraying. Let it dry up. Sanitization is done by dehydration.
4. It is not recommended to culture HEK 293 cells in excess of 20 passages for the cells will exhibit extremely slow and minimal proliferation. After 20 passages, it is of best interest to discard these cells and thaw fresh cells from frozen stock according to the aforementioned procedure.
5. It is of utmost importance to process a frozen stock of HEK 293 cells in the early stages of culture.
6. In generating the pShuttle vector, ensure that the gene of interest does not have PmeI and PacI recognition site(s), since these enzymes will be used in the process of homologous recombination and in the process of transfection, respectively.
7. Unless there is no alternative choice(s), avoid use of the NotI recognition site on MCS. Creating a NotI recognition site for the gene of interest via PCR is not easily achievable since high GC-containing primers (NotI recognition sequence: GC/GGCCGC) tend to form a secondary structure especially when the template recognition site has high GC contents. When there is no alternative choice of action, add 0.5–1 % of DMSO to the PCR reaction which can prevent a formation of a secondary structure.
8. If E1 and E3 deleted backbone (such as AdEasy) are used, the maximum size of the inserted gene of interest can be approximately 7.5 kb.

9. It is not especially easy to make a large volume of LB media to be absorbed into a LB agar plate. You can transfer 500 μL and 300 μL of the transformed mixture into 1.5 mL tube(s), centrifuge at 14,000 rpm for 1 min, and remove most of the LB broth, leaving 100 μL of LB broth in volume. Then, homogenize the centrifuged, transformed bacteria with the remaining LB and plate on the LB agar plate.
10. It is highly likely that either only small or isolated colonies that stand alone are the correctly homologous recombined DNA containing cells. pShuttle (~7 kb) containing cells can grow much faster than homologous recombined DNA (~40 kb) containing cells due to the size of plasmid DNAs they produce. Since the size of satellite colonies and homologous recombined DNA containing colonies are very similar, small colonies that are in the close proximity of others may be satellite colonies that do not contain any plasmid DNA.
11. The use of a mini-prep kit is only suitable for less than the 10 kb size of plasmid DNAs. Therefore, use the previously mentioned protocol (Subheading 3.3) for isolation of any adenoviral backbone plasmid DNAs.
12. Plasmid DNAs with too large a size that are isolated from BJ5813 cells are known to be unstable DNAs, and, thus, BJ5813 cells are not suitable to produce an appropriate amount of plasmid DNAs for cell transfections. Also, chemical-competent cells are not suitable to transform a large size of plasmid DNAs, so re-transformation of homologous recombined DNAs has to be done in electro-competent cells.
13. Overnight digestion with 1× BSA is recommended when using PmeI and PacI. This overnight incubation with BSA can maximize enzyme digestion (12–18 h) and minimize nonspecific cutting (BSA)
14. Adding 80 % volume of ethanol to your sample is enough to precipitate DNAs at the bottom of the microfuge tube. However, adding 400 μL makes the DNA precipitate more towards the side of the tube and facilitates aspiration of the supernatant.
15. Use pre-warmed dH_2O to dissolve DNA as opposed to any buffers. Most of plasmid DNAs produced through this protocol are byproducts for virus production rather than for DNA stocks. Some commercially provided buffers are not applicable for some steps during the adenoviral production processes (e.g., TE (Tris and EDTA) buffer: lower an efficiency of enzyme digestion and cell transfection when DNA is in TE buffer).
16. In the case of generating a replication-competent adenovirus, the A549 (adenocarcinomic human alveolar basal epithelial cells) cell line can be used.

17. There are other approaches, which may be found in other resources for isolating virus particles from the microfuge tube (e.g., collecting them from the bottom of the tube by puncturing the bottom or collecting them from the top of the tube by passing through the other layers). Find the best approach to suit your needs.
18. SDS-treated virus particles obtained after titration can be used to isolate viral DNA using a Blood and Tissue Kit such as those available from Qiagen.

References

1. Majhen D, Ambriovic-Ristov A (2006) Adenoviral vectors—how to use them in cancer gene therapy? Virus Res 119: 121–133
2. Green NK, Seymour LW (2002) Adenoviral vectors: systemic delivery and tumor targeting. Cancer Gene Ther 9:1036–1042
3. He TC et al (1998) A simplified system for generating recombinant adenoviruses. Proc Natl Acad Sci U S A 95:2509–2514
4. Graham FL (1987) Growth of 293 cells in suspension culture. J Gen Virol 68:937–940
5. Lavappa KS, Macy ML, Shannon JE (1976) Examination of ATCC stocks for HeLa marker chromosomes in human cell lines. Nature 259:211–213
6. Stulberg CS et al (1970) The animal cell culture collection. In Vitro 5:1–16
7. Graham FL et al (1977) Characteristics of a human cell line transformed by DNA from human adenovirus type 5. J Gen Virol 36: 59–74
8. Coecke S et al (2005) Guidance on good cell culture practice. A report of the second ECVAM task force on good cell culture practice. Altern Lab Anim 33:261–287
9. Graham FL, Van der Eb AJ, Heijneker HL (1974) Size and location of the transforming region in human adenovirus type 5 DNA. Nature 251:687–691
10. Graham FL, Prevec L (1991) Manipulation of adenovirus vectors. Methods Mol Biol 7: 109–128
11. Sambrook J (2001) Molecular cloning: a laboratory manual. In: Russell DW (ed) Cold Spring Harbor Laboratory Press, Cold Spring Harbor, NY
12. He TC (2004) Adenoviral vectors. Curr Protoc Hum Genet Chapter 12: Unit 12.4
13. Kim JW et al (2013) An adenovirus vector incorporating carbohydrate binding domains utilizes glycans for gene transfer. PLoS One 8: e55533
14. Hanahan D, Gluzman Y (1984) Rescue of functional replication origins from embedded configurations in a plasmid carrying the adenovirus genome. Mol Cell Biol 4:302–309
15. Imler JL et al (1995) An efficient procedure to select and recover recombinant adenovirus vectors. Gene Ther 2:263–268
16. Chartier C et al (1996) Efficient generation of recombinant adenovirus vectors by homologous recombination in Escherichia coli. J Virol 70:4805–4810
17. Krougliak V, Graham FL (1995) Development of cell lines capable of complementing E1, E4, and protein IX defective adenovirus type 5 mutants. Hum Gene Ther 6:1575–1586
18. Ng P, Graham FL (2002) Construction of first-generation adenoviral vectors. Methods Mol Med 69:389–414
19. Kanegae Y, Makimura M, Saito I (1994) A simple and efficient method for purification of infectious recombinant adenovirus. Jpn J Med Sci Biol 47:157–166
20. Maizel JV, White DO Jr, Scharff MD (1968) The polypeptides of adenovirus. I. Evidence for multiple protein components in the virion and a comparison of types 2, 7A, and 12. Virology 36:115–125

Part IV

Viral Vector Tropism

Chapter 10

Controlling AAV Tropism in the Nervous System with Natural and Engineered Capsids

Michael J. Castle, Heikki T. Turunen, Luk H. Vandenberghe, and John H. Wolfe

Abstract

More than one hundred naturally occurring variants of adeno-associated virus (AAV) have been identified, and this library has been further expanded by an array of techniques for modification of the viral capsid. AAV capsid variants possess unique antigenic profiles and demonstrate distinct cellular tropisms driven by differences in receptor binding. AAV capsids can be chemically modified to alter tropism, can be produced as hybrid vectors that combine the properties of multiple serotypes, and can carry peptide insertions that introduce novel receptor-binding activity. Furthermore, directed evolution of shuffled genome libraries can identify engineered variants with unique properties, and rational modification of the viral capsid can alter tropism, reduce blockage by neutralizing antibodies, or enhance transduction efficiency. This large number of AAV variants and engineered capsids provides a varied toolkit for gene delivery to the CNS and retina, with specialized vectors available for many applications, but selecting a capsid variant from the array of available vectors can be difficult. This chapter describes the unique properties of a range of AAV variants and engineered capsids, and provides a guide for selecting the appropriate vector for specific applications in the CNS and retina.

Key words AAV, Adeno-associated virus, Capsid, Virus, Vector, Serotype, Gene therapy, Gene delivery

1 Introduction

The first recombinant adeno-associated virus (AAV) vectors were generated using both the protein capsid and the inverted terminal repeat (ITR) DNA sequence of AAV serotype 2 [1–4]. AAV2 vectors remain widely used today, and nearly all recombinant AAV genomes carry the AAV2 ITR sequence. However, a large number of alternative capsid variants have been identified from humans, baboons, chimpanzees, and rhesus, pigtailed, and cynomolgus macaques [5, 6]. These alternative capsids often have distinct tropism and antigenic profiles, although many have yet to be thoroughly studied. Traditionally, capsid variants are categorized

Fredric P. Manfredsson (ed.), *Gene Therapy for Neurological Disorders: Methods and Protocols*, Methods in Molecular Biology, vol. 1382, DOI 10.1007/978-1-4939-3271-9_10,

according to their serotype. Each serotype is defined as an antigenically distinct viral capsid, as determined by serum cross-neutralization. In addition, several strategies can be used to alter the tropism of the AAV capsid, including chemical modification of the virus capsid, production of hybrid capsids, peptide insertion, capsid shuffling, directed evolution, and rational mutagenesis. The large array of capsid variants generated by these strategies, along with techniques that can enhance or alter their native tropisms, provides a rapidly expanding toolkit for gene transfer to the central nervous system (CNS). However, the number of options can be overwhelming, making it difficult to select the appropriate AAV vector for a specific application.

The recombinant AAV vectors described in this chapter can all be prepared using the same technique, described in Chapter 7, with the unique capsid sequence provided in *trans* during production. The basic T = 1 icosahedral architecture of the viral capsid does not differ among these serotypes and engineered vectors, although the proteins encapsidating the recombinant DNA are slightly different, resulting in limited structural changes. For many AAV serotypes cellular surface receptors or binding determinants have been identified, including sialic acid for AAVs 1, 4, 5, and 6 [7, 8], heparan sulfate proteoglycan (HSPG) for AAV2 [9], the laminin receptor for AAV8 [10], and galactose for AAV9 [11, 12]. In addition, human fibroblast growth factor receptor 1 and alphaV-beta5 integrin have both been proposed as co-receptors for AAV2 [13, 14], as has platelet-derived growth factor receptor for AAV5 [15]. These differences in receptor binding among capsid serotypes contribute to differences in tropism within the brain and other tissues. However, while differences in receptor affinity can drive variability among AAV serotypes, most, if not all, AAVs demonstrate broad tropism without absolute specificity, in part due to the wide presence of AAV receptors throughout the body. Different AAV variants can, however, differ in absolute levels of gene transfer to a specific tissue, as well as in their relative transduction strength among multiple tissues.

Several techniques can be used to generate novel AAV capsids with unique, targeted tropism. Chemical modification of the viral capsid with receptor-binding moieties can confer enhanced tropism, and chemical masking of native receptor-binding moieties can alter the normal tropism of AAV and shield the capsid from neutralizing antibodies. Hybrid capsids can be generated by co-expressing *cap* genes from different serotypes during production, combining the unique properties of both parental serotypes. Peptide insertion of novel receptor-binding elements on the capsid surface can alter the native tropism of AAV, and insertion of fluorescent proteins can be used to tag vector particles. Capsid shuffling and directed evolution can be used to create and screen a library of unique capsid variants for a desired trait, such as tropism for a specific cell type. Finally, rational modification of the viral capsid via site-directed mutagenesis can alter tropism, confer evasion of neutralizing antibodies, and increase transduction efficiency.

In this chapter, we describe the differing tropisms of AAV serotypes in the CNS and retina, the various factors that can influence AAV tropism, the techniques which can be used to alter the tropism of the vector, and the engineered variants that have been developed for use in the nervous system. This will provide an in-depth guide for selecting the optimal capsidserotype or engineered variant for specific experimental or therapeutic applications in the CNS.

2 Selection of the Capsid Serotype

1. Nervous cell tropism varies among AAVcapsid serotypes. In primary cultures of rat nervous cells, AAV5 appears to possess a strong glial tropism, and gene expression rarely colocalizes with the neuronal marker NeuN [16]. AAV serotypes 1, 2, 6, 7, 8, and 9 transduce both neurons and astrocytes in primary culture [16, 17]. AAVs 1, 6, and 7 appear to have the strongest neuronal tropism in vitro, with 75 % or more of transduced cells representing neurons [17]. AAV9, however, has relatively weak neuronal tropism in vitro, with less than 50 % of transduced cells representing neurons [17]. AAV5 is therefore recommended for transduction of cultured astrocytes, and AAVs 1, 6, and 7 are recommended for transduction of cultured neurons.
2. Following intraparenchymal brain injection, AAVs 1, 2, 5, 7, 8, 9, and rh.10 all exhibit strong neuronal tropism, as gene expression rarely colocalizes with markers of astrocytes or oligodendrocytes [18–21]. However, others have observed astroglial transduction with AAVs 1, 2, 5, 6, and 8 [22–24], and AAV8 has also been observed to transduce oligodendrocytes within the cortex [24]. AAV4 possesses strong glial tropism in vivo, and primarily drives gene expression within glial fibrillary acidic protein (GFAP)-positive astrocytes [25]. In addition, AAVrh.43 appears to possess stronger glial tropism in vivo than AAV8 [26]. Thus, while most AAV serotypes exhibit strong neuronal tropism following direct intraparenchymal brain injection, glial transduction has been observed in some cases, and AAVs 4 and rh.43 appear to possess stronger astroglial tropism than most AAV serotypes.
3. While most AAV serotypes appear to preferentially transduce neurons within the brain, the relative strength of neuronal transduction varies greatly. When compared against other serotypes, AAVs 2 and 4 typically mediate weaker and less widespread neuronal gene expression [19, 22, 27–31]. Thus, AAVs

2 and 4 are not recommended for widespread transduction of neurons. AAV2 diffuses less readily through both the brain and spinal cord parenchyma when compared against other serotypes, and therefore mediates transduction over a smaller area [19, 28, 32, 33]. This property can be harnessed for the targeting of small nuclei. The strongest and most widespread neuronal transduction is observed with AAV serotypes 1, 9, and rh.10 [19, 21, 23, 30, 34, 35]. Several novel serotypes, including pi.2, rh.8, hu.11, hu.32, and hu.37, also appear to mediate strong and widespread neuronal transduction in the brain [36]. However, these serotypes have not yet been extensively tested, and can also transduce glia within the white matter [36]. AAV serotypes 1, 9, and rh.10 are therefore recommended for targeting of neurons via intraparenchymal brain injection. Furthermore, AAV2 is recommended for the targeting of small brain regions due to its reduced diffusion in brain tissue.

4. Tropism also varies among AAV serotypes when administered to the cerebrospinal fluid (CSF), either via intrathecal or intracerebroventricular injection. AAV4 strongly transduces ependymal cells when administered to the ventricles, and demonstrates greater ependymal cell tropism than AAVs 2 or 5 [22]. Intracerebroventricular injection of AAV4 is therefore recommended for targeting of ependymal cells. In contrast, AAV7 and AAV9 can bypass the ependymal layer following intrathecal injection, penetrating the parenchyma and transducing neurons throughout the cortex, cerebellum, and spinal cord [37, 38]. AAVs 1, 2, 4, 5, 6, and 8 do not appear to extensively penetrate the brain parenchyma when administered to the CSF [22, 39, 40]. However, AAV6 mediates widespread transduction of spinal motor neurons following intrathecal injection [40]. Further, AAV8 possesses a unique tropism for large-diameter neurons of the dorsal root ganglion (DRG), and drives specific gene expression within these cells following intrathecal injection [39]. Thus, CSF injection of AAV4 can mediate transduction of epithelial cells, CSF injection of AAV7 or 9 can mediate widespread transduction of cortical, cerebellar, and spinal neurons, CSF injection of AAV6 can mediate transduction of spinal motor neurons, and CSF injection of AAV8 can mediate transduction of large-diameter DRG neurons.

5. When the spinal cord is targeted directly via intraparenchymal injection, AAVs 1, 5, and 9 demonstrate the strongest neuronal tropism, while AAVs 2, 6, and 8 demonstrate weaker neuronal tropism [19, 40]. AAV8 retains its tropism for large-diameter DRG neurons after intraparenchymal injection [39]. AAVs 1, 5, and 9 are therefore recommended for intraparenchymal targeting of discrete populations of spinal

motor neurons, while AAV8 is recommended for intraparenchymal targeting of large-diameter DRG neurons.

6. When administered intravenously, AAVs 9, rh.10, rh.8, and rh.43 can penetrate the blood–brain barrier and drive gene expression throughout the nervous system [41–44]. These serotypes possess both neuronal and glial tropism when administered systemically, but transduction is primarily neuronal in neonatal animals and primarily glial in adults [41–43]. Penetration of the intact blood–brain barrier and transduction of brain tissue is limited with other serotypes, including AAVs 1, 2, 5, 6, and 8 [44–47]. *See* also Chapters 16 and 17 for discussion of systemic AAV administration.

7. Tropism also differs among AAV serotypes following ocular administration via subretinal injection, which is generally an efficient method for outer retina transduction. AAVs 1, 2, 4, 5, 7, 8, and 9 transduce cells of the retinal pigmented epithelium (RPE) [48–52]. AAVs 1, 2, and 5 transduce these RPE cells with similar efficiency when directly compared [48]. In addition, AAVs 1, 2, 5, 7, 8, and 9 transduce photoreceptors (PRs), while AAV4 does not [48–53]. AAV8 also possesses tropism for ganglion cells and cells of the inner nuclear layer [50, 54], and demonstrates greater tropism for PRs than AAV2 [54]. In nonhuman primate subretinal injections, AAV8 efficiently targets rod PRs [54], while AAV9 is superior for cone PRs and retains tropism for rod PRs [53]. AAV5 has also been shown to target both rod and cone PRs, but has not been compared directly in this setting against other serotypes [55].

8. Surprisingly, nervous cell tropism can also vary among vector preparations of the same serotype, even when injected under identical conditions. For example, CsCl-purified AAV8 exhibited strong astroglial tropism following intraparenchymal brain injection, while iodixanol-purified AAV8, injected under identical conditions, transduced only neurons [23]. Variability among vector preps or among injection conditions may therefore explain the differences in tropism that are frequently observed among experiments.

9. As a result of this variability, it is not possible to confidently restrict gene expression to neuronal or glial populations based solely on the capsidserotype. Thus, a cell type-specific promoter should be utilized if astrocyte-, oligodendrocyte-, or neuron-specific transduction is desired. However, most serotypes preferentially transduce neurons following intraparenchymal brain injection, and therefore a pancellular promoter can be used to drive neuronal gene expression, so long as the potential transduction of glia is not problematic. On the other hand, if strong glial expression is desired, a cell type-specific promoter should be utilized.

10. In some cases it may be desirable for AAV to undergo axonal transport, either to increase the spread of gene transfer, or to retrogradely target a specific subpopulation of projection neurons. For example, anterograde transport of AAV9 injected into the ventral tegmental area of the brain can greatly enhance transgene distribution [34], and retrograde transport of AAV1 injected into muscle or sciatic nerve can specifically label discrete pools of motor neurons [56]. Axonal transport is a fundamental property of AAV vectors, and thus any vector that is endocytosed at high levels by projection neurons is likely to transduce distal brain regions [57, 58]. AAV1 appears to be more effective than AAVs 2, 3, 4, 5, or 6 for retrograde transduction of motor neurons following muscle or sciatic nerve injection [56]. Further, AAVs 1 and 5 demonstrate greater retrograde transduction of brainstem than AAVs 2, 8, or 9 following injection of the transected spinal cord [59]. Within the brain, AAV9 is most frequently observed to undergo axonal transport, and is the recommended choice if distal transduction is desired [21, 23, 34, 36, 57, 60]. In addition, AAVs 1, 8, and rh.10 also undergo axonal transport within the brain and can be used to drive distal transduction [21, 28–30, 34, 57, 60].

3 Modification of the Capsid

To further improve the utility of AAV as a gene therapy vector, research has concentrated on altering capsid properties such as tropism, targeting specificity, and antigenicity. This can be achieved by (1) chemical modification of the capsid; (2) assembling mosaic capsids consisting of subunits from two or more different serotypes; (3) peptide insertion; (4) capsid shuffling; or (5) rational design.

1. Chemical modifications to improve the tropic properties of adenoviral and lentiviral vectors have resulted in moderate success (reviewed in refs. [61, 62]). Similar strategies have been applied to AAV, although to a lesser extent. Bispecific antibodies capable of binding both AAV2 and $\alpha_{IIb}\beta_3$ integrin can increase the transduction of $\alpha_{IIb}\beta_3$ integrin-expressing cells by 70-fold [63]. In another study, linkage of biotin-coated AAV to an EGF-streptavidin fusion protein increased the transduction efficiency of EGFR-expressing SKOV3.ip1 cells more than 100-fold [64]. However, despite these promising results, enhancement of AAV tropism has not been achieved in vivo. Chemical capsid modifications can also be used to mask receptor-binding domains on the AAV capsid, de-targeting the virus from its native receptors, allowing infection through alternate receptors, and shielding

the capsid from neutralizing antibodies. Indeed, moderate success has been achieved by coating capsids with poly(ethylene) glycol [65], poly-[*N*-(2-hydroxypropyl) methacrylamide] [66], and α-dicarbonyl compounds [67]. However, chemically modified capsids have yet to be widely used in vivo, and their utility in the CNS remains limited.

2. AAV capsids are assembled as icosahedral particles from 60 subunits of the VP1, VP2, and VP3 structural proteins. Hybrid capsids are designed to harness the structural similarity among AAVs, combining beneficial properties from two or more different serotypes by co-expressing their capsid proteins during vector production [68]. Hybrid capsids of AAV1 and AAV2 (AAV1/2) can mediate stronger transgene expression in lung and muscle in vivo than either of the parental serotypes alone [69]. Further, AAV1/2 appears to combine the tropism of AAV2 for TH-positive dopamine neurons with the ability of AAV1 to diffuse more widely through brain tissue, mediating strong transduction of dopamine neurons in the substantia nigra, and has been used to model Parkinson's disease in the rat [70]. Hybrid capsids can also be used to transfer binding affinities from their parental serotypes, such as HSPG binding (AAV2 or AAV3) or mucin binding (AAV4 or AAV5) [71]. However, although the composition of hybrid capsids can be influenced by expressing the parental *cap* genes at specific ratios, the composition of individual capsids cannot be directly controlled, and undesired capsid arrangements are likely to occur. Furthermore, direct genetic manipulation of the *cap* gene provides a more precise method by which the properties of different serotypes can be combined, and thus hybrid AAV capsids are rarely utilized.
3. The earliest successful alterations of AAV tropism by capsid engineering relied on insertion of short peptides into the AAV capsid [72]. Inserted peptides are displayed on the capsid surface and provide affinity for a receptor specifically expressed by the target cell type. Simultaneous disruption of the native capsid tropism can increase the likelihood of specific interaction with the novel target receptor. However, early attempts to provide AAV5 with the ability to bind HSPG indicated that conferring efficient receptor binding does not necessarily confer efficient transduction of the target tissue. Mutant AAV5 virions, despite being able to bind HSPG as efficiently as AAV2, lost their native infectivity and thus did not demonstrate increased tropism for HSPG-expressing cells [73]. In addition to specific insertions of known receptor binding peptides, insertion screens of random peptide libraries have also been utilized. In both cases, it must be ensured that the insertion does not negatively affect vector production, infectivity, or other properties required for gene transfer. Several regions of the capsid are amenable to

insertions, including the N-termini of VP1 and VP2, as well as the various loop regions shared by all VP proteins [72, 74–78]. A peptide inserted into the common C-terminal domain shared by all three VP proteins will be displayed on every capsid subunit (60 copies per viral particle), whereas a VP1 or VP2 insertion will only be present in up to 6 or 12 copies per capsid, respectively. This is an important consideration, as the density of receptor-binding peptides on a capsid can affect its tropism [79]. Although initial experiments focused on the insertion of small peptides (5–15 amino acids), later studies have shown that insertions of full length proteins, such as GFP or mCherry, can also be tolerated without loss of virus function [78, 80]. However, despite these promising in vitro proof-of-principle studies, few AAV mutants with enhanced tropism in vivo have been published. To date, improved targeting of skeletal muscle [81, 82], cardiac muscle [83], vasculature [84–86], lung [87, 88], diseased brain endothelial cells [89], retina [90], ovarian cancer cells [91] and breast cancer cells [88] has been reported. Most of these peptide insertions are targeted between amino acids 587 and 588 of AAV2, the region that mediates HSPG binding [73, 92], in order to disrupt the function of this region and de-target AAV2 from its native tropism. The AAV2-7m8 mutant, which was generated via random insertion of a seven amino acid sequence, can efficiently target most retinal cell types following intravitreal injection [90].

4. In addition to peptide insertion, the development of capsid shuffling [93] and directed evolution [94], discussed in Chapter 11, has generated many promising novel AAV variants. Briefly, *cap* genes of different AAV serotypes are nuclease digested, mixed together, and randomly reassembled to produce mutated chimeric genomes, which are subsequently selected for a specific function or tropism via directed evolution screening. Capsids with improved transduction of heart [95], lung [96, 97], Müller glia in the retina [98], CNS [99–101], and neural and pluripotent stem cells [102, 103] have been described, with numerous others yet unpublished. Directed evolution was also used to screen the random insertion of short peptides into AAV2 VP3, resulting in the AAV2-7m8 mutant capsid, which is capable of transducing all retinal layers after intravitreal injection [90]. In addition to AAV2-7m8, ShH10 can specifically transduce Müller cells from the vitreous [98].
5. An improved understanding of AAV structure and biology has enabled researchers to modify vector function by rationally targeting mutations of amino acids on the viral capsid, rather than selecting clones with the desired property from a library of mutants. Some rationally designed mutants combine the desired functions of different serotypes, while others disrupt

the domains responsible for unwanted characteristics. The former group includes AAV2i8, a chimera of AAV2 and AAV8 [104], AAV2.5, a chimera of AAV2 and AAV1 [105], and chimeras of AAV1 and AAV6 differing by single amino acid changes [106]. These studies indicate that changing only a small number of amino acids is sufficient to generate capsids with the characteristics of both parental serotypes. In addition, AAV2i8 and AAV2.5 possess unique antigenic properties [104, 105]. Similarly, AAV6.2, a novel vector with improved transduction in mouse airways, was generated via targeted single amino acid changes to AAV6 [107, 108]. Disruption of native AAV properties by targeted amino acid mutations has primarily focused on masking the capsid from neutralizing antibodies, which can inhibit AAV-mediated gene transfer. Several mutations that alter serum antibody recognition and neutralization of AAV2 while retaining normal vector function have been identified [109, 110]. Rational mutations have also been designed to increase transduction efficiency, reducing the vector dose required for clinically relevant transgene expression and avoiding the immune response associated with large viral loads. It is hypothesized that phosphorylation of the AAV capsid leads to ubiquitination and subsequent proteasome-mediated degradation of the vector particle, reducing transduction efficiency [111]. Indeed, mutating several surface exposed tyrosine and threonine residues significantly increased the transduction efficiency of AAVs 2, 5, and 8 [111, 112]. As hypothesized, proteosomal degradation of these capsid mutants was reduced, resulting in increased viral nuclear transport and transgene expression [111, 113, 114]. In addition, novel transduction patterns are observed when AAV2 tyrosine mutants are applied to the mouse retina, in particular following intravitreal injection, which may eliminate the need for surgically challenging subretinal vector administration [115, 116]. Tyrosine mutations of different serotypes also demonstrate improved transduction of mesenchymal stem cells [113] and the mouse brain [117, 118]. A similar strategy to disrupt AAV2 phosphorylation by targeting serine, threonine, or lysine residues can increase liver transduction in mice [119]. Although tyrosine mutations improve retinal transduction following intravitreal injection, in the only comparison published to date, AAV2-7m8 demonstrated more efficient transduction from the vitreous [90]. Finally, double tyrosine-mutant AAV9 vectors containing AAV3 ITRs and the neuron-specific synapsin promoter appear to possess stronger neuronal tropism than AAV9 in the murine CNS following systemic delivery, although these mutant vectors were not compared directly against AAV9 or other variants [117].

4 Additional Methods to Refine Gene Targeting

1. The broad natural tropism of the AAVcapsid can be enhanced or made more specific by harnessing cell type-specific promoters. For example, the 1.3 kb CaMK2a promoter can drive transgene expression in glutamatergic excitatory neurons with high specificity [120]. Further, the 1.8 kb neuron-specific enolase promoter [121], the 470 bp human synapsin-1 promoter [120, 122], the 229 bp MeCP2 promoter [123], and the 2 kb herpes simplex virus 1 latency associated transcript promoter [124] can all drive neuron-specific gene expression. The GFAP promoter can drive astrocyte-specific expression, and the myelin basic protein (MBP) promoter can drive oligodendrocyte-specific expression [26, 125]. However, in order for these promoters to be effective, the AAV capsid must possess tropism for the target cell type. For example, AAV4 carrying a CaMK2a promoter is unlikely to drive strong expression within excitatory neurons, as AAV4 does not transduce this cell type efficiently [22]. Similarly, AAV2 carrying a GFAP promoter was found to drive expression primarily in neurons following intraparenchymal brain injection, likely due to the limited astroglial tropism of AAV2 [121]. *See* also Chapter 6 for discussion of cell type-specific promoters.
2. Woodchuck hepatitis virus posttranscriptional regulatory elements (WPREs) are frequently included in the recombinant AAV genome, and can increase the strength of transgene expression [121, 126]. However, WPREs drive greater expression not only in the target cell type, but also in off-target cells that endocytose a lesser number of AAV particles. This can decrease the specificity of an engineered vector or a cell type-specific promoter. Although this effect has not been thoroughly studied, WPRE elements are not recommended when high specificity for a single cell type is desired. WPREs have also been implicated as a contributing or causal factor in oncogenesis in preclinical studies, likely via an ORF within the WPRE [127]. Modified versions that eliminate this ORF have been developed [128].
3. The injected dose and volume can also influence AAV tropism. Raising the injected dose increases the number of AAV particles that are endocytosed by all cells local to the injection site, driving stronger gene expression within off-target cells. For example, when 1.2×10^{11} genome copies (GC) of AAV1 carrying a human synapsin 1 promoter were intraparenchymally injected, gene expression was highly specific for inhibitory neurons [120]. However, raising the injected dose to 1.7×10^{12} GC resulted in similar levels of gene expression within excitatory and inhibitory neurons, and further raising the dose to 8.4×10^{12}

GC resulted in gene expression primarily within excitatory neurons [120]. This is likely due to increased uptake of AAV1 by excitatory neurons at higher injected doses. Decreasing the injected volume while maintaining the injected dose is likely to have a similar effect, as this will apply AAV more focally, driving stronger gene expression within a smaller number of cells. On the other hand, decreasing the injected dose, or applying AAV more diffusely by increasing the injected volume, is likely to increase specificity by reducing the number of vector particles that are endocytosed per cell. Low doses of AAV are therefore recommended when cell-specific expression is desired. If strong or widespread transduction is required, a dose escalation experiment can be performed to identify the injection parameters that result in the strongest gene expression without loss of specificity. *See* also Chapter 14 for discussion of intraparenchymal injection. Similar findings were observed in the retina with AAV8, as increased dose shifted tropism from RPE alone to both RPE and PRs [54].

4. AAV tropism can also be modified via the inclusion of microRNA (miRNA) target sites within the AAV genome [129–131]. By utilizing miRNAs that are expressed only in certain cell types, gene expression can be specifically reduced within these target cells. For example, subretinally injected AAV5 typically transduces both RPE cells and PRs, as described in Subheading 2, **item** 7. However, binding sites for the RPE-specific miR-204 can block AAV5-mediated gene expression in RPE cells, resulting in PR-specific expression [130]. Further, binding sites for the PR-specific miR-124 can block gene expression in PRs, resulting in RPE-specific expression [130]. Thus, gene expression can also be restricted from specific populations via miRNA binding sites within the recombinant AAV genome.

Acknowledgments

The AAV work in our laboratories is supported by NIH grants R01-NS038690, R01-DK063973 and U01-HD079066 to J.H.W. and DP1-OD008267 to L.H.V.; additional support was provided to J.H.W. from the Foerderer Fdn and to L.H.V. from Curing Kids Fund, Foundation for Retina Research, Foundation Fighting Blindness, Research to Prevent Blindness, and Corinne and Wyc Grousbeck. M.J.C. was supported in part by NIH T32-NS007413. L.H.V. is an inventor on patents related to AAV gene therapy; has served as a consultant; is inventor on technologies licensed to biotechnology and pharmaceutical industry; and is cofounder and consultant to GenSight Biologics.

References

1. Hermonat PL, Muzyczka N (1984) Use of adeno-associated virus as a mammalian DNA cloning vector: transduction of neomycin resistance into mammalian tissue culture cells. Proc Natl Acad Sci U S A 81: 6466–6470
2. Samulski RJ, Berns KI, Tan M, Muzyczka N (1982) Cloning of adeno-associated virus into pBR322: rescue of intact virus from the recombinant plasmid in human cells. Proc Natl Acad Sci U S A 79:2077–2081
3. Samulski RJ, Chang LS, Shenk T (1987) A recombinant plasmid from which an infectious adeno-associated virus genome can be excised in vitro and its use to study viral replication. J Virol 61:3096–3101
4. Tratschin JD, West MH, Sandbank T, Carter BJ (1984) A human parvovirus, adeno-associated virus, as a eucaryotic vector: transient expression and encapsidation of the procaryotic gene for chloramphenicol acetyltransferase. Mol Cell Biol 4:2072–2081
5. Gao G, Vandenberghe LH, Alvira MR, Lu Y, Calcedo R, Zhou X et al (2004) Clades of adeno-associated viruses are widely disseminated in human tissues. J Virol 78: 6381–6388
6. Gao G, Vandenberghe LH, Wilson JM (2005) New recombinant serotypes of AAV vectors. Curr Gene Ther 5:285–297
7. Wu Z, Miller E, Agbandje-McKenna M, Samulski RJ (2006) Alpha2,3 and alpha2,6 N-linked sialic acids facilitate efficient binding and transduction by adeno-associated virus types 1 and 6. J Virol 80:9093–9103
8. Kaludov N, Brown KE, Walters RW, Zabner J, Chiorini JA (2001) Adeno-associated virus serotype 4 (AAV4) and AAV5 both require sialic acid binding for hemagglutination and efficient transduction but differ in sialic acid linkage specificity. J Virol 75:6884–6893
9. Summerford C, Samulski RJ (1998) Membrane-associated heparan sulfate proteoglycan is a receptor for adeno-associated virus type 2 virions. J Virol 72:1438–1445
10. Akache B, Grimm D, Pandey K, Yant SR, Xu H, Kay MA (2006) The 37/67-kilodalton laminin receptor is a receptor for adeno-associated virus serotypes 8, 2, 3, and 9. J Virol 80:9831–9836
11. Bell CL, Vandenberghe LH, Bell P, Limberis MP, Gao GP, Van Vliet K et al (2011) The AAV9 receptor and its modification to improve in vivo lung gene transfer in mice. J Clin Invest 121:2427–2435
12. Shen S, Bryant KD, Brown SM, Randell SH, Asokan A (2011) Terminal N-linked galactose is the primary receptor for adeno-associated virus 9. J Biol Chem 286:13532–13540
13. Qing K, Mah C, Hansen J, Zhou S, Dwarki V, Srivastava A (1999) Human fibroblast growth factor receptor 1 is a co-receptor for infection by adeno-associated virus 2. Nat Med 5:71–77
14. Summerford C, Bartlett JS, Samulski RJ (1999) AlphaVbeta5 integrin: a co-receptor for adeno-associated virus type 2 infection. Nat Med 5:78–82
15. Pilz IH, Di Pasquale G, Rzadzinska A, Leppla SH, Chiorini JA (2012) Mutation in the platelet-derived growth factor receptor alpha inhibits adeno-associated virus type 5 transduction. Virology 428:58–63
16. Howard DB, Powers K, Wang Y, Harvey BK (2008) Tropism and toxicity of adeno-associated viral vector serotypes 1, 2, 5, 6, 7, 8, and 9 in rat neurons and glia in vitro. Virology 372:24–34
17. Royo NC, Vandenberghe LH, Ma JY, Hauspurg A, Yu L, Maronski M et al (2008) Specific AAV serotypes stably transduce primary hippocampal and cortical cultures with high efficiency and low toxicity. Brain Res 1190:15–22
18. Bartlett JS, Samulski RJ, McCown TJ (1998) Selective and rapid uptake of adeno-associated virus type 2 in brain. Hum Gene Ther 9:1181–1186
19. Burger C, Gorbatyuk OS, Velardo MJ, Peden CS, Williams P, Zolotukhin S et al (2004) Recombinant AAV viral vectors pseudotyped with viral capsids from serotypes 1, 2, and 5 display differential efficiency and cell tropism after delivery to different regions of the central nervous system. Mol Ther 10:302–317
20. Passini MA, Watson DJ, Vite CH, Landsburg DJ, Feigenbaum AL, Wolfe JH (2003) Intraventricular brain injection of adeno-associated virus type 1 (AAV1) in neonatal mice results in complementary patterns of neuronal transduction to AAV2 and total long-term correction of storage lesions in the brains of beta-glucuronidase-deficient mice. J Virol 77:7034–7040
21. Cearley CN, Wolfe JH (2006) Transduction characteristics of adeno-associated virus vectors expressing cap serotypes 7, 8, 9, and Rh10 in the mouse brain. Mol Ther 13:528–537
22. Davidson BL, Stein CS, Heth JA, Martins I, Kotin RM, Derksen TA et al (2000)

Recombinant adeno-associated virus type 2, 4, and 5 vectors: transduction of variant cell types and regions in the mammalian central nervous system. Proc Natl Acad Sci U S A 97:3428–3432
23. Klein RL, Dayton RD, Tatom JB, Henderson KM, Henning PP (2008) AAV8, 9, Rh10, Rh43 vector gene transfer in the rat brain: effects of serotype, promoter and purification method. Mol Ther 16:89–96
24. Hutson TH, Verhaagen J, Yanez-Munoz RJ, Moon LD (2012) Corticospinal tract transduction: a comparison of seven adeno-associated viral vector serotypes and a non-integrating lentiviral vector. Gene Ther 19:49–60
25. Liu G, Martins IH, Chiorini JA, Davidson BL (2005) Adeno-associated virus type 4 (AAV4) targets ependyma and astrocytes in the subventricular zone and RMS. Gene Ther 12:1503–1508
26. Lawlor PA, Bland RJ, Mouravlev A, Young D, During MJ (2009) Efficient gene delivery and selective transduction of glial cells in the mammalian brain by AAV serotypes isolated from nonhuman primates. Mol Ther 17:1692–1702
27. Klein RL, Dayton RD, Leidenheimer NJ, Jansen K, Golde TE, Zweig RM (2006) Efficient neuronal gene transfer with AAV8 leads to neurotoxic levels of tau or green fluorescent proteins. Mol Ther 13:517–527
28. Reimsnider S, Manfredsson FP, Muzyczka N, Mandel RJ (2007) Time course of transgene expression after intrastriatal pseudotyped rAAV2/1, rAAV2/2, rAAV2/5, and rAAV2/8 transduction in the rat. Mol Ther 15:1504–1511
29. Taymans JM, Vandenberghe LH, Haute CV, Thiry I, Deroose CM, Mortelmans L et al (2007) Comparative analysis of adeno-associated viral vector serotypes 1, 2, 5, 7, and 8 in mouse brain. Hum Gene Ther 18: 195–206
30. Sondhi D, Hackett NR, Peterson DA, Stratton J, Baad M, Travis KM et al (2007) Enhanced survival of the LINCL mouse following CLN2 gene transfer using the rh.10 Rhesus macaque-derived adeno-associated virus vector. Mol Ther 15:481–491
31. Vite CH, Passini MA, Haskins ME, Wolfe JH (2003) Adeno-associated virus vector-mediated transduction in the cat brain. Gene Ther 10:1874–1881
32. Passini MA, Watson DJ, Wolfe JH (2004) Gene delivery to the mouse brain with adeno-associated virus. Methods Mol Biol 246:225–236
33. Peng SP, Kugler S, Ma ZK, Shen YQ, Schachner M (2011) Comparison of AAV2 and AAV5 in gene transfer in the injured spinal cord of mice. Neuroreport 22:565–569
34. Cearley CN, Wolfe JH (2007) A single injection of an adeno-associated virus vector into nuclei with divergent connections results in widespread vector distribution in the brain and global correction of a neurogenetic disease. J Neurosci 27:9928–9940
35. Li SF, Wang RZ, Meng QH, Li GL, Hu GJ, Dou WC et al (2006) Intra-ventricular infusion of rAAV1-EGFP resulted in transduction in multiple regions of adult rat brain: a comparative study with rAAV2 and rAAV5 vectors. Brain Res 1122:1–9
36. Cearley CN, Vandenberghe LH, Parente MK, Carnish ER, Wilson JM, Wolfe JH (2008) Expanded repertoire of AAV vector serotypes mediate unique patterns of transduction in mouse brain. Mol Ther 16:1710–1718
37. Gray SJ, Nagabhushan KS, McCown TJ, Jude SR (2013) Global CNS gene delivery and evasion of anti-AAV-neutralizing antibodies by intrathecal AAV administration in non-human primates. Gene Ther 20:450–459
38. Samaranch L, Salegio EA, San SW, Kells AP, Bringas JR, Forsayeth J et al (2013) Strong cortical and spinal cord transduction after AAV7 and AAV9 delivery into the cerebrospinal fluid of nonhuman primates. Hum Gene Ther 24:526–532
39. Jacques SJ, Ahmed Z, Forbes A, Douglas MR, Vigenswara V, Berry M et al (2012) AAV8(gfp) preferentially targets large diameter dorsal root ganglion neurones after both intra-dorsal root ganglion and intrathecal injection. Mol Cell Neurosci 49:464–474
40. Snyder BR, Gray SJ, Quach ET, Huang JW, Leung CH, Samulski RJ et al (2011) Comparison of adeno-associated viral vector serotypes for spinal cord and motor neuron gene delivery. Hum Gene Ther 22:1129–1135
41. Foust KD, Nurre E, Montgomery CL, Hernandez A, Chan CM, Kaspar BK (2009) Intravascular AAV9 preferentially targets neonatal neurons and adult astrocytes. Nat Biotechnol 27:59–65
42. Gray SJ, Matagne V, Bachaboina L, Yadav S, Ojeda SR, Samulski RJ (2011) Preclinical differences of intravascular AAV9 delivery to neurons and glia: a comparative study of adult mice and nonhuman primates. Mol Ther 19:1058–1069
43. Yang B, Li S, Wang H, Guo Y, Gessler DJ, Cao C et al (2014) Global CNS transduction of adult mice by intravenously delivered rAAVrh.8 and rAAVrh.10 and nonhuman primates by rAAVrh.10. Mol Ther 22:1299–1309

44. Zhang H, Yang B, Mu X, Ahmed SS, Su Q, He R et al (2011) Several rAAV vectors efficiently cross the blood-brain barrier and transduce neurons and astrocytes in the neonatal mouse central nervous system. Mol Ther 19:1440–1448
45. Towne C, Raoul C, Schneider BL, Aebischer P (2008) Systemic AAV6 delivery mediating RNA interference against SOD1: neuromuscular transduction does not alter disease progression in fALS mice. Mol Ther 16:1018–1025
46. Fu H, Muenzer J, Samulski RJ, Breese G, Sifford J, Zeng X et al (2003) Self-complementary adeno-associated virus serotype 2 vector: global distribution and broad dispersion of AAV-mediated transgene expression in mouse brain. Mol Ther 8:911–917
47. Chen SJ, Sanmiguel J, Lock M, McMenamin D, Draper C, Limberis MP et al (2013) Biodistribution of AAV8 vectors expressing human low-density lipoprotein receptor in a mouse model of homozygous familial hypercholesterolemia. Hum Gene Ther Clin Dev 24:154–160
48. Acland GM, Aguirre GD, Bennett J, Aleman TS, Cideciyan AV, Bennicelli J et al (2005) Long-term restoration of rod and cone vision by single dose rAAV-mediated gene transfer to the retina in a canine model of childhood blindness. Mol Ther 12:1072–1082
49. Le Meur G, Stieger K, Smith AJ, Weber M, Deschamps JY, Nivard D et al (2007) Restoration of vision in RPE65-deficient Briard dogs using an AAV serotype 4 vector that specifically targets the retinal pigmented epithelium. Gene Ther 14:292–303
50. Stieger K, Colle MA, Dubreil L, Mendes-Madeira A, Weber M, Le Meur G et al (2008) Subretinal delivery of recombinant AAV serotype 8 vector in dogs results in gene transfer to neurons in the brain. Mol Ther 16:916–923
51. Lebherz C, Maguire A, Tang W, Bennett J, Wilson JM (2008) Novel AAV serotypes for improved ocular gene transfer. J Gene Med 10:375–382
52. Lei B, Zhang K, Yue Y, Ghosh A, Duan D (2009) Adeno-associated virus serotype-9 efficiently transduces the retinal outer plexiform layer. Mol Vis 15:1374–1382
53. Vandenberghe LH, Bell P, Maguire AM, Xiao R, Hopkins TB, Grant R et al (2013) AAV9 targets cone photoreceptors in the nonhuman primate retina. PLoS One 8:e53463
54. Vandenberghe LH, Bell P, Maguire AM, Cearley CN, Xiao R, Calcedo R et al (2011) Dosage thresholds for AAV2 and AAV8 photoreceptor gene therapy in monkey. Sci Transl Med 3:88ra54
55. Boye SE, Alexander JJ, Boye SL, Witherspoon CD, Sandefer KJ, Conlon TJ et al (2012) The human rhodopsin kinase promoter in an AAV5 vector confers rod- and cone-specific expression in the primate retina. Hum Gene Ther 23:1101–1115
56. Hollis ER 2nd, Kadoya K, Hirsch M, Samulski RJ, Tuszynski MH (2008) Efficient retrograde neuronal transduction utilizing self-complementary AAV1. Mol Ther 16:296–301
57. Castle MJ, Gershenson ZT, Giles AR, Holzbaur EL, Wolfe JH (2014) Adeno-associated virus serotypes 1, 8, and 9 share conserved mechanisms for anterograde and retrograde axonal transport. Hum Gene Ther 25(8):705–20
58. Castle MJ, Perlson E, Holzbaur EL, Wolfe JH (2014) Long-distance axonal transport of AAV9 is driven by dynein and kinesin-2 and is trafficked in a highly motile Rab7-positive compartment. Mol Ther 22:554–566
59. Klaw MC, Xu C, Tom VJ (2013) Intraspinal AAV injections immediately rostral to a thoracic spinal cord injury site efficiently transduces neurons in spinal cord and brain. Mol Ther Nucleic Acids 2, e108
60. Masamizu Y, Okada T, Ishibashi H, Takeda S, Yuasa S, Nakahara K (2010) Efficient gene transfer into neurons in monkey brain by adeno-associated virus 8. Neuroreport 21:447–451
61. Schaffer DV, Koerber JT, Lim KI (2008) Molecular engineering of viral gene delivery vehicles. Annu Rev Biomed Eng 10:169–194
62. Morizono K, Chen IS (2011) Receptors and tropisms of envelope viruses. Curr Opin Virol 1:13–18
63. Bartlett JS, Kleinschmidt J, Boucher RC, Samulski RJ (1999) Targeted adeno-associated virus vector transduction of non-permissive cells mediated by a bispecific F(ab'gamma)2 antibody. Nat Biotechnol 17:181–186
64. Ponnazhagan S, Mahendra G, Kumar S, Thompson JA, Castillas M Jr (2002) Conjugate-based targeting of recombinant adeno-associated virus type 2 vectors by using avidin-linked ligands. J Virol 76: 12900–12907
65. Le HT, Yu QC, Wilson JM, Croyle MA (2005) Utility of PEGylated recombinant adeno-associated viruses for gene transfer. J Control Release 108:161–177
66. Carlisle RC, Benjamin R, Briggs SS, Sumner-Jones S, McIntosh J, Gill D et al (2008)

Coating of adeno-associated virus with reactive polymers can ablate virus tropism, enable retargeting and provide resistance to neutralising antisera. J Gene Med 10:400–411

67. Horowitz ED, Weinberg MS, Asokan A (2011) Glycated AAV vectors: chemical redirection of viral tissue tropism. Bioconjug Chem 22:529–532
68. Rabinowitz JE, Samulski RJ (2000) Building a better vector: the manipulation of AAV virions. Virology 278:301–308
69. Hauck B, Chen L, Xiao W (2003) Generation and characterization of chimeric recombinant AAV vectors. Mol Ther 7:419–425
70. Koprich JB, Johnston TH, Reyes MG, Sun X, Brotchie JM (2010) Expression of human A53T alpha-synuclein in the rat substantia nigra using a novel AAV1/2 vector produces a rapidly evolving pathology with protein aggregation, dystrophic neurite architecture and nigrostriatal degeneration with potential to model the pathology of Parkinson's disease. Mol Neurodegener 5:43
71. Rabinowitz JE, Bowles DE, Faust SM, Ledford JG, Cunningham SE, Samulski RJ (2004) Cross-dressing the virion: the transcapsidation of adeno-associated virus serotypes functionally defines subgroups. J Virol 78:4421–4432
72. Yang Q, Mamounas M, Yu G, Kennedy S, Leaker B, Merson J et al (1998) Development of novel cell surface CD34-targeted recombinant adenoassociated virus vectors for gene therapy. Hum Gene Ther 9:1929–1937
73. Opie SR, Warrington KH Jr, Agbandje-McKenna M, Zolotukhin S, Muzyczka N (2003) Identification of amino acid residues in the capsid proteins of adeno-associated virus type 2 that contribute to heparan sulfate proteoglycan binding. J Virol 77:6995–7006
74. Girod A, Ried M, Wobus C, Lahm H, Leike K, Kleinschmidt J et al (1999) Genetic capsid modifications allow efficient re-targeting of adeno-associated virus type 2. Nat Med 5:1438
75. Shi W, Arnold GS, Bartlett JS (2001) Insertional mutagenesis of the adeno-associated virus type 2 (AAV2) capsid gene and generation of AAV2 vectors targeted to alternative cell-surface receptors. Hum Gene Ther 12:1697–1711
76. Wu P, Xiao W, Conlon T, Hughes J, Agbandje-McKenna M, Ferkol T et al (2000) Mutational analysis of the adeno-associated virus type 2 (AAV2) capsid gene and construction of AAV2 vectors with altered tropism. J Virol 74:8635–8647
77. Grifman M, Trepel M, Speece P, Gilbert LB, Arap W, Pasqualini R et al (2001) Incorporation of tumor-targeting peptides into recombinant adeno-associated virus capsids. Mol Ther 3:964–975
78. Warrington KH Jr, Gorbatyuk OS, Harrison JK, Opie SR, Zolotukhin S, Muzyczka N (2004) Adeno-associated virus type 2 VP2 capsid protein is nonessential and can tolerate large peptide insertions at its N terminus. J Virol 78:6595–6609
79. Gigout L, Rebollo P, Clement N, Warrington KH Jr, Muzyczka N, Linden RM et al (2005) Altering AAV tropism with mosaic viral capsids. Mol Ther 11:856–865
80. Judd J, Wei F, Nguyen PQ, Tartaglia LJ, Agbandje-McKenna M, Silberg JJ et al (2012) Random insertion of mCherry Into VP3 domain of adeno-associated virus yields fluorescent capsids with no loss of infectivity. Mol Ther Nucleic Acids 1:e54
81. Yu CY, Yuan Z, Cao Z, Wang B, Qiao C, Li J et al (2009) A muscle-targeting peptide displayed on AAV2 improves muscle tropism on systemic delivery. Gene Ther 16:953–962
82. Lee NC, Falk DJ, Byrne BJ, Conlon TJ, Clement N, Porvasnik S et al (2012) An acidic oligopeptide displayed on AAV2 improves axial muscle tropism after systemic delivery. Genet Vaccines Ther 10:3
83. Ying Y, Muller OJ, Goehringer C, Leuchs B, Trepel M, Katus HA et al (2010) Heart-targeted adeno-associated viral vectors selected by in vivo biopanning of a random viral display peptide library. Gene Ther 17:980–990
84. White SJ, Nicklin SA, Buning H, Brosnan MJ, Leike K, Papadakis ED et al (2004) Targeted gene delivery to vascular tissue in vivo by tropism-modified adeno-associated virus vectors. Circulation 109:513–519
85. Work LM, Buning H, Hunt E, Nicklin SA, Denby L, Britton N et al (2006) Vascular bed-targeted in vivo gene delivery using tropism-modified adeno-associated viruses. Mol Ther 13:683–693
86. White K, Buning H, Kritz A, Janicki H, McVey J, Perabo L et al (2008) Engineering adeno-associated virus 2 vectors for targeted gene delivery to atherosclerotic lesions. Gene Ther 15:443–451
87. Michelfelder S, Kohlschutter J, Skorupa A, Pfennings S, Muller O, Kleinschmidt JA et al (2009) Successful expansion but not complete restriction of tropism of adeno-associated virus by in vivo biopanning of random virus display peptide libraries. PLoS One 4:e5122
88. Michelfelder S, Varadi K, Raupp C, Hunger A, Korbelin J, Pahrmann C et al (2011)

Peptide ligands incorporated into the three-fold spike capsid domain to re-direct gene transduction of AAV8 and AAV9 in vivo. PLoS One 6:e23101

89. Chen YH, Chang M, Davidson BL (2009) Molecular signatures of disease brain endothelia provide new sites for CNS-directed enzyme therapy. Nat Med 15:1215–1218
90. Dalkara D, Byrne LC, Klimczak RR, Visel M, Yin L, Merigan WH et al (2013) In vivo-directed evolution of a new adeno-associated virus for therapeutic outer retinal gene delivery from the vitreous. Sci Transl Med 5:189ra176
91. Shi W, Bartlett JS (2003) RGD inclusion in VP3 provides adeno-associated virus type 2 (AAV2)-based vectors with a heparan sulfate-independent cell entry mechanism. Mol Ther 7:515–525
92. Kern A, Schmidt K, Leder C, Muller OJ, Wobus CE, Bettinger K et al (2003) Identification of a heparin-binding motif on adeno-associated virus type 2 capsids. J Virol 77:11072–11081
93. Li W, Asokan A, Wu Z, Van Dyke T, DiPrimio N, Johnson JS et al (2008) Engineering and selection of shuffled AAV genomes: a new strategy for producing targeted biological nanoparticles. Mol Ther 16:1252–1260
94. Maheshri N, Koerber JT, Kaspar BK, Schaffer DV (2006) Directed evolution of adeno-associated virus yields enhanced gene delivery vectors. Nat Biotechnol 24:198–204
95. Yang L, Jiang J, Drouin LM, Agbandje-McKenna M, Chen C, Qiao C et al (2009) A myocardium tropic adeno-associated virus (AAV) evolved by DNA shuffling and in vivo selection. Proc Natl Acad Sci U S A 106: 3946–3951
96. Excoffon KJ, Koerber JT, Dickey DD, Murtha M, Keshavjee S, Kaspar BK et al (2009) Directed evolution of adeno-associated virus to an infectious respiratory virus. Proc Natl Acad Sci U S A 106:3865–3870
97. Li W, Zhang L, Johnson JS, Zhijian W, Grieger JC, Ping-Jie X et al (2009) Generation of novel AAV variants by directed evolution for improved CFTR delivery to human ciliated airway epithelium. Mol Ther 17: 2067–2077
98. Klimczak RR, Koerber JT, Dalkara D, Flannery JG, Schaffer DV (2009) A novel adeno-associated viral variant for efficient and selective intravitreal transduction of rat Muller cells. PLoS One 4:e7467
99. Koerber JT, Klimczak R, Jang JH, Dalkara D, Flannery JG, Schaffer DV (2009) Molecular evolution of adeno-associated virus for enhanced glial gene delivery. Mol Ther 17:2088–2095
100. Maguire CA, Gianni D, Meijer DH, Shaket LA, Wakimoto H, Rabkin SD et al (2010) Directed evolution of adeno-associated virus for glioma cell transduction. J Neurooncol 96:337–347
101. Gray SJ, Blake BL, Criswell HE, Nicolson SC, Samulski RJ, McCown TJ et al (2010) Directed evolution of a novel adeno-associated virus (AAV) vector that crosses the seizure-compromised blood-brain barrier (BBB). Mol Ther 18:570–578
102. Jang JH, Koerber JT, Kim JS, Asuri P, Vazin T, Bartel M et al (2011) An evolved adeno-associated viral variant enhances gene delivery and gene targeting in neural stem cells. Mol Ther 19:667–675
103. Asuri P, Bartel MA, Vazin T, Jang JH, Wong TB, Schaffer DV (2012) Directed evolution of adeno-associated virus for enhanced gene delivery and gene targeting in human pluripotent stem cells. Mol Ther 20:329–338
104. Asokan A, Conway JC, Phillips JL, Li C, Hegge J, Sinnott R et al (2010) Reengineering a receptor footprint of adeno-associated virus enables selective and systemic gene transfer to muscle. Nat Biotechnol 28:79–82
105. Bowles DE, McPhee SW, Li C, Gray SJ, Samulski JJ, Camp AS et al (2012) Phase 1 gene therapy for Duchenne muscular dystrophy using a translational optimized AAV vector. Mol Ther 20:443–455
106. Wu Z, Asokan A, Grieger JC, Govindasamy L, Agbandje-McKenna M, Samulski RJ (2006) Single amino acid changes can influence titer, heparin binding, and tissue tropism in different adeno-associated virus serotypes. J Virol 80:11393–11397
107. Vandenberghe LH, Breous E, Nam HJ, Gao G, Xiao R, Sandhu A et al (2009) Naturally occurring singleton residues in AAV capsid impact vector performance and illustrate structural constraints. Gene Ther 16:1416–1428
108. Limberis MP, Vandenberghe LH, Zhang L, Pickles RJ, Wilson JM (2009) Transduction efficiencies of novel AAV vectors in mouse airway epithelium in vivo and human ciliated airway epithelium in vitro. Mol Ther 17:294–301
109. Lochrie MA, Tatsuno GP, Christie B, McDonnell JW, Zhou S, Surosky R et al (2006) Mutations on the external surfaces of adeno-associated virus type 2 capsids that affect transduction and neutralization. J Virol 80:821–834
110. Li C, Diprimio N, Bowles DE, Hirsch ML, Monahan PE, Asokan A et al (2012) Single

amino acid modification of adeno-associated virus capsid changes transduction and humoral immune profiles. J Virol 86:7752–7759

111. Zhong L, Li B, Mah CS, Govindasamy L, Agbandje-McKenna M, Cooper M et al (2008) Next generation of adeno-associated virus 2 vectors: point mutations in tyrosines lead to high-efficiency transduction at lower doses. Proc Natl Acad Sci U S A 105: 7827–7832
112. Kay CN, Ryals RC, Aslanidi GV, Min SH, Ruan Q, Sun J et al (2013) Targeting photoreceptors via intravitreal delivery using novel, capsid-mutated AAV vectors. PLoS One 8: e62097
113. Li M, Jayandharan GR, Li B, Ling C, Ma W, Srivastava A et al (2010) High-efficiency transduction of fibroblasts and mesenchymal stem cells by tyrosine-mutant AAV2 vectors for their potential use in cellular therapy. Hum Gene Ther 21:1527–1543
114. Zhong L, Li B, Jayandharan G, Mah CS, Govindasamy L, Agbandje-McKenna M et al (2008) Tyrosine-phosphorylation of AAV2 vectors and its consequences on viral intracellular trafficking and transgene expression. Virology 381:194–202
115. Petrs-Silva H, Dinculescu A, Li Q, Min SH, Chiodo V, Pang JJ et al (2009) High-efficiency transduction of the mouse retina by tyrosine-mutant AAV serotype vectors. Mol Ther 17:463–471
116. Petrs-Silva H, Dinculescu A, Li Q, Deng WT, Pang JJ, Min SH et al (2011) Novel properties of tyrosine-mutant AAV2 vectors in the mouse retina. Mol Ther 19:293–301
117. Iida A, Takino N, Miyauchi H, Shimazaki K, Muramatsu S (2013) Systemic delivery of tyrosine-mutant AAV vectors results in robust transduction of neurons in adult mice. Biomed Res Int 2013:974819
118. Zolotukhin I, Luo D, Gorbatyuk O, Hoffman B, Warrington K Jr, Herzog R et al (2013) Improved adeno-associated viral gene transfer to Murine glioma. J Genet Syndr Gene Ther 4
119. Gabriel N, Hareendran S, Sen D, Gadkari RA, Sudha G, Selot R et al (2013) Bioengineering of AAV2 capsid at specific serine, threonine, or lysine residues improves its transduction efficiency in vitro and in vivo. Hum Gene Ther Methods 24:80–93
120. Nathanson JL, Yanagawa Y, Obata K, Callaway EM (2009) Preferential labeling of inhibitory and excitatory cortical neurons by endogenous tropism of adeno-associated virus and lentivirus vectors. Neuroscience 161:441–450
121. Xu R, Janson CG, Mastakov M, Lawlor P, Young D, Mouravlev A et al (2001) Quantitative comparison of expression with adeno-associated virus (AAV-2) brain-specific gene cassettes. Gene Ther 8:1323–1332
122. Kugler S, Lingor P, Scholl U, Zolotukhin S, Bahr M (2003) Differential transgene expression in brain cells in vivo and in vitro from AAV-2 vectors with small transcriptional control units. Virology 311:89–95
123. Gray SJ, Foti SB, Schwartz JW, Bachaboina L, Taylor-Blake B, Coleman J et al (2011) Optimizing promoters for recombinant adeno-associated virus-mediated gene expression in the peripheral and central nervous system using self-complementary vectors. Hum Gene Ther 22:1143–1153
124. Husain T, Passini MA, Parente MK, Fraser NW, Wolfe JH (2009) Long-term AAV vector gene and protein expression in mouse brain from a small pan-cellular promoter is similar to neural cell promoters. Gene Ther 16:927–932
125. von Jonquieres G, Mersmann N, Klugmann CB, Harasta AE, Lutz B, Teahan O et al (2013) Glial promoter selectivity following AAV-delivery to the immature brain. PLoS One 8:e65646
126. Paterna JC, Moccetti T, Mura A, Feldon J, Bueler H (2000) Influence of promoter and WHV post-transcriptional regulatory element on AAV-mediated transgene expression in the rat brain. Gene Ther 7:1304–1311
127. Kingsman SM, Mitrophanous K, Olsen JC (2005) Potential oncogene activity of the woodchuck hepatitis post-transcriptional regulatory element (WPRE). Gene Ther 12:3–4
128. Schambach A, Bohne J, Baum C, Hermann FG, Egerer L, von Laer D et al (2006) Woodchuck hepatitis virus post-transcriptional regulatory element deleted from X protein and promoter sequences enhances retroviral vector titer and expression. Gene Ther 13:641–645
129. Geisler A, Jungmann A, Kurreck J, Poller W, Katus HA, Vetter R et al (2011) microRNA122-regulated transgene expression increases specificity of cardiac gene transfer upon intravenous delivery of AAV9 vectors. Gene Ther 18:199–209
130. Karali M, Manfredi A, Puppo A, Marrocco E, Gargiulo A, Allocca M et al (2011) MicroRNA-restricted transgene expression in the retina. PLoS One 6:e22166
131. Qiao C, Yuan Z, Li J, He B, Zheng H, Mayer C et al (2011) Liver-specific microRNA-122 target sequences incorporated in AAV vectors efficiently inhibits transgene expression in the liver. Gene Ther 18:403–410

Chapter 11

Altering Tropism of rAAV by Directed Evolution

Damien Marsic and Sergei Zolotukhin

Abstract

Directed evolution represents an attractive approach to derive AAV capsid variants capable of selectively infect specific tissue or cell targets. It involves the generation of an initial library of high complexity followed by cycles of selection during which the library is progressively enriched for target-specific variants. Each selection cycle consists of the following: reconstitution of complete AAV genomes within plasmid molecules; production of virions for which each particular capsid variant is matched with the particular capsid gene encoding it; recovery of capsid gene sequences from target tissue after systemic administration. Prevalent variants are then analyzed and evaluated.

Key words AAV, Gene therapy, Directed evolution, Capsidlibrary, In vivo selection

1 Introduction

AAV-derived vectors are among the most promising tools for human gene therapy because of their safety profile (absence of pathogenicity, episomal localization) and their ability to express transgenes in a sustained manner [1]. However, their promiscuity and the lack of convenient methods to alter their tropism have limited their clinical applications. As tropism is determined by the nature of amino acid residues exposed at the surface of the viral capsid, efforts have been made to develop capsid variants capable of selectively infecting target tissues. As opposed to rational design, which is limited by current knowledge of virus biology and molecular interactions between viral capsid and receptors at the surface of the targeted cells, directed evolution represents an attractive approach that does not require such extensive understanding. Instead, it relies on the generation of large numbers of different sequence variants, somewhat similarly to an organism's adaptive immune system mounting a response to novel antigens by producing a large diversity of antibodies. In fact, the likelihood of directed

Fredric P. Manfredsson (ed.), *Gene Therapy for Neurological Disorders: Methods and Protocols*, Methods in Molecular Biology, vol. 1382, DOI 10.1007/978-1-4939-3271-9_11, © Springer Science+Business Media New York 2016

evolution success is directly related to the initial library size (also known as complexity). Another crucial feature is the availability of an efficient selection method that allows the recovery of desirable variants.

Typically, altering AAV tropism by directed evolution involves the generation of a large initial capsid library followed by several rounds of in vivo selection during which the library is enriched for target-specific variants. The three main methods used to generate the initial library, either alone or in combination, are error-prone PCR [2–4], DNA shuffling [5–9], and random display peptide insertion [10–13]. More recently, a gene synthesis-based method, combining rational design and virtual shuffling at the nucleotide level, was introduced in an attempt to increase library complexity [14]. Due to such a diversity of approaches, it would fall beyond the scope of this chapter to go into the details on how to design and generate the permutated capsid gene fragment for an initial library. For simplicity, it will be assumed that AAV2 is used as a backbone and that the starting material is DNA (obtained through any of the four methods mentioned above, or through a combination of any of them) encoding positions 235–735 of the VP1 capsid protein (ideally, the ends should match the oligonucleotide primers described in **item 3** of Subheading 2.8). Also, it will be assumed that the selection will be done in mice.

An outline of the directed evolution process is shown in Fig. 1. The DNA encoding the initial capsid gene variants is first inserted into a vector encoding a complete AAV genome from which the corresponding portion of the *cap* gene had been removed, thus generating a plasmid library of functional reconstituted AAV genomes (Subheading 3.1). The plasmid library is then amplified in *E. coli* (Subheading 3.2) and purified (Subheading 3.3) before being converted into a viral library in which every virion carries the genome variant that encodes its own variant capsid (Subheading 3.4). The resulting viral population is then purified from the host cells and culture medium (Subheading 3.5), quantified (Subheading 3.6), and injected into the bloodstream of the animal in which selection is to take place (Subheading 3.7). After the target tissue has been harvested, episomal DNA is purified and the viral capsid gene sequences (now enriched for variants targeting that particular tissue) are amplified (Subheading 3.8). The whole process is then repeated several times (except that the first step now uses DNA from the last step instead of the initial library), producing an increasingly enriched library at each new cycle. After 3 or 4 cycles, selected variants are analyzed (Subheading 3.9) and evaluated (Subheading 3.10) as illustrated in Fig. 2.

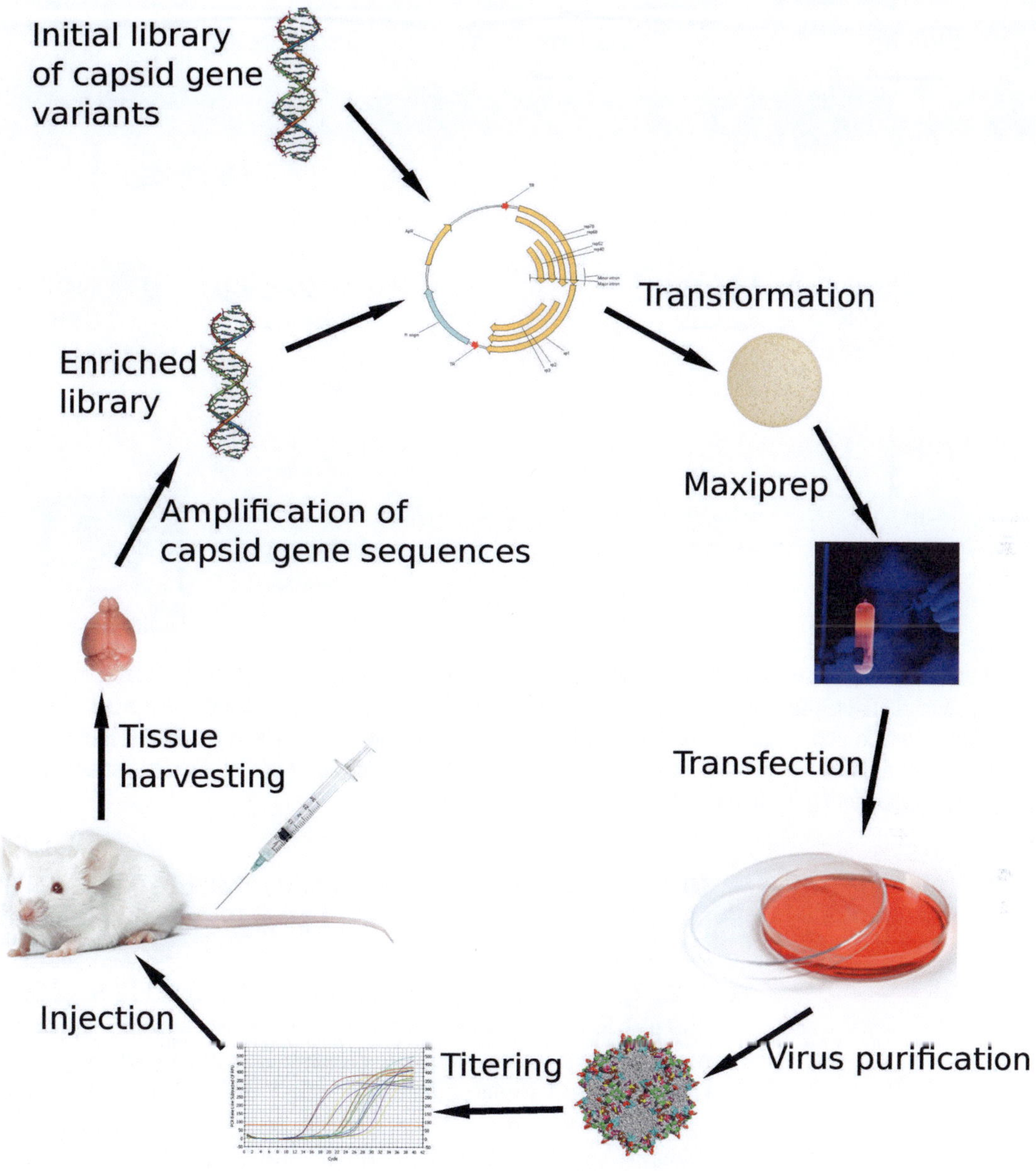

Fig. 1 Library enrichment for target-specific variants. The DNA encoding the initial capsid library (*top left*) is cloned into the vector and enters the cycle through which target-specific variants are enriched. At every new cycle, the complexity of the new enriched library decreases while the prevalence of target-specific variants increases

2 Materials

2.1 Plasmid Assembly

1. pSubEagApa plasmid (*see* **Note 1**) that has been linearized with restriction enzymes EagI and ApaI and gel-purified.
2. 2.5× Isothermal DNA Assembly (IDA) buffer (*see* **Note 2**): for 1.44 ml, mix 664.62 μl H_2O, 500 μl 1 M Tris–HCl pH 7.5,

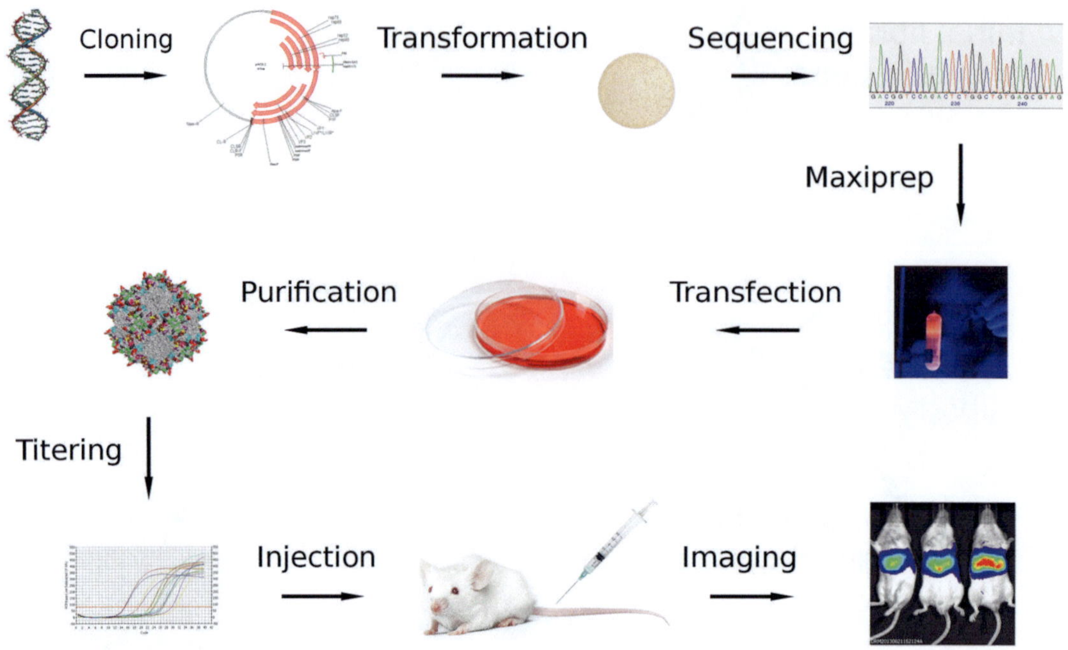

Fig. 2 Variant analysis and evaluation. Enriched capsid gene variants (*top left*) are cloned into a vector that does not contain ITRs. After transformation, clones are analyzed by sequencing to determine the most interesting variants, which capsid genes are then used to generate recombinant AAV that express luciferase. After injecting equal amounts in animals, specificities and transduction efficiencies of the different variants can be compared by monitoring luciferase activity using in vivo imaging

10.38 µl 4.82 M magnesium chloride, 100 µl 10 mM dNTP, 50 µl 1 M Dithiothreitol, 625 µl 40 % PEG 8000, 50 µl 100 mM NAD. Store at −20 °C.

3. 1.33× IDA master mix (*see* **Note 2**): for 375 µl, mix 118.55 µl H_2O, 200 µl 2.5× IDA buffer, 0.2 µl 10 U/µl T5 exonuclease (Epicentre, Madison, WI, USA), 6.25 µl Phusion 2 U/µl DNA polymerase (New England Biolabs, Ipswich, MA, USA), 50 µl 40 U/µl Taq ligase (New England Biolabs, Ipswich, MA, USA). Aliquot in PCR tubes at 7.5, 15, and 30 µl per tube, store at −20 °C.
4. DNA Clean & Concentrator kit (Zymo Research, Irvine, CA, USA).

2.2 Transformation

1. Electrocompetent *E. coli*: E. cloni 10G SUPREME (Lucige, Middleton, WI, USA) or equivalent (*see* **Note 3**).
2. Electroporation cuvettes: 2 mm gap.
3. Electroporation instrument: Bio-Rad Gene Pulser Xcell or equivalent.
4. LB (Lysogeny Broth, also known as Luria Broth or Luria-Bertani medium): 10 g/L sodium chloride, 10 g/L tryptone, 5 g/L yeast extract. Sterilize by autoclaving.

5. LB-Carbenicillin agar plates: 15 g/L agar in LB, 100 mg/L carbenicillin.
6. LB with 100 mg/L carbenicillin.

2.3 Maxiprep

1. Solution I: 50 mM Tris–HCl pH 7.5, 10 mM EDTA.
2. Solution II: 0.2 N NaOH, 1 % SDS.
3. Solution II: 3 M cesium chloride, 1 M potassium acetate, 0.67 M acetic acid (for 1 L: 505.08 g cesium chloride, 98.4 g potassium acetate, 38.46 ml glacial acetic acid).
4. 10 mg/ml RNase A solution.
5. 250 ml flat-bottom centrifuge tubes.
6. Gauze sponges.
7. Isopropanol.
8. TE buffer: 10 mM Tris–HCl pH 8.0, 1 mM EDTA.
9. Cesium chloride.
10. 10 mg/ml ethidium bromide solution.
11. 4.9 ml ultracentrifuge tubes and NVT90 rotor (Beckman, Indianapolis, IN, USA) or equivalent.
12. 18 G needles and 5 ml syringes.
13. Isoamyl alcohol.
14. Corex centrifuge tubes.
15. Ethanol.
16. Phenol-chloroform: phenol/chloroform/isoamyl alcohol solution (25:24:1).
17. Chloroform.
18. 3 M sodium acetate, pH 5.2.
19. 75 % ethanol.
20. SmaI restriction enzyme.

2.4 Transfection

1. Human Embryonic Kidney (HEK) 293 cells (Sigma-Aldrich, St. Louis, MO, USA) grown in 150 mm tissue culture dishes.
2. Culture medium: Dulbecco's Modified Eagle Medium (DMEM) supplemented with 5–10 % Fetal Bovine Serum (FBS) and Antibiotic-Antimycotic mix (final concentrations: 100 Units/ml each penicillin and streptomycin, 250 ng/ml Fungizone).
3. 2×HBS: 280 mM sodium chloride, 50 mM Hepes, 1.5 mM sodium phosphate dibasic. Adjust pH to 7.17 using 10 M sodium hydroxide (approximately 1.6 ml for 1 L). Sterilize by filtration through a 0.22 μm filter. Aliquot in 50 ml tubes and store at −20 °C.

4. 2.5 M calcium chloride. Sterilize using a 0.22 μm syringe filter. Store at −20 °C.
5. pHelper plasmid (Agilent Technologies, Santa Clara, CA, USA).
6. Standard cell culture equipment (CO_2 incubator, sterile hood, centrifuge, water bath) and reagents (PBS, Trypsin).
7. Lysis buffer: 150 mM sodium chloride, 50 mM Tris–HCl, pH 8.5.

2.5 Virus Purification

1. 0.22 μm vacuum filter.
2. Tangential flow filtration system such as Minim II with a 100 kDa membrane such as 100 K Omega Centramate cassette (Pall Corporation, Port Washington, NY, USA).
3. Benzonase (Sigma Aldrich, St. Louis, MO, USA) or equivalent nuclease.
4. Saturated magnesium chloride solution: 4.82 M magnesium chloride. Store at room temperature.
5. 60 % Iodixanol solution: OptiPrep (Sigma-Aldrich, St. Louis, MO, USA).
6. PBS-MK buffer: 1× phosphate- buffered saline (PBS), 1 mM magnesium chloride, 2.5 mM potassium chloride. Prepare as a 10× concentrated stock solution.
7. 40 % Iodixanol in PBS-MK.
8. 25 % Iodixanol in PBS-MK.
9. 15 % Iodixanol in PBS-MK and 1 M sodium chloride.
10. Multichannel peristaltic pump such as Watson Marlow 205S or equivalent.
11. 29.9 ml ultracentrifuge tubes and 70 Ti rotor (Beckman or equivalent).
12. 18 G needles and 10 ml syringes.
13. Lactated Ringer's solution.
14. Centrifugal quantitative concentrators: Apollo 20 ml, 150 kDa (Orbital Biosciences, Topsfield, MA, USA).
15. 0.22 μm sterile filters.
16. Sterile low retention 1.5 ml microcentrifuge tubes.

2.6 Determination of Viral Titer

1. DNase I and its 10× reaction buffer.
2. Maxiprep Solution II (**item 2** from Subheading 2.3).
3. DNA Clean & Concentrator kit (Zymo Research, Irvine, CA, USA).
4. qPCR instrument, such as Bio-Rad MyiQ (Bio-Rad, Hercules, CA, USA).

5. qPCR reagent, such as ABsolute Blue qPCR SYBR Green Fluorescein Mix (Thermo Fisher Scientific, Waltham, MA, USA).
6. qPCR standard: linearized plasmid containing the target sequence, at known concentration, aliquoted in low retention tubes and stored at −20 °C.
7. qPCR primers targeting AAV2 *rep* gene (to quantify capsid library titers): 5′-gcaagaccggatgttcaaat (forward) and 5′-cctcaaccacgtgatccttt (reverse).
8. qPCR primers targeting the CBA promoter (to quantify capsid variant titers): 5′-tcccatagtaacgccaatagg (forward) and 5′-cttggcatatgatacacttgatg (reverse).

2.7 In Vivo Selection

1. Young adult (6–10 weeks old) male mice.
2. 1 ml syringes with 25–28 Gauge needles.
3. Induction chamber.
4. Isoflurane.
5. Scissors. Forceps.
6. PBS.
7. Maxiprep solutions I and III (*see* Subheading 2.3).
8. 1.2 % sodium dodecyl sulfate (SDS) solution.
9. Proteinase K (lyophilized).
10. 56 °C shaking incubator or water bath.
11. 10 mg/ml RNase A solution.
12. 0.45 μm syringe filters.
13. Plasmid midiprep kit: EZNA Plasmid Midi kit (Omega Biotek, Norcross, GA, USA) or equivalent.
14. DNA Clean & Concentrator kit (Zymo Research, Irvine, CA, USA).

2.8 Capsid Gene Amplification and Purification

1. High fidelity DNA polymerase such as Q5 or Phusion (New England Biolabs, Ipswich, MA, USA).
2. Standard PCR components and accessories: dNTP, PCR-grade water, PCR tubes, thermal cycler.
3. PCR primers: 5′-ggatgggcgacagagtcatc (forward) and 5′-caagcaattacagattacgagtcagg (reverse).
4. Standard DNA electrophoresis equipment: electrophoresis apparatus, power supply, agarose, TAE, buffer, loading dye, DNA ladder.
5. Gel extraction kit (Omega Biotek, Norcross, GA, USA) or equivalent.

2.9 Variant Analysis

1. pACG2-m56 plasmid (*see* **Note 4**) that has been linearized with restriction enzymes EagI and ApaI and gel-purified.
2. Plasmid miniprep kit.
3. PstI restriction enzyme.
4. Sequencing primers 5′-gtctaggaactggcttcctg and 5′-ccattgaggtggtacttggtag.

2.10 In Vivo Evaluation

1. pTR-UF50-BC plasmid (*see* **Note 5**).
2. Modified Dulbecco's Phosphate Buffered Saline (mDPBS) without calcium chloride and magnesium chloride (Sigma-Aldrich, St. Louis, MO, USA).
3. Luciferin solution: 15 mg/ml firefly luciferin sodium salt in mDPBS. Sterilize by filtering through a 0.22 μm syringe filter, aliquot in 2 ml tubes and store at −20 °C.
4. In vivo imaging instrument such as IVIS Imaging System (Perkin-Elmer, Waltham, MA, USA).

3 Methods

3.1 Plasmid Assembly

See **Note 6** for background information and explanations on why this particular method is used.

1. Assemble reaction on ice by adding vector and insert (collection of capsid gene variants) DNA to one or more IDA master mix tubes (*see* **Note 7**) and adjust volume with water. *See* Table 1 for

Table 1
IDA reaction assembly and purification. Recommended parameters for IDA reaction assembly and purification depending on application, from initial library through enriched libraries and finally to variant analysis

	Initial library	First enriched library	Second and subsequent enriched libraries	Variant analysis
Number and type of IDA master mix tubes	>10 × 30 μl	1 × 30 μl	1 × 15 μl	1 × 7.5 μl
Vector identity	pSubEagApa	pSubEagApa	pSubEagApa	pACG2-m56
Vector amount per tube	200 ng	200 ng	100 ng	25 ng
Insert amount per tube	150 ng	150 ng	75 ng	19 ng
Final reaction volume per tube	40 μl	40 μl	20 μl	10 μl
Elution volume after DNA purification	500 μl	100 μl	50 μl	30 μl

suggested amounts. Centrifuge briefly to collect reaction mixture to the bottom of the tubes.

2. Incubate 2 h (*see* **Note 8**) at 50 °C, preferably in a thermal cycler with a heated lid.
3. Purify DNA from reactions using DNA Clean & Concentrator kit. Use at least 200 μl binding buffer. Elute with sterile water (not elution buffer), *see* Table 1 for recommended amount.

3.2 Transformation

1. Place competent cells on ice (*see* **Note 9**). The total volume of competent cells, expressed in microliter, should be at least 4 times the amount of vector DNA used in plasmid assembly, expressed in ng.
2. Place appropriate number of electroporation cuvettes on ice at least 15 min in advance.
3. Add a volume of LB equal to 5 times the total volume of competent cells to a sterile flask and place on ice. Alternatively, aliquot the LB into several smaller flasks or tubes (the number of electroporation cuvettes should be a multiple of the number of flasks or tubes). Choose the size of the containers so that the LB does not fill more than 20 % of their total volume.
4. Once the competent cells are thawed (if previously frozen), add purified DNA from Subheading 3.1 to the cells, mix gently, and aliquot into the electroporation cuvettes.
5. Electroporate at 2.8 kV or other optimal voltage (*see* **Note 10**) and immediately transfer to the LB.
6. Incubate 1 h at 37 °C in a shaking incubator (220 rpm).
7. Place an LB-carbenicillin agar plate under a sterile hood to warm up to room temperature and dry excess moisture.
8. Transfer the transformation to a volume of LB-carbenicillin at least equal to 5 times the volume of the transformation and mix. If necessary, aliquot into several flasks so that each flask is filled at 20 % of its maximal volume.
9. Spread 50 μl into the LB-carbenicillin agar plate.
10. Incubate the agar plate at 37 °C overnight.
11. Incubate the flask(s) at 30 °C (*see* **Note 11**) in a shaking incubator (220 rpm) overnight.
12. Next morning, count colonies on the agar plate. Knowing the total volume of medium from which the aliquot was plated, calculate the complexity (extrapolated total number of colony forming units). Knowing the amount of vector DNA that was used, calculate the transformation efficiency (cfu/μg).
13. If needed, Sanger or PacBio sequencing may be performed from colonies and from the liquid culture respectively.

3.3 Maxiprep

The following instructions assume an overnight culture volume of 1 L. Adjust the values according to the actual volume.

1. Save 2 × 1 ml aliquots as glycerol stocks at −80 °C.
2. Centrifuge at 5,000 × *g* for 15 min and discard supernatant.
3. Completely resuspend pellet in 50 ml Solution I and 500 μl RNase A solution using a 10 ml pipette.
4. Transfer to a 250 ml Beckman centrifuge tube.
5. Add 50 ml Solution II and gently mix.
6. Incubate on ice for 5 min.
7. Add 50 ml Solution III and immediately mix, using the pipette in a stirring motion.
8. Incubate on ice for 5 min.
9. Centrifuge at 12,000 × *g* for 10 min.
10. Transfer supernatant to a new 250 ml Beckman tube by filtering through a gauze sponge.
11. Add 90 ml isopropanol and mix well.
12. Incubate on ice for 5 min.
13. Centrifuge at 12,000 × *g* for 10 min.
14. Discard supernatant and let the tube dry upside down on a paper towel.
15. Dissolve pellet in TE and transfer to a 15 ml conical tube, in two or three steps, using a total volume of 3.8 ml TE for each 4.9 ml ultracentrifugation tube that you plan to use (1 or 2 tubes for each liter of culture, depending on the DNA yield).
16. Add 4.2 g cesium chloride and 240 μl ethidium bromide solution for each 3.9 ml TE that you used in the previous step.
17. When cesium chloride is completely dissolved, transfer to 4.9 ml Beckman ultracentrifuge tubes, add TE if necessary to fill the tubes completely.
18. Ultracentrifuge at 210,000 × *g* (55,000 rpm with NVT 90 rotor) overnight.
19. Puncture tube with needle, aspirate plasmid band (if more than one band is visible, the plasmid should be in the lower band, check using a UV lamp if necessary to locate the high intensity band) and place in a 2 ml microcentrifuge tube.
20. Extract 3 times (or more if solution is still colored) with equal volume of isoamyl alcohol (vortex, spin 2 min at maximum speed and discard upper phase).
21. Transfer to a 30 ml Corex tube, add 2.5 volumes of water and mix.
22. Add 2 volumes of ethanol and mix by inverting several times.
23. Incubate at −20 °C for 30 min.

24. Centrifuge at 10,000 ×*g* for 15 min at 4 °C.
25. Discard supernatant and dissolve pellet in appropriate volume of TE (typically 400 μl for each ultracentrifugation tube from **step 17**).
26. Transfer to a microcentrifuge or centrifuge tube of appropriate size if necessary and add 1 volume of phenol-chloroform.
27. Vortex, centrifuge 2 min at maximum speed, transfer aqueous (top) phase to a new tube.
28. Repeat **steps 26** and **27** using chloroform instead of phenol-chloroform.
29. Add 0.1 volume of sodium acetate solution and mix well.
30. Add 2.5 volumes of ethanol and mix by inversion.
31. Centrifuge for 10 min at maximum speed (>13,000 ×*g* if using 1.5 ml tubes) at 4 °C.
32. Under a sterile hood (*see* **Note 12**), discard supernatant and wash pellet with smaller volume of 75 % Ethanol.
33. Centrifuge 5 min at maximum speed at 4 °C.
34. Discard supernatant and dry pellet.
35. Add 1 ml sterile TE, place on ice and allow to dissolve slowly for several hours.
36. Vortex, collect an aliquot (10–20 μl) for analysis, aliquot if necessary into 1.5 ml tubes and store at −20 °C.
37. Use the aliquot to quantitate DNA and perform restriction analysis with SmaI according to the manufacturer's instructions (*see* **Note 13**).

3.4 Transfection

1. Culture HEK293 cells until they reach approximately 90 % confluence. The number of tissue culture dishes depends on the expected library complexity and desired yield. For the initial library, 50–100 dishes (150 mm in diameter) are recommended. For subsequent enriched libraries, 20–25 dishes are sufficient.
2. Aliquot 2×HBS into the minimum number of 50 ml tubes for which each tube is no more than 50 % full. The total volume of 2×HBS that is needed is 1.44 ml multiplied by the number of 150 mm tissue culture dishes.
3. In a separate tube, add the following amounts multiplied by the number of dishes to transfect (numbers indicated are per 150 mm dish): 144 μl calcium chloride solution, 70 μg pHelper, 10 ng library plasmid (*see* **Note 14**) and sterile water to reach a final volume of 1.44 ml.
4. Add an equal volume of the mix from **step 3** to each tube of 2×HBS, and mix gently by inverting the tubes several times.

5. Wait for the solution to become slightly cloudy, while occasionally inverting the tubes (typically no more than 15 min).
6. Add 2.88 ml to each dish drop by drop.
7. Rock gently back and forth and side to side, then return dishes to incubator for 6 h.
8. Discard culture medium from each dish and replace with culture medium containing only 3 % FBS. Return to incubator for 72 h.
9. Place 250 ml conical tubes on ice (the number of tubes should be sufficient to fit the total volume of the transfected dishes). Perform **steps 10–12** for one dish at a time.
10. Transfer 10 ml of the culture medium into the flipped over lid.
11. Using a 10 ml pipette, detach the cells from the dish surface by pipetting in and out; transfer to the conical tubes from **step 9**. Alternatively, use a cell scraper.
12. Use the medium from the lid to rinse the dish and collect leftover cells; transfer to the same conical tubes.
13. Centrifuge at 800 × *g* for 15 min at 4 °C.
14. Transfer supernatant into another tube and centrifuge at 5,000 × *g* for 10 min at 4 °C (*see* **Note 15**).
15. Transfer supernatant to a plastic bottle and store at –80 °C.
16. Resuspend cell pellet (from **step 13**) in 350 μl lysis per dish (typically 7 ml for 20 dishes) and store at –80 °C.

3.5 Virus Purification

Steps 1–8 should be performed in parallel, as cell lysate and concentrated culture medium will be mixed together at **step 9** (*see* **Note 16**). **Steps 10** and **11** can be performed while waiting for **step 6** to complete.

1. Lyse cells by subjecting the resuspended cell pellet to three freeze–thaw cycles using a 37 °C water bath and either a –80 °C freezer or liquid nitrogen. Make sure to vortex vigorously after each thaw.
2. Add 1 μl of Benzonase and 1 μl of saturated magnesium chloride per 10 ml of lysate and incubate for 30 min at 37 °C.
3. Centrifuge at 3,000 × *g* for 10 min at 4 °C and keep on ice.
4. Prepare tangential filtration system by processing 3 L water with equal flow rates for the permeate and retentate lines.
5. Filter clarified medium (from **step 15**, Subheading 3.4) through a 0.22 μm vacuum filter.
6. Transfer medium to the tangential filtration system reservoir and equilibrate the membrane for 10 min with the permeate line closed and the retentate fed back to the reservoir, with a pump speed of 25–30 rpm.

7. Open the permeate line (connected to a waste container) and allow medium to concentrate until the minimum volume is reached.
8. Close the permeate line and let circulate for 15 min to allow AAV particles to detach from the membrane.
9. Open the retentate line and collect concentrated medium into a 50 ml tube.
10. Add supernatant from **step 3** to the tube containing the concentrated medium, mix and keep on ice.
11. Prepare the multichannel pump. Insert tubings. Attach disposable microcapillary pipettes to each ends of tubings. Calibrate if necessary, so that equal volumes will be dispensed through all channels used.
12. Rinse tubings by pumping 50 ml water through them, then continue pumping to flush the water out.
13. Knowing that each 10 ml of unpurified AAV (from **step 9**) requires one ultracentrifuge tube and 5.5 ml, 5.5 ml, 4 ml and 5 ml respectively of 15, 25, 40 and 60 % Iodixanol, prepare appropriate amounts of the four Iodixanol solutions in separate 50 ml tubes.
14. Transfer 10 ml unpurified AAV into each ultracentrifuge tube using a 5 ml pipette.
15. Place all the microcapillary pipettes on the aspirating end of the tubings together into the 50 ml tube containing the 15 % Iodixanol solution and start pumping.
16. As soon as the solution reaches the opposite end, pause the pump and place the microcapillary pipettes on the dispensing end of the tubings into the ultracentrifuge tubes (1 pipette per tube), at the bottom.
17. Resume pumping until just before the tube with Iodixanol becomes empty, then transfer the pipettes on the aspirating end into the next Iodixanol tube.
18. Repeat **steps 16–17** with remaining Iodixanol tubes (the order is 15, 25, 40, 60 %) except that with the last one, continue pumping until either the ultracentrifuge tubes are full or air is about to come out from the dispensing end, whichever happens first.
19. Top the tubes off with lysis buffer if necessary and plug them.
20. Centrifuge at 350,000 × g (69,000 rpm with type 70 Ti rotor) for 1 h at 18 °C.
21. Rinse the pump tubings by pumping water, then dry by pumping air and discard capillaries.
22. Puncture the side of the tubes below the 40–60 % junction and aspirate approximately 4 ml which should consist of about 1 ml of the 60 % and 3 ml of the 40 % fractions.

23. Transfer to a 50 ml tube, add at least 5 volumes of cold Lactated Ringer's solution, mix and place on ice.
24. Concentrate through repeated filling, centrifuging (at 1,500 × *g*, 4 °C) and discarding flow-through, using an Apollo concentrator until the retentate volume reaches about 1 ml.
25. Dilute again with cold Lactated Ringer's solution to fill the concentrator.
26. Continue concentrating down to about 0.5–1 ml.
27. Aspirate with a needle to collect all the retentate and sterilize through a 0.22 μm syringe filter.
28. Add 200–500 μl of Lactated Ringer's solution to the concentrator, rinse the membrane by pipetting in and out to collect leftover AAV, and repeat **step 27**. Combine the filtrates and homogenize by mixing.
29. Transfer 10 μl into a PCR tube for quantitation, aliquot the rest in low retention 1.5 ml tubes and store at −80 °C.

3.6 Determination of Viral Titer

1. Add 34.5 μl water, 5 μl DNase buffer and 0.5 μl DNase I to the 10 μl virus sample and incubate for 1 h at 37 °C.
2. Add 17 μl of maxiprep Solution II, mix and incubate at 65 °C for 30 min.
3. Chill on ice for at least 5 min.
4. Purify DNA with the DNA Clean & Concentrator kit following kit instructions except: use 700 μl of binding buffer (*see* **Note 17**); for the last step, elute twice into a low-retention tube using 100 μl elution buffer (final volume is therefore 200 μl).
5. Store at −20 °C or place on ice until ready for qPCR.
6. Prepare standard solutions at 400, 40, 4, 0.4, and 0.04 pg/μl using linearized plasmid containing the target sequence (such as pSubEagApa, which has the *rep* gene) (*see* **Note 18**).
7. Prepare reaction mix (qPCR reagent, forward and reverse primers, water) following the qPCR kit instructions. Typically, reaction volume is 25 μl including 22.5 μl reaction mix and 2.5 μl template.
8. Distribute 22.5 μl of reaction mix per tube into qPCR tubes.
9. Add 2.5 μl template (negative control, standard dilutions, AAV samples) per well, in duplicate or triplicate depending on your instrument reproducibility.
10. Run qPCR according to qPCR reagent instructions. Typical cycling parameters using ABsolute Blue qPCR SYBR Green Fluorescein Mix are: 15 min at 95 °C initial denaturation, followed by 40 cycles of 15 s at 95 °C denaturation, 25 s at 60 °C annealing and 15 s at 72 °C extension.

11. From qPCR results (amount of DNA per sample, as calculated from the standard curve constructed using the standard dilutions), calculate viral titers, expressed as viral genomes per ml (vg/ml), knowing the length of the standard (size of the linearized plasmid in bp).

3.7 In Vivo Selection

1. Inject up to 250 μl (typically 1E10–1E12 viral genomes) of the purified AAV per mouse into the tail vein. This step should be performed by a trained technician.
2. 3–5 days after injection, place mice, one by one, in an induction chamber and anesthetize them with isoflurane according to your institution's protocol.
3. Immediately after anesthesia has been induced, euthanize the mice by cervical dislocation.
4. Harvest the tissue or organ that is the target for selection and immediately place on ice.
5. Rinse with PBS and transfer to a Petri dish of appropriate size.
6. Add small amount of maxiprep solution I (*see* **Note 19**).
7. Dissociate tissue with sterile scissors into as small pieces as possible.
8. Transfer to 50 ml tube.
9. Add solution I if necessary so that all tissue is in suspension.
10. Add 1 volume of 1.2 % SDS and mix.
11. Add proteinase K at 200 μg/ml final concentration.
12. Incubate at 56 °C while shaking for 2–6 h or until lysis is complete.
13. Cool down to room temperature.
14. Add RNase A at 200 μg/ml final concentration.
15. Mix well and incubate for 10 min at room temperature.
16. Add 0.7 volume of maxiprep Solution III.
17. Immediately mix gently and place on ice for 15 min.
18. Centrifuge for 15 min at maximum speed (typically 4,000 × *g*).
19. Transfer supernatant to a fresh tube. If necessary, filter through a gauze sponge (if solid material is present at the surface).
20. Filter through a 0.45 μm syringe filter into a new tube (*see* **Note 20**).
21. Load into a midiprep column, centrifuge according to the midiprep kit instructions and discard flow-through. Repeat until all sample has been processed through the column.
22. Follow midiprep kit protocol for washing and eluting (*see* **Note 20**).

23. Further purify and concentrate using DNA Clean & Concentrator kit according to the kit instructions, except use 7 volumes binding buffer before loading into the column.

3.8 Capsid Gene Amplification and Purification

1. Perform PCR to amplify capsid gene sequences from purified episomal DNA (final product of previous section) using a high-fidelity DNA polymerase. Include a negative control (reaction in which template is replaced with water) and three template dilutions (for example 1×, 3× and 10×). Example of cycling parameters for Q5 polymerase: 30 s at 98 °C initial denaturation, followed by 30 cycles of 10 s at 98 °C denaturation, 20 s at 67 °C annealing and 35 s at 72 °C extension, followed by 1 min at 72 °C final extension.
2. Perform DNA electrophoresis of the PCR products on a 1.2 % agarose gel, along with a DNA ladder.
3. Briefly illuminate using a long wavelength portable UV lamp, cut visible bands with a clean scalpel and transfer each band into a separate 1.5 ml tube.
4. Bring the gel to an imaging system, take a picture and verify the absence of band in the negative control lane and the size of the cut bands (*see* **Note 21**).
5. Purify DNA from the gel using a gel extraction kit according to the kit instructions and quantitate.
6. If more cycles of selection are needed, go to Subheading 3.1, otherwise (after at least three cycles have been completed) go to Subheading 3.9.

3.9 Variant Analysis

1. Insert purified DNA from previous section into pACG2-m56 vector (*see* **Note 22**) following instructions from Subheading 3.1. *See* Table 1 for parameters to be used in the case of variant analysis.
2. Transform competent cells with assembly product following instructions from Subheading 3.2 with the following exceptions: use 50 μl of competent cells only, and skip **steps 8** and **11–13**.
3. On the next day, pick at least 20 colonies and grow overnight in 3 ml LB-ampicillin (or LB-carbenicillin) in a shaking incubator (37 °C/220 rpm).
4. Isolate plasmidDNA using a miniprep kit according to kit instructions and quantitate. Store leftover cultures at 4 °C.
5. Digest 750 ng of each plasmid with PstI restriction enzyme according to the manufacturer's instructions.
6. Perform agarose gel electrophoresis of digested plasmids along a DNA ladder and verify the presence of the three expected bands (4,413, 2,300, and 1,463 bp).

7. Sequence all positive clones (those with the three expected bands with correct size).
8. Assemble both sequences for each clone using the overlap region and translate into a protein sequence.
9. Compare protein sequences and identify those that are most prevalent and that are not wild type.

3.10 In Vivo Evaluation

1. For each capsid variant that needs to be evaluated as well as for each control (wild-type AAV2 capsid gene, possibly others as well depending on the target tissue), purify plasmidDNA according to Subheading 3.3, starting from 1 L of overnight culture (*see* **Note 23**).
2. For each variant from the previous step (including controls), perform transfection according to Subheading 3.4 with the following exceptions: 20 tissue culture dishes are sufficient (**step 1**); three plasmids are used (instead of two) in equimolar amounts with a total of 70 μg per dish (**step 3**); the three plasmids are the pACG2 derivative containing the capsid gene variant (or the control), pHelper, and pTR-UF50-BC (*see* **Note 24**).
3. Purify all viral preparations according to Subheading 3.5.
4. Determine titer of each preparation according to Subheading 3.6.
5. Dilute all viral preparations except the one with the lowest titer to bring them to the same titer, using sterile Lactated Ringer's solution.
6. Inject at least three animals with the same amount of virus for each preparation.
7. Monitor luciferase activity over a period of several weeks by in vivo imaging following your institution's protocol. Briefly: inject intraperitoneally 150 μg of luciferin per gram of body weight and image at plateau phase (determined in advance by doing a kinetics curve) under anesthesia.

4 Notes

1. pSubEagApa is a pSub201 [15] derivative containing a deletion between ApaI sites at 3,764 and 4,049 and including an EagI site at position 4,373 (a silent mutation). It allows reconstitution of a full length *cap* gene after insertion, between the ApaI and EagI sites, of a capsid gene fragment that has been amplified using appropriate primers.
2. A commercial version of the isothermal DNA assembly [16] master mix is available from New England Biolabs under the name Gibson Assembly Master Mix.

3. A compromise needs to be achieved between high transformation efficiency (essential for the library complexity) and stability of the inverted terminal repeats (ITRs) which are crucial for AAV production. In many *E. coli* strains, ITRs are unstable and get deleted during culturing. Differences in both transformation efficiency and ITR stability can also exist between clones of the same strain. We suggest testing several clones by making electrocompetent cells from individual colonies and saving the best clone as a glycerol stock from which new electrocompetent cells are prepared as needed. In this case, we had better reproducibility by starting cultures directly from a few microliters of that glycerol stock rather than from an individual colony derived from the same stock. Here are examples of protocols on how to make *E. coli* cells electrocompetent:
http://mcb.berkeley.edu/labs/krantz/protocols/electrocomp_cells.pdf
http://openwetware.org/wiki/Knight:Preparing_electrocompetent_cells
http://2012.igem.org/wiki/images/5/57/Making_Ecoli_Electrocompetent.pdf
4. The vector pACG2-m56 is derived from pACG-2 [17] and has the same modifications as pSubEagApa (deletion between 2 ApaI sites and introduction of an EagI site allowing cloning of amplified capsid fragments between the ApaI and EagI sites).
5. pTR-UF50-BC (GenBank Accession #KF926476) is a luciferase-expressing barcoded vector that allows to evaluate the distribution of AAV variants by in vivo imaging.
6. This method is a modified version of isothermal DNA assembly [16], now also called Gibson assembly. It is particularly suited to assemble libraries because of its superior efficiency compared to traditional cloning. Typically, over 95 % of randomly selected clones are positive (correct insert size in the correct orientation). For convenience, it can also be used in cases in which such high efficiency is not needed, such as Subheading 3.9; in that case, reaction volumes as low as 10 μl can be used with reproducible results.
7. The total reaction size depends on the expected complexity. Depending on the method used to generate the capsid variations, plasmid assembly might be the limiting step, in which case a larger reaction volume will produce higher library complexity. For enriched libraries, the complexity is expected to decrease with each round of selection, therefore the reaction volume can be decreased.
8. Incubation time can vary, but yield will be higher with 2 h. If only a small number of clones are needed, 30 min is sufficient. However, for initial and first enriched library assemblies, 2 h are strongly recommended.

9. Transformation efficiency is usually higher with freshly made competent cells. For the initial library as well as for the first enriched library, we recommend making competent cells on the same day and keeping them on ice, so that they are ready no more than 1 h before transformation. For the subsequent enriched libraries, frozen competent cells can be used instead.
10. Optimal voltage may vary depending on electroporation instrument, type of cuvettes and batch of competent cells. Optimization is especially important for the initial library, for which transformation efficiency needs to be as high as possible. A lower voltage (e.g., 2.5 kV) will decrease the number of transformants, while a higher voltage (e.g., 3 kV) will increase the risk of arcing. Tests should be performed using small aliquots of the same batch of cells and the same model of cuvette.
11. An incubation at 30 °C is recommended in order to limit possible recombination of the ITRs (*see* **Note 3**). Note that at this stage a high yield is not needed. The culture volume is only related to the transformation size, which is dictated by a need for complexity.
12. From this step on, the tube containing plasmidDNA should only be opened under a sterile hood. Although previous steps were not performed under sterile conditions, the addition of 2.5 volumes of ethanol in **step 29** effectively sterilizes the content (after inverting the tube several times). As the plasmid is intended to transfect mammalian cells, potential contamination by airborne microorganisms is to be avoided.
13. Restriction analysis is a convenient way to assess the integrity of the ITRs, which are crucial to the effectiveness of the viral library production (both intact ITRs are needed for the viral DNA to be packaged into infectious virions). Each ITR contains two SmaI restriction sites. A complete digestion of the library plasmid with the SmaI restriction enzyme should generate two large fragments, 4,586 and 3,702 bp in size respectively (the two other fragments are only 11 bp long each, too small to be visible on a gel). If one of the two ITR has been deleted, the SmaI digestion will produce a 8.2 kb fragment instead. In practice, the presence of a fraction of plasmid molecules with rearranged ITRs is unavoidable, and is acceptable as long as it is not predominant. An example of SmaI digestion of a successful library plasmid maxiprep is shown in Fig. 3. In cases where the 4,586 and 3,702 bp bands are absent or of lower intensity than the 8.2 kb band, a new transformation is required, using a different strain of *E. coli*, or a different clone of the same strain, or a lower incubation temperature.
14. A crucial aspect of any library, for it to have any usefulness, is a strong physical coupling between each viral particle's

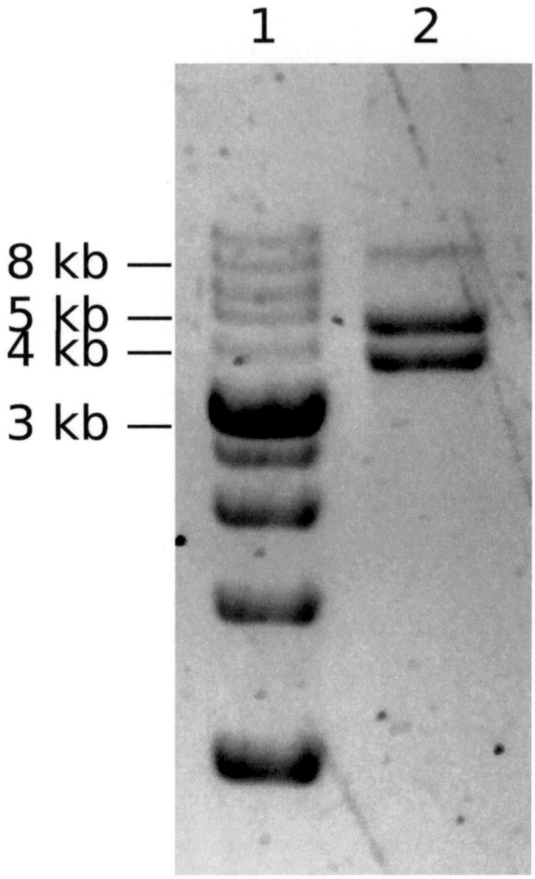

Fig. 3 Example of gel electrophoresis of a plasmid library restriction digest. 1: DNA ladder. 2: AAVcapsid library plasmid DNA digested with SmaI. The expected bands at 4,586 and 3,702 bp are clearly visible. A band corresponding to single-cut plasmid is also present but at lower intensity

phenotype (the capsid surface) and genotype (the viral genome encoding that particular capsid sequence). We assume this to be achieved by highly diluting the library plasmid with helper plasmid (molar ratio 1:5,000) during transfection. According to previous studies [18], no more than 5 % of total DNA used in transfection is taken up by cells, and only 10 % of that DNA reaches the nucleus where it can be transcribed. Under our experimental conditions (5.6E12 total plasmid molecules for 2.7E7 cells), it would mean that only 0.2 copy of the library plasmid reaches a cell nucleus on average. Even under the more optimistic assumptions of 50,000 plasmid molecules entering each cell [3], no more than 1 copy of the library plasmid would end up being transcribed per cell.

15. While the first centrifugation (at lower speed, **step 12**) aims at collecting intact cells for purification of intracellular AAV from

the cell pellet, the goal of the second centrifugation (high speed, **step 13**) is to clear the medium by removing particles that could interfere with efficient filtration (first step of virus purification from the culture medium). Two separate centrifugations are necessary because excessive speed might disrupt cells prematurely, while insufficient speed might not allow effective clarification.

16. Purifying AAV from both cells and culture medium allows not only a higher yield but also potentially higher complexity, compared with purifying from only cells or only medium. During AAV production in 293 cells, virions are released from cells and are re-entering cells depending on interactions with cell surface receptors [19]. As a result, the intracellular and extracellular distributions of capsid variants are likely different. The method described in Subheading 3.5 is a combination of AAV purification methods developed by Zolotukhin [20] and Vendenberghe [19].

17. Single-stranded DNA does not bind to the column as efficiently as double-stranded DNA unless a much larger volume of binding buffer is used. As the AAV genome is single-stranded, a large amount of binding buffer is crucial to recover viral DNA and allow accurate quantitation.

18. Dilutions should not be stored. Instead, they should be prepared when needed from a stock solution of linearized plasmid. The stock solution should be stored at –20 °C, and its concentration should be verified from time to time.

19. **Steps 6–22** of Subheading 3.7 are adapted from Arad's modified Hirt DNA extraction method [21].

20. Aliquots can be saved after **steps 20** and **22** and used as PCR template in the case the purified episomal DNA (at the end of **step 23**) fails to generate a PCR product.

21. If the negative control lane shows a band at the same position as the bands in the template lanes, identify the source of contamination (water, primers, dNTP, buffer, DNA polymerase) by performing PCR with no template using various combinations of new and previously used reagents. Once the contaminated reagent has been identified, replace it and repeat PCR. If no band of expected size (1,515 bp using primers described in **step 3** of Subheading 2.8) is present in template lanes, try more dilutions, different templates (*see* **Note 20**), different DNA polymerases and/or more PCR cycles. Proceed to **step 5** only if the negative control lane is empty and a band of expected size is present in a template lane. If a band of expected size is visible in lanes corresponding to more than one template dilution, select the one corresponding to the lowest dilution, as the PCR product will be more representative of the capsid variant population present in the original tissue sample.

22. The purpose of cloning the enriched library into pACG2-m56 is to have capsid variants ready to be assembled into transgenic vectors harboring reporter genes, in the case that sequencing analysis reveals the presence of enriched variants.
23. In the case of the selected capsid variants, leftover cultures from minipreps can be used to inoculate the 1 L cultures for the maxipreps. For the controls, if sufficient amount of purified plasmidDNA is not available, the culture can be started from a glycerol stock if available, or from a colony obtained by transforming competent cells with an aliquot of the plasmid. In all cases, the plasmid should include a complete functional copy of the same *rep* gene as well as the variant-specific *cap* gene, but no ITR.
24. As opposed to transfections during selection where the goal was to produce viral capsid libraries, in this case the goal is to make a large number of copies of a single AAV variant. Furthermore, the viral genome packaged in each virion is no longer an AAV genome encoding that particular virion's capsid. Instead, the viral genome (encoded in the pTR-UF50-BC plasmid) now encodes two reporter genes (firefly luciferase and mApple GFP derivative) between AAV2 ITRs. The capsid variant gene is expressed but not packaged, as pACG2-derived vectors do not have ITRs. Note that pTR-UF50-BC is available in 46 different versions, each with a different 6-nt barcode. If biodistribution study of viral genomes is intended, each variant can be produced with a different barcode and a mixture of all variants then injected to a small number of animals (instead of that small number of animal for each variant separately). The barcode region of the viral genomes can then be amplified and sequenced from different tissue samples while total AAV titer and concentration of genomic DNA are determined as well, allowing to calculate the prevalence of each variant in each tissue sample.

References

1. Wu Z, Asokan A, Samulski RJ (2006) Adeno-associated virus serotypes: vector toolkit for human gene therapy. Mol Ther 14:316–327
2. Perabo L, Endell J, King S et al (2006) Combinatorial engineering of a gene therapy vector: directed evolution of adeno-associated virus. J Gene Med 8:155–162
3. Maheshri N, Koerber JT, Kaspar BK et al (2006) Directed evolution of adeno-associated virus yields enhanced gene delivery vectors. Nat Biotechnol 24:198–204
4. Koerber JT, Maheshri N, Kaspar BK et al (2006) Construction of diverse adeno-associated viral libraries for directed evolution of enhanced gene delivery vehicles. Nat Protoc 1:701–706
5. Maguire CA, Gianni D, Meijer DH et al (2010) Directed evolution of adeno-associated virus for glioma cell transduction. J Neurooncol 96:337–347
6. Li W, Asokan A, Wu Z et al (2008) Engineering and selection of shuffled AAV genomes: a new strategy for producing targeted biological nanoparticles. Mol Ther 16:1252–1260
7. Grimm D, Lee JS, Wang L et al (2008) In vitro and in vivo gene therapy vector evolution via

multispecies interbreeding and retargeting of adeno-associated viruses. J Virol 82:5887–5911
8. Gray SJ, Blake BL, Criswell HE et al (2010) Directed evolution of a novel adeno-associated virus (AAV) vector that crosses the seizure-compromised blood-brain barrier (BBB). Mol Ther 18:570–578
9. Koerber JT, Jang J-H, Schaffer DV (2008) DNA shuffling of adeno-associated virus yields functionally diverse viral progeny. Mol Ther 16:1703–1709
10. Müller OJ, Kaul F, Weitzman MD et al (2003) Random peptide libraries displayed on adeno-associated virus to select for targeted gene therapy vectors. Nat Biotechnol 21:1040–1046
11. Michelfelder S, Lee M-K, deLima-Hahn E et al (2007) Vectors selected from adeno-associated viral display peptide libraries for leukemia cell-targeted cytotoxic gene therapy. Exp Hematol 35:1766–1776
12. Sellner L, Stiefelhagen M, Kleinschmidt JA et al (2008) Generation of efficient human blood progenitor-targeted recombinant adeno-associated viral vectors (AAV) by applying an AAV random peptide library on primary human hematopoietic progenitor cells. Exp Hematol 36:957–964
13. Naumer M, Ying Y, Michelfelder S et al (2012) Development and validation of novel AAV2 random libraries displaying peptides of diverse lengths and at diverse capsid positions. Hum Gene Ther 23:492–507
14. Marsic D, Govindasamy L, Currlin S et al (2014) Vector design tour de force: integrating combinatorial and rational approaches to derive novel adeno-associated virus (AAV) variants. Mol Ther 22:1900–9
15. Samulski RJ, Chang LS, Shenk T (1987) A recombinant plasmid from which an infectious adeno-associated virus genome can be excised in vitro and its use to study viral replication. J Virol 61:3096–3101
16. Gibson DG, Young L, Chuang R-Y et al (2009) Enzymatic assembly of DNA molecules up to several hundred kilobases. Nat Methods 6:343–345
17. Li J, Samulski RJ, Xiao X (1997) Role for highly regulated rep gene expression in adeno-associated virus vector production. J Virol 71:5236–5243
18. Batard P, Jordan M, Wurm F (2001) Transfer of high copy number plasmid into mammalian cells by calcium phosphate transfection. Gene 270:61–68
19. Vandenberghe LH, Xiao R, Lock M et al (2010) Efficient serotype-dependent release of functional vector into the culture medium during adeno-associated virus manufacturing. Hum Gene Ther 21:1251–1257
20. Zolotukhin S, Byrne BJ, Mason E et al (1999) Recombinant adeno-associated virus purification using novel methods improves infectious titer and yield. Gene Ther 6:973–985
21. Arad U (1998) Modified Hirt procedure for rapid purification of extrachromosomal DNA from mammalian cells. Biotechniques 24: 760–762

Chapter 12

Altering Entry Site Preference of Lentiviral Vectors into Neuronal Cells by Pseudotyping with Envelope Glycoproteins

Kenta Kobayashi, Shigeki Kato, Ken-ichi Inoue, Masahiko Takada, and Kazuto Kobayashi

Abstract

A lentiviral vector system provides a powerful strategy for gene therapy trials against a variety of neurological and neurodegenerative disorders. Pseudotyping of lentiviral vectors with different envelope glycoproteins not only confers the neurotropism to the vectors, but also alters the preference of sites of vector entry into neuronal cells. One major group of lentiviral vectors is a pseudotype with vesicular stomatitis virus glycoprotein (VSV-G) that enters preferentially cell body areas (somata/dendrites) of neurons and transduces them. Another group contains lentiviral vectors pseudotyped with fusion envelope glycoproteins composed of different sets of rabies virus glycoprotein and VSV-G segments that enter predominantly axon terminals of neurons and are transported through axons retrogradely to their cell bodies, resulting in enhanced retrograde gene transfer. This retrograde gene transfer takes a considerable advantage of delivering the transgene into neuronal cell bodies situated in regions distant from the injection site of the vectors. The rational use of these two vector groups characterized by different entry mechanisms will further extend the strategy for gene therapy of neurological and neurodegenerative disorders.

Key words Gene transfer, Lentiviral vector, Neuron, Entry site, Retrograde transport, Pseudotyping, Vesicular stomatitis virus glycoprotein, Rabies virus glycoprotein, Fusion glycoprotein

1 Introduction

A lentiviral vector system offers a useful approach for gene therapy trials against a variety of neurological and neurodegenerative disorders, such as Parkinson's disease and motor neuron diseases [1–3]. In particular, human immunodeficiency virus type 1 (HIV-1)-based lentiviral vectors have been applied extensively as gene therapy vehicles in animal models for these diseases [4–10]. The lentiviral vector enters the host cells through receptor-mediated endocytosis, and the viral genome is integrated into the host genome, resulting in stable and long-lasting expression of the transgene [11, 12]. The HIV-1-based vector consists of a replication-defective, self-inactivating

Fredric P. Manfredsson (ed.), *Gene Therapy for Neurological Disorders: Methods and Protocols*, Methods in Molecular Biology, vol. 1382, DOI 10.1007/978-1-4939-3271-9_12, © Springer Science+Business Media New York 2016

form of the virus, and the cis- and trans-acting elements required for production of functional vector particles are encoded separately by different plasmids [13–15]. The transfer plasmid encodes the viral genome including the transgene, whereas the two packaging plasmids contain the genes (*gag/pol* and *rev*) necessary for the vector packaging, reverse transcription, nuclear transport, and integration of viral DNA into the host genome. Another plasmid encodes the gene for viral envelope glycoprotein that coats the vector particles. The HIV-1 vector has a large packaging capacity consisting of ~8 kilobases in length [16].

Viral envelope glycoproteins interact with the receptor molecules situated on the host cell surface. The host range of lentiviral vectors is altered through pseudotyping with different types of envelope glycoproteins [17]. The majority of HIV-1 vectors is pseudotyped with vesicular stomatitis virus glycoprotein (VSV-G), which transduces a broad range of host cells [18, 19]. In the nervous system, the VSV-G-pseudotyped vector efficiently transduces neuronal cells [20–24]. Pseudotyping with envelope glycoproteins derived from Mokola virus, lymphocytic choriomeningitis virus, and Moloney murine leukemia virus also confers tropism to the nervous system [25, 26]. In addition, some viral vectors enter synaptic terminals of neuronal cells and are transported, through retrograde axonal transport, to their cell bodies situated in regions remote from the injection sites of the vectors [27–31]. This retrograde gene transfer confers great advantage as the transgene can be delivered into neuronal cell bodies that project to the injection sites. Pseudotyping of lentiviral vectors with rabies virus glycoprotein (RV-G) increases the efficiency of retrograde gene transfer [32–37], but the gene transfer by RV-G-pseudotyped vector is still less efficient. To resolve this issue, HIV-1-based vectors showing highly efficient retrograde gene transfer (HiRet) and neuron-specific retrograde gene transfer (NeuRet) have been developed by pseudotyping with novel fusion envelope glycoproteins that consist of different combinations of RV-G and VSV-G segments [38–42]. Both the HiRet and the NeuRet vectors display high efficiency of retrograde gene delivery into neuronal populations, although they differ in gene transduction pattern of the particular cell types surrounding the injection site.

In this chapter, we summarize the property of gene transduction of the representative lentiviral vector pseudotypes with neurotropism, and classify these vectors into two groups based on the preference of vector entry sites into neuronal cells, including (1) VSV-G-pseudotyped vector that preferentially enters cell body areas (somata/dendrites) of neurons and transduces them; and (2) HiRet/NeuRet vectors that predominantly enter axonal terminals of neurons, showing enhanced retrograde gene transfer (*see* Fig. 1). In addition, we describe the application of the HiRet/NeuRet vectors for retrograde gene delivery into the target neurons for gene therapy trials.

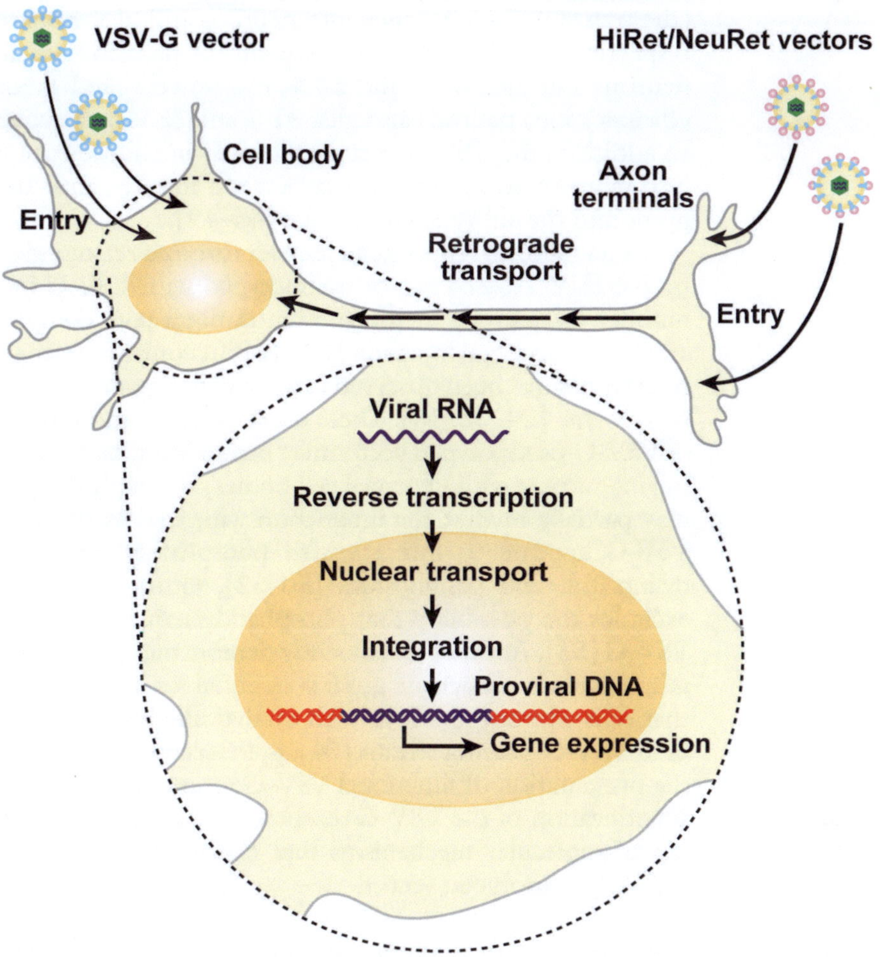

Fig. 1 Entry site preference of pseudotyped lentiviral vectors into neuronal cells. Lentiviral vector pseudotyped with VSV-G mainly enters the cell body areas of neurons, whereas the HiRet/NeuRet vectors with FuG-B/B2 or FuG-C enter nerve terminals and are transported within axons through retrograde transport to the cell body. Viral RNA is reverse-transcribed and the DNA intermediates are transported into the nucleus and integrated into the host genome, resulting in the expression of the transgene

2 Gene Transduction Property of Lentiviral Vector Pseudotypes

2.1 VSV-G-Pseudotyped Vectors

The pattern of gene transduction in the nervous system by VSV-G-pseudotyped HIV-1 lentiviral vector has been characterized in some animal species. Injection of the VSV-G-pseudotyped vector into the striatum in mice transduces both neuronal and glial cells around the injection site [43]. The VSV-G pseudotype preferentially transduces neuronal cells over glial cells in the mouse striatum, and the transduction efficiency into neurons and glia, defined as the percentage of the number of transduced neurons or glia divided by the total number of transduced cells, is 90 % and 7 %, respectively [36]. The preferential gene transduction of the VSV-G pseudotype into neuronal cells is observed in the brain stem of rats (the transduction efficiency into neurons and glia: 56 % and 26 %, respectively) [44] and in the striatum of rats (the efficiency into neurons and glia: 68 % and 22 %, respectively) and monkeys (the efficiency into neurons and glia: 51 % and 38 %, respectively) [45]. In addition, the VSV-G vector efficiently introduces the transgene into neural stem/progenitor cells when injected into the dentate gyrus and the subventricular zone [46–49].

A few studies report gene transfer through retrograde transport by the HIV-1-based vector pseudotyped with VSV-G into a small number of neurons in the substantia nigra pars compacta (SNc) after an intrastriatal injection [22, 45]. In contrast, retrograde gene transfer has not been observed in most experiments with the VSV-G pseudotype [24, 36, 43]. These observations suggest that the entry of VSV-G-pseudotyped vector may be mediated principally through the interaction with neuronal cell bodies, although the vector entry may partially involve the interaction with the dendritic structures. VSV-G appears to interact with phosphatidylserine, phosphatidylinositol, and gangliosides [50–52], although a recent study excludes the possibility that phosphatidylserine is the receptor for VSV-G [53]. Another recent study demonstrates that the endoplasmic reticulum chaperone gp96 is essential for infection of VSV and that cells without gp96, or with catalytically inactive gp96, do not bind VSV-G-bearing virions [54]. gp96 seems to be responsible for the presentation of functional VSV-G receptors on the cell surface. Identification of the VSV-G receptor will facilitate the understanding of molecular mechanisms that explain the localized entry of VSV-G-pseudotyped vector.

2.2 HiRet/NeuRet Vectors

The HiRet vector is a pseudotype of the HIV-1-based vector containing the fusion glycoprotein type B (FuG-B), which consists of the extracellular and transmembrane domains of RV-G (derived from the challenge virus standard strain) and the cytoplasmic domain of VSV-G (*see* Fig. 2) [38]. Injection of the HiRet vector into the dorsal striatum of mice produces efficient retrograde gene transfer into neuronal populations that innervate the striatum, such as the cerebral cortical areas, intralaminar thalamic nuclei, and ventral midbrain (SNc), showing a 8- to 14-fold greater efficiency in each brain region than that obtained after injection of the RV-G pseudotyped vector. Injection of the HiRet vector into the ventral striatum generates retrograde gene transfer into the piriform cortex, subiculum, basolateral nucleus of the amygdala, and lateral hypothalamus; and the vector injection into the medial prefrontal cortex leads to gene transduction of the cingulated cortex, hippocampus, and mediodorsal and ventromedial thalamic nuclei. Around the injection site in the dorsal striatum, the HiRet vector transduces a large number of astroglial cells (~75 %) and a much smaller number of neurons (~20 %), and this transduction pattern around the injection site is similar to that of the RV-G-pseudotyped vector. In

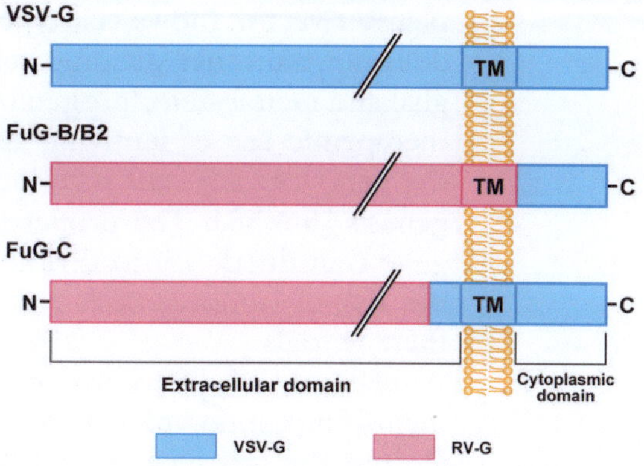

Fig. 2 Structure of viral envelope glycoproteins. The extracellular, transmembrane (TM), and cytoplasmic domains are shown. FuG-B/B2 consist of the extracellular and transmembrane domains of RV-G fused to the cytoplasmic domain of VSV-G. The amino acid sequences of RV-G segments in FuG-B and FuG-B2 are those of glycoproteins derived from the challenge virus standard and Pasteur virus strains, respectively. FuG-C contains the N-terminal region of the extracellular domain of RV-G and the membrane-proximal region and the transmembrane/cytoplasmic domains of VSV-G

addition, the RV-G sequence in FuG-B can be replaced with the sequence derived from another rabies virus strain: Pasteur virus (termed FuG-B2); a HiRet vector pseudotyped with FuG-B2 produces a greater level of retrograde gene transfer, showing an approximately 1.3-fold increase than the original FuG-B-pseudotyped vector [39].

The NeuRet vector is another pseudotype of the HIV-1-based vector showing high efficiency of retrograde gene transfer with fusion glycoprotein type C (FuG-C), which is composed of the N-terminal region of the extracellular domain (439 amino acids) of RV-G, and a short C-terminal region of the extracellular domain or membrane-proximal region (16 amino acids), and the transmembrane/cytoplasmic domains of VSV-G (*see* Fig. 2) [40]. Intrastriatal injection of the NeuRet vector results in efficient retrograde gene transfer into neuronal populations innervating the striatum in a similar fashion to the gene transfer by the HiRet vector. However, the efficiency of retrograde gene delivery into different neural pathways varies between these two vectors; for instance, the gene transfer into the corticostriatal pathways is relatively higher in the NeuRet vector than in the HiRet vector, whereas the transfer into the thalamostriatal pathway is the opposite [55]. More interestingly, the NeuRet vector transduces only a small number of neurons (~6 %) around the injections site, but the transduction of glial cells is less efficient (~0.3 %). Gene transfer into neural stem/progenitor cells by the NeuRet vector also shows a low level compared with VSV-G- or RV-G-pseudotyped vector [40]. Thus, the

NeuRet vector shows cell-type specificity for gene transfer and it does not transduce dividing cells in the nervous system, including glial and neural stem/progenitor cells. One significant issue on the therapeutic use of lentiviral vectors is transgene integration into the gene loci adjacent to cellular oncogenes that lead to tumorigenesis [56–58]. The property of NeuRet vector that suppresses gene transduction into dividing cells will be beneficial to reduce the risk of tumorigenesis and improve the safety of future gene therapy trials.

The use of different combinations of RV-G and VSV-G segments for pseudotyping affects the extent of retrograde gene transfer and the cell type specificity of gene transduction around the injection site. Pseudotyping with FuG-B/B2, as compared with RV-G, increases the transduction into cultured cells, whereas it does not appear to influence the yield of vector particles. This pseudotyping results in enhanced retrograde gene transfer into various neural pathways. Substitution of the cytoplasmic domain of RV-G with the corresponding part of VSV-G may change the interaction with host cells or the transduction level of the pseudotyped vector. Pseudotyping with FuG-C, in addition to its enhanced retrograde gene transfer, suppresses gene transduction into dividing cells around the injection site. These results suggest that the N-terminal region of the RV-G extracellular domain of 439 amino acids is responsible for retrograde gene transfer, probably by promoting the entry into axonal terminals of neuronal cells. Amino acid residues involved in rabies virus virulence are reported to be localized in this extracellular domain [59, 60]. In addition, the results of FuG-C pseudotyping suggest that the membrane-proximal region of the extracellular domain may be implicated in the determination of the cell-type specificity of gene transduction around the injection site.

The entry of the pseudotyped vectors into nerve terminals may be mediated via the receptor for RV-G. Previous studies suggest that RV-G interacts with highly sialylated gangliosides [61] and certain receptor proteins, including the nicotinic acetylcholine receptor (nAChR) α-subunit in the neuromuscular junction [62, 63], the low-affinity nerve growth factor receptor or p75NTR [64], and the neural cell adhesion molecule [65]. However, nAChR is located at the postsynaptic muscle membrane, but not at the presynaptic terminals, suggesting the presence of other receptor molecules that mediate viral entry into neuronal cells [66]. Furthermore, there is evidence that p75NTR is not necessary for rabies virus infection [67]. Identification of the RV-G receptor will promote the understanding of mechanisms that mediate the entry of the HiRet/NeuRet vectors into the nerve terminals.

3 Delivery of Transgene into Target Neurons Through Retrograde Gene Transfer

3.1 Gene Transfer into Nigrostriatal Dopamine Neurons

Parkinson's disease is the most common motor system disorder resulting from selective degeneration of nigrostriatal dopamine neurons. The HIV-1 vector pseudotyped with VSV-G has been used for gene therapy trials in rodent and nonhuman primate models to deliver transgenes required for survival and protection of dopamine neurons such as glial cell line-derived neurotrophic factor, neurturin, and parkin into the striatum or the SNc [4–10]. In this review, we describe a model experiment to evaluate the capability of the NeuRet vector for gene transfer via retrograde transport into the nigrostriatal dopamine system in nonhuman primates [40]. Injection of the NeuRet vector carrying the gene for green fluorescent protein (GFP) into the striatum (putamen) of crab-eating monkeys results in the occurrence of a large number of GFP-positive cells in the SNc, and transgene expression is observed in the majority of SNc dopamine neurons (~70 %) (Fig. 3). Thus, the NeuRet vector system makes it possible to efficiently introduce transgenes involved in neuronal survival and protection for Parkinson's disease therapy through retrograde gene transfer after intrastriatal injection of the vector.

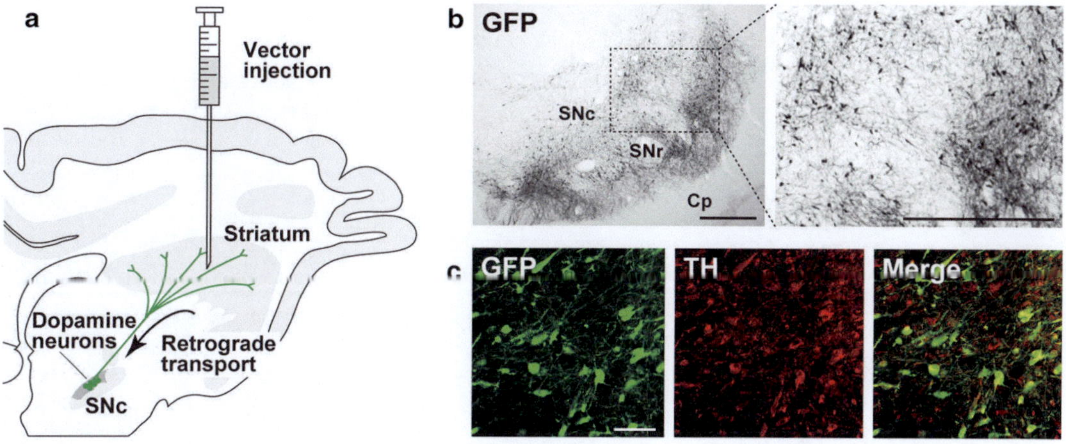

Fig. 3 Retrograde gene transfer into nigrostriatal dopamine neurons after intrastriatal injection. (**a**) Schematic illustration of intrastriatal injection of a lentiviral vector for retrograde gene transfer in the macaque monkeys. The NeuRet vector encoding the GFP transgene (1.2×10^{10} copies/ml, 5.0 μl × 10 sites) was injected into the striatum (putamen) of the crab-eating monkeys. (**b**) GFP immunohistochemistry. After 4 weeks, sections through the SNc were prepared, and stained by GFP immunohistochemistry. A number of GFP-positive neurons are visualized in the SNc of the injected monkeys. *Cp* cerebral peduncle, *SNr* substantia nigra pars reticulata. (**c**) Double immunohistochemistry for GFP and tyrosine hydroxylase (TH). The majority of GFP-positive neurons exhibited TH immunoreactivity, indicating efficient retrograde gene transfer into the nigrostriatal dopamine neurons. Scale bars: 500 μm (**b**) and 50 μm (**c**). (The data are modified from ref. [40])

3.2 Gene Delivery into Motor Neurons

Motor neuron diseases, including amyotrophic lateral sclerosis and spinal muscular atrophy, are characterized by progressive muscle weakness and paralysis resulting from degeneration of motor neurons in the spinal cord and brain. Intramuscular injection of an adenoviral vector encoding genes required for neuronal survival and protection, such as brain-derived neurotrophic factor, glial cell line-derived neurotrophic factor, and neuronal apoptosis inhibitory protein, prevents motor neuron death in axotomy-induced injury models [27–31]. Adeno-associated virusserotype 6 and 9 vectors also deliver desired transgenes into motor neurons following an intramuscular injection [68–70]. Recently, we reported the gene transfer capability of the HiRet vector into motor neurons in rodents via retrograde transport [71]. Injection of the HiRet vector (pseudotyped with FuG-B2) encoding GFP into the hindlimb muscles in mice leads to the transduction of motor neurons in the spinal cord at the lumbar level (Fig. 4). The gene transfer efficiency of the HiRet vector is much higher than that of the RV-G-pseudotyped vector, showing a 4.6-fold increase. Injection of the HiRet vector into the tongue muscle also produces a large number of GFP-positive cells in the hypoglossal nucleus, showing a 14.8-fold increase in efficiency compared with the RV-G vector. The efficiency of retrograde gene transfer of the NeuRet vector into motor neurons is not comparable to that of the HiRet vector.

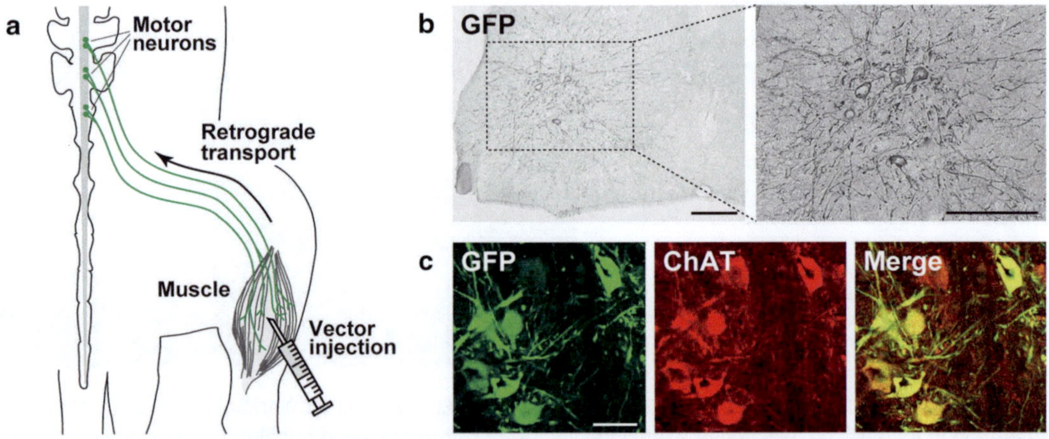

Fig. 4 Retrograde gene transfer into spinal cord motor neurons after intramuscular injection. (**a**) Schematic illustration of intramuscular injection of a lentiviral vector for retrograde gene transfer in mice. The HiRet vector (FuG-B2 pseudotype) encoding the GFP transgene (5.0×10^{11} copies/ml, 5.0 μl × 6 sites) was injected into the hindlimb muscle of mice. (**b**) GFP immunohistochemistry. After 4 weeks, sections through the lumbar spinal cords were prepared, and stained by GFP immunohistochemistry. GFP-positive neurons are seen in the ventral horn of the spinal cord. (**c**) Double immunohistochemistry for GFP and choline acetyltransferase (ChAT). GFP-positive neurons express ChAT showing the retrograde gene transfer into spinal cord motor neurons. Scale bars: 200 μm (**b**) and 50 μm (**c**). (The data are modified from ref. [71])

Therefore, the HiRet vector system provides a useful approach for efficient introduction of the genes involved in the survival of motor neurons through retrograde gene transfer.

4 Concluding Remarks

Pseudotyping with different viral envelope glycoproteins not only confers neurotropism to lentiviral vectors, but also converts the preference of vector entry sites into neuronal cells. The pseudotyped HIV-1 vectors with neurotropism are distinguishable based on the vector entry site preference: One group is the VSV-G-pseudotyped vector that preferentially enters cell body areas (somata/dendrites) of neurons and transduces them, and another group contains the HiRet/NeuRet vectors pseudotyped with FuG-B/B2 or FuG-C that predominantly enter axon terminals of neurons and are transported retrogradely into their cell bodies, displaying enhanced retrograde gene delivery. In addition to the general use of the VSV-G-pseudotyped vector, the application of our NeuRet vector results in gene transfer into a large number of nigrostriatal dopamine neurons that are the major target for gene therapy of Parkinson's disease. Furthermore, intramuscular injection of the HiRet vector achieves gene transfer into motor neurons in the brain and spinal cord that are the target for gene therapy of motor neuron diseases. Recently, the efficiency of retrograde gene transfer by the NeuRet vector was further improved by optimizing the junction of RV-G and VSV-G segments in the membrane-proximal region of fusion envelope glycoproteins [72]. A better understanding of the molecular mechanisms underlying vector entry into neuronal cells will contribute to the continuous development of genetic therapeutic approaches for the treatment of intractable neural diseases.

References

1. Azzouz M, Kingsman SM, Mazarakis ND (2004) Lentiviral vectors for treating and modeling human CNS disorders. J Gene Med 6:951–962
2. Wong LF, Goodhead L, Prat C et al (2006) Lentivirus-mediated gene transfer to the central nervous system: therapeutic and research applications. Hum Gene Ther 17:1–9
3. Lundberg C, Björklund T, Carlsson T et al (2008) Applications of lentiviral vectors for biology and gene therapy of neurological disorders. Curr Gene Ther 8:461–473
4. Kordower JH, Emborg ME, Bloch J et al (2000) Neurodegeneration prevented by lentiviral vector delivery of GDNF in primate models of Parkinson's disease. Science 290:767–773
5. Palfi S, Leventhal L, Chu Y et al (2002) Lentivirally delivered glial cell line-derived neurotrophic factor increases the number of striatal dopaminergic neurons in primate models of nigrostriatal degeneration. J Neurosci 22:4942–4954
6. Georgievska B, Jakobsson J, Persson E et al (2004) Regulated delivery of glial cell line-derived neurotrophic factor into rat striatum, using a tetracycline-dependent lentiviral vector. Hum Gene Ther 15:934–944

7. Lo Bianco C, Schneider BL, Bauer M et al (2004) Lentiviral vector delivery of parkin prevents dopaminergic degeneration in an α-synuclein rat model of Parkinson's disease. Proc Natl Acad Sci U S A 101:17510–17515
8. Fjord-Larsen L, Johansen JL, Kusk P et al (2005) Efficient in vivo protection of nigral dopaminergic neurons by lentiviral gene transfer of a modified Neurturin construct. Exp Neurol 195:49–60
9. Vercammen L, Van der Perren A, Vaudano E et al (2006) Parkin protects against neurotoxicity in the 6-hydroxydopamine rat model for Parkinson's disease. Mol Ther 14: 716–723
10. McCoy MK, Ruhn KA, Martinez TN et al (2008) Intranigral lentiviral delivery of dominant-negative TNF attenuates neurodegeneration and behavioral deficits in hemiparkinsonian rats. Mol Ther 16:1572–1579
11. Kafri T (2004) Gene delivery by lentivirus vectors an overview. Methods Mol Biol 246:367–390
12. Cockrell AS, Kafri T (2007) Gene delivery by lentivirus vectors. Mol Biotechnol 36: 184–204
13. Dull T, Zufferey R, Kelly M et al (1998) A third-generation lentivirus vector with a conditional packaging system. J Virol 72: 8463–8471
14. Miyoshi H, Blömer U, Takahashi M et al (1998) Development of a self-inactivating lentivirus vector. J Virol 72:8150–8157
15. Zufferey R, Dull T, Mandel RJ et al (1998) Self-inactivating lentivirus vector for safe and efficient in vivo gene delivery. J Virol 72: 9873–9880
16. Thomas CE, Ehrhardt A, Kay MA (2003) Progress and problems with the use of viral vectors for gene therapy. Nat Rev Genet 4: 346–358
17. Cronin J, Zhang XY, Reiser J (2005) Altering the tropism of lentiviral vectors through pseudotyping. Curr Gene Ther 5:387–398
18. Akkina RK, Walton RM, Chen ML et al (1996) High-efficiency gene transfer into CD34+ cells with a human immunodeficiency virus type 1-based retroviral vector pseudotyped with vesicular stomatitis virus envelope glycoprotein G. J Virol 70:2581–2585
19. Hanawa H, Kelly PF, Nathwani AC et al (2002) Comparison of various envelope proteins for their ability to pseudotype lentiviral vectors and transduce primitive hematopoietic cells from human blood. Mol Ther 5:242–251
20. Naldini L, Blömer U, Gallay P et al (1996) In vivo gene delivery and stable transduction of nondividing cells by a lentiviral vector. Science 272:263–267
21. Naldini L, Blömer U, Gage FH et al (1996) Efficient transfer, integration, and sustained long-term expression of the transgene in adult rat brains injected with a lentiviral vector. Proc Natl Acad Sci U S A 93:11382–11388
22. Blömer U, Naldini L, Kafri T et al (1997) Highly efficient and sustained gene transfer in adult neurons with a lentivirus vector. J Virol 71:6641–6649
23. Mochizuki H, Schwartz JP, Tanaka K et al (1998) High-titer human immunodeficiency virus type 1-based vector systems for gene delivery into nondividing cells. J Virol 72:8873–8883
24. Kordower JH, Bloch J, Ma SY et al (1999) Lentiviral gene transfer to the nonhuman primate brain. Exp Neurol 160:1–16
25. Watson DJ, Kobinger GP, Passini MA et al (2002) Targeted transduction patterns in the mouse brain by lentivirus vectors pseudotyped with VSV, Ebola, Mokola, LCMV, or MuLV envelope proteins. Mol Ther 5:528–537
26. Cannon JR, Sew T, Montero L et al (2011) Pseudotype-dependent lentiviral transduction of astrocytes or neurons in the rat substantia nigra. Exp Neurol 228:41–52
27. Gravel C, Götz R, Lorrain A et al (1997) Adenoviral gene transfer of ciliary neurotrophic factor and brain-derived neurotrophic factor leads to long-term survival of axotomized motor neurons. Nat Med 3:765–770
28. Baumgartner BJ, Shine HD (1998) Permanent rescue of lesioned neonatal motoneurons and enhanced axonal regeneration by adenovirus-mediated expression of glial cell-line-derived neurotrophic factor. J Neurosci Res 54: 766–777
29. Perrelet D, Ferri A, MacKenzie AE et al (2000) IAP family proteins delay motoneuron cell death *in vivo*. Eur J Neurosci 12:2059–2067
30. Sakamoto T, Kawazoe Y, Shen JS et al (2003) Adenoviral gene transfer of GDNF, BDNF and TGF beta 2, but not CNTF, cardiotrophin-1 or IGF1, protects injured adult motoneurons after facial nerve avulsion. J Neurosci Res 72:54–64
31. Barkats M, Horellou P, Colin P et al (2006) 1-methyl-4-phenylpyridinium neurotoxicity is attenuated by adenoviral gene transfer of human Cu/Zn superoxide dismutase. J Neurosci Res 83:233–242
32. Mazarakis ND, Azzouz M, Rohll JB et al (2001) Rabies virus glycoprotein pseudotyping of lentiviral vectors enables retrograde axonal transport and access to the nervous system after

peripheral delivery. Hum Mol Genet 10: 2109–2121
33. Azzouz M, Ralph GS, Storkebaum E et al (2004) VEGF delivery with retrogradely transported lentivector prolongs survival in a mouse ALS model. Nature 429:413–417
34. Wong LF, Azzouz M, Walmsley LE et al (2004) Transduction patterns of pseudotyped lentiviral vectors in the nervous system. Mol Ther 9:101–111
35. Mentis GZ, Gravell M, Hamilton R et al (2006) Transduction of motor neurons and muscle fibers by intramuscular injection of HIV-1-based vectors pseudotyped with select rabies virus glycoproteins. J Neurosci Methods 157:208–217
36. Kato S, Inoue K, Kobayashi K et al (2007) Efficient gene transfer via retrograde transport in rodent and primate brains using a human immunodeficiency virus type 1-based vector pseudotyped with rabies virus glycoprotein. Hum Gene Ther 18:1141–1151
37. Federici T, Kutner R, Zhang XY et al (2009) Comparative analysis of HIV-1-based lentiviral vectors bearing lyssavirus glycoproteins for neuronal gene transfer. Genet Vaccines Ther 7:1
38. Kato S, Kobayashi K, Inoue K et al (2011) A lentiviral strategy for highly efficient retrograde gene transfer by pseudotyping with fusion envelope glycoprotein. Hum Gene Ther 22:197–206
39. Kato S, Kuramochi M, Kobayashi K et al (2011) Selective neural pathway targeting reveals key roles of thalamostriatal projection in the control of visual discrimination. J Neurosci 31:17169–17179
40. Kato S, Kuramochi M, Takasumi K et al (2011) Neuron-specific gene transfer through retrograde transport of lentiviral vector pseudotyped with a novel type of fusion envelope glycoprotein. Hum Gene Ther 22:1511–1523
41. Kato S, Kobayashi K, Kuramochi M et al (2011) Highly efficient retrograde gene transfer for genetic treatment of neurological diseases. In: Xu K (ed) Viral gene therapy, chapter 17. InTech, Rijeka, pp 371–380
42. Kato S, Kobayashi K, Inoue K et al (2013) Vectors for highly efficient and neuron-specific retrograde gene transfer for gene therapy of neurological diseases. In: Molina FM (ed) Gene therapy—tools and potential applications, chapter 15. InTech, Rijeka, pp 387–398
43. Baekelandt V, Claeys A, Eggermont K et al (2002) Characterization of lentiviral vector-mediated gene transfer in adult mouse brain. Hum Gene Ther 13:841–853
44. Duale H, Kasparov S, Paton JF et al (2004) Differences in transductional tropism of adenoviral and lentiviral vectors in the rat brainstem. Exp Physiol 90:71–78
45. Kitagawa R, Miyachi S, Hanawa H et al (2007) Differential characteristics of HIV-based versus SIV-based lentiviral vector systems: gene delivery to neurons and axonal transport of expressed gene. Neurosci Res 57:550–558
46. Englund U, Fricker-Gates RA, Lundberg C et al (2002) Transplantation of human neural progenitor cells into the neonatal rat brain: extensive migration and differentiation with long-distance axonal projections. Exp Neurol 173:1–21
47. Consiglio A, Gritti A, Dolcetta D et al (2004) Robust *in vivo* gene transfer into adult mammalian neural stem cells by lentiviral vectors. Proc Natl Acad Sci U S A 101:14835–14840
48. Geraerts M, Eggermont K, Hernandez-Acosta P et al (2006) Lentiviral vectors mediate efficient and stable gene transfer in adult neural stem cells in vivo. Hum Gene Ther 17: 635–650
49. Capowski EE, Schneider BL, Ebert AD et al (2007) Lentiviral vector-mediated genetic modification of human neural progenitor cells for ex vivo gene therapy. J Neurosci Methods 163:338–349
50. Schlegel R, Tralka TS, Willingham MC et al (1983) Inhibition of VSV binding and infectivity by phosphatidylserine: is phosphatidylserine a VSV-binding site? Cell 32:639–646
51. Sinibaldi L, Goldoni P, Seganti L et al (1985) Gangliosides in early interactions between vesicular stomatitis virus and CER cells. Microbiologica 8:355–365
52. Mastromarino P, Conti C, Goldoni P et al (1987) Characterization of membrane components of the erythrocyte involved in vesicular stomatitis virus attachment and fusion at acidic pH. J Gen Virol 68:2359–2369
53. Coil DA, Miller AD (2004) Phosphatidylserine is not the cell surface receptor for vesicular stomatitis virus. J Virol 78:10920–10926
54. Bloor S, Maelfait J, Krumbach R et al (2010) Endoplasmic reticulum chaperone gp96 is essential for infection with vesicular stomatitis virus. Proc Natl Acad Sci U S A 107: 6970–6975
55. Kato S, Kobayashi K, Kobayashi K (2013) Dissecting circuit mechanisms by genetic manipulation of specific neural pathways. Rev Neurosci 24:1–8
56. De Palma M, Montini E, Santoni de Sio FR et al (2005) Promoter trapping reveals significant differences in integration site selection

between MLV and HIV vectors in primary hematopoietic cells. Blood 105:2307–2315
57. Themis M, Waddington SN, Schmidt M et al (2005) Oncogenesis following delivery of a nonprimate lentiviral gene therapy vector to fetal and neonatal mice. Mol Ther 12:763–771
58. Montini E, Cesana D, Schmidt M et al (2006) Hematopoietic stem cell gene transfer in a tumor-prone mouse model uncovers low genotoxicity of lentiviral vector integration. Nat Biotechnol 24:687–696
59. Prehaud C, Coulon P, LaFay F et al (1988) Antigenic site II of the rabies virus glycoprotein: structure and role in viral virulence. J Virol 62:1–7
60. Coulon P, Ternaux JP, Flamand A et al (1998) An avirulent mutant of rabies virus is unable to infect motoneurons in vivo and in vitro. J Virol 72:273–278
61. Superti F, Seganti L, Tsiang H et al (1986) Role of phospholipids in rhabdovirus attachment to CER cells. Brief report. Arch Virol 81:321–328
62. Hanham CA, Zhao F, Tignor GH (1993) Evidence from the anti-idiotypic network that the acetylcholine receptor is a rabies virus receptor. J Virol 67:530–542
63. Gastka M, Horvath J, Lentz TL (1996) Rabies virus binding to the nicotinic acetylcholine receptor α subunit demonstrated by virus overlay protein binding assay. J Gen Virol 77: 2437–2440
64. Thoulouze MI, Lafage M, Schachner M et al (1998) The neural cell adhesion molecule is a receptor for rabies virus. J Virol 72: 7181–7190
65. Tuffereau C, Bénéjean J, Blondel D et al (1998) Low-affinity nerve-growth factor receptor (P75NTR) can serve as a receptor for rabies virus. EMBO J 17:7250–7259
66. Albertini AA, Baquero E, Ferlin A et al (2012) Molecular and cellular aspects of rhabdovirus entry. Viruses 4:117–139
67. Tuffereau C, Schmidt K, Langevin C et al (2007) The rabies virus glycoprotein receptor p75NTR is not essential for rabies virus infection. J Virol 81:13622–13630
68. Towne C, Schneider BL, Kieran D et al (2010) Efficient transduction of non-human primate motor neurons after intramuscular delivery of recombinant AAV serotype 6. Gene Ther 17: 141–146
69. Towne C, Setola V, Schneider BL et al (2011) Neuroprotection by gene therapy targeting mutant SOD1 in individual pools of motor neurons does not translate into therapeutic benefit in fALS mice. Mol Ther 19:274–283
70. ElMallah MK, Falk DJ, Lane MA et al (2012) Retrograde gene delivery to hypoglossal motoneurons using adeno-associated virus serotype 9. Hum Gene Ther Methods 23:148–156
71. Hirano M, Kato S, Kobayashi K et al (2013) Highly efficient retrograde gene transfer into motor neurons by a lentiviral vector pseudotyped with fusion glycoprotein. PLoS One 8, e75896
72. Kato S, Kobayashi K, Kobayashi K (2014) Improved transduction efficiency of a lentiviral vector for neuron-specific retrograde gene transfer by optimizing the junction of fusion envelope glycoprotein. J Neurosci Methods 227:151–158

Chapter 13

Directed Evolution of Adenoviruses

Jason G. Smith

Abstract

The ability to evolve viruses in cell culture in the face of selective pressure is an invaluable method to elucidate the molecular mechanisms of synthetic or natural antivirals, expand tropism, or alter virulence. Recently, mutations to the human adenovirus polymerase that reduce replicative fidelity were described, and we have incorporated one of these mutations into the *pol* gene of a conditionally replicating human adenovirus serotype 5 (HAdV-5)-based vector (Uil et al., Nucleic Acids Res 39:e30, 2011; Myers et al., J Virol 87:6047–6050, 2013). Here, we describe methods to evolve this virus (HAdV-5.polF421Y) under selective pressure from antivirals to identify their mechanisms of action.

Key words Adenovirus, Mutator, Evolution, Antiviral

1 Introduction

The in vitro evolution of viruses has been used as a powerful method for many aspects of virology. This approach is particularly suited to RNA viruses, capitalizing on the inherently high mutational rate of RNA polymerases. Although this approach has been used successfully with DNA viruses including herpesviruses, the adenovirus polymerase has a prohibitively high fidelity ill suited to this method [1]. We recently described the creation of a mutator adenovirus through the introduction of previously identified point mutations in the *pol* gene of a HAdV-5-based vector [2]. This protocol describes in practical detail the process through which diversity in the starting population of this vector is generated, the process by which the virus is evolved under selective pressure, and analysis and validation of the resulting evolved viruses. This protocol presupposes understanding and availability of mammalian cell culture, end point PCR, and adenovirus purification by ultracentrifugation through cesium chloride gradients (*see* Chapter 9). Methods to sequence the viral

Fredric P. Manfredsson (ed.), *Gene Therapy for Neurological Disorders: Methods and Protocols*, Methods in Molecular Biology, vol. 1382, DOI 10.1007/978-1-4939-3271-9_13, © Springer Science+Business Media New York 2016

genome and to evaluate the contribution of genetic changes to antiviral resistance are discussed but not described in detail. Although focused on antivirals, this general strategy could be extended to experiments to alter the host range or tropism of adenovirus or to uncover viral protein functions. In addition, mutator viruses of alternative adenovirus serotypes could be engineered.

2 Materials

1. 293 cells (ATCC, Manassas, VA, USA): More adhesive 293 variants have been described (e.g., 293β5), which facilitate the use of these cells for adenovirus studies [3].
2. HAdV-5.polF421Y: An E1/E3-deleted HAdV-5-based vector encoding enhanced green fluorescent protein (eGFP) driven by a cytomegalovirus promoter in E1. In addition, a single amino acid change, F421Y, has been introduced in polymerase to reduce replication fidelity (*see* **Note 1**). Available from the Smith laboratory [2].
3. Complete DMEM: DMEM supplemented with 10 % fetal bovine serum, 100 units/mL penicillin, 100 μg/mL streptomycin, 4 mM l-glutamine, and 0.1 mM nonessential amino acids.
4. Liquid nitrogen (LN_2).
5. T162 or similarly sized cell culture flasks. HYPER*Flasks* (Corning, Corning, NY, USA) are a space saving alternative.
6. Proteinase K solution: 2 mg/mL proteinase K, 1 mM EDTA, 0.5 % SDS, 50 mM Tris–HCl pH 7.4.
7. Reagents for phenol/chloroform extraction: phenol/chloroform/isoamyl alcohol, chloroform, TE (10 mM Tris–HCl, 1 mM EDTA, pH 8.0).
8. Oligonucleotides: F421YsequF (5′-TTTGGAACGCGGATTTGGC-3′).

   F421YsequR (5′-GCTTCCAACCCTCATCTTGCAG-3′).
9. PCR reagents: Any commercially available high-fidelity PCR kit such as Roche Expand High Fidelity PCR System (Roche, Basel, Switzerland) or equivalent.
10. Flow cytometer with analysis software, Typhoon 9400, 9410, or Trio (GE Healthcare, Little Chalfont, UK) or equivalent variable mode imager with ImageJ software.
11. Prism software (GraphPad Software, Inc., La Jolla, CA, USA) for nonlinear regression analysis.

3 Methods

Carry out all open procedures with cell lines and infectious virus in a biosafety cabinet with sterile technique under BSL-2 guidelines.

3.1 Generating Diversity in the Starting Population

1. The original HAdV-5.polF421Y virus was created from a bacterial artificial chromosome (BAC) containing adenoviral genomic DNA into which the F421Y mutation was introduced by recombineering [2]. DNA from a single BAC clone was transfected into 293 cells and minimally expanded. This section describes serial passage of this relatively homogeneous virus stock to generate diversity in the starting viral population that will be used for selection.
2. Culture 293 cells in a T162 flask to confluency, approximately 2×10^7 cells/flask, in 25 mL complete DMEM.
3. In a 50 mL conical tube, dilute 2×10^6 infectious units of purified HAdV-5.polF421Y virus in 25 mL of fresh, warm (25–37 °C) complete DMEM.
4. Add diluted virus to the T162 flask of 293 cells to achieve a multiplicity of infection (MOI) of 0.1 and incubate the infected flask at 37 °C, 5 % CO_2 until most of the cells exhibit cytopathic effect (CPE). CPE manifests as rounding of the cells and partial or complete detachment from the flask and will generally progress to completion at 48–54 h post-infection (p.i.).
5. Keeping the flask closed and held horizontally, bang the side of the flask several times with your hand to dislodge the cells. Excessive force should not be required. Collect the dislodged cells in their media and transfer to a 50 mL conical tube.
6. Centrifuge at 335 × *g* for 5 min at 4 °C.
7. Reserve 2 mL of the supernatant for **step 9**. Divide the remaining supernatant into two equal aliquots in 15 mL conical tubes. Resuspend the cell pellet in 0.5 mL of PBS or supernatant. Transfer the resuspended cell pellet to a sterile Eppendorf tube.
8. Snap freeze both aliquots of supernatant and the resuspended cell pellet in LN_2 and store at −80 °C as P1 (passage 1).
9. Dilute the 2 mL of supernatant from **step 7** into 23 mL of fresh, warm complete DMEM (*see* **Note 2**).
10. Add diluted virus to a new, confluent T162 flask of 293 cells and incubate at 37 °C, 5 % CO_2 until complete CPE, as in **step 4**. This will take approximately 2–4 days (*see* **Note 3**).
11. Perform **steps 5** through **10**, increasing the passage number by 1 through each round of infection, until P9 (or greater) is reached (*see* **Note 4**).

12. To create a stock of virus for CsCl purification, dilute 20 mL of supernatant into a volume of complete DMEM sufficient to infect 10–20 T162 flasks (20 mL per flask). Alternatively, a HYPER*Flask* (Corning) requires a total of 550 mL of diluted virus and has a surface area equivalent to ten T175 flasks. Complete CPE should be observed 2–4 days p.i., and virus can be purified from the lysate following protocols in Chapter 9. Snap freeze purified virus in small aliquots (10–25 μL) in LN_2 and store at −80 °C.

3.2 Confirm Maintenance of the F421Y Mutation

1. Once a viral stock for selection is generated, it is important to verify that the F421Y mutation has not reverted. This step can also be performed periodically throughout the selection process (during Subheading 3.5, *see* **Note 5**).
2. Dilute 10 μL of purified virus to a final volume of 25 μL with ddH_2O.
3. Add 25 μL proteinase K solution and incubate at 37 °C for 1–2 h.
4. Add 50 μL TE, extract with 100 μL phenol/chloroform/isoamyl alcohol, and then extract the aqueous layer with 100 μL chloroform.
5. A 300 bp PCR product from 1 to 2 μL of the extracted DNA can be generated from oligonucleotides F421YsequF and F421YsequR using a high-fidelity polymerase. The reaction Tm is 53 °C.
6. Once gel purified, the PCR product can be sequenced using F421YsequR to verify that the F421Y mutation is still present in the viral population (Fig. 1).

3.3 Optimize Infection Conditions for the Selection Assay

1. We have performed selection using a tissue culture infectious dose 80 % ($TCID_{80}$) of virus. Thus, this section identifies the dilution factor of a stock concentration of virus needed to achieve 80 % infection of cells. We then try to maintain ~90 % inhibition for each round of selection (Subheading 3.5) using inhibitor concentrations determined in Subheading 3.4.
2. Culture 293 cells to confluency in 12-well culture dishes.

GATCTCGTC**GTA**GCCGTTGAT
CTAGAGCAG**CAT**CGGCAACTA
IleGluAsp<u>Tyr</u>GlyAsnIle

Fig. 1 The viral genomic sequence immediately surrounding the F421Y mutation (*bold*) is indicated. Note that the *pol* gene is on the complementary strand (translated) as the adenovirus genome is conventionally depicted

3. Dilute 5 μL purified HAdV-5.polF421Y from Subheading 3.1, **step 12**, into 1.5 mL complete DMEM.
4. Create ten serial threefold dilutions of this virus by adding 0.5 mL of the previous dilution to 1.0 mL complete DMEM and mixing thoroughly by gentle vortexing.
5. Remove media from each well of 293 cells and replace with 1 mL/well serially diluted virus. Replace one well with fresh complete DMEM without virus as a negative control. Incubate at 37 °C, 5 % CO_2.
6. Quantitate eGFP expression 18–24 h p.i. by using flow cytometry to determine the percent of eGFP-positive cells in each well compared to uninfected cells or by gently washing the 12-well plate with PBS, adding 1 mL PBS/well, and scanning the plate with a 488 nm laser using the 520 bp 40 filter on a Typhoon 9400, 9410, or Trio variable mode imager. For Typhoon images, we use ImageJ software to quantify the integrated density of relative fluorescence units (RFUs) for each well above background fluorescence in the uninfected control well.
7. Plot either the percent positive cells from flow cytometry or RFUs from the Typhoon scan against a log transformation of the dilution factor for each well.
8. Perform nonlinear regression analysis using Prism software using the function "log(agonist) vs. response—Find ECanything," constraining "F" to 80. The resulting ECF value will be the dilution of the starting virus stock that will yield 80 % maximal infection of the culture (*see* **Note 6**).

3.4 Optimize Inhibitor Concentration for the Selection Assay

1. This assay determines an inhibitor concentration that results in 90 % inhibition of infection (IC_{90}).
2. Culture 293 cells to confluency in 12-well culture dishes.
3. Dilute purified HAdV-5.polF421Y from Subheading 3.1 **step 12** into 6 mL complete DMEM by a dilution factor equal to half of the value of ECF obtained in Subheading 3.3 **step 8**. This will yield a 2× virus stock.
4. Create ten serial threefold dilutions of the inhibitor by first diluting it into a total of 0.75 mL complete DMEM and then adding 0.25 mL of the previous dilution to 0.5 mL complete DMEM and mixing thoroughly by gentle vortexing.
5. Mix 0.5 mL of each inhibitor dilution with 0.5 mL 2× virus stock (*see* **Note 7**).
6. Remove media from each well of 293 cells and replace with 1 mL/well virus plus serially diluted inhibitor. Replace one well with fresh complete DMEM without virus as a negative control. Replace one well with 2× virus diluted to 1× with complete DMEM alone as a positive control. Incubate at 37 °C, 5 % CO_2.

7. Quantitate eGFP expression 18–24 h p.i as in Subheading 3.3 **step 6**.
8. Plot either the percent positive cells from flow cytometry or RFUs from the Typhoon scan against a log transformation of inhibitor concentration for each well.
9. Perform nonlinear regression analysis using Prism software as in Subheading 3.3, **step 8**, constraining "F" to 10. The resulting ECF value will be the IC_{90} of the inhibitor.

3.5 Passaging Virus under Selective Pressure by Bulk Selection (*See* Note 8)

1. Culture 293 cells to confluency in two 12-well culture dishes.
2. Dilute purified HAdV-5.polF421Y from Subheading 3.1, **step 12**, into 2 mL complete DMEM by a dilution factor equal to half of the value of ECF obtained in Subheading 3.3 **step 8**. This will yield a 2× virus stock. Dilute 0.1 mL of 2× virus with 0.9 mL complete DMEM to yield 0.2× virus (*see* **Note 9**).
3. Dilute inhibitor in 0.5 mL complete DMEM to a concentration equal to twice the IC_{90}. This will yield a 2× inhibitor stock.
4. Mix 0.5 mL 2× virus with 0.5 mL complete DMEM (*no inhibitor control*). Mix 0.5 mL 2× virus with 0.5 mL 2× inhibitor (*treated sample*).
5. Mix 0.5 mL 0.2× virus with 0.5 mL complete DMEM to create a sample in the absence of inhibitor that will mimic the infection level of the inhibited sample (*passaging control*).
6. Incubate all samples under the conditions used in Subheading 3.4, **step 5**, (*see* **Note** 7).
7. On one plate, remove media from three wells of 293 cells and replace with 1 mL/well *treated sample, passaging control*, or fresh complete DMEM without virus as a negative control. To guard against cross-contamination, the treated sample and the passaging control are placed on opposite sides of the 12-well plate. On a separate plate, replace media in one well with the *no inhibitor control*. Incubate at 37 °C, 5 % CO_2.
8. If a Typhoon is available, the plates can be scanned 18–24 h p.i (*see* **Note 10**). The treated sample should be infected tenfold less than the no-inhibitor control and approximately equally to the passaging control. At this point, the no-inhibitor control can be discarded.
9. Examine the wells daily until complete CPE is observed in the treated sample and passaging control. This should occur 3–4 days p.i.
10. Carefully harvest the cells and media from the treated sample and passaging control wells into separate 1.5 mL Eppendorf tubes on ice. This can be accomplished by forcefully pipetting the media onto the monolayer. A cell scraper can be used but should not be required. Care must be taken to avoid cross-contamination of the samples.

11. Centrifuge at 350–500 ×*g* for 5 min at 4 °C.
12. For each sample, create two equal aliquots of supernatant (~500 μL each), resuspend the cell pellet in 300 μL PBS, and snap freeze all samples by submerging in LN_2 until frozen solid.
13. To create a cell lysate containing virus, freeze/thaw each cell pellet three times alternating between LN_2 and a 37 °C water bath.
14. Centrifuge the cell pellets at 5000 ×*g* for 10 min at 4 °C.
15. Collect the supernatant (lysate) from each sample separately and divide into five to six aliquots of 50 μL each. Snap freeze all samples in LN_2. Store lysate and supernatant aliquots at 80 °C. These are round 1 samples (R1).
16. Thaw one aliquot of lysate from the treated sample and passaging control and determine the dilution of each sample needed to infect 80 % of the cells by following the instructions in Subheading 3.3. Note that this assay can be scaled down to a smaller culture size, and the initial lysate dilution can be decreased, if the apparent titer of the lysate is low.
17. Repeat the steps in this section, substituting lysate from the previous round of selection for the purified virus that was used in **step 2**, which is diluted based on the results of **step 16**. Lysate from the treated sample of each previous round of selection is used to create the no inhibitor control and treated sample for the next round. Lysate from the passaging control of each previous round of selection is used to create the passaging control for the next round. The idea is to maintain a consistent level of infection in all rounds of infection.

3.6 Evaluation of Resistance

1. For each round of selection, the treated sample should be ~10-fold less infected than the no inhibitor control when evaluated 18–24 h p.i. If this ratio shrinks, which is evidence of resistance, it may be necessary to reestablish the IC_{90} of the inhibitor as in Subheading 3.4 using the lysate from the round of selection where resistance becomes apparent (*see* **Note 11**). Resistance may also manifest as a shortened time to CPE in the treated wells.

3.7 Identifying and Validating the Genetic Basis for Resistance

Once the degree of resistance that is desired for the particular application has been reached, the genetic basis for resistance can be uncovered. There are several options:

1. Option 1: *Deep sequence the selected pool of viruses.* The strength of this approach is that most of the genetic diversity in the population can be captured. We have amplified virus from every five rounds of selection. The virus was expanded on 293 cells in the absence of inhibitor and purified on CsCl gradients. We then purified DNA from the virus as in Subheading 3.2 and used an

Illumina MiSeq for sequencing. Depending on the level of multiplexing of the sample, read depths in excess of 10,000 per base pair can be achieved. The disadvantage of this approach is that mutations could revert during viral amplification in the absence of inhibitor and that the short read lengths limit the identification of mutations that occur simultaneously in the same genome, although isolation of viral DNA straight from lysate without amplification of the virus might also be possible.

2. Option 2: *Plaque purify individual viruses from the bulk population and deep sequence.* Plaque purification [4] allows a more straightforward association between genotype and phenotype, if the plaque-purified viruses are also assayed for inhibitor resistance. Moreover, the higher purity allows for a better assessment of mutations that occur within the same genome. However, plaque purification is laborious, and only a small fraction of the genetic content of the bulk population could be analyzed.
3. Option 3: *PCR amplify a particular part of the genome and sequence.* If the target of the inhibitor is known in advance (e.g., hexon for an anti-hexon neutralizing antibody), then only that part of the genome need be sequenced. This can be from either the bulk population or from individual purified plaques.
4. To validate the identified mutations and to assign function to individual mutations if many are identified, it is useful to engineer them into a stable, non-mutator genome. There are a number of techniques to accomplish this, and our method of choice is recombineering [5, 6]. Virus from these stable clones can be amplified and assessed for phenotype and are less likely to revert.

4 Notes

1. In Uil et al., 2011, two polymerase mutations produced the highest mutation rate while supporting replication, T286I and F421Y [1]. Therefore, use of virus with the T286I mutation in polymerase is expected to be equivalent to our results with F421Y, while the double mutant is likely a stronger mutator. We have engineered both of these constructs but have not thoroughly evaluated them.
2. Although the titer of the virus in the supernatant will be two- to fourfold less, the 2 mL of supernatant can also be snap frozen in LN_2 and thawed prior to dilution, if there is a need to stop the protocol in between passages. In addition, the frozen supernatant from **step 7** serves as a reserve in the event of mishap during passaging. A lysate can also be generated from the cell pellet by following Subheading 3.5, **steps 13–15**.

3. In general, the inoculum can be left on when passaging virus from supernatant. Occasionally, we observe cytotoxicity within a few hours of adding unpurified virus to cells, most commonly when passaging lysate. In that event, the virus can be added, incubated for 1–2 h at 37 °C, 5 % CO_2, removed, washed gently with warm complete DMEM, and incubated with fresh DMEM until CPE is observed.
4. Ten passages in the absence of selection were used in Uil et al. and our previous study [1, 2]. Additional passaging will introduce more variability in the starting population; however, the virus will also continue to evolve under selective pressure in Subheading 3.5.
5. In place of proteinase K digestion of purified virus to obtain viral genomic DNA, we have used the QIAamp Viral RNA Mini Kit (Qiagen) to isolate viral genomic DNA from supernatant or lysate that contains virus.
6. When the viral stock is diluted by the factor derived from this method, infection will be in the linear range and is expected to be dose-responsive to inhibition. The MOI will be ~1.6. If appropriate software is not available, this value can be estimated from any graph of percent positive cells or RFUs against dilution factor.
7. Depending on the mode of inhibition, it may be important to incubate the virus with the inhibitor in the absence of cells at this point (e.g., 1 h on ice). In our previous studies with the 9C12 anti-hexon neutralizing antibody, this step permitted binding of the antibody to its epitope on the virus [2, 7]. If the inhibitor targets the cell or a viral component that is exposed or produced only after interaction with the cell, then preincubation can be skipped and the mixture can be immediately added to cells. If serum interferes with the antiviral, then this step can be performed in part in serum free media, although complete media will need to be added to the cells prior to overnight incubation.
8. Bulk selection may result in a mixed population of viruses with different mutations that convey resistance. Alternatively, viruses bearing a single genetic change may dominate the population over successive rounds of selection. Therefore, an alternative approach to uncover multiple genetic changes that result in antiviral resistance is to infect cells at a very low MOI, overlay with agar containing the inhibitor, and pick resistant plaques at every round of selection. Propagating individual plaques is much more laborious but also makes genotype/phenotype correlations more straightforward.
9. If the antiviral compound is not limiting, it is advantageous to perform multiple bulk selections in parallel. Utmost care must

be taken to avoid cross-contamination. We generally keep infected wells on separate plates and handle infected wells under selection prior to and separate from control wells infected in the absence of selection.

10. Alternatively, parallel wells can be analyzed by flow cytometry if a Typhoon is not available.
11. A single genetic change may generate a completely resistant virus. Alternatively, multiple changes may need to accumulate to confer resistance. It is the latter case where consistently increasing the inhibitor concentration to maintain an IC_{90} may be beneficial.

Acknowledgement

This work was supported by Royalty Research Fund Grant 65-6165 from the University of Washington and by R01 AI104920 from the National Institute for Allergy and Infectious Diseases.

References

1. Uil TG, Vellinga J, de Vrij J, van den Hengel SK, Rabelink MJ, Cramer SJ, Eekels JJ, Ariyurek Y, van Galen M, Hoeben RC (2011) Directed adenovirus evolution using engineered mutator viral polymerases. Nucleic Acids Res 39:e30
2. Myers ND, Skorohodova KV, Gounder AP, Smith JG (2013) Directed evolution of mutator adenoviruses resistant to antibody neutralization. J Virol 87:6047–6050
3. Nguyen EK, Nemerow GR, Smith JG (2010) Direct evidence from single-cell analysis that human alpha-defensins block adenovirus uncoating to neutralize infection. J Virol 84:4041–4049
4. Green M, Loewenstein PM (2006) Human adenoviruses: propagation, purification, quantification, and storage. Curr Protoc Microbiol Chapter 14:Unit 14C.11
5. Smith JG, Silvestry M, Lindert S, Lu W, Nemerow GR, Stewart PL (2010) Insight into the mechanisms of adenovirus capsid disassembly from studies of defensin neutralization. PLoS Pathog 6:e1000959
6. Warming S, Costantino N, Court DL, Jenkins NA, Copeland NG (2005) Simple and highly efficient BAC recombineering using galK selection. Nucleic Acids Res 33:e36
7. Smith JG, Cassany A, Gerace L, Ralston R, Nemerow GR (2008) Neutralizing antibody blocks adenovirus infection by arresting microtubule-dependent cytoplasmic transport. J Virol 82:6492–6500

Part V

Delivery Methods

Chapter 14

Intraparenchymal Stereotaxic Delivery of rAAV and Special Considerations in Vector Handling

Matthew J. Benskey and Fredric P. Manfredsson

Abstract

Stereotaxic surgery enables precise and consistent microinjections to discrete neural nuclei. Using stereotaxic surgery to deliver viral vectors is a powerful tool that provides the ability to manipulate gene expression in specific regions, or even specific cell types in the brain. Here, we describe the proper handling and stereotaxic delivery of recombinant adeno-associated virus to various neuroanatomical structures of the rodent brain.

Key words Stereotaxic surgery, Adeno-associated virus (rAAV), Gene delivery, Brain

1 Introduction

Stereotaxic surgery is a commonly used and minimally invasive surgical technique that utilizes a three-dimensional Cartesian (or polar) coordinate system to localize central neural nuclei. Since the advent of the stereotaxic frame in the early twentieth century, stereotaxic surgery has dramatically improved the accuracy and consistency of neurosurgical microinjections [1]. Stereotaxic surgery takes advantage of the relatively constant location of neuroanatomical structures in reference to specific features of the skull. Using either the bregmatic or lambdoidal sutures of the skull as a reference point (Fig. 1), a set of three coordinates in the X (medial-lateral, ML), Y (anterior-posterior, AP), and Z (dorsal-ventral, DV) plane will describe the exact location of neuroanatomical structures.

Combining the precision of stereotaxic surgery with the highly efficacious viral vectors now available, researchers can easily manipulate gene expression within discrete neural nuclei. Specifically, recombinant adeno-associated virus (rAAV) vectors are favored for their high biosafety profile, low immunogenicity, and ability to transduce and achieve long-term gene expression in non-dividing

Fredric P. Manfredsson (ed.), *Gene Therapy for Neurological Disorders: Methods and Protocols*, Methods in Molecular Biology, vol. 1382, DOI 10.1007/978-1-4939-3271-9_14, © Springer Science+Business Media New York 2016

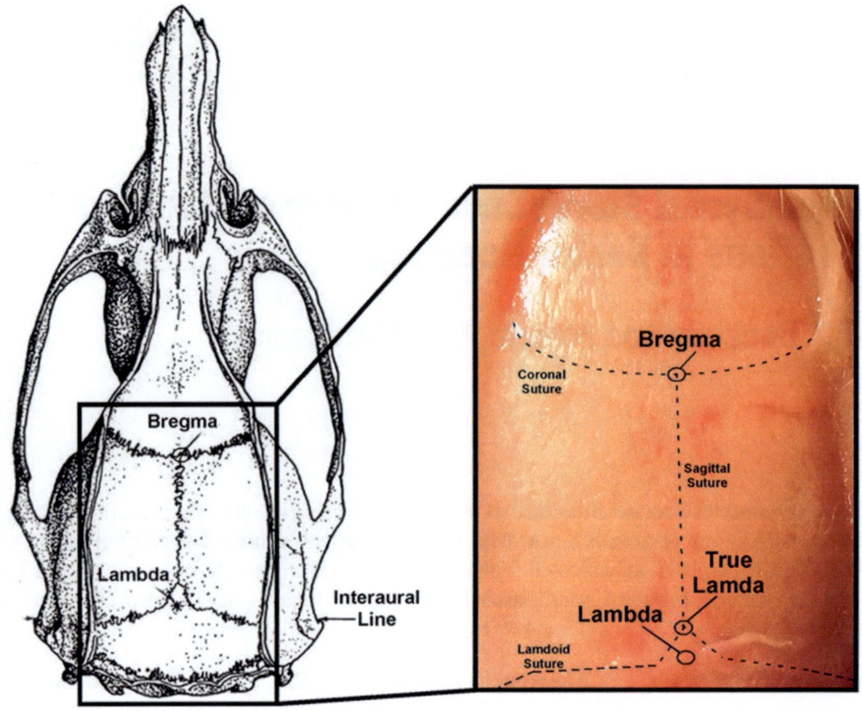

Fig. 1 Skull surface sutures and reference points. Stereotaxic surgery utilizes reference points created by the intersection of skull sutures to pinpoint the location of neuroanatomical structures in the brain. The most common reference points used are Bregma and Lambda. Bregma is the point at which the coronal suture intersects the sagittal suture. Lambda is the point that is in line with both the sagittal suture and the interaural line. Alternatively, True Lambda is the point at the intersection of the lambdoid suture and the sagittal suture. Shown is a diagram of the sutures and reference points found on the rat skull as well as a representative picture of a rat skull (corresponding to the area in the *box* on the diagram to the *right*) with the locations of the skull sutures and reference points demarcated

cells of adult animals. Using rAAV, transduction can be achieved in structures as large as the striatum [2] or limited to structures as small as nuclei of the hypothalamus (Fig. 2) [3, 4]. Transduction kinetics and tropism can be optimized using particular rAAV pseudotypes and cell-specific promoters, to achieve tightly controlled transgene expression solely in cells of interest (Fig. 3) [2, 5–7].

This chapter describes a protocol for attaining accurate and consistent delivery of rAAV vectors to the brain. Provided herein are detailed instructions for placing an animal in a stereotaxic frame, determination of skull surface reference points, localization of stereotaxic coordinates, and injection of viral vectors using a motorized micropump. Finally, due to the complications associated with utilizing viral vectors, special consideration will be given to the proper handling of virus in order to maximize transduction potential.

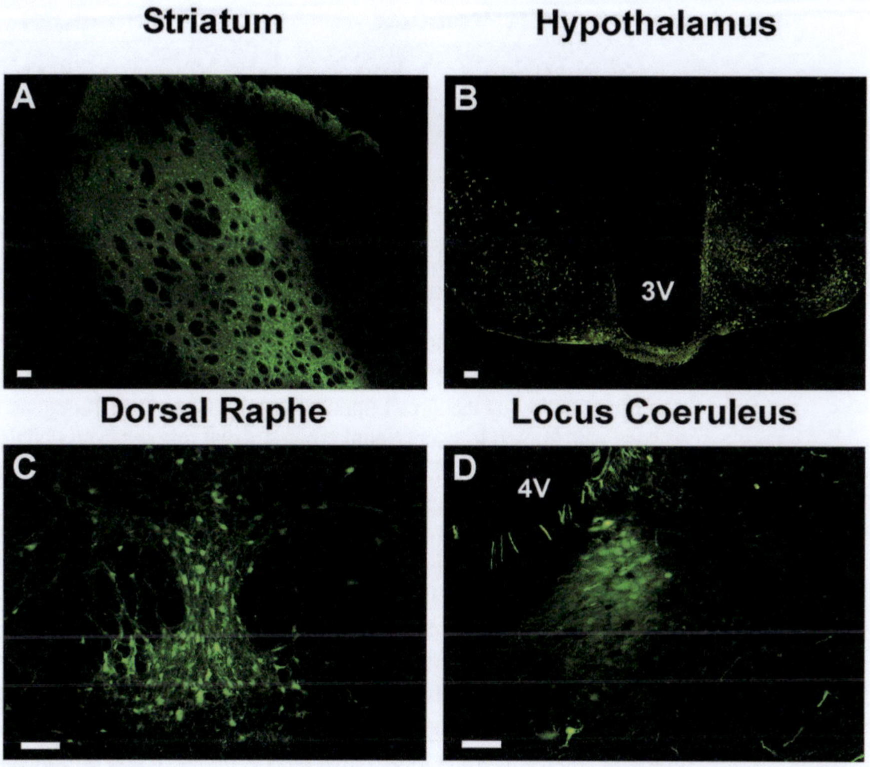

Fig. 2 Stereotaxic delivery of rAAV to discrete neural nuclei. rAAV 2/5 expressing GFP under control of the cytomegalovirus/chicken β-actin promoter was stereotaxically injected into various brain regions of the rat or mouse to demonstrate gene delivery and expression in discrete nuclei of the brain. Successful viral delivery, transduction and transgene expression is demonstrated in the rat striatum (**a**), the mouse mediobasal hypothalamus (**b**), the rat dorsal raphe nucleus (**c**) and the rat locus coeruleus (**d**). All coordinates, volumes of injection and flow rates for the shown stereotaxic injections are available in Table 1 (except the mouse hypothalamus which were bilateral, 250 nl injections performed at a 10° lateral angle at −1.75 anterior–posterior, ±1.25 mediolateral, and −6.2 dorsal ventral). Scale bars in all panels represent 100 μm. Abbreviations: *3 V* third ventricle, *4 V* fourth ventricle

2 Materials

1. Adult Sprague–Dawley rat (*see* **Note 1**).
2. rAAV (*see* **Note 2**).
3. Stereotaxic atlas (e.g., Paxinos and Watson).
4. Isoflurane vaporizer with induction box and nose cone adapters.
5. Isoflurane (Abbott Animal Health, Abbott Park, IL, USA).
6. Programmable motorized micropump (e.g., QSI by Stoelting or equivalent).

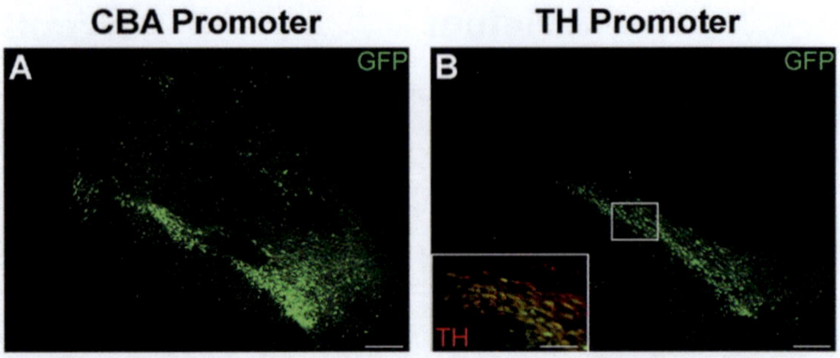

Fig. 3 Cell-specific promoters increase specificity of transgene expression following stereotaxic delivery of rAAV. The precision of stereotaxic gene delivery can be improved by using cell-specific promoters to limit transgene expression to cells with known transcriptional profiles. Adult rats received unilateral stereotaxic injection of rAAV 2/5 into the midbrain. The expression cassette consisted of a GFP transgene under the control of either: (**a**) the ubiquitously expressed cytomegalovirus/chicken β-actin (CBA) hybrid promoter, or (**b**) the tyrosine hydroxylase (TH) promoter, which is active solely in catecholaminergic neurons. Under the control of the CBA promoter, GFP expression is observed throughout the majority of the midbrain, within many different cell types. In contrast, GFP expression under the control of the TH promoter is limited solely to the dopaminergic neurons of the substantia nigra and ventral tegmental area. *Yellow* cells in the *inset* of panel **b** represent the restricted expression of GFP (*green*) within TH immunoreactive neurons (*red*). Scale bars in panels **a** and **b** represent 200 μm. Scale bar within *inset* in panel **b** represents 50 μm

7. 10 μL Hamilton syringe with 26-gauge needle (Hamilton, Reno, NV, USA).
8. 50 μL microcapillary pipette (Drummond Scientific, Broomall, PA, USA).
9. Surgical drill (e.g., Osada electric Xl-30 W or equivalent).
10. Drill Bit (Brasseler, Savannah, GA, USA; *see* **Note 3**).
11. Intramedic polyethylene tubing (I.D. × O.D.: 0.588 × 0.965 mm; Clay Adams Brand, Sparks, MD, USA).
12. Sigmacote siliconizing reagent (Sigma-Aldrich, St. Louis, MO, USA).
13. Stereotaxic frame (Kopf Instruments, Tujunga, CA, USA; *see* **Note 4**).
14. Buprenorphine.
15. Betadine surgical scrub (Thermo Fisher Scientific, Waltham, MA, USA).
16. 70 % ethanol.
17. Sterile adhesive plastic surgical drape (e.g., Bioclusive by Johnson & Johnson, or equivalent).

Table 1
Established stereotaxic coordinates of various neural nuclei

Brain region	Reference point	Anterior–posterior	Medio-lateral	Dorsal-ventral	Injection volume	Flow rate (μL/min)	Notes
Striatum (single injection)	Bregma	0	+3.0	−4	2 μL	0.5	Lower needle to site, inject immediately (4 min), wait 1 min retract needle 1 mm, wait 4 min before full retraction = *total 9 min*
Striatum (double injection)[a]	Bregma	First site: 0 Second site: −1.6	First site: +3.0 Second site: +4.5	First site: −4.0 Second site: −5.5	2 μL per injection	0.5	Lower needle to site, inject immediately (4 min), wait 1 min retract needle 1 mm, wait 4 min before full retraction = *total 9 min*
Substantia nigra	Bregma	−5.4	+2.0	−7.2 from dura	2 μL	0.5	Lower needle to site, inject immediately (4 min), wait 5 min retract needle = *total 9 min*
Hippocampus	Bregma	−3.8	+2.0	−2.6 from dura	2 μL	0.5	Lower needle to site, inject immediately (4 min), wait 5 min before full retraction = *total 9 min*
Dorsal raphe	Bregma	−7.8	−3.1	−7.5 from skull	2 μL	0.5	Stereotaxic arm at a 30° lateral angle. Lower needle to site, inject immediately (4 min), wait 5 min before full retraction = *total 9 min*

(continued)

Table 1
(continued)

Brain region	Reference point	Anterior–posterior	Medio-lateral	Dorsal-ventral	Injection volume	Flow rate (μL/min)	Notes
Locus coeruleus	True Lambda	−3.7	−1.2	−6.1 from skull	1.5 μL	0.25	Stereotaxic arm at a 12° angle in the anterior-posterior direction. Lower to site, inject immediately (6 min), wait 5 min before full retraction = total 11 min
Hypothalamus	Bregma	−1.8	+1.0	−9.0 from Dura	0.25 μL	0.125	Lower needle to site, inject immediately (2 min), wait 5 min before full retraction = *total 7 min*
Motor cortex	Bregma	+1.6	−2.6	−1.6 from dura	1.5 μL	0.25	Lower needle to site, inject immediately (6 min), wait 5 min before full retraction = *total 11 min*

Listed are the coordinates, reference points, injection volumes, and flow rates for several neural nuclei that have been tested and validated. All coordinates are in millimeters. The notes section lists the injection paradigm used for the listed nuclei, including the time the needle is left in place before full retraction, and any special notes pertaining to the specific surgery

[a]The single striatal injection paradigm is used to deliver virus to the approximate center of the striatum, while the double injection paradigm is used to transduce the entirety of the striatum. Animals are pre-treated with mannitol (0.75 g/kg) to improve convection-enhanced delivery and spread with large injections such as the double striatal injection. The volume of injection also depends upon the spread and depth of penetration into the parenchyma desired. It is best practice to empirically validate all parameters of the surgery

18. Sterile surgical gauze (Thermo Fisher Scientific, Waltham, MA, USA).
19. Long cotton swabs.
20. Scalpel with # 10 blade.
21. 11.5 cm straight fine scissors (F.S.T., Foster City, CA, USA).
22. Fine curved forceps (e.g., Medical #7S forceps; F.S.T., Foster City, CA, USA).
23. Serrated bulldog serrefines (F.S.T., Foster City, CA, USA).
24. Autoclip stapler (F.S.T., Foster City, CA, USA).

25. 9 mm Autoclip surgical staples (Braintree Scientific, Braintree, MA, USA; *see* **Note 5**).
26. Lidocaine.
27. Puralube vet ophthalmic ointment (Dechra Veterinary Products, Overland Park, KS, USA).
28. Hair clippers.
29. Isothermal heating pad.
30. Parafilm (Neenah, WI, USA).
31. Pipette and pipette tips (p20 and p200).
32. Balanced salt solution (Alcon Laboratories, Fort Worth, TX, USA).
33. 1 and 10 mL syringe.
34. 25-gauge needle.
35. 1.5 mL Eppendorf tubes.
36. 3 % hydrogen peroxide.
37. Sterile 0.9 % Saline.
38. Calculator.
39. Timer.
40. Paper towel.

3 Methods

3.1 Preoperative Preparation

1. Determine stereotaxic coordinates, volume of injection, and flow rate for the brain region to be injected (*see* **Note 6**). Table 1 shows a list of confirmed coordinates, volumes, and flow rates for various brain regions.
2. Siliconize microcapillary pipettes, Eppendorf tubes, and pipette tips. To siliconize the microcapillary pipettes, place the pipettes in a beaker and use a handheld pipettor or transfer pipette to drip Sigmacote into the microcapillaries. Be sure fill the entire microcapillary with Sigmacote. Repeat process until the entire interior surface of the microcapillary is thoroughly coated with Sigmacote. To siliconize Eppendorf tubes, drip Sigmacote down all interior surfaces of the Eppendorf. Repeat until surface is thoroughly coated with Sigmacote. To siliconize pipette tips, infuse enough Sigmacote to fill the entire pipette tip, then expel contents. Repeat until entire interior surface of pipette tip is coated with Sigmacote. Drain any excess Sigmacote from microcapillary pipettes, Eppendorf tubes, and pipette tips and air-dry overnight in a hood (*see* **Note 7**).

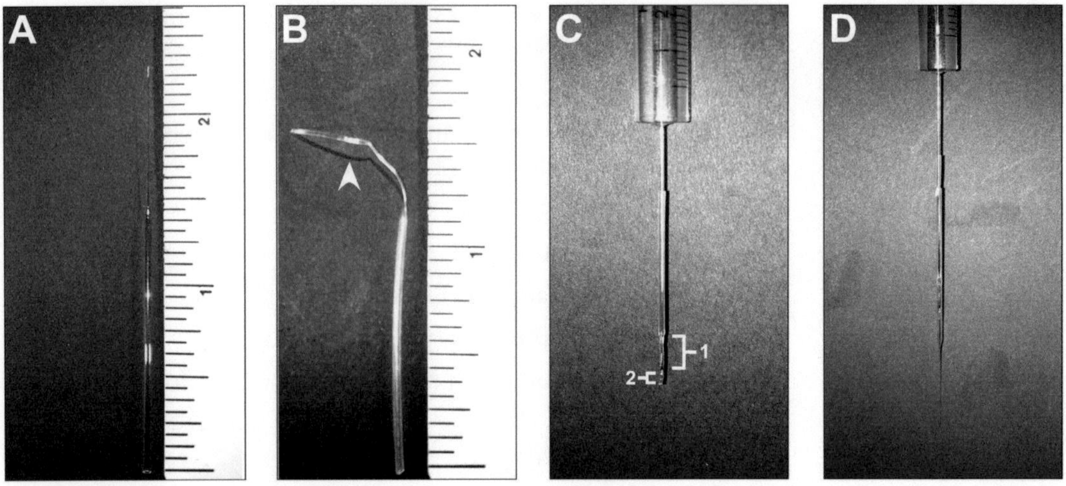

Fig. 4 Fitting the microcapillary pipette to the Hamilton Syringe. Siliconize the microcapillary pipette and pull the pipette to a total length of 4–6 cm with a 2 cm tip and a 60–80 μm diameter (**a**). Pull a 1–2 cm piece of intramedic tubing with a twisting motion to create a taper (*arrowhead* in panel **b**) at the end of the tubing (**b**). Pull the tubing onto the Hamilton needle without puncturing or tearing the tubing. The taper will tightly encase the needle (*1* in panel **c**). The tubing should extend from the tip of the Hamilton needle to approximately 0.5–1 in. from the barrel of the syringe. Trim the tubing approximately 1 mm (*2* in panel **c**) below the tip of the Hamilton needle (**c**). Finally, pull the microcapillary pipette over the tubing onto the Hamilton syringe. Trim the very end (≈1 mm) of the microcapillary pipette tip once successfully secured on the Hamilton needle tip. It is essential that the microcapillary tip forms an airtight seal over the Hamilton needle (**d**). The numbered graduations on the vertical rulers in panels **a** and **b** represent inches

3. Using a pipette puller, pull the siliconized microcapillary pipettes to a total length of 4–6 cm with a 2 cm tip and a 60–80 μm tip diameter (Fig 4a).
4. All surgical tools should be autoclaved prior to use, including: Scissors, forceps, bulldog serrefines, autoclip stapler and staples, drill bit, cotton swabs and gauze. Thoroughly sterilize other surgical supplies that cannot be autoclaved. Disinfect and sterilize surgical station (*see* **Note 8**).
5. Secure the microcapillary pipette (previously siliconized/pulled) to the needle of the Hamilton syringe. Pull a small (1–2 in.) section of intramedic tubing off of the roll (Fig. 4b). Pulling will stretch the tubing and create a taper on one end of the tube. Cut the end of the tubing opposite the taper to create a clean opening. Pull the tubing over the needle of the Hamilton syringe, with the taper facing the tip of the needle. Cut the tapered tip of the tubing ≈ 1 mm below the tip of the Hamilton needle (Fig. 4c). Pull the microcapillary pipette tip over the tubing and the Hamilton syringe needle (Fig. 4d). Check to ensure that the pipette forms an airtight seal with the tubing and needle (*see* **Note 9**).

6. Program the micropump (for the specific Hamilton syringe) to inject the desired volume at the desired flow rate (*see* **Note 10**).
7. Secure an anesthesia adapter to the stereotaxic frame. Place the end of the anesthesia adapter where the animal's nose will rest once in the frame.
8. Set a timer for the total time of the injection (*see* Table 1). This includes the time needed to perform the actual injection and the time the needle will be left in place after the injection is finished (*see* **Note 11**).
9. Prepare the working titer of rAAV. Anything coming into contact with the virus should be siliconized (*see* **step 2**, Subheading 3.1). If necessary dilute virus to desired titer using balanced salt solution (*see* **Note 12**).
10. Prepare a 1.5 mL Eppendorf tube containing 3 % hydrogen peroxide and another containing sterile saline.

3.2 Stereotaxic Injection

1. Turn on isothermal heating pad.
2. Turn on isoflurane vaporizer and set at 2 % isoflurane. Allow the gas to fill the induction chamber (*see* **Note 13**).
3. Place animal in the induction chamber until fully anesthetized (*see* **Note 14**).
4. Administer preemptive analgesic (e.g., buprenorphine; 0.05–0.1 mg/kg, subcutaneous).
5. Use clippers to thoroughly shave the head of the animal (Fig. 5a; *see* **Note 15**).
6. Turn on isoflurane flow to the adapter on the stereotaxic frame. Place the animal in the ear bars of the stereotaxic frame. Keep one ear bar clamped in place and loosen the other ear bar so it moves freely. Holding the head of the animal, guide the head toward the clamped ear bar and insert the ear bar into the animal's ear. Gently move the head onto the tip of the ear bar until you feel the bar enter the bony ear canal in the skull. Hold the head in place and slide the other ear bar into the bony ear canal of the opposite side of the head. Secure the second ear bar in place (Fig. 5a; *see* **Note 16**).
7. Bring the bite bar toward the animal. Insert the animal's incisors into the bite bar and secure the bite bar into position. Secure the nose clamp (*see* **Note 17**).
8. Place heating pad beneath the animal with paper towel in-between the animal and the heating pad.
9. Disinfect the surgical site. Using the sterile gauze, scrub the shaved scalp with betadine followed by 70 % ethanol, wiping from the center of the surgical site outward. Repeat three times. Be careful to avoid the animal's eyes (Fig. 5b).

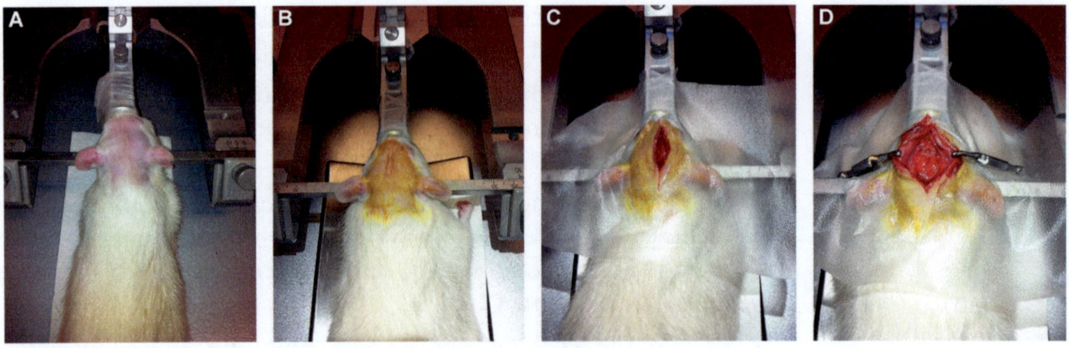

Fig. 5 Exposing the skull. Once the animal is fully anesthetized, thoroughly shave the head and secure the skull in the stereotaxic frame, ensuring constant isoflurane flow through the anesthesia nose adapter (**a**). Disinfect the surgical site by scrubbing the scalp three times with betadine surgical scrub (**b**). Place the heating pad under the animal and isolate the surgical site with the plastic surgical drape. Make a vertical incision in the surgical drape to allow access to the incision site. Make a vertical incision in the scalp, from between the eyes to between the ears (**c**). Clamp either side of the incision with bulldog serrefines and lay them to their respective sides to expose the skull (**d**)

10. Coat the animal's eyes with puralube vet ointment to avoid ocular dehydration.
11. Drape the head with the sterile adhesive plastic and make an incision in the surgical drape to provide isolated access to the surgical site.
12. Confirm animal is fully anesthetized (*see* **Note 14**). Make an anterior/posterior incision in the center of the scalp with the scalpel. The incision should extend from the rear of the skull to just in-between the eyes (Fig. 5c).
13. Using a 1 mL syringe, apply lidocaine to the incision on the scalp.
14. Clamp either side of the incision with bulldog serrefines and lay them to their respective sides to keep the incision open. Use cotton swabs to remove blood and fascia and dry the skull (Fig. 5d).
15. Ensure that the skull is centered and level in the stereotaxic frame (*see* **Note 18**).
16. Secure the motorized pump to the stereotaxic arm and secure the Hamilton syringe (with the sealed microcapillary pipette tip) into the pump (Fig. 6a).
17. Using the motorized micropump manifold, load the syringe with the desired amount of virus and ensure flow (*see* **Note 19**).
18. Find bregma (or the skull surface landmark used) and position the tip of the needle directly over the center of bregma (Fig. 1). Read the vernier scales on the AP guide bar and ML arm of the stereotaxic frame and record the AP and ML coordinates of bregma (*see* **Note 20**).

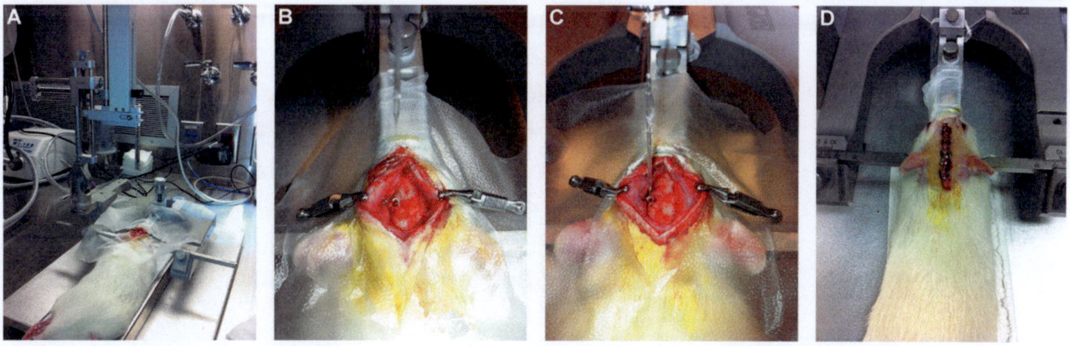

Fig. 6 Stereotaxic injection. Secure the motorized pump to the stereotaxic arm. Secure the Hamilton syringe with the siliconized, pulled microcapillary pipette in the motorized pump. Program the pump to the desired infusion volume and flow rate, and aspire the virus into the syringe (**a**). Locate skull surface reference point and place the tip of the needle directly over reference point. Record the position of the reference point in the AP and ML planes, and calculate the corresponding coordinates of your brain nuclei of interest. Place the needle over the calculated coordinates. If your stereotaxic coordinates are "from skull," carefully touch the tip of the needle to the skull and denote the DV coordinates. Carefully drill a hole in the skull at the calculated coordinates (**b**). If the stereotaxic coordinates are "from dura," lower the tip of the needle until it is just touching the dura mater. Record the DV coordinates and calculate the correct DV coordinates, from skull or dura, corresponding to your region of interest. Cut through the dura with an 18-gauge needle, lower the needle to the correct DV plane and begin injection (**c**). Once the full injection is complete, raise the needle and wash the tip with hydrogen peroxide and saline. Apply lidocaine to the incision and staple the incision closed (**d**)

19. Using the previously determined stereotaxic coordinates (obtained from stereotaxic atlas and pilot surgeries), calculate the correct coordinates from bregma, corresponding to your region of interest. Use the calculator to add or subtract your coordinates from bregma.
20. Using the scales on the AP guide bar and ML arm, position the tip of the needle to the calculated coordinates corresponding to your brain region of interest.
21. Lower the needle to just above the skull. If your stereotaxic coordinates are "from skull," carefully touch the tip of the needle to the skull and denote the DV coordinates. Make a mark on the skull where the needle tip will penetrate the skull (*see* **Note 21**).
22. Carefully drill through the depth of the skull (Fig. 6b; *see* **Note 21**).
23. If the stereotaxic coordinates are "from dura," lower the tip of the needle until it is just touching the dura mater. Record the DV coordinates on the vertical arm of the stereotaxic frame.
24. Calculate the correct DV coordinates, from skull or dura, corresponding to your region of interest. Use the calculator to subtract your coordinates from bregma.

25. Use the 25-gauge needle to very carefully make a small incision in the dura mater (*see* **Note 22**).
26. Lower the needle to the calculated DV coordinates using the scale on the DV arm of the stereotaxic frame (Fig. 6c).
27. Start the injection using the micropump manifold. Start the timer (*see* **Note 23**).
28. After the full injection is finished (including the actual injection and leaving the needle in place), slowly retract the needle out of the skull.
29. Remove the Hamilton syringe from the motorized pump. Fully depress the plunger to expel any remaining contents in the syringe. Dip the tip of the needle into the Eppendorf tube containing hydrogen peroxide. Next, rinse the tip of the needle in the Eppendorf tube containing saline. Finally, remove the plunger and use the 10 mL syringe containing saline to flush the needle and ensure proper flow. Place the plunger back into the Hamilton (*see* **Note 24**).
30. Remove the serrefines. Apply bone wax to fill the burr hole in the skull.
31. Using a 1 mL syringe, apply lidocaine to the incision on the scalp.
32. Use forceps to pull the skin away from the skull and squeeze the skin from the left and right side of the incision together. Staple along the entire axis of the incision using the Autoclip stapler (Fig. 6d).
33. Place the animal on a heating pad in a temporary holding cage until awake and ambulatory. Replace animal to home cage.
34. When performing multiple surgeries disinfect surgical tools with a glass bead sterilizer between surgeries.

3.3 Postoperative Care

1. Closely monitor the animals twice daily for the next 48 h post-surgery, for signs of distress or infection. Monitor once daily for the next week (*see* **Note 25**).
2. Provide postoperative analgesic for the next 24 h.
3. Remove staples 10–14 days post-surgery. Anesthetize the animal and use a surgical staple remover to remove all remaining wound clips.

4 Notes

1. All protocols using live animals must be reviewed and approved by an Institutional Animal Care and Use Committee (IACUC) and must follow officially approved procedures for the care and use of laboratory animals.

2. The spread, tropism, and level of transgene expression in different brain structures and different cell types will be affected by the titer and pseudotype/serotype of rAAV used. These parameters should be tailored to the specific needs of the experiment and empirically validated in pilot experiments.
3. Dental drill bits (Brasseler USA, Savannah, GA, USA) work well for stereotaxic surgeries. We have determined that a number 8 round-bur drill bit is suitable for rat surgeries and a number 6 round-bur drill bit is suitable for mouse surgeries.
4. Ensure that the frame and accessories (ear bars and bite bars) are appropriate for the type of animal to receive surgery.
5. The 9 mm autoclip staples are appropriate size for rats but may be too large for mice. It may be prudent to use a 7 mm staple for mice or alternatively use a tissue adhesive such as Gluture (Abbott Animal Health, Abbot Park, IL, USA).
6. Stereotaxic coordinates found in an atlas usually correspond to an average size adult male animal. For example, most rat atlases entail coordinates for an adult male rat weighing approximately 290 g. Accordingly, if the experimental animal is significantly smaller or larger than the animal described in the atlas, stereotaxic coordinates should be obtained empirically. In general, it is best practice to use the stereotaxic atlas to obtain an initial set of coordinates and perform pilot surgeries. Performing pilot injections with a diluted dye is a very useful way to determine the localization of the injection. Injected brains can be quickly sectioned post-mortem and the location and spread of the dye can be visualized immediately. Coordinates and volume of injection can then be adjusted until probe placement and spread is within the nuclei of interest.
7. Due to the inherently "sticky" nature of the viral capsid, it is imperative to siliconize every surface that the virus will come into contact with (this includes the pipette tips and tubes used to move and store the virus). Siliconizing will prevent the adherence of virus to contact surfaces and subsequent loss of viral particles.
8. Perform surgery using aseptic technique. Anything coming into contact with the animal during the surgical procedure must be sterilized and aseptic. Further, the surgery should be performed on a decontaminated surgical station within a sterile surgical room (preferably under a hood with positive filtered airflow) wearing lab coats, sterile gloves and filter masks. Following strict sterile technique will significantly improve surgical outcome and decrease postoperative complications or discomfort.
9. Pull the tubing from the role with a twisting motion in order to create a tight taper (Fig. 4b). Pull the tubing onto the

Hamilton needle to the limit of the elastic resistance of the tubing without puncturing or tearing the tubing. The taper will tightly encase the needle. The tubing should extend from the tip of the Hamilton needle to approximately 0.5–1 in. from the barrel of the syringe. Trim the tubing using fine scissors approximately 1 mm below the tip of the needle. When pulling the pulled microcapillary pipette tip over the tubing and needle, the tubing will bunch and create a seal. Avoid sheering the tubing with the microcapillary pipette. Trim the very end (≈1 mm) of the microcapillary pipette tip once successfully secured on the Hamilton needle tip. It is essential that the microcapillary tip forms an airtight seal over the Hamilton needle. To ensure an airtight seal, fill a 10 mL syringe with saline. Make a temporary gasket by covering the needle of the 10 mL syringe needle with a short piece of intramedic tubing. Remove the plunger from the Hamilton syringe. Place the 25-gauge needle of the 10 mL syringe in the plunger hole of the Hamilton syringe and force saline through the Hamilton. If any saline leaks from the interface of the microcapillary pipette, intramedic tubing, and Hamilton needle, the seal is not airtight. Repeat the process of securing the microcapillary pipette tip until an airtight seal is achieved. Once the seal is obtained, replace the plunger into the Hamilton slowly, expelling saline out of the tip of the microcapillary. Some saline should remain in the syringe after the plunger is replaced.

10. Although rAAV produces high levels of transgene expression within the CNS, one limiting factor is the distribution of virus within the brain. The protocol described herein utilizes constant hydrostatic pressure to produce convection-enhanced delivery and increase the distribution of rAAV in the CNS. For more information on convection-enhanced delivery, *see* Chapter 21. When injecting larger brain structures, convection-enhanced delivery can be improved by pre-treating the animals with mannitol to enhance the spread of the virus [8, 9].
11. The needle is left in place after the injection is finished to allow for the convection-enhanced spread of the virus and to prevent reflux along the needle track when retracting the needle.
12. Although all contact surfaces (of containers and pipettes) have been siliconized, the virus could still potentially stick to these surfaces and result in loss of viral particles when transferring. In order to maintain titer when transferring virus to a new container, it is necessary to allow the virus to coat all surfaces it will come into contact with. Using a pipettor, slowly aspire the desired volume of virus into a siliconized pipette tip. Allow virus to sit in the tip for 1 min. Expel virus back into original container. Next, aspire the desired volume of virus and expel into the new container. Allow to sit for 1 min, and again,

replace virus from the new container back into the original container. Finally, transfer desired volume of virus from the original container to the new container.

13. The 2 % isoflurane is a general guideline. Animals should be monitored throughout the entire surgery to ensure that they are fully anesthetized but showing no signs of overdose (shallow or irregular breathing, cyanosis on footpads) and anesthesia should be adjusted accordingly.
14. When fully anesthetized the animal will exhibit loss of pedal withdrawal reflex and tail pinch response, decreased or absent limb muscle tone, and slow regular breathing. To test pedal withdrawal reflex extend the animal's hind limb and use a pair of forceps to lightly pinch the foot. If the limb withdrawals or muscles twitch, the animal is not sufficiently anesthetized. The lack of a response indicates deep anesthesia.
15. The animal may wake up during or after shaving. If so, place the animal back into the induction chamber prior to placing the animal in the stereotaxic frame. This will ensure proper anesthesia and prevent any struggle that may harm the animal or the surgeon.
16. It is absolutely essential that the animal be secured in the ear bars properly. Check to make sure that the ear bars are in the ear canals and not pinching the neck or skull. If the animal is placed in the ear bars properly, the nose will be able to move up and down but not side to side.
17. Keep the bite bar and nose clamp loose prior to putting the animal in the stereotaxic frame. Use the blunt end of a long cotton swab to open the lower jaw of the animal. With the jaw open pull the bite bar into place and insert the animal's incisors into the hole in the bite bar. When securing the nose clamp, tighten just enough to hold the nose in place. Be sure that the isoflurane adapter is flowing into the nose of the animal. It may be helpful to wrap a piece of parafilm around the animal's nose and the anesthetic adapter, creating a small anesthesia microenvironment. This will ensure the animal remains fully anesthetized throughout the entire surgery.
18. All stereotaxic coordinates, and the accuracy of stereotaxic surgery, are predicated on the animal being secured in the stereotaxic frame properly, with a flat and centered skull. To ensure that the skull is centered lay the scalpel (or anything with a straight edge) in line with the sagittal suture of the skull. The axis of the sagittal suture should be in parallel with the axis of the nose clamp. If it is not, the animal may not be in the ear bars properly. If the animal is in the ear bars properly, it may be necessary to carefully unclamp the ear bars and slide the skull (still in the ear bars) until it is centered in the frame. To confirm

that the skull is flat measure the DV coordinates of bregma and lambda. If bregma and lambda are not in the same plane, adjust the anterior mount until they are in level with each other.

19. When loading the syringe, retract the amount of virus needed for injections plus 1 μL extra. Due to the small size of the microcapillary needle, clogging is common. To ensure proper flow, use the pump manifold to carefully expel a small amount of virus from the tip of the needle after the syringe is loaded. If the tip is clogged use fine forceps to trim a very small amount off of the tip of the microcapillary needle.
20. It can be difficult to accurately visualize the exact location of bregma or lambda with the naked eye. Accordingly, it is extremely beneficial to use a surgery microscope when localizing the skull surface landmarks, drilling, or obtaining the DV coordinates of the skull/dura. Although a surgical scope is highly recommended and will greatly enhance accuracy, it is not absolutely essential.
21. If using a surgical microscope it is not necessary to make a mark on the skull for drilling. Raise the tip of the needle approximately 1 in. from the surface of the skull, then while looking through the surgical scope, bring the drill bit in directly below the tip of the needle and carefully drill. If you do not have access to a surgical scope use pencil or a fine tip marker to make a small mark on the surface of the skull and carefully drill there. Drill through a fraction of the depth of the skull, then stop drilling and lower the tip of the needle toward the skull to ensure you are drilling in the correct position. Complete drilling until the skull begins to crack and splinter. Then, using the fine forceps, carefully remove the splintered pieces of skull from the drill site, leaving a clean hole. Do not drill all the way through the skull into the cranial cavity or you may destroy the dura matter or damage the brain. Drilling may cause effusive bleeding. If bleeding is severe and cotton swabs are insufficient to stop the bleeding, use cellulose-based sponge such as a weck-cel (BVI, Waltham, MA, USA) to absorb excess fluid.
22. The dura mater covering the rat brain is too thick for the microcapillary pipette tip to penetrate, so it is necessary to make an incision in the dura prior to lowering the needle in place. It is not necessary to cut the dura of smaller rodents such as mice. It is very beneficial to have a surgical scope when performing the incision in the dura mater.
23. Again, the timer should be set to include the total time needed for the injection as well as the time the needle will be left in place after the injection. Some injection protocols stipulate

that the needle should be partially retracted 1 mm after a given amount post injection (prior to completely retracting the needle). If this is the case, set two timers, one timer for the partial retraction and one timer for the complete retraction.

24. The hydrogen peroxide will destroy any remaining virus on the Hamilton needle tip and also sterilize the needle. Rinse in sterile saline to remove any remaining hydrogen peroxide prior to loading the needle with virus for the next surgery. Finally, rinsing the entire syringe with saline will ensure proper flow and help prevent clogging.
25. Monitor animals for symptoms of pain and distress and administer analgesic as necessary. Closely monitor wound healing. If necessary anesthetize the animal, disinfect the wound and reapply surgical staples. Finally, monitor for signs of dehydration or weight loss. If necessary administer warm sterile saline (1–5 cc; i.p. and 5–10 cc; s.c.) or high fat supplements in order to minimize discomfort and maximize animal health.

Acknowledgement

This work was supported by the Saint Mary's Foundation.

References

1. Gildenberg PL, Krauss JK (2009) History of stereotactic surgery. Stereotactic and functional neurosurgery. Springer, Berlin
2. Reimsnider S, Manfredsson FP, Muzyczka N, Mandel RJ (2007) Time course of transgene expression after intrastriatal pseudotyped rAAV2/1, rAAV2/2, rAAV2/5, and rAAV2/8 transduction in the rat. Mol Ther 15:1504–1511
3. de Backer MWA, Brans MAD, Luijendijk MC, Garner KM, Adan RAH (2010) Optimization of adeno-associated viral vector-mediated gene delivery to the hypothalamus. Hum Gene Ther 21:673–682
4. Benskey MJ, Manfredsson FP, Lookingland KJ, Goudreau JL (2014) The role of parkin in the differential susceptibility of tuberoinfundibular and nigrostriatal dopamine neurons to acute toxicant exposure. Neurotoxicology 46C:1–11
5. Burger C (2004) Recombinant AAV viral vectors pseudotyped with viral capsids from serotypes 1, 2, and 5 display differential efficiency and cell tropism after delivery to different regions of the central nervous system. Mol Ther 10:302–317
6. Gray SJS et al (2011) Optimizing promoters for recombinant adeno-associated virus-mediated gene expression in the peripheral and central nervous system using self-complementary vectors. Hum Gene Ther 22:1143–1153
7. Shevtsova Z, Malik JMI, Michel U, Bähr M, Kügler S (2005) Promoters and serotypes: targeting of adeno-associated virus vectors for gene transfer in the rat central nervous system in vitro and in vivo. Exp Physiol 90:53–59
8. Burger C, Nguyen FN, Deng J, Mandel RJ (2005) Systemic mannitol-induced hyperosmolality amplifies rAAV2-mediated striatal transduction to a greater extent than local co-infusion. Mol Ther 11:327–331
9. Carty N et al (2010) Convection-enhanced delivery and systemic mannitol increase gene product distribution of AAV vectors 5, 8, and 9 and increase gene product in the adult mouse brain. J Neurosci Methods 194:144–153

Chapter 15

MRI-Guided Delivery of Viral Vectors

Ernesto A. Salegio, John Bringas, and Krystof S. Bankiewicz

Abstract

Gene therapy has emerged as a potential avenue of treatment for many neurological disorders. Technological advances in imaging techniques allow for the monitoring of real-time infusions into the brain of rodents, nonhuman primates, and humans. Here, we discuss the use of magnetic resonance imaging (MRI) as a tool in the delivery of adeno-associated viral (AAV) particles into brain of nonhuman primates.

Key words MRI, Gene therapy, AAV, Nonhuman primates, Preclinical

1 Introduction

Treatment of central nervous system (CNS) disorders is challenging due to the protected compartmentalization of the brain and spinal cord by the blood-brain barrier (BBB). A barrier harbored by tight endothelial junctions with selective permeability. Temporary penetrability of the BBB has been explored by the peripheral/systemic administration of chemical, biological, and physical interventions [1]. However, while these approaches provide certain benefits, they also have their limitations. For instance, while there are CNS disorders that would require systemic consideration, neurological conditions such as Parkinson's disease would benefit more from precise local targeting. The importance of direct, focal parenchymal drug delivery was recently highlighted as a key determinant for the low efficacy encountered in a Phase 2 clinical trial in Parkinson's patients. Evidence for this came from a prospective study conducted in nonhuman primates, which replicated the clinical protocol used in the Phase 2 clinical trial and found that the methods for drug delivery were suboptimal [2]. In addition, based on clinical outcomes [3–6], it is clear that parenchymal delivery, dosage, protein expression and drug distribution in situ will continue to be major determinant of therapeutic benefit. Hence, our commitment to developing technological tools that are highly translatable.

Fredric P. Manfredsson (ed.), *Gene Therapy for Neurological Disorders: Methods and Protocols*, Methods in Molecular Biology, vol. 1382, DOI 10.1007/978-1-4939-3271-9_15, © Springer Science+Business Media New York 2016

With these challenges in mind, we have previously described the use of multiple interventional magnetic resonance imaging (iMRI) targeting systems, including one that allows for the delivery of clinically relevant volumes with a target error of less than 1 mm [7]. This novel delivery system integrates an MRI-compatible skull-mounted aiming device (SmartFrame®), a sophisticated software platform (ClearPoint®) and a reflux-resistant cannula specifically designed for the delivery of pressurized infusates [8]. This delivery platform was validated preclinically in nonhuman primates confirming its accuracy, predictability and overall safety [8, 9]. Optimization of such delivery systems has been accomplished by ongoing modifications and improvements of previous cannula-guided designs such as the chimney-design array [10]. In this chapter we will describe two styles of MRI compatible guides and a newly developed ball-joint port system and discuss their application in real-time MRI-guided delivery of adeno-associated viral (AAV) vector into the nonhuman primate brain.

2 Materials

2.1 Animal Guidelines

1. As per any other experimental investigation involving the use of animals, animal protocols are submitted and approved by the appropriate Institutional Animal Care and Use Committees.
2. When working with adult rhesus or cynomolgus monkeys, they are housed individually, in rooms set to a 12-h light/dark cycle and with room temperature ranging between 64 and 84 °F.
3. Once animals have been purchased from the appropriate vendor, there is a 31-day quarantine period mandated by the Centers for Disease Control (CDC).
4. Upon arrival and prior to allocating animal to different studies, they should undergo a detailed examination by a qualified veterinarian.

2.2 Adeno-Associated Virus

1. AAV preparations were purchased from the Children's Hospital of Philadelphia (CHOP, *see* **Note 1**). It is important that each vector be manufactured according to the investigator's needs, as this will determine selection of the promoter, transgene, and serotype of the AAV. AAV serotype 2 is the most common vector used in clinical trials and we have recently published a detailed protocol describing AAV2 production from the HEK 293 cell line [11].

2.3 Contrast Agent

1. An essential part of performing real-time iMRI-guided infusions is the ability to visualize the infusate during the procedure. To do this, AAV preparations were mixed with the MRI contrast agent, Gadoteridol (ProHance®), to a final concentra-

tion of 1–2 mM in phosphate-buffered saline (PBS, pH 7.4) and Pluronic F-68 (0.001 % v/v; Invitrogen, Carlsbad, CA, USA). Gadoteridol is commercially available (Bracco Diagnostics Inc., Monroe Township, NJ, USA) and each mL contains 279.3 mg of Gadoteridol with 0.23 mg Calteridol Calcium and 1.21 mg of Tromethamine.

2.4 Analgesics

1. Ketamine (Ketaset, 7 mg/kg, administered intramuscularly) and xylazine (Rompun, 3 mg/kg, administered intramuscularly).

2.5 List of Tools and Equipment

1. Periosteal elevator (Fine Scientific Tools, Foster City, CA, USA), rongeurs calipers (Fine Scientific Tools, Foster City, CA, USA), gelfoam (Baxter, Deerfield, IL, USA), dental acrylic, gauze, syringes (5 and 50 mL), latex gloves, stopwatch timer.
2. Reflux-resistant infusion cannula (Upchurch Scientific, West Berlin, NJ, USA; *see* **Note 2**).
3. Teflon tubing for secondary and loading lines (1.57 mm outer diameter, 0.76 mm inner diameter; Upchurch Scientific, West Berlin, NJ, USA; *see* **Note 3**).
4. Plastic cannula guide ports (three types available, *see* **Note 4**).
5. AAV2 vector (CHOP; *see* Subheading 2.2).
6. Gadoteridol (ProHance®; Bracco Diagnostics Inc., Monroe Township, NJ, USA; *see* Subheading 2.3).
7. Skull-mounted aiming device (SmartFrame®, MRI Interventions Inc., Memphis, TN, USA) and software (ClearPoint®, MRI Interventions Inc., Memphis, TN, USA).
8. Sterile hardware: Plastic screws, pens, rulers, screwdriver, dummy catheter (Upchurch Scientific, West Berlin, NJ, USA), large animal MRI-compatible stereotaxic frame (Kopf Instruments, Tujunga, CA, USA), 3500 Medfusion pump (Strategic Applications Inc., Lake Villa, IL, USA), Tefzel ferrule connectors and Luer-Lock adapters (Upchurch Scientific, West Berlin, NJ, USA), impaction drill (3.5 mm round drill bit; Stryker, Portage, MI, USA).
9. 1.5-T MRI scanner (Signa LX; GE Medical Systems, Waukesha, WI, USA), 5-in. circular surface MRI coil (MR Instruments Inc., Hopkins, MN, USA).
10. OsiriX® software (v5.5.2; Pixmeo, Bernex, Switzerland).

3 Methods

3.1 Calculation of Coordinates

Prior to port implantation, baseline MR images should be acquired to establish optimum trajectory to the region of interest (ROI).

Placement of skull-mounted ports and predicted cannula trajectory (e.g., into a gray matter structure) should avoid (1) puncturing through major cortical blood vessels; (2) lateral ventricles; (3) inaccurate targeting of ROI; and (4) white matter tracts. These factors are known to affect the distribution and confinement of the delivered infusate [12].

1. Sedate the animal with a mixture of ketamine and xylazine (*see* Subheading 2.4), test for reflexes and intubate.
2. Continue anesthesia with isoflurane and monitor vital signs.
3. Establish an intravenous line to delivery sedatives and/or administer other drugs as necessary.
4. Place animal in a stereotactic frame, preferably a frame that is MRI compatible and can be used during infusion in the magnet.
5. Position the animal inside the MRI bore, center the animal using the laser crosshairs and reposition if needed.
6. Advance the MRI bed into the bore and begin scanning (*see* Subheading 3.6).
7. MR images should be acquired in an axial plane and/or 3-dimensions (3D) to allow for the visualization and confirmation of the ear bar position along the anterior-posterior and mediolateral axis.
8. Cannula trajectories are determined by the position of the skull-mounted port relative to the location of subcortical target sites from the ear bars (i.e., applicable to the chimney and ball-joint ports). Additional steps are required when using the SmartFrame/ClearPoint platform (*see* **Note 5**).

3.2 Surgery: Port Implantation

Each surgical procedure is customized to each animal and comparison is drawn between three ports.

1. Browse through the baseline MR images, locate the image where the ear bars are most noticeable and determine the relative anterior-posterior (AP) distance to the ROI (*see* **Note 6**). Note that these (AP) coordinates can also be generated using an anatomy atlas, which provides 3D coordinates for all regions within the brain [13].
2. Select a trajectory the trajectory that safely reaches the targeted ROI through a burr hole. The size of the craniotomy is governed by the size of the skull-mounted plastic port (*see* **Note 7**).
3. Port/s (chimney-array and SmartFrame) should be surgically implanted onto the skull at approximately 2 weeks prior to infusion, whereas the ball-joint port is implanted and removed on the same day of the procedure (*see* **Note 4**).
4. Sedate the animal (*see* **steps 1–5**, Subheadings 3.1).
5. Record ear bar and eye bar measurements.

6. Under aseptic conditions, make a midline incision in the skin and dissect the underlying muscle layers. Make sure to remove all connective tissue from the skull prior to drilling.
7. Based on AP coordinates, place marks on the skull using a sterile pen outlining the port placement (*see* **Note 8**). Double check measurements and coordinates.
8. Commence drilling (Stryker drill, 3.5 mm round drill bit) the burr hole/s and be cautious not to damage the sagittal sinus.
9. Once burr holes are drilled, remove excess bone fragments and make sure the skull is clean and dry to ensure the dental acrylic binds to the skull. Note that this step is not necessary with the ball-joint port.
10. Prior to adding the dental acrylic, secure three plastic screws (equally spaced) into the skull to provide extra support to the port.

    Next, position the port into the burr hole and cover all exposed regions where the dura is visible. Gelfoam can be added to protect the dural surface that is not covered by the port (*see* **Note 9**).
11. Apply the dental acrylic around the port and screws (*see* **Note 10**). Allow the first layer to dry before adding more acrylic. Ensure that the acrylic covers the base of the port, the screws and any small bone apertures (reinforced with gelfoam).
12. Check patency of the dental acrylic in the ports, screws, and skull.
13. Suture the muscle, connective tissue, and skin in anatomical layers. Allow the animal to recover for 2 weeks before conducting an infusion procedure through the ports (*see* **Note 11**). As mentioned in Subheading 3.2, **step 9**, the ball-joint port does not require a 2-week waiting period prior to the infusion procedure (*see* **Note 10**).

3.3 Setup of Infusion System

There are three components that will be required for a successful infusion (*see* **Note 3**): a step-design infusion cannula, a loading line (sterile tubing; *see* Subheading 3.7), and a secondary line (non-sterile tubing; *see* Subheading 3.4) that is directly connected to the infusion pump (Fig. 1).

3.4 Secondary Line Setup

1. Start with the non-sterile secondary line, which is connected to a 5 mL syringe mounted on the infusion pump.
2. Measure the distance from the pump to the MRI bore and the approximate the distance of the loading line from the cannula.
3. Cut the Teflon tubing (1.57 mm outer diameter, 0.76 mm inner diameter) to the desired distance.
4. Add Tefzel ferrule connectors and Luer-Lock adapters to both ends of the tubing.

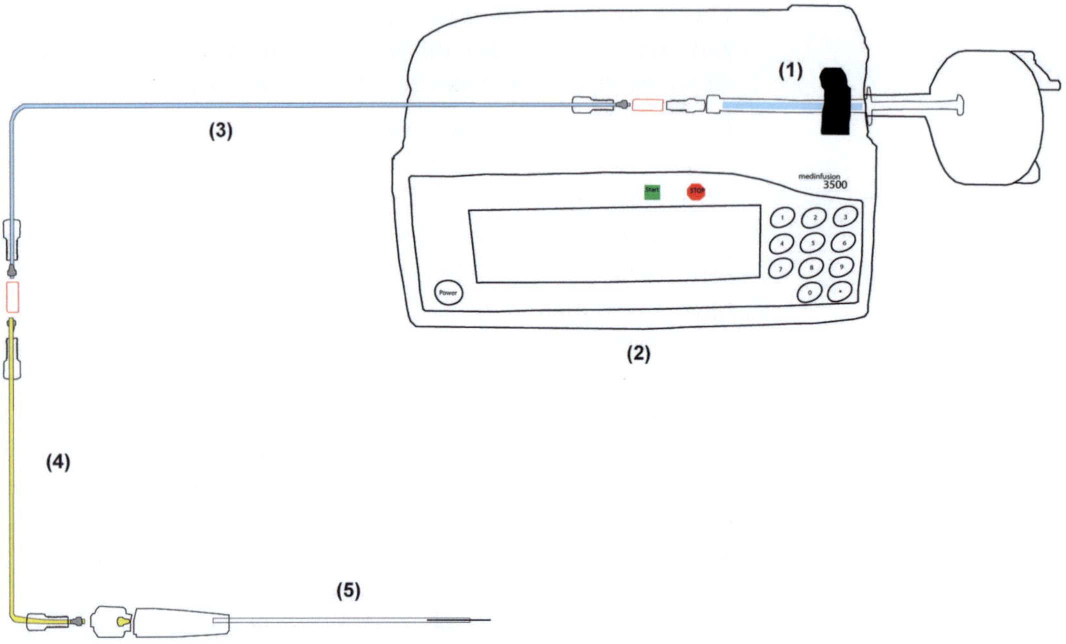

Fig. 1 Infusion system setup: This schematic representation demonstrates all of the components in the infusion system. Starting with a 5 mL syringe (*1*) mounted onto an infusion pump (*2*), a secondary non-sterile line primed with sterile saline and appropriate connectors at the distal end (*3*), followed by a sterile loading line containing the infusate (AAV + Gd) with connectors at each end of the line (*4*) and terminating at a sterile infusion cannula (*5*). Note illustrations are not up to scale

5. Fill the tubing with sterile saline using a 50 mL syringe.
6. Fill a 5 mL syringe with saline, attach to the secondary line and mount the syringe directly onto the pump.
7. Check for leaks before use and ensure that there are no air bubbles in the system.
8. Begin setting up the loading line.

3.5 Loading Line Setup

1. Create a sterile field and layout the sterile materials.
2. Place the infusion cannula onto the sterile field, attach it to a 5 mL syringe on the proximal end, flush with sterile saline and check for patency.
3. Add connectors to each end of the loading line.
4. Prime the loading line with sterile saline prior to loading the infusate (i.e., AAV mixed with MRI tracer).
5. Load the infusate into the sterile line and connect to the proximal end of the cannula.
6. Continue pushing the infusate until the cannula is filled.

7. Make sure *NO* air bubbles are present in the loading line and/ or cannula (*see* **Note 12**).
8. Connect the secondary line to the loading line and cannula.
9. Check for leaks in the entire system.
10. Transfer the system to the MRI bore and introduce the cannula through an identified aperture on the infusion port.
11. Begin infusion (*see* Subheading 3.7).

3.6 MRI Protocol

The MRI parameters are specific to visualizing the Gadoteridol contrast agent during infusion (Fig. 2). Note that this protocol has been written for use in at 1.5 T Signa LX scanner and should be modified when using different types of scanners, coils, and hardware.

1. Start with a 3D high-resolution MP-RAGE (magnetization prepared-rapid acquisition gradient echo) scan (for additional parameters, *see* **Note 13**).
2. Continue with T2-weighted images to determine trajectory of the cannula from the top of the port (*see* **Note 14**).
3. Advance the lancet rod through the selected aperture and penetrate the dura without damaging the cortical surface.
4. Using a sterile ruler, measure the desired depth on the cannula and place a mark. Slide a depth stopper over the tip of the cannula and secure it to the desired length.
5. Insert the cannula through the opening in the dura and advance to the desired depth.
6. Monitor vital signs during cannula insertion.
7. Begin T1-weighted image acquisition with a spoiled GRASS (gradient recalled acquisition in stead state) sequence.
8. Set the slice thickness to 1 mm and determine the number of slices to visualize the ROI. Depending on the number of slices, the scanning time should be range between 9 and 12 min.

3.7 Infusion Protocol

1. Double-check all lines and connectors (*see* Subheadings 3.4 and 3.5).
2. Calculate the total amount of infusate required to fill the dead space in the lines, connectors, cannulae, and the dose volume.
3. Start the pump at 0.5 μL/min and ensure that infusate exits the tip of the catheter prior to insertion into the selected aperture.
4. After the catheter is advanced to the target site, set the total volume delivered on the pump to zero. Document the start time and keep track of the infusion time by using a stopwatch timer.
5. Acquire baseline T1- and T2-weighted images to ensure accuracy of cannula penetration to reach the target structure.

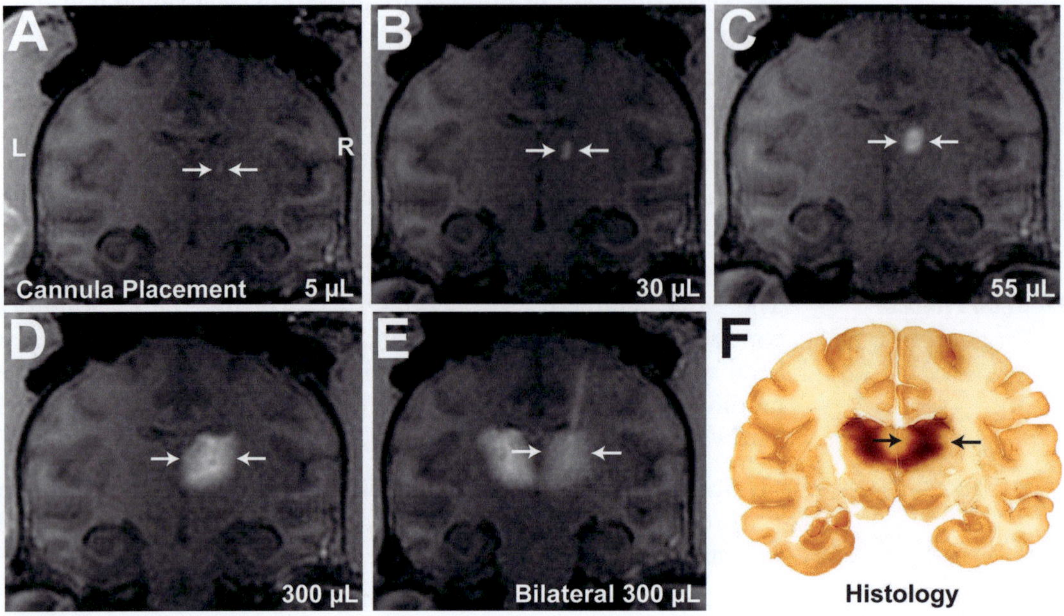

Fig. 2 Real-time MRI infusion into the nonhuman primate thalamus and comparative histology section: Infusion of AAV2-GDNF/Gd visualized as a contrast demarcation on the MR images indicates placement of the cannula tip within the thalamus (**a**; *right* side; *white arrow*). As the infusate volume increases during the procedure, the size of the demarcation continues to grow until reaching 300 μL (**b–d**). Note that this was a bilateral infusion completed sequentially and the infusate was visualized in the contralateral side (**e**). A matching histology section processed 5 weeks after infusion demonstrates expression of the infused transgene, closely representing original MRI observations (**f**; *black arrows*). *AAV2* adeno-associated virus, *GDNF* glial-derived neurotrophic factor, *Gd* gadoteridol (MRI tracer)

6. Once the infusate is visualized in the MR images, the rate can be progressively increased at 0.5 increments to 1.0, 1.5, 2.0, 2.5, 3.0, 3.5, 4.0, 4.5, and 5.0 μL/min (*see* **Note 15**).
7. Continue scanning during the infusion procedure.
8. Monitor MR images to ensure infusate is confined within the ROI.

3.8 Calculating Volume of Distribution Versus Volume Infused

Linearity between the volume of distribution (Vd) versus volume infused (Vi) is a determining factor on how well an infusion was performed. This type of analysis is conducted after the infusion procedure using open-source OsiriX software and involves manually tracing the ROI on each slice. The software uses an automated calculation feature to determine the ROI volume.

1. Upload DICOM (Digital Imaging and Communications in Medicine)-formatted MR images to a software capable of performing 3D reconstructions and volumetric analysis, such as OsiriX®.

2. Locate the last MRI sequence where the total intended volume has been infused (*see* **Note 16**). Either a T1- or T2-weighted scans can be used; however, this is dependent on the scanning protocol and type of images acquired (*see* **Note 17**).
3. On the mouse button function menu select the pencil icon and begin tracing the ROI (i.e., infusate) throughout the MRI series. Remember to trace regions where the infusate can be visualized (*see* **Note 18**).
4. After tracing all ROIs, go to the "ROI" menu, select ROI volume and compute volume. This tool also provides the option of generating missing ROIs or deleting generated ROIs.
5. By using the volume of distribution generated in OsiriX and referencing the total volume infused from the study notes, the Vd/Vi ratio can now be calculated (*see* **Note 19**).

4 Notes

1. Material transfer agreement (MTA) will need to be organized between the Primary Investigator, Institution and Provider, in addition to adequate protocol approvals. When working with AAV preparations it is important to prepare them on the day of the procedure to avoid vector aggregation and/or lysis due to frequent thawing and re-freezing of aliquots. If infusions are conducted at a location remote from the MRI facility, it is recommended to transport the infusate on dry ice prior to aliquoting and/or if already aliquoted, it can be placed on wet ice.
2. Infusion cannulas are specifically made to have a step design in which the height of the step, measured from the distal part of the tip, can vary depending on the dimensions of the targeted structure. In larger animal models, for example, the step measures 3 mm from the tip, whereas, in smaller species, the step is normally 1 mm. Note that the steplike design is produced by having a thicker diameter sheath (outer diameter 0.53 mm, inner diameter 0.45 mm) over a thinner diameter cannula (outer diameter 0.43 mm, inner diameter 0.32 mm). This is an important configuration of the delivery platform and has been consistently used in our experiments [14].
3. The actual length of Teflon tubing is proportional to the distance of the pump from the cannula. For instance, MRI-compatible pumps can be positioned close to the magnet bore, whereas, conventional pumps must be operated from a considerable distance outside the magnetic field. There are three components required for infusion: (1) step-design infusion cannula, (2) a loading line containing the AAV/MRI tracer, and (3) a secondary line filled with sterile saline that is con-

nected to a syringe and mounted onto the infusion pump. Therefore, the distance of the pump to the infusion cannula will dictate the length of tubing. It is important to note that the dead volume of the infusion system should be calculated in advance to ensure that an adequate dose volume is administered into the ROI.

4. The most commonly used port is the chimney-designed array with 27 access holes that is threaded into the port (Fig. 3a) [10]. A neuronavigational platform that has been clinically approved consists of a skull-mounted SmartFrame® (Fig. 3b), a reflux-resistant cannula and software package (ClearPoint®) that communicates with the MRI console [8, 15]. Recent developments in our laboratory have included the use of a "ball-joint" style (Hayes Manufacturing Services Inc., USA) that involves a simpler procedure for skull mounting and provides a broad range of movement for targeting, as compared to the static chimney array (Fig. 3c). Briefly, this port is implanted on the same day of the infusion procedure by securing three screws to the skull, suturing the muscles and transferring the animal to the scanner. Once the infusion is completed, the sedated animal is returned to the surgical room, the port is removed by reopening through the anatomical layers and closing the wound site.
5. The SmartFrame (i.e., skull-mounted aiming device) has a built-in fluid stem and fiducials that are registered to neuronavigational software (ClearPoint). MR images are acquired on the MRI console and sent directly to an external computer containing the Clearpoint system. Surgical coordinates are calculated from a defined reference point, the anterior and posterior commissure (AC-PC), to the target site. Cannula trajectory is adjusted/aligned using a hand-handled controller with an X-Y (pitch and roll) translational stage [8].
6. During baseline scanning the animal must be centered in the stereotactic frame using the ear and eyebar attachments. Points of reference used to determine the location of the burr hole and implanted port will be dictated by visualizing the ear bars on the MR images. Based on MRI slice thickness, the appropriate anterior-posterior coordinates can be generated to best target the ROI.
7. These access ports are made of plastic, are hollow and threaded on the inside (14 mm diameter × 12 mm height). Chimney-designed arrays with 27 access holes are threaded into the port (12 mm diameter × 14 mm height).
8. Note that this is not a complete craniotomy but rather a small burr hole on the skull over the targeted anatomical region. To ensure accuracy, all coordinates must be generated using sterile stereotactic equipment.

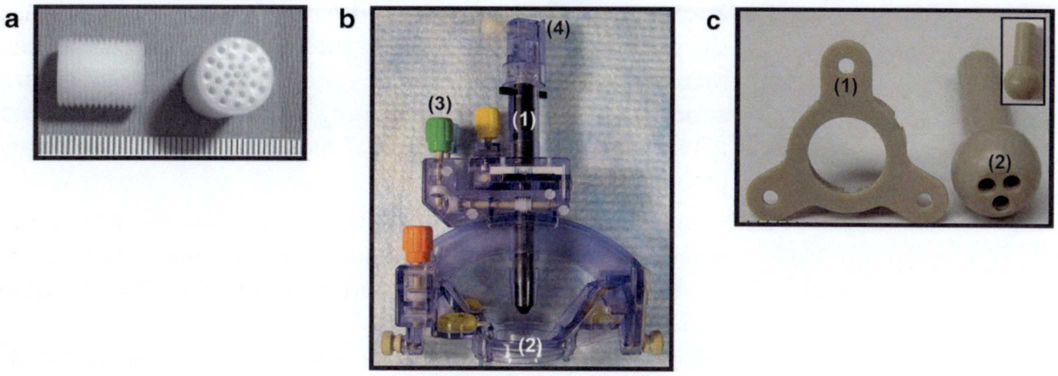

Fig. 3 Types of cannula guides for parenchymal infusions. (**a**) Chimney-style port designed with threaded exterior and 27 access holes. The use of this cannula-guide requires placement of three plastic screws onto the skull around the port base. Dental acrylic is applied over the screws and proximal to the base to secure the port onto the skull. (**b**) SmartFrame device consisting of a fluid-filled stem (*1*), an internal thread used for attachment to a skull-mounted base (*2*), color-coded adjustment knobs used for aligning the cannula trajectory (*3*) and a single access hole for the cannula (*4*). The Smartframe device requires using Clearpoint neuronavigational software, a handheld controller attachment and dental acrylic to secure the base. (**c**) Novel ball-joint cannula guide cannula system with built-in screw holes at the base (*1*) and a snap-on ball socket allowing for simple orientation of cannula trajectories. This port does not require dental acrylic and unlike the other two ports, the ball-joint can be used on the same day of the infusion procedure

9. This is an important step to ensure that the application of dental acrylic will not leak through the hole and bind to the dural surface.
10. The ball-joint style port is secured to the skull by placing three screws into the assigned holes located on the base of the port. These holes provide stability when inserting the ball into the base and do not require the use of dental acrylic as compared to the chimney and the SmartFrame designs (Fig. 3).
11. The 2-week recovery period before an infusion procedure is an arbitrary time point. It simply allows the animal to recover from surgery, acclimate to the implant, and ensure port integration onto the skull.
12. When flushing the system with sterile saline and/or the infusate it is important to avoid the introduction and/or formation of air bubbles within the cannula, loading line or secondary line. Air that exits the cannula tip can impact infusate distribution within the ROI, damage parenchymal tissue and may lead to complications during anesthesia recovery. Therefore, air bubbles within the infusion system should be avoided at all costs.
13. Repetition time (TR): 2110 ms; echo time (TE): 3.6 ms; flip angle: 15°; number of excitations (NEX): 1 (repeated three times); matrix: 240×240; field of view (FOV): $240 \times 240 \times 240$.

14. Based on the type of port used to guide the cannula through the parenchymal tissue, T2-weighted images can provide information regarding optimum trajectories for target access, approximate depth for positioning the cannula tip, and identifying anatomical structures that may lead to potential leakage pathways.
15. Confirmation of catheter tip placement within the ROI is verified by the initial MRI sequences after cannula placement and infusion commences. Depending on the targeted structure, infusion rates should start at 0.5 μL/min and increased at 0.5–1.0 μL/min increments up to 5.0 μL/min. Infusions should be constantly monitored in the acquired MR images to prevent leakage of the infusate into adjacent regions.
16. There is a delay in image acquisition and the actual time of infusion completion that must be adjusted. Depending on the MRI sequence, acquisition times can be 5–20 min after infusion completion. Investigators must ensure that consecutive sequences are acquired during the infusion session.
17. Co-administration of an MRI contrast agent with the test article is essential to monitor real-time distribution of infusate in a targeted structure. T1-weighted scans will have white demarcations representative of the introduced agent. In studies that do not involve co-administration of the contrast agent, infusate distribution can be assessed using T2-weighted images [7].
18. To improve visualization of the ROI, select the gradient tool on the default toolbar. This will adjust the brightness and contrast on the MR image to reduce non-specific background noise.
19. The example provided in Fig. 4, a bilateral infusion of 300 μL into the left and right thalamus generated a volume of distribution of 733–738 mm^3, respectively. This resulted in Vd/Vi ratio of 2.4 and 2.5 with a linear correlation of $R = 0.98$. The higher the ratio, the better the spread of the infusate in relation to cannula placement. Complete 3D reconstruction of the infusate within the ROI can also be conducted using iPlan Cranial® (BrainLab, Germany).

Acknowledgement

This work was supported by a grant to K.S.B. from NIH-NINDS (R01NS073940-01).

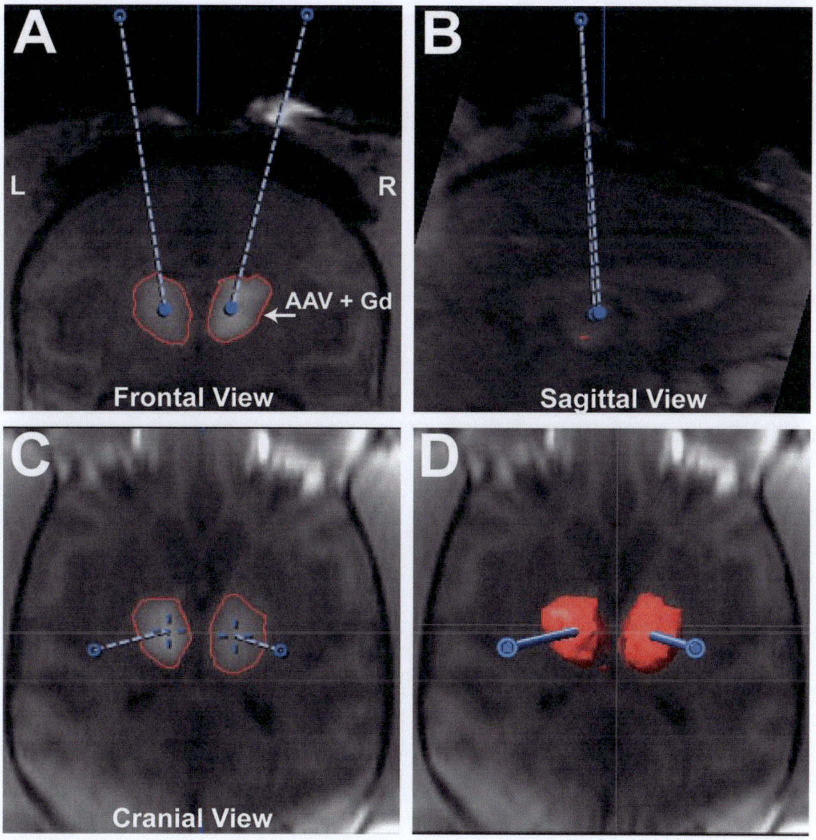

Fig. 4 Three-dimensional reconstruction of a bilateral infusion into the nonhuman primate thalamus: Infusion of (AAV + Gd) delivered into the thalami can be seen as a *white demarcation* within both ROIs. (**a**) Attenuation of Gadoteridol signal intensity between the left and right thalami are commonly observed during bilateral infusions, particularly when one side (*left* side) is completed before the other (*right* side). (**a–d**) Cannula trajectory has been included in all orientations to demonstrate the path taken to initiate infusion and reconstruct the total volume infused. Note that no signs of leakage can be seen refluxing along the cannula and/or outside the ROI

References

1. Chen Y, Liu L (2012) Modern methods for delivery of drugs across the blood-brain barrier. Adv Drug Deliv Rev 64(7):640–665
2. Salvatore MF, Ai Y, Fischer B et al (2006) Point source concentration of GDNF may explain failure of phase II clinical trial. Exp Neurol 202(2):497–505
3. Marks WJ Jr, Ostrem JL, Verhagen L et al (2008) Safety and tolerability of intraputaminal delivery of CERE-120 (adeno-associated virus serotype 2-neurturin) to patients with idiopathic Parkinson's disease: an open-label, phase I trial. Lancet Neurol 7(5):400–408
4. Eberling JL, Kells AP, Pivirotto P et al (2009) Functional effects of AAV2-GDNF on the dopaminergic nigrostriatal pathway in parkinsonian rhesus monkeys. Hum Gene Ther 20(5):511–518
5. Bartus RT, Herzog CD, Chu Y et al (2011) Bioactivity of AAV2-neurturin gene therapy

(CERE-120): differences between Parkinson's disease and nonhuman primate brains. Mov Disord 26(1):27–36

6. Marks WJ Jr, Bartus RT, Siffert J et al (2010) Gene delivery of AAV2-neurturin for Parkinson's disease: a double-blind, randomised, controlled trial. Lancet Neurol 9(12):1164–1172
7. Richardson R, Gimenez F, Salegio EA et al (2011) T2-imaging in monitoring of intraparenchymal real-time convection enhanced delivery. Neurosurgery 69(1):154–163
8. Richardson RM, Kells AP, Martin AJ et al (2011) Novel platform for MRI-guided convection-enhanced delivery of therapeutics: preclinical validation in nonhuman primate brain. Stereotact Funct Neurosurg 89(3):141–151
9. San Sebastian W, Richardson RM, Kells AP et al (2012) Safety and tolerability of magnetic resonance imaging-guided convection-enhanced delivery of AAV2-hAADC with a novel delivery platform in nonhuman primate striatum. Hum Gene Ther 23(2):210–217
10. Salegio E, Kells AP, Richardson RM et al (2010) Magnetic resonance imaging-guided delivery of adeno-associated virus type 2 to the primate brain for the treatment of lysosomal storage disorders. Hum Gene Ther 21:1–11
11. Hadaczek P, Kohutnicka M, Krauze MT et al (2006) Convection-enhanced delivery of adeno-associated virus type 2 (AAV2) into the striatum and transport of AAV2 within monkey brain. Hum Gene Ther 17(3):291–302
12. Yin D, Richardson R, Fiandaca M et al (2010) Cannula placement for effective convection-enhanced delivery in the nonhuman primate thalamus and brainstem: implications for clinical delivery of therapeutics. J Neurosurg 113:240–248
13. Paxinos G, Huang X, Petrides M, Toga A (2008) The rhesus monkey brain. Elsevier Inc, USA
14. Yin D, Forsayeth J, Bankiewicz K (2010) Optimized cannula design and placement for convection-enhanced delivery in rat striatum. J Neurosci Methods 178:46–51
15. Krauze MT, Saito R, Noble C et al (2005) Reflux-free cannula for convection-enhanced high-speed delivery of therapeutic agents. J Neurosurg 103(5):923–929

Chapter 16

Systemic Gene Therapy for Targeting the CNS

Sara E. Gombash and Kevin D. Foust

Abstract

Systemic gene delivery is useful for modeling and treatment of a body-wide disease. Recently, it has been shown that certain agents, when delivered systemically, can efficiently target the central nervous system. This technique has been used to model and treat rodent models of neurological disease with unprecedented success. Here, we describe intravenous delivery in neonate and adult mice. These techniques are easily learned and have minimal equipment requirements.

Key words Intravenous injection, Superficial temporal vein, Tail vein, Adeno-associated virus type 9, Systemic gene therapy

1 Introduction

The feasibility of systemic gene therapy for neurological disease is starting to be realized [1–3]. The ability to rapidly create chimeric organisms has proven to be powerful for clinical development and as a laboratory tool [4]. Indeed, systemically delivering corrective genes to mouse models of neurological disease has produced unprecedented preclinical results [5–8]. An advantage of systemic delivery is the ease that the gene therapy can be administered. No expensive, special equipment or surgical procedure is required. The injection procedures in mice are minimally invasive and quick to perform, often less than 5 min per animal. A second advantage is the robust widespread CNS transgene expression that can be achieved following systemic delivery of novel adeno-associated viral serotypes [9–11] (Figs. 1 and 2). The broad transduction has allowed for delivery of intracellular transgenes for diseases with global pathology [12–14]. Surprisingly, data from preclinical studies suggest that gene restoration in a fraction of cells can correct severe neurological phenotypes [7, 8]. Potential disadvantages of systemic gene therapy for CNS disease are the relatively high viral doses required for efficient brain and spinal cord targeting.

Fredric P. Manfredsson (ed.), *Gene Therapy for Neurological Disorders: Methods and Protocols*, Methods in Molecular Biology, vol. 1382, DOI 10.1007/978-1-4939-3271-9_16, © Springer Science+Business Media New York 2016

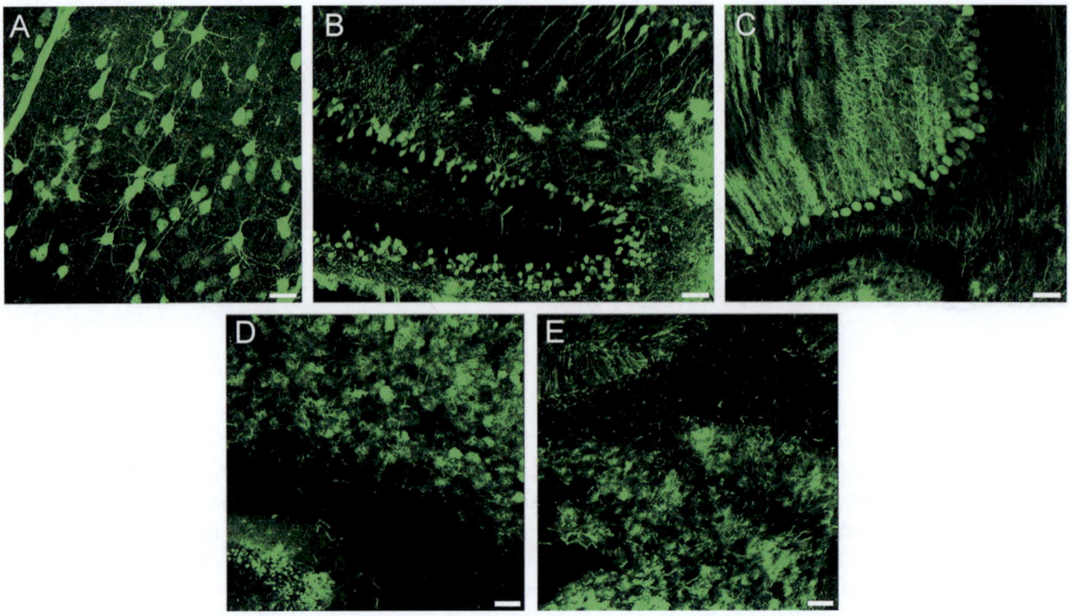

Fig. 1 Representative images of mouse brain following systemic injection of AAV9 CB GFP. GFP immunofluorescence shows extensive GFP expression in neurons throughout multiple structures including the cortex (**A**), hippocampus (**B**) and cerebellum following injection into neonate mice. Sections from the cortex (**D**) and medulla (**E**) show that brains of animals that received intravenous AAV9 injections as adults have GFP expression primarily in astrocytes. Scale bars A-C = 50μm; D-E = 100μm

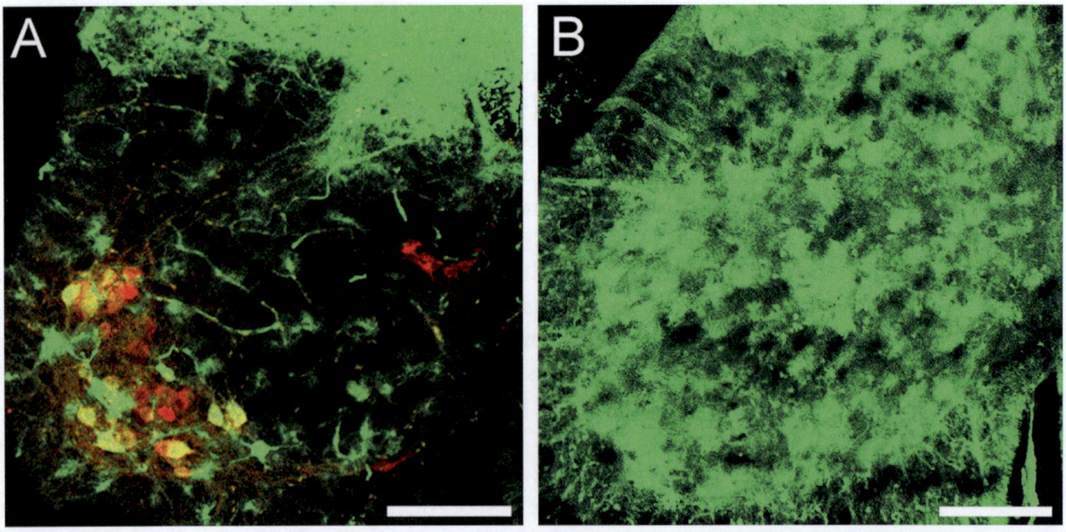

Fig. 2 Representative images of mouse lumbar spinal cord following systemic injection of AAV9 CB GFP. GFP immunofluorescence from lumbar spinal cord sections following intravenous injections into neonate (**A**) or adult (**B**) mice show differences in viral transgene expression. A) Injection into neonate mice results in abundant motor neuron (ChAT, red) and dorsal root ganglia cell transduction. GFP expression is also observed in endothelial cells and astrocytes. B) Intravenous injection in adult mice results in robust GFP expression within spinal grey matter astrocytes. GFP expression in motor neurons and dorsal root ganglia is reduced. All scale bars = 200μm

These may pose challenges for vector manufacturing and raise the potential for immune responses [15–17]. Strategies are being devised to overcome these obstacles with new manufacturing platforms and clever molecular biology techniques to limit infection and gene expression such as molecular evolution for tissue detargeting, and by incorporating microRNA binding sites into the vector expression cassette to restrict cell type expression [18–20].

Our experience is with the systemic delivery of adeno-associated viruses (AAV) in mice and non-human primates, though others have shown that similar crossing of the blood–brain barrier exists in rats and cats as well [7, 10, 21, 22]. Systemic delivery for targeting the CNS has also been reported in neonate mice using lenti-, retro-, and adenovirus vectors [23–26]. Currently AAV9 is the most widely used, but newer AAV serotypes are emerging that show similar properties [11]. Further characterization remains to determine the niche each new serotype can fill. Clinically, systemic AAV gene therapy for treatment of CNS disease has not yet been tested in humans, but the US Food and Drug Administration recently granted investigational new drug approval for IV delivery of AAV9 in 0- to 9-month-old children with the lower motor neuron disease, spinal muscular atrophy. A phase 1 clinical trial is slated to begin in the second half of 2014. Here, we describe the techniques we use to systemically deliver adeno-associated viruses to neonate and adult mice.

2 Materials

2.1 Neonate Injections

1. Container of wet ice.
2. Mouse pups on their first or second postnatal day.
3. Thinpro Insulin Syringe (Terumo, Sommerset, NJ, USA), 3/10 cc, 3/8″ needle, 30 g, 1 per mouse.
4. Cotton Tipped Applicators.
5. Fiber Optic Light Source.
6. Dissecting Microscope.
7. For training, 1 % Evans Blue Dye solution, made with 1× Phosphate Buffered Saline.
8. For experiments, viral solution.

2.2 Adult Injections

1. Mouse tail vein injection device (Braintree Scientific, Braintree, MA, USA).
2. Mice, >21 days of age.
3. Thinpro Insulin Syringe (Terumo, Sommerset, NJ, USA), 3/10 cc, 3/8″ needle, 30 g, 1 per mouse.

4. For training, 1× Phosphate Buffered Saline.
5. For experiments, viral solution.
6. Alcohol wipes to clean the tail.

3 Methods

3.1 Neonate Injections

1. Gather wet ice to anesthetize the mouse pups, an empty cage to segregate the dam from the litter, a dissecting microscope, a light source that can be positioned at an angle to the injection (use of a light source at a 90° angle to the injection site obscures the vein), a clean surface to place the animal for the injection, cotton swabs, 3/10 cc insulin syringe with 3/8″ 30 g needle (one per animal) and 1 % Evans Blue Dye (made with phosphate buffered saline (PBS)) solution for training (*see* **Note 1**).
2. Remove the dam to a separate cage while manipulating the pups (*see* **Note 2**).
3. Place a single pup directly on the wet ice for 30–60 s to anesthetize the animal.
4. While the animal is on ice, load your syringe with 30 μl of Evans Blue Dye.
5. When the animal is fully anesthetized, confirmed by lack of movement on the ice while still breathing, move it under the microscope. For a right-handed injection, face the animal's muzzle to the right. Place your left index finger on the muzzle and your left middle finger caudal to the ear bud. The ear bud should be between the index and middle fingers (*see* **Note 3**).
6. Examine just anterior to the ear bud for a superficial capillary that moves when the skin is manipulated. This capillary is NOT the target, however identification is important for temporal vein identification. Next, locate a dark, shadowy vein inferior to the capillary that remains fixed regardless of skin position. The temporal vein appears shadowy, runs dorsal to ventral, and feeds into the jugular vein.
7. Enter the temporal vein with the needle bevel up. The needle bevel will fill with blood through the skin. Then depress the plunger slowly and note blanching of the vein down the side of the face.
8. Allow the needle to remain within the vein for an added 10–15 s to prevent backflow in the injectant.
9. After a proper injection, the pup should turn blue almost immediately. Remove the needle and apply a cotton swab to the injection site until the blood clots.
10. Monitor the pup for signs of distress. Allow the pup 2–3 min to recover and re-warm, recognized when the pup is conscious,

upright and moving, before returning to the cage. Pups are often cupped in the investigator's gloved hands to provide appropriate warmth to aid in recovery.

11. Place the pup back into the home cage and ensure the pup is coated with bedding and/or nestlet to ensure reacceptance by the dam.
12. A new syringe and cotton swab should be used for each pup to maintain sterility.
13. Return the mother to the cage with the pups.

3.2 Adult Tail Vein Injections

1. Load the syringe with the viral solution.
2. Secure the animal into the restrainer.
3. Clean the tail with the alcohol wipe.
4. Warm the tail with warm water (if not using the Braintree restrainer) or the trough in the mouse tail illuminator.
5. Target the vein starting closer to the tip of the tail. Enter the vein. If you miss the vein, move closer to the base of the tail and try again.
6. Inject viral solution in a bolus. Injection volumes range from 100 to 300 μl (*see* **Note 4**).
7. Return the animal to its housing.

4 Notes

1. Evans blue dye is only used in training. Viral solutions are not mixed with dye. Injection volume in neonate mice is restricted to 50 μl. For neonate injections of self-complementary AAV viruses, viral doses range from 1 to 5×10^{11} vg for efficient CNS targeting.
2. A danger of this injection technique is the mother rejecting the treated pups, or the litter entirely. Although a rare event, the risk can be minimized by the selection of mouse strain and dam choice. Some inbred strains are observed to be "better" mothers, which impacts the tendency to reject the treated litters. In our experience, FVB mice are very tolerant of the injection procedure. Mice on the C57/Bl6 background also tolerate manipulation of the pups but loss of treated pups is more frequent compared to FVB mice, though these are personal observations and have not been quantified. First-time mothers also tend to be less accepting of handling the pups. Choosing dams that have had at least one litter prior to the pups for treatment is also advantageous though not required. Finally, the disease model can also influence how well the neonate injection procedure is tolerated. For example, a model of an autism

spectrum disorder had a much higher rate of pup rejection than expected based on the background strain.

3. The superficial temporal vein is visible on either side of the head. Left-handed individuals may find that facing the animal to the investigator's left may be ergonomically better and should be determined for each individual investigator. Should the vein not be visible or is mistargeted on initial attempts, targeting the vein on the opposite side of the head is appropriate. If the vein is punctured on both sides of the head, some leakage may occur through the first site during injection into the second site.

4. For delivery of self-complementary adeno-associated virus, injections are routinely performed into 21-day-old mice. Animals are dosed at 1.5×10^{11} vector genomes/gram body weight which is generally delivered in a volume of 200 μl. Injection volume is capped at 100 μl for 14-day-old mice. As mice pass 20 g in body weight, we will administer injections up to 300 μl. We have not performed studies examining the effect of injection volume on viral spread, and have chosen to maintain a constant injection volume at each age group by mixing virus with PBS.

References

1. Dayton RD, Wang DB, Klein RL (2012) The advent of AAV9 expands applications for brain and spinal cord gene delivery. Expert Opin Biol Ther 12(6):757–766. doi:10.1517/14712598.2012.681463
2. Silva GA (2008) Nanotechnology approaches to crossing the blood-brain barrier and drug delivery to the CNS. BMC Neurosci 9(Suppl 3):S4. doi:10.1186/1471-2202-9-S3-S4
3. Blumling Iii JP, Silva GA (2012) Targeting the brain: advances in drug delivery. Curr Pharm Biotechnol 13(12):2417–2426
4. Ramesh T, Lyon AN, Pineda RH, Wang C, Janssen PM, Canan BD, Burghes AH, Beattie CE (2010) A genetic model of amyotrophic lateral sclerosis in zebrafish displays phenotypic hallmarks of motoneuron disease. Dis Model Mech 3(9–10):652–662. doi:10.1242/dmm.005538
5. Adachi H, Katsuno M, Minamiyama M, Sang C, Pagoulatos G, Angelidis C, Kusakabe M, Yoshiki A, Kobayashi Y, Doyu M, Sobue G (2003) Heat shock protein 70 chaperone overexpression ameliorates phenotypes of the spinal and bulbar muscular atrophy transgenic mouse model by reducing nuclear-localized mutant androgen receptor protein. J Neurosci 23(6):2203–2211
6. Foust KD, Salazar DL, Likhite S, Ferraiuolo L, Ditsworth D, Ilieva H, Meyer K, Schmelzer L, Braun L, Cleveland DW, Kaspar BK (2013) Therapeutic AAV9-mediated suppression of mutant SOD1 slows disease progression and extends survival in models of inherited ALS. Mol Ther 21(12):2148–2159. doi:10.1038/mt.2013.211
7. Foust KD, Wang X, McGovern VL, Braun L, Bevan AK, Haidet AM, Le TT, Morales PR, Rich MM, Burghes AH, Kaspar BK (2010) Rescue of the spinal muscular atrophy phenotype in a mouse model by early postnatal delivery of SMN. Nat Biotechnol 28(3):271–274. doi:10.1038/nbt.1610
8. Garg SK, Lioy DT, Cheval H, McGann JC, Bissonnette JM, Murtha MJ, Foust KD, Kaspar BK, Bird A, Mandel G (2013) Systemic delivery of MeCP2 rescues behavioral and cellular deficits in female mouse models of Rett syndrome. J Neurosci 33(34):13612–13620. doi:10.1523/JNEUROSCI.1854-13.2013
9. Foust KD, Nurre E, Montgomery CL, Hernandez A, Chan CM, Kaspar BK (2009) Intravascular AAV9 preferentially targets neonatal neurons and adult astrocytes. Nat Biotechnol 27(1):59–65. doi:10.1038/nbt.1515

10. Duque S, Joussemet B, Riviere C, Marais T, Dubreil L, Douar AM, Fyfe J, Moullier P, Colle MA, Barkats M (2009) Intravenous administration of self-complementary AAV9 enables transgene delivery to adult motor neurons. Mol Ther 17(7):1187–1196. doi:10.1038/mt.2009.71
11. Zhang H, Yang B, Mu X, Ahmed SS, Su Q, He R, Wang H, Mueller C, Sena-Esteves M, Brown R, Xu Z, Gao G (2011) Several rAAV vectors efficiently cross the blood-brain barrier and transduce neurons and astrocytes in the neonatal mouse central nervous system. Mol Ther 19(8):1440–1448. doi:10.1038/mt.2011.98
12. Dufour BD, Smith CA, Clark RL, Walker TR, McBride JL (2014) Intrajugular vein delivery of AAV9-RNAi prevents neuropathological changes and weight loss in Huntington's disease mice. Mol Ther. doi:10.1038/mt.2013.289
13. Ahmed SS, Li H, Cao C, Sikoglu EM, Denninger AR, Su Q, Eaton S, Liso Navarro AA, Xie J, Szucs S, Zhang H, Moore C, Kirschner DA, Seyfried TN, Flotte TR, Matalon R, Gao G (2013) A single intravenous rAAV injection as late as P20 achieves efficacious and sustained CNS Gene therapy in canavan mice. Mol Ther 21(12):2136–2147. doi:10.1038/mt.2013.138
14. Iwata N, Sekiguchi M, Hattori Y, Takahashi A, Asai M, Ji B, Higuchi M, Staufenbiel M, Muramatsu S, Saido TC (2013) Global brain delivery of neprilysin gene by intravascular administration of AAV vector in mice. Sci Rep 3:1472. doi:10.1038/srep01472
15. Samaranch L, Sebastian WS, Kells AP, Salegio EA, Heller G, Bringas JR, Pivirotto P, Dearmond S, Forsayeth J, Bankiewicz KS (2014) AAV9-mediated expression of a non-self protein in nonhuman primate central nervous system triggers widespread neuroinflammation driven by antigen-presenting cell transduction. Mol Ther 22(2):329–337. doi:10.1038/mt.2013.266
16. Ciesielska A, Hadaczek P, Mittermeyer G, Zhou S, Wright JF, Bankiewicz KS, Forsayeth J (2013) Cerebral infusion of AAV9 vector-encoding non-self proteins can elicit cell-mediated immune responses. Mol Ther 21(1):158–166. doi:10.1038/mt.2012.167
17. Mingozzi F, High KA (2013) Immune responses to AAV vectors: overcoming barriers to successful gene therapy. Blood 122(1):23–36. doi:10.1182/blood-2013-01-306647
18. Pulicherla N, Shen S, Yadav S, Debbink K, Govindasamy L, Agbandje-McKenna M, Asokan A (2011) Engineering liver-detargeted AAV9 vectors for cardiac and musculoskeletal gene transfer. Mol Ther 19(6):1070–1078. doi:10.1038/mt.2011.22
19. Xie J, Xie Q, Zhang H, Ameres SL, Hung JH, Su Q, He R, Mu X, Seher Ahmed S, Park S, Kato H, Li C, Mueller C, Mello CC, Weng Z, Flotte TR, Zamore PD, Gao G (2011) MicroRNA-regulated, systemically delivered rAAV9: a step closer to CNS-restricted transgene expression. Mol Ther 19(3):526–535. doi:10.1038/mt.2010.279
20. Brown BD, Venneri MA, Zingale A, Sergi Sergi L, Naldini L (2006) Endogenous microRNA regulation suppresses transgene expression in hematopoietic lineages and enables stable gene transfer. Nat Med 12(5):585–591. doi:10.1038/nm1398
21. Bevan AK, Duque S, Foust KD, Morales PR, Braun L, Schmelzer L, Chan CM, McCrate M, Chicoine LG, Coley BD, Porensky PN, Kolb SJ, Mendell JR, Burghes AH, Kaspar BK (2011) Systemic gene delivery in large species for targeting spinal cord, brain, and peripheral tissues for pediatric disorders. Mol Ther 19(11):1971–1980. doi:10.1038/mt.2011.157
22. Tatom JB, Wang DB, Dayton RD, Skalli O, Hutton ML, Dickson DW, Klein RL (2009) Mimicking aspects of frontotemporal lobar degeneration and Lou Gehrig's disease in rats via TDP-43 overexpression. Mol Ther 17(4):607–613. doi:10.1038/mt.2009.3
23. Kobayashi H, Carbonaro D, Pepper K, Petersen D, Ge S, Jackson H, Shimada H, Moats R, Kohn DB (2005) Neonatal gene therapy of MPS I mice by intravenous injection of a lentiviral vector. Mol Ther 11(5):776–789. doi:10.1016/j.ymthe.2004.10.006
24. Carbonaro DA, Jin X, Petersen D, Wang X, Dorey F, Kil KS, Aldrich M, Blackburn MR, Kellems RE, Kohn DB (2006) In vivo transduction by intravenous injection of a lentiviral vector expressing human ADA into neonatal ADA gene knockout mice: a novel form of enzyme replacement therapy for ADA deficiency. Mol Ther 13(6):1110–1120. doi:10.1016/j.ymthe.2006.02.013
25. Kyosen SO, Iizuka S, Kobayashi H, Kimura T, Fukuda T, Shen J, Shimada Y, Ida H, Eto Y, Ohashi T (2010) Neonatal gene transfer using lentiviral vector for murine Pompe disease: long-term expression and glycogen reduction. Gene Ther 17(4):521–530. doi:10.1038/gt.2009.160
26. Hu C, Cela RG, Suzuki M, Lee B, Lipshutz GS (2011) Neonatal helper-dependent adenoviral vector gene therapy mediates correction of hemophilia A and tolerance to human factor VIII. Proc Natl Acad Sci U S A 108(5):2082–2087. doi:10.1073/pnas.1015571108

Chapter 17

Widespread Neuronal Transduction of the Rodent CNS via Neonatal Viral Injection

Ji-Yoen Kim, Stacy D. Grunke, and Joanna L. Jankowsky

Abstract

The rapid pace of neuroscience research demands equally efficient and flexible methods for genetically manipulating and visualizing selected neurons within the rodent brain. The use of viral vectors for gene delivery saves the time and cost of traditional germline transgenesis and offers the versatility of readily available reagents that can be easily customized to meet individual experimental needs. Here, we present a protocol for widespread neuronal transduction based on intraventricular viral injection of the neonatal mouse brain. Injections can be done either free-hand or assisted by a stereotaxic device to produce lifelong expression of virally delivered transgenes.

Key words Adeno-associated virus, Neonatal brain, Intracerebroventricular injection, Viral transduction, Transgenic mouse

1 Introduction

The ease of viral delivery permits genetic manipulation of the rodent brain without the time and cost of traditional germline manipulations. The widespread tropism of adeno-associated virus (AAV) facilitates robust somatic transfer of DNA to a broad range of neuronal subtypes. While AAV has long been used for targeted stereotaxic injection into the adult mammalian brain, Passini and Wolfe were the first to demonstrate that this vector could also be successfully delivered to the lateral ventricles of the neonatal brain [1]. Rather than the spatially limited transduction obtained by stereotaxic injection of the adult, intracerebroventricular injection into the neonate achieves widespread distribution of viral particles [1–4]. Optimal transduction is achieved when the virus is delivered within 24 h of birth, when the immature ependymal layer lining the ventricles allows passage of AAV particles from the CSF into the brain parenchyma [1, 2, 4–6]. Expression of virally delivered transgenes usually begins within 2–3 days after injection and persists for well over a year [1, 2].

Fredric P. Manfredsson (ed.), *Gene Therapy for Neurological Disorders: Methods and Protocols*, Methods in Molecular Biology, vol. 1382, DOI 10.1007/978-1-4939-3271-9_17,

Here, we provide an optimized technique for free-hand injection of AAV into the neonatal mouse brain based on our own experience and that of past publications [1, 2, 4]. We also introduce a complementary approach for viral injection based on stereotaxic targeting of the lateral ventricles. The only major equipment required for free-hand injection is a microsyringe fitted with a fine bore needle. In contrast, stereotaxic injection requires a dedicated apparatus but can improve reproducibility of the injections for novice investigators. This benefit is worth noting, as accurate targeting of the lateral ventricles is the most critical factor for successful viral expression when using these techniques. By including a non-toxic dye (2–4 % trypan blue) in the viral solution, the injection site and ventricular spread can be visualized to provide an immediate read-out of injection accuracy. With successful targeting of the lateral ventricles, both free-hand and stereotaxic techniques provide widespread neuronal transduction throughout the brain from the olfactory bulb to the cerebellum.

While free-hand and stereotaxic injection produce similar patterns of neuronal transduction, they are not identical. The angle at which the needle approaches the brain and the precise location of viral injection within the ventricle differ between the two techniques. The method we describe for free-hand injection targets the caudal portion of the ventricle and results in highest expression within the posterior and lateral regions of cortex. In contrast, the stereotaxic coordinates provided for single-site injection (as well as the alternative free-hand injection site we describe) target a more rostral location within the ventricle, leading to strongest expression in anterior regions. Both injection sites generate transgene expression throughout the cortex and beyond, but the bias of maximal expression toward rostral or caudal forebrain may influence which of the two approaches is more appropriate for a particular experiment. If desired, the stereotaxic manipulator can be used to perform two injections into each ventricle, one rostral and one caudal, to attain a more even distribution of viral transduction throughout the cortex.

The transduction pattern resulting from intraventricular viral injection is also significantly influenced by the serotype and dilution of AAV particles. In our hands, AAV8 displays the greatest transduction efficiency, although other groups have reported success with AAV1, AAV2, AAV6, and AAV9 [1, 3, 4, 7–11]. The density of neuronal transduction can be controlled by adjusting the viral titer, with high titer injections producing very dense viral expression and dilute injections resulting in sparse transduction of isolated cells [2]. Moreover, multiple viruses can be co-injected to express distinct proteins, using the serotype of each to bias transduction toward distinct or overlapping populations of neurons [2]. There are endless combinations of promoter, serotype, and titer than can be exploited to tailor viral expression for experimental needs [12]. This inherent flexibility has been advantageous for studies ranging from in vivo imaging to gene therapy of neurological disorders [2, 8, 13–19].

2 Materials

2.1 Preparation of Viral Solution and Decontamination Afterwards

1. Personal protective equipment, including lab coat or disposable gown, gloves, and facemask.
2. Aliquot of recombinant AAV8 (*see* **Notes 1** and **2**).
3. Ice-cold Dulbecco's phosphate-buffered saline.
4. Trypan blue stain.
5. 2 % bleach.
6. 70 % ethanol.
7. Biohazard containers.

2.2 Postnatal Day 0 (P0) Mice

1. Neonatal mouse pups born less than 6–12 h earlier, along with the female that delivered them (*see* **Note 3**).
2. *Optional*: an ICR or FVB female with offspring less than 4 days of age to serve as foster mother to the injected pups (*see* **Note 4**).
3. High fat rodent chow for nursing females (Purina Mills, St. Louis, MO, USA) (*see* **Note 5**).

2.3 Intraventricular Injection of Neonatal Mice

1. Warming pad suitable for laboratory animals.
2. Ice bucket.
3. Cotton swab soaked in 70 % ethanol.
4. Ten-microliter injection syringe (Hamilton, Reno, NV, USA).
5. 32-gauge needle (Hamilton, Reno, NV, USA).
6. Raw aluminum plate (4 × 6, 0.25 in. thick).

2.4 Additional Materials for Stereotaxic Injection

1. Stereotaxic apparatus with digital display (David Kopf Instruments, Tujunga, CA, USA).
2. Universal syringe holder with needle support foot (David Kopf Instruments, Tujunga, CA, USA).
3. Neonatal rat adaptor for stereotaxic device (Stoelting, Wood Dale, IL, USA).
4. 100 % ethanol.
5. Crushed dry ice.

3 Methods

3.1 Dilute Viral Preparations

Wear personal protective equipment (lab coat, gloves, and face mask) when working with viruses. Check with your institute's veterinary and environmental safety staff to learn what additional precautions are required for viral use at your research site. Regulations regarding biohazard isolation, cage identification, and waste disposal may vary across institutions.

1. Remove viral aliquots from the −80 °C freezer and place on ice to thaw (*see* **Note 6**).
2. Dilute viral stocks in ice-cold, sterile DPBS to the desired titer (*see* **Note 7**).
3. Add trypan blue to a final concentration of 0.05–0.1 % (*see* **Note 8**).
4. Fit an injection syringe with a clean needle. Load 5 μl of diluted viral suspension into the syringe. Remove air and bubbles in the viral suspension by holding the syringe needle up and tapping the syringe until the bubbles can be dispensed from the needle.

3.2 Prepare Neonatal Mice for Intracranial Injection

Plan to perform viral injections as soon as the pups are nursing, as evidenced by visible milk spots, or within 6 h of birth, whichever comes first (*see* **Note 9**).

1. Place the aluminum plate on ice. Cover the plate with a dry tissue to protect the skin from the cold metal (Fig. 1).
2. Transfer neonates from the cage to the warming pad (*see* **Note 10**).
3. Take one neonate from the warming pad and place it onto the cold metal block to induce hypothermic anesthesia. Wait 2–3 min for cessation of movement and respiration, indicating full anesthesia (*see* **Note 11**).
4. Gently wipe skin overlying the skull with a cotton swab soaked in 70 % ethanol.

3.3 Free-Hand Intraventricular Injection of AAV

1. Identify the desired site of injection, which is located at 2/5 of the distance from the lambda suture to each eye (Fig. 1, *see* **Note 12**).
2. Hold the syringe so that the volume scale is visible. Make sure your thumb can comfortably reach the top of the plunger. Put your elbow on the bench table and lean your arm on the ice bucket to stabilize your injection hand.
3. With your free hand, place the neonate on its side with its skull directly under the syringe. Rotate the head of the neonate until the injection site is perpendicular to the syringe.
4. Insert the needle at the marked injection site to a depth of approximately 3 mm (*see* **Note 13**).
5. Infuse up to 2 μl of the viral suspension over ~3 s (*see* **Note 14**).
6. Monitor the dye spread within the lateral ventricle during the infusion (*see* **Note 15**).
7. Remove your thumb from the plunger and hold the needle in place for a few seconds.

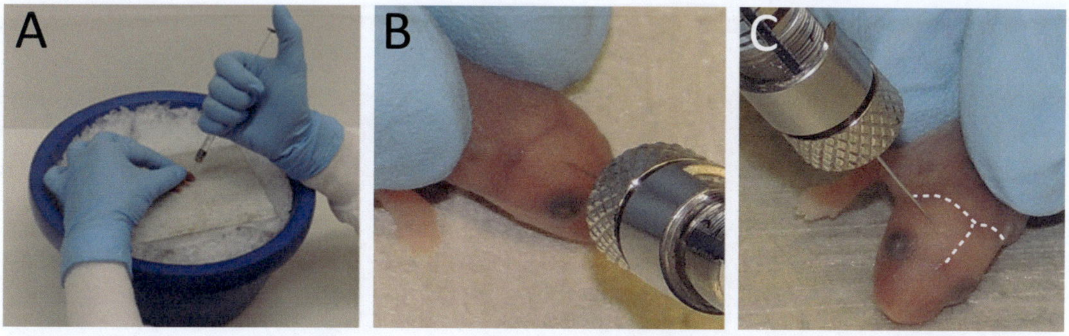

Fig. 1 Setup for free-hand intracerebroventricular injections. (**a**) Brace arms against the ice bucket for support and stability. Comfortably hold the syringe in one hand with the thumb positioned above the plunger. (**a** and **b**) The primary site for free-hand injection is located 2/5 of the way between the eye and the lambda suture (*dotted line* in **c**). Insert the needle perpendicular to the skull at the injection site

8. Slowly withdraw the needle.
9. Wait a few seconds for the injection site to close (*see* **Note 16**).
10. Inject the contralateral ventricle using the same procedure.

3.4 Stereotaxic Injection of AAV

1. Setup the stereotaxic stage with the neonatal frame and syringe holder. Insert the syringe fitted with a 32-gauge needle into the syringe holder (Fig. 2).
2. Cool the neonatal frame to 4–8 °C by adding 100 % ethanol and dry ice to the reservoir at the front end of the block (*see* **Note 17**).
3. Gently place the head of the neonate between the ear bars of the neonatal frame. Make sure the head is level in the *Y*-axis (front to back) by checking that the line between lambda and bregma is parallel to the stage. Make sure the head is leveled in the *X* axis (side to side) by checking that an imagined line between the ears, or a line between the eyes, is parallel to the stage (*see* **Note 18**).
4. Use the stereotaxic manipulator to position the injection needle above lambda and then zero the *X* and *Y* coordinates.
5. Move the stereotaxic arms to (*X*, *Y*) = (0.8, 1.5 mm) (*see* **Note 19**).
6. Slowly lower the syringe until the needle penetrates the skin. Zero the Z coordinate with the needle at the surface of the skull and the bevel of the needle at the surface of the skin (*see* **Note 20**).
7. Insert the needle until Z = −1.7 mm and then retract to −1.5 mm (*see* **Note 21**).
8. Infuse the viral suspension by depressing the plunger over ~3 s. Monitor the volume dispensed from the syringe and the spread of the dye (*see* **Notes 14** and **15**).

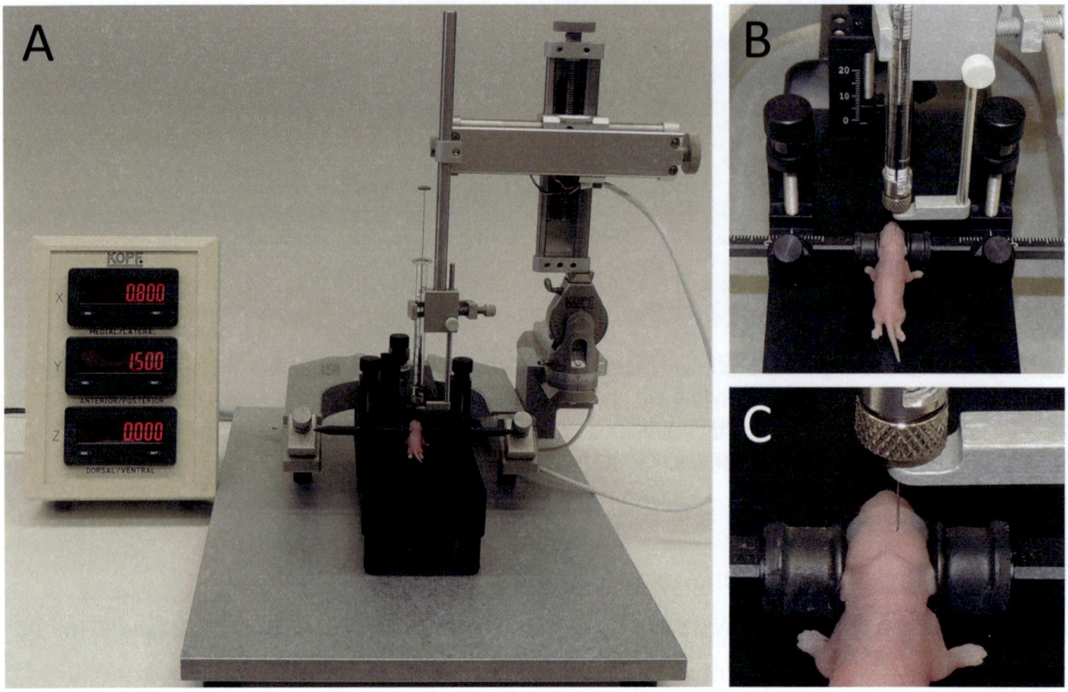

Fig. 2 Setup for stereotaxic intracerebroventricular injections. (**a**) Attach the neonatal adaptor to the stereotaxic frame with digital readout for coordinates. (**b**) Position the neonate between the ear bars with its head parallel to the stage and the injection site centered between the ear bars. Use the manipulator arm to position the needle above lambda, zero the coordinates, and then move the needle to the injection site. (**c**) The injection site is located 0.8 mm lateral to the sagittal suture, halfway between lambda and bregma

9. Leave the needle in place for 30 s after completing the infusion. Then slowly withdraw the needle over 1–2 min (*see* **Note 16**).
10. Repeat for the contralateral hemisphere, using negative X coordinates ($X = -0.8$ mm) for the injection site.

3.5 Postinjection Care and Clean-Up

1. Place the injected pup on a warming pad until its body temperature and skin color return to normal and the pup starts moving again.
2. Once all of the pups have been injected, disinfect the injection needle and syringe with 2 % bleach followed by repeated rinses in deionized water. Clean the working area with 2 % bleach followed by 70 % ethanol.
3. Collect all disposable items including pipette tips, tubes, mask, and gloves in a plastic biohazard bag. Dispose as required by local law (i.e., autoclave or incinerate).
4. Return the injected neonates to their biological mother or foster mother once they have fully recovered (*see* **Note 22**).

5. Label the cage as a biohazard and quarantine in an institutionally approved area.
6. After 3 days, transfer the mother and neonates to a clean cage before returning them to the animal colony (*see* **Note 23**).
7. At an appropriate experimental endpoint, collect mouse brains for analysis of viral transduction efficiency. If fluorescent reporters are included in the viral construct, transduction can be directly visualized within brain sections using the appropriate filter on a fluorescence microscope (Fig. 3).

4 Notes

1. Many universities have core laboratories on site that specialize in viral packaging. Alternatively, large facilities at the University of North Carolina and the University of Pennsylvania offer high quality off-the-shelf reagents in a variety of serotypes at reduced cost. These facilities also provide custom packaging for vectors that are not available as pre-packaged stocks.
2. The quality of the viral preparation can significantly influence which cells will be transduced by this approach. High-quality AAV8 injected at postnatal day 0 as described here will produce strong neuronal transduction [2]. Poor-quality viral preparations with high levels of non-infectious particles produce significantly more astrocytic uptake. The most common means of determining viral titer, using genome amplification, will not distinguish between infectious and non-infectious preparations. If this is a concern, viral titer can be done using the more traditional, but also more expensive, serial dilution infection to determine if the actual infectious titer differs significantly from that determined by genomic amplification.
3. We find it useful to set up breeding cages by adding one or two healthy females to the cage of a single resident male. Each morning, check the females for a mating plug and remove them from the cage if a plug is present.
4. Foster moms will be needed to raise the pups after viral injection if the experimental animals are maintained on a genetic background with poor maternal characteristics (for example, C57BL/6 or C3HeJ). In this case, set up a separate mating cage using ICR or FVB females alongside the experimental breeders so that the two sets of females deliver at approximately the same time. Ideally, the foster strain should have distinct coloration from the experimental strain so that the pups can be easily distinguished (i.e., pink-eyed FVB or ICR for use with dark-eyed C57BL/6). Foster moms are best used to raise pups born within 4 days of their own litter.

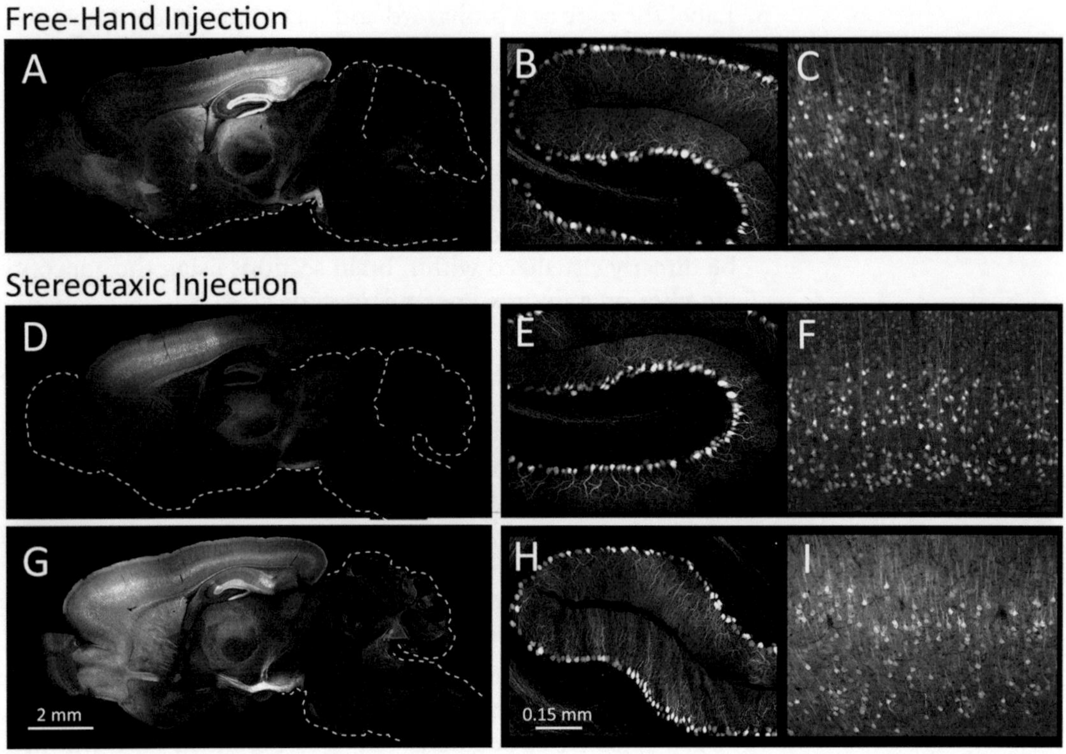

Fig. 3 Neonatal AAV injection produces widespread neuronal transduction. The pattern of viral transduction pattern was analyzed 3–4 weeks after injection using the free-hand technique (**a–c**) or stereotaxic manipulator (single injection: **d–f**, dual injection: **g–i**). The injected AAV8 encodes a chicken β-actin (CBA) promoter controlling expression of yellow fluorescent protein (YFP). Native YFP fluorescence was visualized in sagittal brain sections to examine the spread of viral transduction (**a**, **d**, **g**). These images illustrate the bias toward maximal transduction of caudal cortex by free-hand injection and toward rostral cortex by single site stereotaxic injection. A more uniform distribution of virus can be achieved through injection into two locations within each ventricle (**g**). Although maximal transduction occurs within the forebrain, neurons are labeled throughout the brain by both free-hand and stereotaxic approaches even though cells may not be visualized when the exposure time is limited by the cortical signal. In areas such as the cerebellum where the density of transduced cells is lower than the cortex, higher magnification images clearly reveal YFP expression within individual Purkinje neurons (**b**, **e**, **h**). Overall, transduction efficiency in Purkinje cells is similar to that of layer V cortical pyramidal neurons (**c**, **f**, **i**)

5. Provide all pregnant and nursing females with high fat chow in order to support the energy demands of breeding and lactation. For strains with poor maternal characteristics, it can be beneficial to start females on high fat chow 1–2 weeks prior to mating. Check with your local veterinarian for a breeder chow available at your institution.
6. All work with viral stocks should be done inside a Level 2 biosafety hood; small aliquots can be handled at the bench.

Repeated freeze–thaw cycles should be avoided as they cause the virus to become less effective. Once thawed, viral aliquots will maintain transduction efficiency for at least 2 months when stored at 4 °C [20].

7. Start with a tenfold serial dilution (10^{10} to 10^{7} viral particles/hemisphere) for initial optimization of the transduction pattern.
8. The addition of dye to the viral solution allows the injection accuracy to be monitored in real time by watching for the ventricles to fill with color.
9. Females will generally deliver in 19 ± 1 days from the plug date, depending on the background strain. Three days prior to delivery, place the pregnant female into a clean cage with fresh nesting material (i.e., Nestlets from Ancare) and some form of covered shelter (i.e., Igloo hut from Bio-Serv or Shepherd shack from Shepherd Specialty Papers). Place the cage in a quiet place to reduce stress caused by noise, light, and other disturbances. Females generally deliver pups in the morning but on rare occasions will deliver in afternoon. Therefore, check for newborn pups twice daily starting 2 days prior to the expected delivery date (17 days after the plug date). To minimize stress, check for pups by peaking through the bottom of the cage for the presence of pink newborns. For highest transduction efficiency, it is critical to inject virus as soon as possible after pups are delivered.
10. Removing all the pups from the cage at once or repeated opening of the cage will cause stress to the mother. If the biological mother will nurse the injected neonates, transfer only half the pups to the warming pad and leave the remaining half in the nest. After completing the first set of injections, return the injected neonates to the cage and transfer the remaining pups to the warming pad to await injection.
11. When first learning this technique, use a timer to monitor the duration of anesthesia. Start the timer when you place the neonate on the cold block. Hypothermic anesthesia has no long-term consequences when the total time is <10 min, but permanent brain damage can occur if neonates are hypothermic for >15 min. If the duration of anesthesia approaches this time limit, immediately retract the needle and place the pup back on the warming pad for recovery. After the remaining pups have been injected, the partially injected pup can be returned to the block for viral delivery into the uninjected ventricle.
12. Prior to postnatal day 3, the lambda and bregma sutures are visible through the skin. It can be helpful to mark the injection site with a non-toxic laboratory pen for reference. An alternative injection site is located approximately 0.8–1 mm lateral

from the sagittal suture, halfway between lambda and bregma. This site is identical to the location used for stereotaxic injections.

13. Make sure to remove your thumb from the plunger while you insert the needle through the skull. This ensures that you do not accidentally dispense any virus before you have positioned the needle. When the needle reaches the correct depth, the resistance will decrease slightly, indicating that it has penetrated into the lateral ventricle.
14. Each ventricle can accept a maximum volume of 2 μl. Carefully monitor the volume dispensed from the syringe so that you do not inject more than 2 μl per hemisphere. Ensure that you hold the syringe rigidly and depress the plunger without inserting the needle further into the brain. If the injection is displaced into the thalamus, viral spread will be severely limited.
15. Precise targeting of the lateral ventricle is critical to maximize viral spread. Accurate targeting also minimizes tissue damage caused by rapid infusion of a relatively large volume of fluid.
16. Pause briefly after the needle is retracted from the brain to minimize backflow of virus from the first injection site during contralateral injection. Do not inject the same site more than once. Reinserting the needle causes virus that has already been injected to leak out, and can injure or kill the neonate.
17. Keep the temperature above 1 °C to avoid frostbite of the pups. Thermometer stickers can be added to the side of the stage to assist in temperature control. Continually monitor the temperature of the stage during the procedure and add or remove dry ice as needed to maintain the stage at the correct temperature. It is best to cool the stage with dry ice to <4 °C prior to securing the neonate. This will ensure that the neonate remains deeply anesthetized throughout the procedure. The stage temperature may rise slightly (up to 8 °C) toward the end of the procedure. The slightly higher ending temperature will speed recovery of normal body temperature when the procedure is finished.
18. Place the mouse on the stage with its legs extending out from the body. Ensure that its belly is flat against the cold stage to maintain hypothermia. Make sure the injection site (halfway between lambda and bregma) is centered between the ear bars. If the injection site is located too far from the ear bars, the head will pivot when the needle is lowered.
19. Coordinates for injection of P0 pups are $(X, Y, Z) = (0.8, 1.5, -1.5$ mm) from lambda for infusion at one site in each hemisphere. If desired, two infusions can be made into each hemisphere using coordinates $(X, Y, Z) = (0.8, 2.0, -1.5$ and $1.2, 1.0, -1.5$ mm), alternating between sides for each injection. If two

sites are used, reduce the volume of each injection to 1 μl so that the final volume per hemisphere remains the same. Be aware that pup size and stereotaxic coordinates may vary by strain and age. Adjust coordinates as needed to target the lateral ventricles by monitoring dye spread during practice injections with pups of the same age and strain as will be used experimentally.

20. The surface of the skull will indent as the needle is slowly lowered. Once the needle has penetrated the skull, the pressure will decrease and the skull will rebound slightly. Retract the needle until the skull recovers its normal concave shape but ensure the bevel of the needle remains under the skin.
21. Retracting the needle from $Z = -1.7$ to -1.5 mm releases pressure caused by inserting the needle into the brain and prevents damage from the increased ventricular volume caused by rapid infusion of viral solution.
22. The survival rate after injection is most affected by maternal care. We find it useful to return the injected neonates to a corner of the cage away from the nest. A good, attentive mother will immediately retrieve the pups and bring them to the nest. If the mother has not collected the pups within 10 min, another mother may be needed. Check again later in the day for the presence of milk spots. If pups do not look well or have not nursed, they should be transferred to a new mother immediately. If a foster mom will be used to nurse the injected neonates, place the foster mother's pups and some of their bedding together on a warming pad with the injected pups in order to transfer the scent of the biological offspring. Cull the biological pups to improve acceptance of the injected pups. We recommend ICR or FVB for fostering because they produce ample milk and readily accept new pups into their litter.
23. After 3 days, the injected mice are no longer considered a biohazard. Transfer the mice inside a Level 2 biosafety cabinet to avoid exposure to virally contaminated bedding. Bring the dirty cage with all bedding and shelter to the appropriate biohazard cage wash area at your institution or place in biohazard bag and sterilize by autoclaving if required by your institution.

Acknowledgements

This research was supported by the Robert A. and Rene E. Belfer Family Foundation, NIA R21 AG038856 (J.L.J.), BrightFocus Foundation Alzheimer's Disease research grant A2010097 (J.L.J.), and NIA Biology of Aging Training grant T32 AG000183 (support for S.D.G.).

References

1. Passini MA, Wolfe JH (2001) Widespread gene delivery and structure-specific patterns of expression in the brain after intraventricular injections of neonatal mice with an adeno-associated virus vector. J Virol 75:12382–12392
2. Kim JY, Ash RT, Ceballos-Diaz C et al (2013) Viral transduction of the neonatal brain delivers controllable genetic mosaicism for visualising and manipulating neuronal circuits in vivo. Eur J Neurosci 37:1203–1220
3. Pilpel N, Landeck N, Klugmann M et al (2009) Rapid, reproducible transduction of select forebrain regions by targeted recombinant virus injection into the neonatal mouse brain. J Neurosci Methods 182:55–63
4. Chakrabarty P, Rosario A, Cruz P et al (2013) Capsid serotype and timing of injection determines AAV transduction in the neonatal mice brain. PLoS One 8, e67680
5. Tramontin AD, Garcia-Verdugo JM, Lim DA et al (2003) Postnatal development of radial glia and the ventricular zone (VZ): a continuum of the neural stem cell compartment. Cereb Cortex 13:580–587
6. Spassky N, Merkle FT, Flames N et al (2005) Adult ependymal cells are postmitotic and are derived from radial glial cells during embryogenesis. J Neurosci 25:10–18
7. Rafi MA, Zhi RH, Passini MA et al (2005) AAV-mediated expression of galactocerebrosidase in brain results in attenuated symptoms and extended life span in murine models of globoid cell leukodystrophy. Mol Ther 11:734–744
8. Passini MA, Watson DJ, Vite CH et al (2003) Intraventricular brain injection of adeno-associated virus type 1 (AAV1) in neonatal mice results in complementary patterns of neuronal transduction to AAV2 and total long-term correction of storage lesions in the brains of β-glucuronidase-deficient mice. J Virol 77:7034–7040
9. Broekman ML, Comer LA, Hyman BT et al (2006) Adeno-associated virus vectors serotyped with AAV8 capsid are more efficient than AAV-1 or -2 serotypes for widespread gene delivery to the neonatal mouse brain. Neuroscience 138:501–510
10. Cearley CN, Vandenberghe LH, Parente MK et al (2008) Expanded repertoire of AAV vector serotypes mediate unique patterns of transduction in mouse brain. Mol Ther 16:1710–1718
11. Dirren E, Towne CL, Setola V et al (2014) Intracerebroventricular injection of adeno-associated virus 6 and 9 vectors for cell type-specific transgene expression in the spinal cord. Hum Gene Ther 25:109–120
12. Betley JN, Sternson SM (2011) Adeno-associated viral vectors for mapping, monitoring, and manipulating neural circuits. Hum Gene Ther 22:669–677
13. Wolf DA, Lenander AW, Nan Z et al (2011) Direct gene transfer to the CNS prevents emergence of neurologic disease in a murine model of mucopolysaccharidosis type I. Neurobiol Dis 43:123–133
14. Chakrabarty P, Ceballos-Diaz C, Beccard A et al (2010) IFN-gamma promotes complement expression and attenuates amyloid plaque deposition in amyloid beta precursor protein transgenic mice. J Immunol 184:5333–5343
15. Chakrabarty P, Jansen-West K, Beccard A et al (2010) Massive gliosis induced by interleukin-6 suppresses Aβ deposition in vivo: evidence against inflammation as a driving force for amyloid deposition. FASEB J 24:548–559
16. Kim J, Miller VM, Levites Y et al (2008) BRI2 (ITM2b) inhibits Aβ deposition in vivo. J Neurosci 28:6030–6036
17. Levites Y, Jansen K, Smithson LA et al (2006) Intracranial adeno-associated virus-mediated delivery of anti-pan amyloid-β, amyloid-β40, and amyloid-β42 single-chain variable fragments attenuates plaque pathology in amyloid precursor protein mice. J Neurosci 26:11923–11928
18. Lee NC, Chien YH, Hu MH et al (2014) Treatment of congenital neurotransmitter deficiencies by intracerebral ventricular injection of an adeno-associated virus serotype 9 vector. Hum Gene Ther 25(3):189–198
19. Gadalla KK, Bailey ME, Spike RC et al (2013) Improved survival and reduced phenotypic severity following AAV9/MECP2 gene transfer to neonatal and juvenile male Mecp2 knockout mice. Mol Ther 21:18–30
20. Croyle MA, Cheng X, Wilson JM (2001) Development of formulations that enhance physical stability of viral vectors for gene therapy. Gene Ther 8:1281–1290

Chapter 18

AAV-Mediated Gene Transfer to Dorsal Root Ganglion

Hongwei Yu, Gregory Fischer, and Quinn H. Hogan

Abstract

Transferring genetic molecules into the peripheral sensory nervous system to manipulate nociceptive pathophysiology is a powerful approach for experimental modulation of sensory signaling and potentially for translation into therapy for chronic pain. This can be efficiently achieved by the use of recombinant adeno-associated virus (rAAV) in conjunction with nociceptor-specific regulatory transgene cassettes. Among different routes of delivery, direct injection into the dorsal root ganglia (DRGs) offers the most efficient AAV-mediated gene transfer selectively into the peripheral sensory nervous system. Here, we briefly discuss the advantages and applications of intraganglionic microinjection, and then provide a detailed approach for DRG injection, including a list of the necessary materials and description of a method for performing DRG microinjection experiments. We also discuss our experience with several adeno-associated virus (AAV) options for in vivo transgene expression in DRG neurons.

Key words Dorsal root ganglion, Chronic pain, Microinjection, Gene therapy, Adeno-associated virus

1 Introduction

Chronic neuropathic pain is common and largely resistant to pharmacological treatments. Gene therapy targeting the peripheral nervous system is an approach that shows great promise for the treatment of chronic pain [1–4]. The intent is to deliver analgesic genetic molecules to the pathophysiological area and affect a positive therapeutic outcome. Disordered cellular mechanisms underlying chronic pain, such as that which follows nerve injury, reside at diverse sites, including in receptive fields in peripheral tissues, in the somata of the injured sensory neurons, and in the dorsal horn of the spinal cord. Many maladaptive gene expressions upon nerve damage are mainly synthesized in the primary sensory neurons and trafficked to central presynaptic terminals as well as peripheral axons to induce central or peripheral hypersensitivity. The dorsal root ganglia (DRGs), which harbor the somata of primary sensory neurons that transmit sensory signals from the peripheral organs

Fredric P. Manfredsson (ed.), *Gene Therapy for Neurological Disorders: Methods and Protocols*, Methods in Molecular Biology, vol. 1382, DOI 10.1007/978-1-4939-3271-9_18, © Springer Science+Business Media New York 2016

toward the appropriate integral sites in the central nervous system, are thus an optimal target for therapeutic gene transfer [5–7].

Vector selection plays a critical role for the success of gene therapy. Currently, the most successful gene therapy strategies rely on recombinant viral vectors (e.g., adeno-associated virus, adenovirus, lentivirus, retrovirus, and herpes simplex virus), although the utility of nonviral vectors is continuing to emerge [8]. Enthusiasm for the recombinant adeno-associated virus (AAV) vector system for in vivo viral gene transfer to the DRG neurons has grown in recent years since this vector provides highly efficient gene transfer into the post-mitotic primary sensory neurons and long-term control of neuropathic pain with minimal toxicity [1, 6, 7, 9–11].

Genetic molecules can be transferred to DRG neurons through direct AAV intraganglionic injection [5, 7, 12], or by intrathecal delivery of AAV vectors into cerebrospinal fluid (CSF) [13–16], as well as retrograde transfer by injection of AAV into sciatic nerve and skeletal muscles [11]. Each delivery route has features that make them applicable to certain pain therapies. Among the various routes, direct intraganglionic injection offers the most efficient AAV-mediated gene transfer selectively into the peripheral sensory nervous system, including sensory neuron somata and their central and peripheral axonal terminals [7, 17, 18]. The loose and porous structure of DRG surrounded by a connective tissue capsule renders it well suited to injection. Due to the skeletal elements surrounding the DRG, a foraminotomy (minimal laminectomy) is necessary to expose the distal pole of the DRG for a reliable and consistent successful intraganglionic injection [19]. The small size of the adult rodent DRG (approximately 1 cm long and 0.5 cm in diameter in rat) determines the injected volumes be kept quite small [5]. Slow and sustained injection through a pulled small-tip glass micropipette attached to a microprocessor-controlled injector enables continuous and predictable filling of the DRG with 2 μl of AAV particle suspension without significant leakage and damage to the DRG [5]. Since this approach targets only the peripheral sensory neurons, some of the problems associated with other pain therapies that also affect the central nervous system can be avoided. The injection procedure produces only transient hypersensitivity to threshold mechanical stimulation without any significant changes in hyperalgesia or other sensory modalities or motor behavior [5]. Direct injection into the DRG has been proven well tolerated in both humans and rodents [5, 7, 12, 20, 21] and is a simple outpatient procedure in humans.

Intrathecal delivery of AAV through an intrathecal pump or by lumbar puncture has been demonstrated to produce effective gene transfer to the DRG and spinal cord. This method also has the advantage of allowing for repeated injections, which can be problematic for DRG injection because scar tissue formation caused by the initial surgery can make subsequent surgical exposure difficult.

However, intrathecal delivery requires far more AAV vector [22], which creates a costly biomanufacturing challenge compared to intraganglionic injection. More importantly, intrathecal delivery cannot achieve a localized effect owing to the spread of the vectors within the CSF of the subarachnoid space, and robust transduction occurs in multiple spinal cord tissues including spinal motor neurons. This paradigm indicates that intrathecal AAV delivery has more therapeutic potential for widespread neurological disease processes that require diffuse gene delivery to spinal motoneurons as well as sensory neurons [23]. Intraneural AAV injection into the sciatic nerve can potentially transfer genes to primary sensory neurons, but this route of AAV delivery also shows preferential transduction of motor neurons, and injections into peripheral nerve carry the risk of damage from intrafasicular injection [24, 25]. Intramuscular AAV causes highly efficient transduction of skeletal muscle fibers, but does not efficiently transduce sensory neurons by retrograde axonal transport [11, 26]. In neonate rodents, intraperitoneal administration of AAV has been shown to efficiently target transgene expression in the DRG neurons [27].

In the subsequent material, we provide a detailed approach for DRG injection in adult rats using AAV vectors carrying GFP reporter as the injectate, including a list of the necessary materials and detailed description of a method for performing DRG microinjection experiments. Development of these techniques would likewise be suitable for intraganglionic application of various other therapeutic injectates, such as pharmacological agents and nonviral vectors.

2 Materials

2.1 Animals

1. Male Sprague-Dawley rats (*see* **Note 1**). We generally use rats weighing between 100 and 150 g to allow easier handling over extended post-injection observation. The surgery can be successfully performed on rats as large as 300 g or more if necessary, although the surgery can become more challenging as the size of the animal increases.

2.2 Reagents

1. High titer and high-pure AAV particles (titer should ideally be >10^{12} GC/ml) should be stored in small aliquots at −80 °C to prevent repeated freeze/thaw cycles, which significantly reduce infection efficiency (*see* **Note 2**).
2. Volatile anesthetic (isoflurane, sevoflurane) and delivery system.
3. Anesthesia chamber.
4. Anesthesia nose cone.

5. Betadine.
6. 70 % (v/v) ethanol in spray bottle.
7. Sterile cotton-tipped applicators.
8. Biohazard bin.

2.3 Equipment

1. Dissecting microscope.
2. Glass micropipette puller.
3. Glass Pasteur pipettes.
4. Oxygen tank.
5. Vaporizer.
6. Induction box.
7. Electric razor.
8. Betadine or other skin disinfectant.
9. Injection system: Micromanipulator-mounted, piston-driven injector employing positive displacement of a noncompressible fluid (i.e., mineral oil) under control of a microprocessor-controlled micro-stepping motor such as Nanoliter 2000 (World Precision Instruments, Sarasota, FL, USA) or equivalent.
10. Three-axis micromanipulator.
11. Magnetic Stand.
12. Surgical tools: #3 Scalpel handle, #10 scalpel blade, No. 2 forceps ×2, No. 5 forceps, microrongeur, spring scissors, small animal retraction kit, absorbable 6-0 suture, skin stapler.
13. Glass bead sterilizer.
14. Method to control bleeding (Gauze, Gelfoam, bone wax).
15. Rolled paper towel or wooden dowel of sufficient diameter to support the animal during surgery.
16. Heating pad to maintain the animal's body temperature.

3 Methods

3.1 Setup

1. Pulled glass pipettes for injection: Pulled to produce a long tip with a sharp point, with minimal regard to initial tip diameter. The tip should be pressed through a tautly held laboratory wipe (e.g., Kimwipe). This reliably produces a final tip diameter between 40 and 60 μm and a slightly "jagged" appearance of the tip under microscopic viewing (*see* **Note 3**). The exact procedure can be different based on the personal preference and on the different equipment for glass needle pulling for

injection. Store glass needles upright with the tips pointed down in an electrode storage jar after pulling.

2. Spinal stabilization clamp: a spring-loaded or screw-driven clamp affixed to metal rods that attach to a magnetic base positioned caudal of the animal as it lies in position for surgery. The rods should be long enough to accommodate the injection system. The clamp should be large enough to grasp the spinous processes of the lumbar vertebra, and should grip with enough force to lift the rat slightly, but not tightly enough to damage the bone or overlying fascia.
3. Microinjection rig setup: The microinjector should be mounted on a horizontal bar attached to the same magnetic base as the stabilization clamp. Sufficient distance should be present to allow the injector to approach the animal from the caudal direction at a shallow angle (approximately 20° from horizontal), as close to parallel with the spine of the animal as the anatomy will allow.

3.2 Anesthetize the Rat

1. Anesthetize the rat in a suitable chamber. We routinely use the volatile anesthetic isoflurane (4–5 % for induction, in $O_2 \approx 2$ L/min). Sevoflurane (with 50 % higher dosing) is also suitable.
2. Confirm achievement of a surgical plane of anesthesia by lack of withdrawal to plantar pinch. Anesthetic depth should be monitored by vigilant observation of spontaneous breathing with lowering anesthetic levels if tidal volumes diminish.
3. When immobility is achieved, shave the surgical area of the back, leaving a wide margin around the planned incision to prevent contamination of the surgical field with hair.
4. Disinfect the surgical area, i.e., with Betadine and rinse three times with 70 % ethanol.

3.3 Surgical Exposure of DRG and Injection

1. Place the rat in ventral recumbency, with rolled paper towels or a wooden rod beneath the hips, to arch the back and elevate the lumbar spine. The animal's face is placed in the anesthetic cone to maintain anesthesia and the isoflurane concentration reduced to 1.5–2.5 % (in $O_2 \approx 2$ L/min).
2. Confirm achievement of a surgical plane of anesthesia by lack of withdrawal to plantar pinch. Anesthetic depth should be monitored by vigilant observation of spontaneous breathing with lowering anesthetic levels if tidal volumes diminish.
3. Make a 3–3.5 cm skin incision just lateral of the dorsal midline, extending rostral from just caudal to the iliac crest (Fig. 1).
4. Make an underlying incision in the fascia of the paravertebral muscles just lateral to the midline.

5. Use forceps to separate two layers of the paravertebral musculature by blunt dissection to expose the lateral aspect of the L6-L4 vertebrae, applying retraction using retractors mounted on magnetic holders as needed to maintain a clear view of the surgical field.
6. Position the rat partially on to its side to achieve a better view of the lateral aspect of the L6-L4 vertebrae.
7. Clean the lateral aspect of the vertebrae by blunt dissection to expose the L4/L5 and L5/L6 intervertebral foramina (IVFs), overlaid by a membrane of connective tissue, to reveal the L4 and L5 spinal nerves just distal (approximately 2 mm) from their DRGs, which are covered by laminar bone (Fig. 1a).
8. Gently separate the membrane from the vertebral bone, exposing the entry of the spinal nerves into the IVFs. Pause as appropriate to control bleeding by compression with small balls of sterilized absorbent tissue paper.
9. Use rongeurs to slightly enlarge the IVF and remove a 1 mm-deep crescent of laminar bone, exposing the distal third of the L4 and L5 DRGs (Fig. 1b) (*see* **Note 4**). Avoid compression of the DRG by using the smallest rongeur, taking small bites with only the tip of the rongeur, and maintaining a lifting force against the bone with the portion of the rongeur that is in contact with the DRG.
10. Remove one aliquot of virus from −80 °C freezer. Thaw, flick the tube to mix, and then do a quick spin-down in a microcentrifuge (<2100 × *g*) to ensure no liquid is in the cap or on the walls of the tube.
11. Load the pipette with injectate by drawing injectate back through the tip, taking care to avoid introducing air bubbles or cavitation, which can affect the accuracy of injection volume.
12. Mount the rat in the clamp (Fig. 1c) and lift it such that the abdomen is just barely touching the table (to reduce venous bleeding). Physical support can then be provided under shoulder and hip with paper towel to stabilize the position.
13. Approach the ganglion at as shallow an angle as possible (i.e., tangential to the surface of the capsule) to facilitate entry into the DRG without compression against the underlying bone.
14. Slowly advance the pipette into the DRG to a depth of approximately 100–150 μm.
15. Wait 2–3 min before beginning injection to allow the DRG capsule to close around the pipette tip.
16. Gradually expel injectate (2 μl of AAV) in small boluses, carefully watching for backflow from around the pipette tip (*see* **Note 5**).

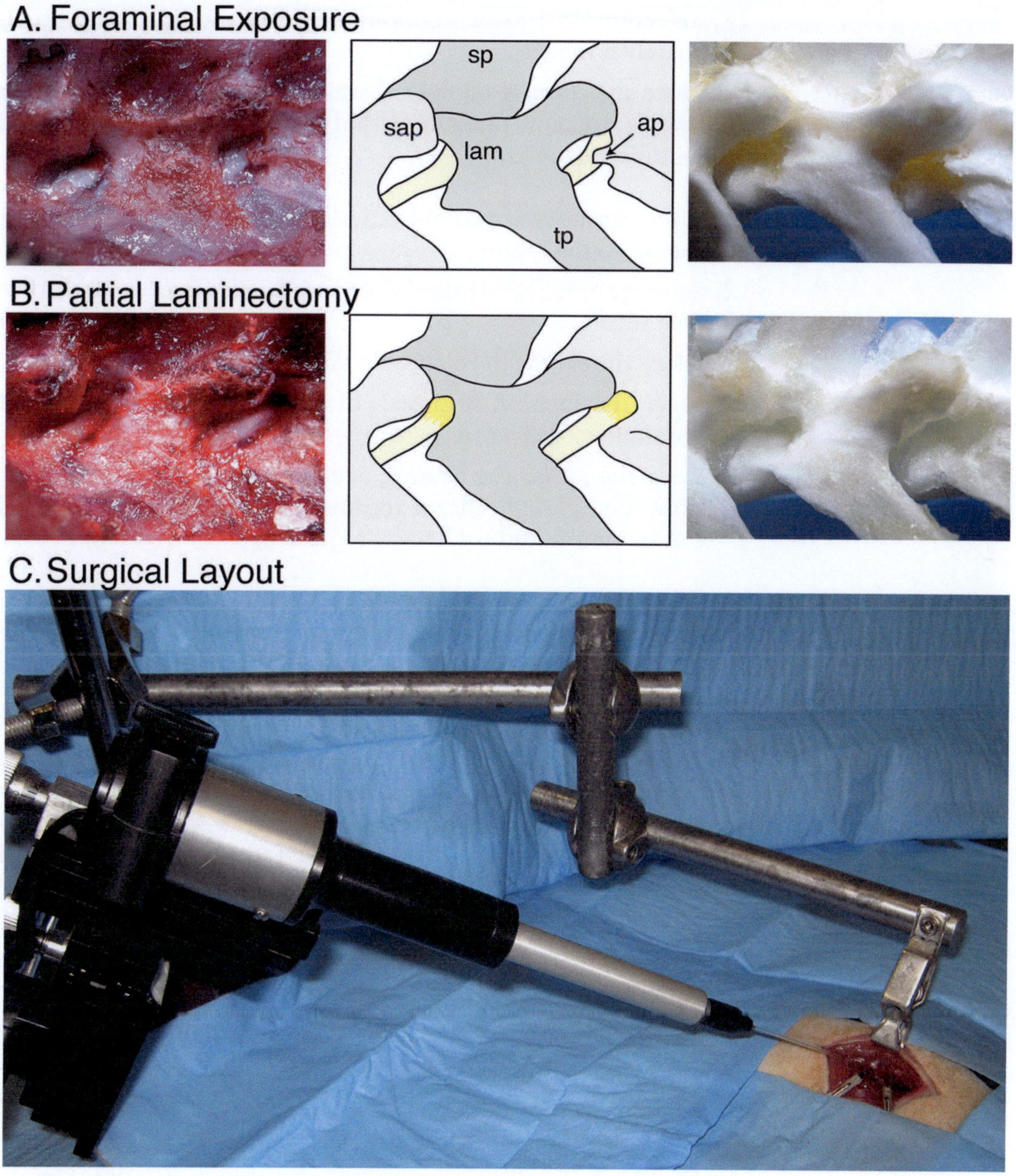

Fig. 1 Intraganglionic injection. Paravertebral surgical exposure for intraganglionic injection, illustrating the operative field (*left* panel), reference diagram (*middle* panel), and cleaned vertebral bones (*right* panel). (**a**) Initial dissection of soft tissues at the level of the fourth lumbar (L4) and L5 spinal nerves (*yellow*) shows the superior articular processes (sap), spinous processes (sp), and transverse processes (tp), as well as the laminar bone (lam) and accessory process on L4 (ap). The dorsal root ganglia are covered by laminar bone. (**b**) Removal of laminar bone superior to the foramen and the L4 accessory process reveals the distal dorsal root ganglion, recognized by its broader diameter and *brownish-orange* color. (**c**) The motorized injection system is mounted on a magnetic stand via a manual micromanipulator. The rat vertebral column is stabilized by clamping the spinous process of L5 using a clip mounted on an articulated arm that is attached to the same magnetic stand. The head of the rat is to the *right*. Reproduced with minor modification from Fischer et al. [5] with permission

17. Injection of 2 μl of solution per DRG should take approximately 5 min.
18. Once injection is complete, wait 5 min before withdrawing pipette, to allow pressure within the ganglion to equalize and minimize backflow.

3.4 Suture and Recovery

1. After injections are completed, disengage the clamp and lower the rat to the table.
2. After ensuring hemostasis, close the muscle fascia using absorbable suture.
3. Close the skin with removable skin staples or wound clips.
4. Once the incision is closed, discontinue the anesthetic, but continue to deliver oxygen.
5. Keep the rat on the heating pad and monitor it continuously until the righting reflex is regained, at which point the animal can be transferred back to its cage.

3.5 Representative Results

1. When following this protocol, a very precise injection with consistently successful transgene expression mediated by AAV is obtained. As a representative result of this method, we injected self-complimentary AAVs of various serotypes (5, 6, 8, and 9) into the L4 and L5 DRGs. Variable levels of EGFP expression were detected in the primary sensory neurons and their axons at 1 week following vector administration. A robust EGFP expression for all vectors tested was observed at 2–4 weeks after injection. EGFP expression was restricted to the sensory neuronal somata and projections, and no EGFP could be identified in satellite glia or other nonneuronal cells (Fig. 2).

4 Notes

1. All protocols using live animals must be reviewed and approved by an Institutional Animal Care and Use Committee (IACUC) and must be performed in accordance with the National Institutes of Health Guidelines for the Care and Use of Laboratory Animals.
2. A highly purified AAV vector with a titer of at least 1×10^9 GC viral particles (injected) and >90 % purity by silver stain is recommended. Injecting less than 10^8 viral particles with low-purity per DRG may not lead to a desired in vivo transduction. AAV is generally considered nonpathogenic and regarded as a Biosafety Level 2 (BSL-2) material, and appropriate practices must be followed during AAV application.
3. A 40–60 μm tip is suitable for dissolved compounds and suspensions of very small particles, such as viruses and nanoparti-

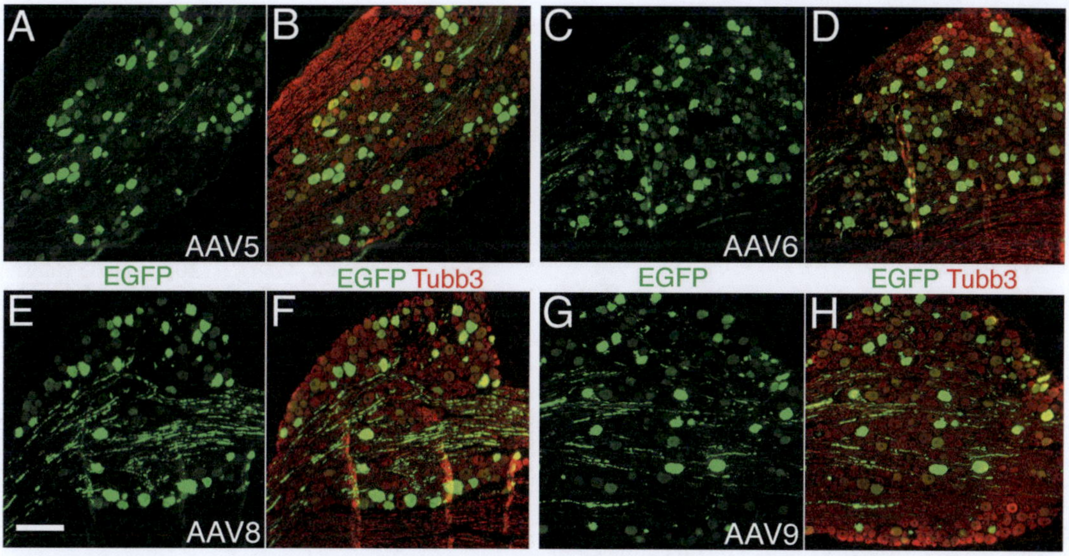

Fig. 2 Typical sensory neuron transduction after DRG injection of AAV5-, 6-, 8-, and 9-EGFP. Paraffin-sections from L5 DRGs were double-immunolabeled with the antibodies against GFP (*green* color) and neuron-specific β3-tubulin (Tubb3, *red* color). High-efficient transgene (EGFP) expression with predominant neuronal tropism is demonstrated in DRGs 4-week after intraganglionic injection of AAV5 (**a** and **b**), AAV6 (**c** and **d**), AAV8 (**e** and **f**), and AAV9 (**g** and **h**). Scale bar for all images: 200 μm

cles. Larger openings may be necessary for suspensions of larger particles, i.e., cell suspensions.

4. Cauterization is not recommended due to the risk of thermal injury to the spinal nerve. In the case of the L4 DRG, foraminotomy will also involve the removal of the small accessory process that projects caudally and partially overlays the foramen.

5. Increasing injectate volume over 2 μl per DRG does not necessarily increase the transduction but will lead to significant leakage [5, 12]. Use of a visible dye (i.e., Fast Green) is especially helpful at this stage to ensure that no leakage occurs. Fast Green at a concentration of 0.1 % w/v is an effective method to track the spread of injectate in real time. DAPI can be used at a concentration of 0.25 μg/μl to measure injectate spread in fixed tissue sections for at least 2 weeks after injection without toxicity [5].

Acknowledgement

This work was supported by grants from the Department of Veterans Affairs Rehabilitation Research and Development (3690-03), the Advancing a Healthier Wisconsin (FP00005706), and the National Institute of Neurological Disorders and Stroke (R01NS079626-01), to Q.H.H.

References

1. Beutler AS (2010) AAV provides an alternative for gene therapy of the peripheral sensory nervous system. Mol Ther 18(4):670–673. doi:10.1038/mt.2010.41
2. Glorioso JC, Fink DJ (2009) Gene therapy for pain: introduction to the special issue. Gene Ther 16(4):453–454. doi:10.1038/gt.2009.18
3. Goins WF, Cohen JB, Glorioso JC (2012) Gene therapy for the treatment of chronic peripheral nervous system pain. Neurobiol Dis 48(2):255–270. doi:10.1016/j.nbd.2012.05.005
4. Handy CR, Krudy C, Boulis N (2011) Gene therapy: a potential approach for cancer pain. Pain Res Treat 2011:987597
5. Fischer G, Kostic S, Nakai H, Park F, Sapunar D, Yu H, Hogan Q (2011) Direct injection into the dorsal root ganglion: technical, behavioral, and histological observations. J Neurosci Methods 199(1):43–55
6. Fischer G, Pan B, Vilceanu D, Hogan QH, Yu H (2013) Sustained relief of neuropathic pain by AAV-targeted expression of CBD3 peptide in rat dorsal root ganglion. Gene Ther. doi:10.1038/gt.2013.56
7. Yu H, Fischer G, Ferhatovic L, Fan F, Light AR, Weihrauch D, Sapunar D, Nakai H, Park F, Hogan QH (2013) Intraganglionic AAV6 results in efficient and long-term gene transfer to peripheral sensory nervous system in adult rats. PLoS One 8(4):e61266. doi:10.1371/journal.pone.0061266
8. Sloane EM, Soderquist RG, Maier SF, Mahoney MJ, Watkins LR, Milligan ED (2009) Long-term control of neuropathic pain in a non-viral gene therapy paradigm. Gene Ther 16(4):470–475. doi:10.1038/gt.2009.21
9. Beutler AS, Reinhardt M (2009) AAV for pain: steps towards clinical translation. Gene Ther 16(4):461–469
10. Eaton MJ, Blits B, Ruitenberg MJ, Verhaagen J, Oudega M (2002) Amelioration of chronic neuropathic pain after partial nerve injury by adeno-associated viral (AAV) vector-mediated over-expression of BDNF in the rat spinal cord. Gene Ther 9(20):1387–1395
11. Towne C, Pertin M, Beggah AT, Aebischer P, Decosterd I (2009) Recombinant adeno-associated virus serotype 6 (rAAV2/6)-mediated gene transfer to nociceptive neurons through different routes of delivery. Mol Pain 5:52
12. Mason MR, Ehlert EM, Eggers R, Pool CW, Hermening S, Huseinovic A, Timmermans E, Blits B, Verhaagen J (2010) Comparison of AAV serotypes for gene delivery to dorsal root ganglion neurons. Mol Ther 18(4):715–724
13. Beutler AS, Banck MS, Walsh CE, Milligan ED (2005) Intrathecal gene transfer by adeno-associated virus for pain. Curr Opin Mol Ther 7(5):431–439
14. Wang X, Wang C, Zeng J, Xu X, Hwang PY, Yee WC, Ng YK, Wang S (2005) Gene transfer to dorsal root ganglia by intrathecal injection: effects on regeneration of peripheral nerves. Mol Ther 12(2):314–320. doi:10.1016/j.ymthe.2005.03.032
15. Vulchanova L, Schuster DJ, Belur LR, Riedl MS, Podetz-Pedersen KM, Kitto KF, Wilcox GL, McIvor RS, Fairbanks CA (2010) Differential adeno-associated virus mediated gene transfer to sensory neurons following intrathecal delivery by direct lumbar puncture. Mol Pain 6:31
16. Storek B, Reinhardt M, Wang C, Janssen WG, Harder NM, Banck MS, Morrison JH, Beutler AS (2008) Sensory neuron targeting by self-complementary AAV8 via lumbar puncture for chronic pain. Proc Natl Acad Sci U S A 105(3):1055–1060. doi:10.1073/pnas.0708003105
17. Glatzel M, Flechsig E, Navarro B, Klein MA, Paterna JC, Bueler H, Aguzzi A (2000) Adenoviral and adeno-associated viral transfer of genes to the peripheral nervous system. Proc Natl Acad Sci U S A 97(1):442–447
18. Xu Y, Gu Y, Wu P, Li GW, Huang LY (2003) Efficiencies of transgene expression in nociceptive neurons through different routes of delivery of adeno-associated viral vectors. Hum Gene Ther 14(9):897–906. doi:10.1089/104303403765701187
19. Puljak L, Kojundzic SL, Hogan QH, Sapunar D (2009) Targeted delivery of pharmacological agents into rat dorsal root ganglion. J Neurosci Methods 177(2):397–402. doi:10.1016/j.jneumeth.2008.10.029
20. Zhao X, Tang Z, Zhang H, Atianjoh FE, Zhao JY, Liang L, Wang W, Guan X, Kao SC, Tiwari V, Gao YJ, Hoffman PN, Cui H, Li M, Dong X, Tao YX (2013) A long noncoding RNA contributes to neuropathic pain by silencing Kcna2 in primary afferent neurons. Nat Neurosci 16(8):1024–1031. doi:10.1038/nn.3438
21. Samad OA, Tan AM, Cheng X, Foster E, Dib-Hajj SD, Waxman SG (2012) Virus-mediated shRNA knockdown of Na(v)1.3 in rat dorsal root ganglion attenuates nerve injury-induced neuropathic pain. Mol Ther. doi:10.1038/mt.2012.169

22. Fagoe ND, Eggers R, Verhaagen J, Mason MR (2013) A compact dual promoter adeno-associated viral vector for efficient delivery of two genes to dorsal root ganglion neurons. Gene Ther. doi:10.1038/gt.2013.71
23. Federici T, Taub JS, Baum GR, Gray SJ, Grieger JC, Matthews KA, Handy CR, Passini MA, Samulski RJ, Boulis NM (2012) Robust spinal motor neuron transduction following intrathecal delivery of AAV9 in pigs. Gene Ther 19(8):852–859. doi:10.1038/gt.2011.130
24. Lu YY, Wang LJ, Muramatsu S, Ikeguchi K, Fujimoto K, Okada T, Mizukami H, Matsushita T, Hanazono Y, Kume A, Nagatsu T, Ozawa K, Nakano I (2003) Intramuscular injection of AAV-GDNF results in sustained expression of transgenic GDNF, and its delivery to spinal motoneurons by retrograde transport. Neurosci Res 45(1):33–40
25. Boulis NM, Noordmans AJ, Song DK, Imperiale MJ, Rubin A, Leone P, During M, Feldman EL (2003) Adeno-associated viral vector gene expression in the adult rat spinal cord following remote vector delivery. Neurobiol Dis 14(3):535–541
26. Zheng H, Qiao C, Wang CH, Li J, Yuan Z, Zhang C, Xiao X (2010) Efficient retrograde transport of adeno-associated virus type 8 to spinal cord and dorsal root ganglion after vector delivery in muscle. Hum Gene Ther 21(1):87–97. doi:10.1089/hum.2009.131
27. Machida A, Kuwahara H, Mayra A, Kubodera T, Hirai T, Sunaga F, Tajiri M, Hirai Y, Shimada T, Mizusawa H, Yokota T (2013) Intraperitoneal administration of AAV9-shRNA inhibits target gene expression in the dorsal root ganglia of neonatal mice. Mol Pain 9:36. doi:10.1186/1744-8069-9-36

Chapter 19

Gene Therapy of the Peripheral Nervous System: The Enteric Nervous System

Matthew J. Benskey and Fredric P. Manfredsson

Abstract

The enteric nervous system (ENS) is a complex network of ganglia embedded in the walls of the gastrointestinal (GI) tract. The highly integrated enteric neural network can act independently of sympathetic and parasympathetic innervation to control all aspects of GI function. Due to the central role the ENS plays in GI physiology, the ability to manipulate genetic expression within the ENS is an invaluable tool for both research and clinical applications. Here, we describe a method for gene delivery to the ENS using direct injections of adeno-associated virus (AAV) into the gut wall. This method is able to achieve transgene expression in both neurons and glia located in discrete areas of the ENS.

Key words Enteric nervous system, Gene delivery, Adeno-associated virus, Myenteric plexus, Submucosal plexus

1 Introduction

The ENS is one of the main branches of the peripheral nervous system and maintains the physiological function of the GI tract. The ENS consists of a highly integrated network of neurons and glia that are embedded in the gut wall, spanning from the esophagus to the anus. The intricate complexity and autonomous nature of the ENS has earned it the title of "the second brain." Within the ENS neurons are organized into ganglia located in either the myenteric plexus (MP) or the submucosal plexus (SMP). The MP is located between the circular and longitudinal muscle of the gut wall, while the submucosal plexus is located between the mucosa and circular muscle of the gut [1]. Abnormalities of the ENS can cause pathology in the GI tract that manifest in functional GI disorders, and significantly decrease quality of life [2]. Accordingly, effectively targeting and delivering genetic material to the ENS holds great promise for both research and therapeutics. Previously, viral vector-based transduction of the ENS has been accomplished

Fredric P. Manfredsson (ed.), *Gene Therapy for Neurological Disorders: Methods and Protocols*, Methods in Molecular Biology, vol. 1382, DOI 10.1007/978-1-4939-3271-9_19, © Springer Science+Business Media New York 2016

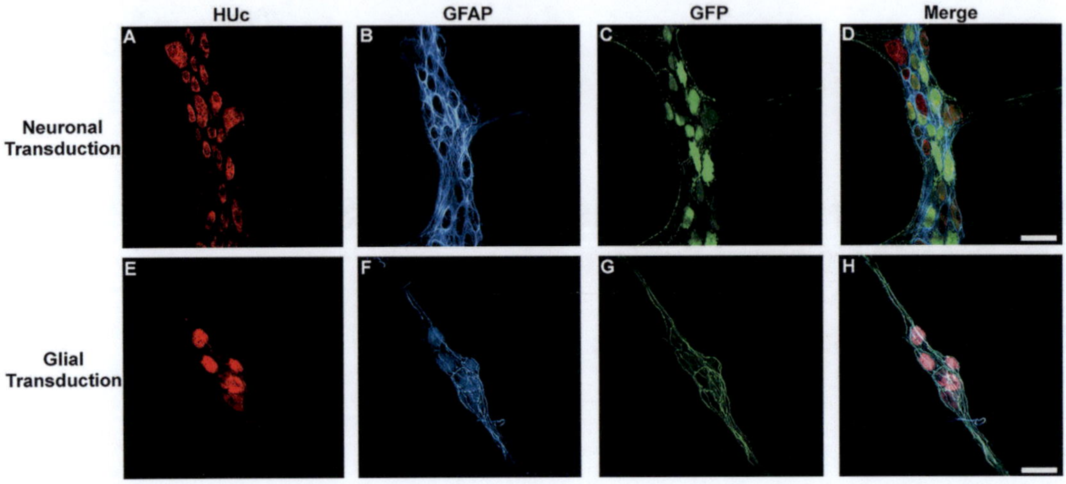

Fig. 1 Transduction of neurons and glia in the myenteric plexus of the ENS. Adult male rats received direct injections of AAV9 (**a–d**) or AAV6 (**e–h**) expressing green fluorescent protein (GFP) into the wall of the descending colon and were sacrificed 1 month later. Tissue from the myenteric or submucosal plexus was stained for the neuronal marker HUc (*red*), and the glia marker, glial fibrillary acidic protein (GFAP; *blue*). AAV9-mediated neuronal transduction (**a–d**) in the myenteric plexus tissue is shown as colocalization of GFP (*green*) expression in HUc immunoreactive cells (colocalization appears *yellow*). Within the submucosal plexus, enteric glia form a sheath around neurons. AAV6-mediated transduction of glia (**e–h**) is shown as GFP expression within GFAP immunoreactive cells (colocalization appears *turquoise*). Scale bar in (**d**) and (**h**) is 100 μm and applies to all panels

through either systemic or peroral delivery of viral vectors [3–5]. Although these delivery methods have shown some success, indirect delivery of viral vectors must overcome significant barriers to achieve successful transduction, and dramatically increase the risk of off-target effects [5, 6].

The protocol detailed herein describes an effective method of gene delivery to the ENS using AAV [7]. Gene delivery is achieved by performing a laparotomy, followed by direct injections of AAV into the wall of the descending colon using a Hamilton syringe connected to a foot-operated pump. Using this technique we demonstrate high levels of transduction of both neurons and glia within discrete areas of the ENS (Fig. 1) [7]. The current chapter will describe the direct injection of AAV to the descending colon of the rat; however, this same technique can be applied to various locations along the GI tract across species [7].

2 Materials

1. Adult male Sprague Dawley rat (*see* **Note 1**).
2. Adeno-associated virus (*see* **Note 2**).
3. Isoflurane vaporizer with induction box and nose cone adapters.

4. Isoflurane (Abbott Animal Health, Abbott Park, IL, USA).
5. Programmable micro syringe pump with foot pedal control (e.g. Harvard Apparatus Pump 11 Elite or equivalent).
6. 50 μL Hamilton syringe with 26 or 31 gauge beveled tip needle (Hamilton, Reno, NV, USA; *see* **Note 3**).
7. Sigmacote siliconizing reagent (Sigma-Aldrich, St. Louis, MO, USA).
8. Heating pad.
9. Hair clippers.
10. Betadine surgical scrub (Thermo Fisher Scientific, Waltham, MA, USA).
11. Sterile Surgical Gauze (Thermo Fisher Scientific, Waltham, MA, USA).
12. Sterile saline (Aspen Veterinary Resources, Liberty, MO, USA).
13. 11.5 cm straight fine scissors (FST, Foster City, CA, USA).
14. Two 12.5 cm serrated halstead hemostats (FST, Foster City, CA, USA).
15. 12 cm adson foreceps with rounded tip (FST, Foster City, CA, USA).
16. 1 and 20 mL syringes (BD, Franklin Lakes, NJ, USA).
17. Sterile adhesive plastic surgical drape (e.g., Bioclusive by Johnson & Johnson, or equivalent).
18. Black animal tattoo pigment and ultra-fine tip tattoo needle (AIMS, Hornell, NY, USA).
19. Perma-Hand 19 mm silk suture (Ethicon, San Lorenzo, Puerto Rico; *see* **Note 2**).
20. Autoclip stapler (FST, Foster City, CA, USA).
21. 9 mm Autoclip surgical staples (Braintree Scientific, Braintree, MA, USA; *see* **Note 3**).
22. Ketofen (ketoprofen; Fort Dodge Animal Health, Fort Dodge, IA, USA).
23. Timentin (ticarcillin disodium and clavulanate potassium; GlaxoSmithKline, Brentford, UK).
24. 3 % hydrogen peroxide.

3 Methods

3.1 Pre-operative Preparation

1. Siliconize the Hamilton syringe using the Sigmacote siliconizing reagent (*see* **Note 4**). Fill an Eppendorf tube with approximately 500 μL of Sigmacote. Using the Hamilton

syringe, withdraw Sigmacote to fill the syringe, allow to sit for 10 s, then expel the Sigmacote back into the Eppendorf tube. Repeat this process approximately 10 times to ensure adequate coating of the syringe. Dry overnight.

2. All surgical tools should be autoclaved prior to use, including: Scissors, forceps, hemostats, autoclip stapler and staples, tattoo needle, and gauze. Thoroughly sterilize other surgical supplies that cannot be autoclaved. Disinfect and sterlilize surgical station (*see* **Note 5**).
3. Prepare working concentrations of Ketofen and Timentin. Determine the total cumulative body weight (in kilograms) of all animals which will receive surgery. Add 10 % to this value to ensure adequate amount of drugs. This value is the total volume (in milliliters) of each respective drug you will need for all surgeries (based on the working concentrations described here). Prepare working dose of Ketofen by diluting Ketofen stock to a final working concentration of 5 mg/mL in the calculated total volume of sterile saline. Dilute Timentin stock to a final working concentration of 60 mg/mL in the calculated total volume of sterile saline (*see* **Note 6**).
4. Prepare a 1.5 mL Eppendorf tube containing 3 % hydrogen peroxide and another containing sterile saline.
5. Fill a 20 mL syringe with sterile saline.
6. Set up the micro syringe pump and secure siliconized Hamilton syringe inside of syringe holder. Program the pump (for the specific syringe) to infuse a volume of 5 μL at a flow rate of 10 μL/min. Load the Hamilton with a total of 30 μL of virus.

3.2 Laparotomy and Viral Injection

1. Turn on isoflurane vaporizer and set at 2 % isoflurane and oxygen set at 0.5 %. Allow the gas to fill the induction chamber (*see* **Note 7**).
2. Place animal in the induction chamber until fully anesthetized (*see* **Note 8**).
3. Use clippers to thoroughly shave the entire abdomen of the animal (*see* **Note 9**).
4. Place the animal on its back on top of the heating pad with the nose of the animal inside of the anesthesia nose cone (*see* **Note 9**).
5. Apply betadine to the shaved surgical area and use the sterile surgical gauze to spread across the entire abdomen (Fig. 2a).
6. Ensure that the animal is fully anesthetized again prior to making initial incision. Using the straight fine scissors cut a small entry incision in the skin overlaying the abdomen (Fig. 2b; *see* **Note 10**).

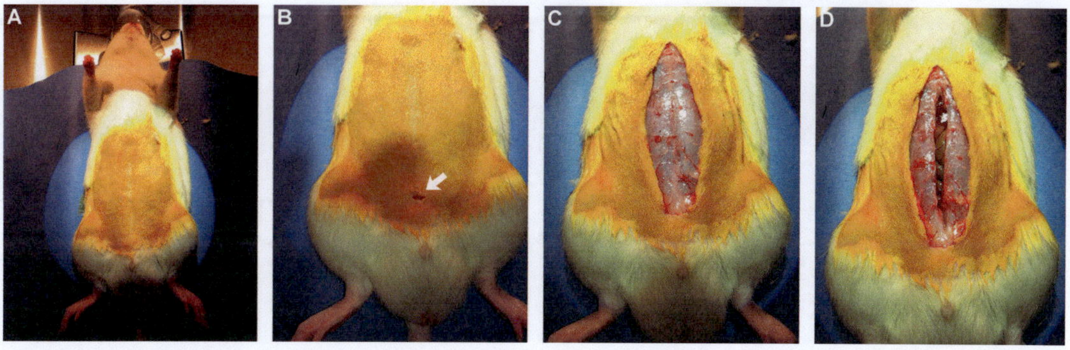

Fig. 2 Performing the laparatomy. Anesthetize the animal, thoroughly shave the abdomen and scrub with betadine to disinfect the surgical site. Place the animal on a heating pad with the abdomen facing up (**a**). Make a small initial incision in the skin at the approximate level of the pelvis (*arrow* in **b**). Continue the incision in the skin vertically up to the level of the sternum (**c**). Make a small initial incision in the abdominal muscle at the level of the pelvis and continue incision in the abdominal wall vertically up to the level of the sternum (**d**)

7. Use the scissors and forceps to delicately continue the vertical incision along the skin up the midline of the abdomen until the approximate level of the sternum. During this time also use the scissors and forceps to remove the fascia connecting the skin to the rectus abdominus muscle below (Fig. 2c; *see* **Note 10**).
8. Using the fine scissors and forceps, cut a small vertical incision in the abdominal wall just rostral to the approximate level of the bladder (*see* **Note 10**).
9. Use the fine scissor and forceps to very carefully continue the vertical incision along the midline of the abdominal wall until the approximate level of the sternum, exposing the peritoneal cavity (Fig. 2d; *see* **Note 11**).
10. Overlay the abdomen with the plastic wrap and cut a vertical incision in the plastic over the incisions in the animals' abdomen (*see* **Note 12**).
11. Use the hemostats to grab the abdominal wall on both sides of the incision and lay them across the plastic wrap to retract the abdominal wall, fully exposing the peritoneal cavity (Fig. 3a).
12. Use the forceps to gently grab the small intestine or cecum. Delicately remove the small intestine and cecum from the peritoneal cavity and place on the plastic wrap overlaying the animals' abdomen. Move the contents of the peritoneal cavity until the descending colon is exposed and accessible (Fig. 3b; *see* **Note 13**).
13. Use the 20 mL syringe containing sterile saline to continuously bathe the exposed small intestine and other contents of the peritoneal cavity that are lying outside of the body (*see* **Note 14**).

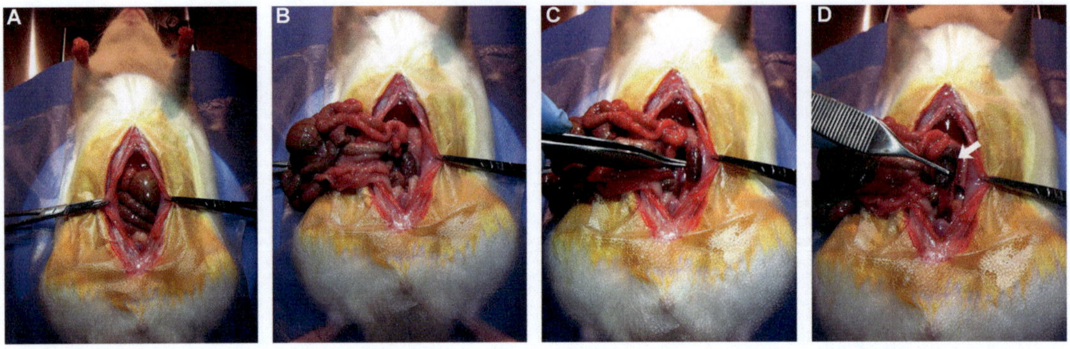

Fig. 3 Retracting the abdominal wall and isolating the colon. Place plastic wrap over the abdomen and cut a vertical incision in the plastic wrap mirroring the incision in the abdomen. Use hemostats to clamp the abdominal wall, and lay the hemostats to either side of the incision to retract the abdomen and expose the peritoneal cavity (**a**). Use forceps to gently remove the small intestine and cecum overlaying the colon (**b**). Use forceps to isolate the descending colon from the peritoneal cavity (**c**). While securing the colon in the forceps, tattoo a horizontal band across the colon (*arrow* in **d**)

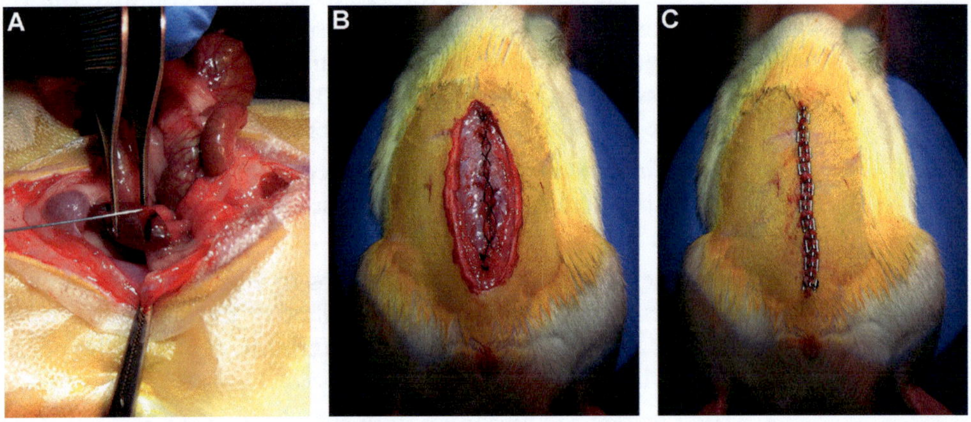

Fig. 4 Injecting the colon and closing the abdominal cavity. Secure the descending colon with the forceps, insert the needle under the serosa and longitudinal muscle of the gut wall and inject (**a**). Following completion of injections, replace the previously removed contents of the peritoneal cavity and suture the abdominal wall closed (**b**). Finally, staple the skin of the abdomen closed (**c**)

14. Once the descending colon is identified and accessible, use the forceps to gently secure it (Fig. 3c). Dip the fine tip tattoo needle in ink and delicately tattoo a horizontal band across the colon. Wash away excess ink with the sterile saline in the 20 mL syringe and repeat the tattooing until a clearly identifiable band remains (Fig. 3d; *see* **Note 15**).
15. Use the forceps to gently grip the colon and insert the Hamilton syringe into the wall of the descending colon just caudal to the level of the tattoo (Fig. 4a). The Hamilton should not puncture the lumen of the colon. Aim to insert the

needle, with the bevel up, just below the longitudinal muscle layer (Fig. 5). Once the Hamilton needle is in place press the foot pedal to inject the virus. Leave the needle in place for 30 s after infusion to avoid reflux (*see* **Note 16**).

16. Repeat the process in **step 15**, five more times, creating a 2×3 grid of injection just caudal to the level of the tattoo (Fig. 6; *see* **Note 17**).
17. After injections are complete, use the forceps to gently replace the small intestines and any other contents of the peritoneal cavity previously removed (*see* **Note 18**).
18. Holding both of the hemostats (still gripped to the abdominal wall) lift the animal and shake in a vertical plane to ensure that the contents of the peritoneal cavity settle back into place and are not twisted or otherwise impeded.
19. Remove the hemostats and the plastic wrap, being careful to avoid dripping the fluid on the plastic wrap into the peritoneal cavity.
20. Use the hemostats to grab the needle of the suture and use the forceps to hold the abdominal wall. Insert the needle of the suture down through the right side of abdominal wall, and then up through the left side. Use the hemostats to pull the suture to the left, through the abdominal wall until approximately 1 cm of suture remains on the right side of the

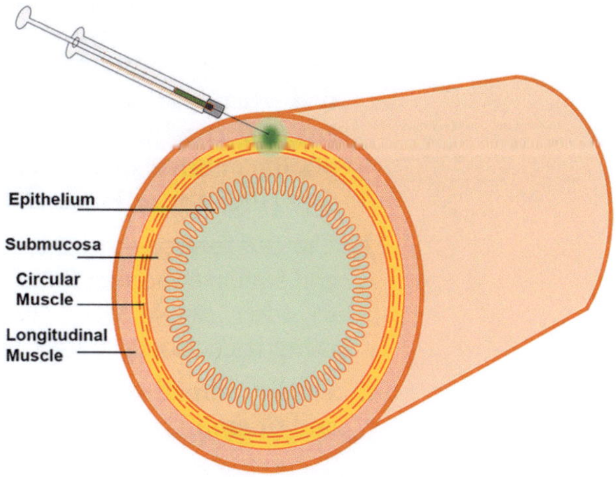

Fig. 5 Needle placement for colonic injections. Aim to insert the needle in a horizontal plane relative to the surface of the gut wall at the level of the tattoo when performing colonic injections. Insert the needle just below the superficial layer of the gut wall, with the bevel of the needle facing up, at the approximate layer of circular muscle. Proper needle placement will allow viral spread to both the myenteric plexus and the submucosal plexus

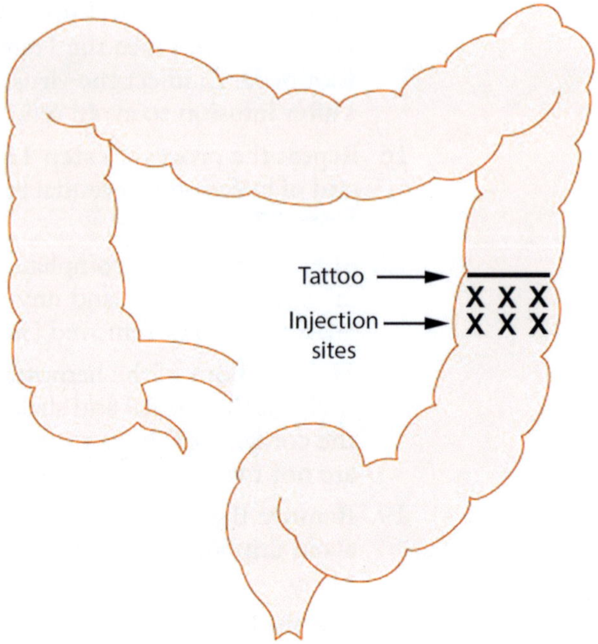

Fig. 6 Colonic injection grid. Perform injections caudal to the tattoo on the descending colon. Place 6 × 5 μL injections in a grid as shown by the X's

incision. Still holding the suture needle in the hemostats, wrap the long end of the suture held in the hemostats around the forceps twice in a clockwise direction, then pull down toward the incision, creating a square knot. Repeat the knotting process again but wrapping the long end of the suture around the forceps in a counterclockwise direction. After the initial knot is tied place a continuous suture up and then back down the entire vertical axis of the incision. Complete the suture by placing another square knot just below the initial point of incision (Fig. 4b; *see* **Note 19**).

21. Use the forceps to pull the skin away from the abdominal wall and squeeze the skin from the left and right side of the incision together. Staple along the entire vertical axis of the incision using the autoclip stapler (Fig. 4c).
22. Finally, administer the calculated dose of ketofen (5 mg/kg) and timentin (60 mg/kg) intraperitoneally.
23. Following completion of the surgery expel any remaining fluid in the Hamilton syringe. Remove the Hamilton from the pump and dip the tip in the 3 % hydrogen peroxide. Next rinse the tip in sterile saline (*see* **Note 20**).

3.3 Post-operative Care

1. Closely monitor the animals twice daily for the next 48 h post-surgery, for signs of distress or infection. Monitor once daily for the next week (*see* **Note 21**).

2. Remove staples 10–14 days post-surgery. Anesthetize the animal and use a surgical staple remover to remove all remaining wound clips.

4 Notes

1. Experiments involving the use live animals must follow officially approved procedures for the care and use of laboratory animals that have been reviewed and approved by an Institutional Animal Care and Use Committee (IACUC).
2. The serotype of AAV will have an impact on the tropism of the virus within the ENS. We have observed high levels of neuronal tropism with AAV9 and high levels of both glial and neuronal tropism with AAV6 [7].
3. We have empirically determined that the 26 gauge needle is suitable for rat colonic injections and the 31 gauge needle is suitable for mouse colonic injections. Similarly, although we have had success with the 9 mm suture in mice, it may be prudent to use a smaller gauge suture and smaller surgical clips (e.g. 7 mm) for mouse or other small rodent surgeries.
4. Due to the inherently "sticky" nature of the viral capsid it is imperative to siliconize every surface that the virus will come into contact with (this includes pipette tips and tubes used to move and store the virus). Siliconizing will prevent the adherence of virus to contact surfaces and subsequent loss of viral particles.
5. The current surgical technique is highly invasive and as such requires very strict sterilization and aseptic technique. Anything coming into contact with the animal during the surgical procedure must be sterilized and aseptic. This includes autoclaving surgical tools, but also using sterile saline, sterilization and decontamination of the Hamilton syringe, sterilization of the abdominal surgical site using an antiseptic scrub, use of sterile sutures and staples, and use of sterile tattoo ink. Further, the surgery should be performed on a decontaminated surgical station within a sterile surgical room (preferably under a hood with positive filtered airflow) wearing lab coats and filter masks. Following strict sterile technique will significantly improve surgical outcome and decrease post-operative complications or discomfort.
6. The working concentration of Timentin can be prepared, aliquoted, and stored at −80 °C until the day of use. Prepare a fresh working concentration of Ketofen on the day of surgery.
7. The 2 % isofluorane is a general guideline. Some anesthesia protocols call for 3–5 % isofluorane for anesthetic induction,

then a reduction to 2 % isofluorane for the surgical procedure. Animals should be monitored throughout the entire surgery to ensure that they are fully anesthetized but showing no signs of overdose (shallow or irregular breathing, cyanosis on footpads) and anesthesia should be adjusted accordingly.

8. When fully anesthetized, the animal will exhibit loss of pedal withdrawal reflex and tail pinch response, decreased or absent limb muscle tone, and slow regular breathing. To test pedal withdrawal reflex extend the animals hind limb and use a pair of forceps to lightly pinch the foot. If the limb withdrawals or muscles twitch the animals is not sufficiently anesthetized. The lack of a response indicates deep anesthesia.
9. The animal may wake up after or during shaving the abdomen. If so, place the animal back into the induction chamber prior to completing the shaving and moving to the heating pad and nose cone. This will ensure proper anesthesia and prevent any struggle that may harm the animal or the surgeon.
10. Keeping the initial incisions into the skin and the abdominal wall as small as possible will allow for easier and cleaner suturing and stapling when closing the animal after the surgery. Removing the fascia connecting the skin to the abdomen will also facilitate easier and cleaner suturing/stapling as well as grant the surgeon increased freedom when manipulating the abdominal wall.
11. When performing the vertical incision up the midline of the abdominal wall, exercise extreme caution not to lacerate any internal organs. Use the forceps to pull the abdominal wall away from the peritoneal cavity and keep the tip of the scissors pointed up and away from the internal organs. The liver will press up against the abdominal wall, so it is necessary to cut carefully when performing the incision.
12. The plastic wrap will provide a sanitary surface to place the small intestines and cecum on when they are removed.
13. When removing the contents overlaying your region of interest pull them out in a straight line and avoid any significant amount of force or torsion. The small intestine and cecum should come out with very little pull. Remove the contents laying on top first and generally try to keep the organs in the order in which they were in the peritoneal cavity. These measures will ensure that the removed organs can be put back in the peritoneal cavity in their original position, obviating the potential for gastrointestinal obstruction produced by torsion. Always handle organs delicately, and avoid directly clamping the organs or using sharp instruments that could poke organs.
14. It is imperative to keep the exposed contents of the peritoneal cavity moist and hydrated. Apply sterile saline frequently throughout the entire surgery to avoid dehydration.

15. The tattoo on the colon is used as a marker to indicate the area of injection at the time of euthanasia and tissue collection. Use the forceps to delicately lift the colon approximately 1 cm from the ventral wall of the peritoneal cavity. Hold the colon just firm enough to keep it from moving but gentle enough to avoid damage. It is helpful to hold an area of the colon that contains a fecal pellet. This provides a firm substrate to grab with the forceps as well as a firm backdrop for the tattooing. The tattooing should be performed delicately, just barely puncturing the superficial layers of the colon.
16. When inserting the syringe, use the same grip as in **step 14**. If possible angle the colon slightly and match the angle of the colon with the needle. When injecting ensue that the needle enters relatively parallel to the colon. This will decrease the probability of puncturing the needle into the lumen of the gut. Be sure that the bevel of the needle is completely under the longitudinal muscle layer so that the virus does not leak out of the injection hole. A bubble of injected fluid will slightly distend the outer layers of the colon visually confirming a successful injection.
17. A single 5ul injection was determined to cover an area of approximately 50 mm^2 [7]. However, the number and volume of injections should be empirically optimized for specific experimental needs. Any fine bleeding produced by injections can be stopped by gently pressing a cotton swab to the site of bleeding.
18. Again, be sure to replace the contents of the peritoneal cavity back in the order in which they originally sat.
19. Be conscious of the depth and angle of the suture needle when suturing. Similar to making the incisions, keep the tip pointed up and away from the internal organs as much as possible. Suture cautiously to avoid puncturing or lacerating the internal organs.
20. The hydrogen peroxide will destroy any remaining virus on the Hamilton needle tip and also sterilize the needle. Rinse in sterile saline to remove any remaining hydrogen peroxide prior to loading the needle with virus for the next surgery.
21. Monitor animals for symptoms of pain and distress and administer analgesic as necessary. Closely monitor wound healing. If necessary anesthetize the animal, disinfect the wound, and reapply surgical staples. Finally, monitor for signs of dehydration or weight loss. If necessary administer sterile saline (1–5 cc; i.p.) or high fat supplements in order to minimize discomfort and maximize animal health.

Acknowledgement

This work was supported by a Strategic Partnership Grant from the Michigan State University Foundation.

References

1. Wood JD (2011) Enteric nervous system: the-brain-in-the-gut. Morgan & Claypool Life Sciences, New Jersey
2. Belkind-Gerson J, Graeme-Cook F, Winter H (2006) Enteric nervous system disease and recovery, plasticity, and regeneration. J Pediatr Gastroenterol Nutr 42:343–350
3. During MJ et al (2000) An oral vaccine against NMDAR1 with efficacy in experimental stroke and epilepsy. Science 287:1453–1460
4. During MJ et al (1998) Peroral gene therapy of lactose intolerance using an adeno-associated virus vector. Nat Med 4:1131–1135
5. Foust KD et al (2009) Intravascular AAV9 preferentially targets neonatal neurons and adult astrocytes. Nat Biotechnol 27:59–65
6. Zincarelli C, Soltys S, Rengo G, Rabinowitz JE (2008) Analysis of AAV serotypes 1-9 mediated gene expression and tropism in mice after systemic injection. Mol Ther 16:1073–1080
7. Benskey MJ et al (2015) Targeted gene delivery to the enteric nervous system using AAV: a comparison across serotypes and capsid mutants. Mol Ther. doi:10.1038/mt.2015.7

Chapter 20

Gene Therapy of the Peripheral Nervous System: Celiac Ganglia

Bradley Hammond and David L. Kreulen

Abstract

Gene therapy has played an integral role in advancing our understanding of the central nervous system. However, gene therapy techniques have yet to be widely utilized in the peripheral nervous system. Critical targets for gene therapy within the PNS are the neurons in sympathetic ganglia, which are the final pathway to end organs. Thus they are the most specific targets for organ-specific neuron modification. This presents challenges because neurons are not viscerotopically organized within the ganglia and therefore cannot be targeted by their location. However, organ-specific neurons have been identified in sympathetic ganglia of some species and this offers an opportunity for targeting and transducing neurons by way of their target. In fact, alterations in sympathetic neurons have had pathological effects, and transducing organ-specific sympathetic neurons offer an exciting opportunity to selectively modify sympathetic pathology. In this chapter, we describe a method to virally transduce the celiac ganglion (CG), a prevertebral sympathetic ganglion that innervates abdominal organs, with AAV serotypes 1 and 6; thereby, providing a potential avenue to modulate specific subsets of neurons within the celiac ganglion.

Key words Celiac ganglia, Gene therapy, Peripheral nervous system, Adeno-associated virus, Viral transduction, Direct injection

1 Introduotion

Adeno-associated viruses (AAV) produce efficient, long-lasting, and nontoxic transductions; making AAVs an ideal vector for gene therapy [1, 2]. Transductions using AAVs have been used to map anatomical pathways, and drive transgene expression in a variety of regions throughout the central nervous system [1, 3–5]. However, few have utilized gene therapy to modulate the peripheral nervous system [6, 7]. Peripheral nervous system (PNS) gene therapy would pave the way for other investigations including manipulation of target genes, utilizing optogenetics to modulate subsets of PNS cell types and anterograde tracing of PNS neurons. The results of these investigations would provide insight into gene function in disease states, function of various cell types throughout the PNS, and

Fredric P. Manfredsson (ed.), *Gene Therapy for Neurological Disorders: Methods and Protocols*, Methods in Molecular Biology, vol. 1382, DOI 10.1007/978-1-4939-3271-9_20,

innervation patterns of PNS neurons, eventually opening up the possibility of treatment of disorders of the autonomic nervous system that are now outside the reach of current therapies.

An unexplored target for gene therapy in the PNS is the celiac ganglion (CG), an important source of innervation to a variety of abdominal organs that are critical for survival. While sympathetic neurons within the CG are not spatially organized, organ-specific neurons have been identified using retrograde tracers [8–10]. Furthermore, there is evidence that some neurons are chemically coded according to their targets [11, 12], which suggests that neurochemical-specific promoters could be used in some species to genetically modify neurons based on their target. In the abdomen, celiac ganglion neurons modulate organ blood flow via the degree of vasoconstriction of arteries and veins; recruitment of these vascular neurons contributes to the elevated blood pressure and if it is chronic, neurogenic hypertension [13]. Apart from the effects of complete removal of the CG, which can attenuate hypertension in animal models [14], little is known about the specific physiologic consequences of gain or loss of function of neurons within the CG. Gene therapy techniques have the potential to resolve the physiological implications of organ-specific innervation by CG neurons. Viral transduction of CG neurons has not been reported. Here, we demonstrate viral transduction of the CG by direct CG injections of AAV serotypes 1 and 6, expressing GFP driven by a CMV promoter.

2 Materials

2.1 Components of AAV Injections

1. Brown/Flaming P-87 micropipette puller (Sutter Instrument Co, Novato, CA, USA).
2. Pulled glass micropipette: pull a glass micropipette (VWR International, Radnor, PA, USA) using the micropipette puller set at Heat = 727, Velocity = 30, Time = 200.
3. Micromanipulator (Warner Instruments, Hamden, CT, USA).
4. Picospritzer (Parker, Hollis, NH, USA).
5. Gut retractor.
6. Dissecting microscope.
7. AAV Serotypes Testing Kit (Vector Biolabs, Eagleville, PA, USA).
8. 5 mM Fast green FCF (Sigma-Aldrich, St. Louis, MO, USA): measure 4.04 mg of fast green FCF and transfer it to a 1 ml microcentrifuge tube containing 1 ml of dH_2O.
9. Dulbecco's Phosphate Buffered Saline (Sigma-Aldrich, St. Louis, MO, USA).
10. 4-0 silk sutures.
11. Electronic heating pad.
12. Oster Golden A5, 1-Speed Clipper.

2.2 Immunohistochemical Components

1. Paraformaldehyde (4 %): on the day of fixation heat 50 ml 0.2 M phosphate buffer (pH 7.4) and 4 g paraformaldehyde powder in a fume hood on a stirring hot plate. Heat the solution until the paraformaldehyde dissolves into solution, but be careful not to exceed 90 °C. Remove the paraformaldehyde solution from heat and bring the solution to 10 °C in an ice bath. Filter the solution using a bottle top filter (Corning, Coming, NY, USA) into a flask large enough to hold 100 ml. Add 50 ml dH_2O to the filtered solution and store at 4 °C.
2. Sucrose (20 %): dissolve 20 g of sucrose in 100 ml 0.1 M phosphate buffer (pH 7.4) at room temperature using a stir bar.
3. Tissue-Tek® O.C.T Compound (VWR, Radnor, PA, USA).
4. Superfrost® plus micro slides (VWR, Radnor, PA, USA).
5. Cryostat.
6. Super PAP pen (Life Technologies, Grand Island, NY, USA).
7. Tris-PBS: Measure 1.211 g Trizma Base, 25 ml 0.4 M Phosphate buffer (pH 7.4), and 9 g NaCl and add them to 800 ml dH_2O. pH the solution to 7.4 with 0.1 M HCL, then bring the solution up to 1 l with dH_2O. Store the solution at 4 °C for a maximum of 2 weeks.
8. Immunobuffer (IB): measure 1 l of Tris-PBS and add 3 ml of Triton X-100. Stir gently on a stir plate until the Triton X-100 dissolves. Store the solution at 4 °C for a maximum of 2 weeks.
9. 10 % Normal goat serum (NGS): dilute 1 ml NGS in 9 ml IB. Do not store, make as needed.
10. Chicken anti-GFP (Aves, Tigard, Oregon, USA) or equivalent. Store per manufacturer's recommendation.
11. NGS (1 %): dilute 0.1 ml NGS in 9.9 ml IB (*see* **step 2.8**). Do not store, make as needed.
12. Fluorescein Goat Anti-Chicken IgY (Aves, Tigard, Oregon, USA) or equivalent. Store per manufacturer's recommendation.
13. ProLong® Gold Antifade (Life Technologies, Grand Island, NY, USA). Store at room temperature.
14. Microscope with filters capable of exciting fluorophores at 495 nm and detecting 519 nm emissions.

3 Methods

For all procedures, use male Sprague-Dawley rats weighing 225–300 g. House the animals two per cage in a temperature- and humidity-controlled room with a 12-h light/dark cycle. Allow animals ad libitum access to rat chow and water for the duration of this protocol.

3.1 Direct Injection of AAVs into the CG

1. Anesthetize the animal with isoflurane (3–4 % during induction, 2 % during surgery) in O_2 (0.75 l/min).
2. Shave the midline abdominal region and then prepare the incision site by alternating three sets of cyclohexidine and ethanol scrubs.
3. Perform a laparotomy and retract the abdominal viscera to reveal the CG (*see* **Note 1**, Fig. 1).
4. Keep the abdomen moist by draping the exposed tissue with sterile gauze soaked in Dulbecco's Phosphate Buffered Saline.
5. Under a dissecting microscope, locate the CG (*see* **Note 2**, Fig. 1) and remove excess connective tissue surrounding the CG (*see* **Note 3**).
6. Fill a glass micropipette pulled according to the specifications above, with a mixture of 1 μl of 1×10^{13} gc/ml AAV6, and 0.1 μl of 5 mM fast green (*see* **Note 4**). Mount the filled micropipette on a micromanipulator that is attached to a Picospritzer set at 15 ms. Keep the glass micropipette a safe distance from the CG until ready to pierce the CG (*see* **Note 5**).
7. Advance the pipette to a depth of 400 μm in the CG (*see* **Notes 6–8**). Wait 2 min, allowing the glass micropipette to make a seal with the CG. After 2 min, inject 25 % of the viral mixture into the CG by activating the Picospritzer as many times as needed (*see* **Note 9**). Wait 5 min before withdrawing the glass micropipette.
8. Move the glass micropipette to different locations around the CG and repeat **step** 7 until the total 1.1 μl volume is injected using four different injection sites. Be careful to inject the viral load evenly throughout the entire CG, using the fast green to track the location of the virus.
9. Remove the gut retractor, replace the abdominal viscera, and close the incision with 4-0 silk sutures.

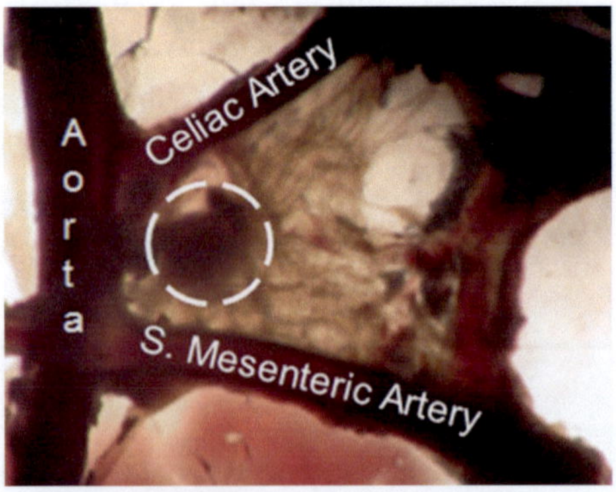

Fig. 1 Photograph of the celiac ganglia under a dissecting microscope. The celiac ganglia (*circled*) can be located by using the aorta, celiac artery, and superior mesenteric artery as landmarks

10. Give the animals analgesic (Carprofen 5 mg/kg) and antibiotic (Enrofloxacin 5 mg/kg) for 3 days post-surgery (House the animals in pairs for 4 weeks until viral expression is confirmed using immunohistochemistry) (*see* **Note 10**).
11. Wipe down the workspace and any equipment used to handle AAVs with a 10 % bleach solution.

3.2 Immunohistochemical Detection of GFP

All washes and incubations are processed on a shaker at room temperature. Reduce light exposure of the samples by keepings the slides in a dark environment during washes and incubations. Each wash is 10 min.

1. 4 weeks after the direct injection of AAV, deeply anesthetize the animals with sodium pentobarbital (50 mg/kg, ip), and then harvest the CG. Submerge the CG in 4 % paraformaldehyde for 1 day at 4 °C.
2. Transfer the CG to 20 % sucrose for 1 day at 4 °C, then freeze the sample at −80 °C in OCT until use.
3. Cryostat section (14–18 μm) the CG in a single series on to Superfrost® plus micro slides.
4. Using a PAP pen outline the slides with a hydrophobic barrier.
5. Wash the slides three times in IB.
6. Incubate for with 10 % NGS (30 mins).
7. Incubate with anti-GFP antibody (1:1000, 1 day), diluted in 10 % NGS.
8. Wash three times with Tris-PBS.
9. Incubate with goat anti-chicken secondary antibody (1:500, 2 h) diluted in 1 % NGS.
10. Coverslip the slides with ProLong® Gold Antifade.
11. Visualize the sections under a microscope capable of detecting the chromogen FITC.
12. To determine specificity of labeling, process control sections in the absence of either primary or secondary antibody. Alternatively, run the full detection protocol on CGs injected with vehicle as negative controls.

4 Notes

1. Animals do not need to be fasted; however, fasting animals overnight will ease viscera retraction and provide a larger working space inside the animal.

2. To locate the CG find the junction of the renal artery and aorta. Move rostrally along the aorta until reaching the superior mesenteric artery branching off of the aorta. Resting on the superior mesenteric artery is the superior mesenteric ganglia (SMG); if the SMG is not located, clear the fat surrounding the superior mesenteric artery. Follow the SMG rostrally until a lobe-like structure is reached. This lobe like structure is the right lobe of the CG, and is immediately caudal to the celiac artery (Fig. 1).
3. When clearing connective tissue avoid damaging nerves and lymph vessels surrounding the CG. Damaging a nerve will compromise the integrity of the CG and potentially reduce transduction rates. Rupturing a lymph vessel will obscure your view of the CG and compromise your injections.
4. AAV serotype 6 was selected for direct injection into the CG over other serotypes based on work done with differentiated pheochromocytoma (PC-12) cells in culture and direct injections of AAVs into the CG. Differentiated PC-12 have properties resembling sympathetic neurons [15], and had higher transduction rates with AAV serotype 6 as compared with AAV serotypes 1, 2, and 5 (Fig. 2). Only serotypes 1 and 6 were tested in vivo, with serotype 6 having the best transduction at a concentration of 1×10^{13} gc/ml (Fig. 3). Therefore, 1×10^{13} gc/ml AAV6 is used to transduce the CG with GFP.
5. Abdominal fluids will enter the glass micropipette via capillary action if the glass micropipette is positioned too close to the CG.
6. Achieving penetration of the CG can be difficult due to several factors: incomplete clearing of the connective tissue covering the CG, or imperfections in or damage to the tip of the glass micropipette. Both of these scenarios lead to below average injection quality and inferior transduction rates. Therefore, take care when removing connective tissue, and while loading the glass micropipette.
7. Due to variability from animal to animal, and injection to injection, the 400 μm depth for injections is an estimate and should be used as such. Achieving an injection depth of 400 μm may be impossible for some injections (*see* **Note 6**). For some injections, the pipette will have to be slowly worked into the CG by advancing and retracting the pipette into the CG several times, making depth tracking onerous.
8. Penetrating the CG at very shallow angles produces the best injections. Poor injections result when the glass micropipette is at an angle close to perpendicular to the CG. At these perpen-

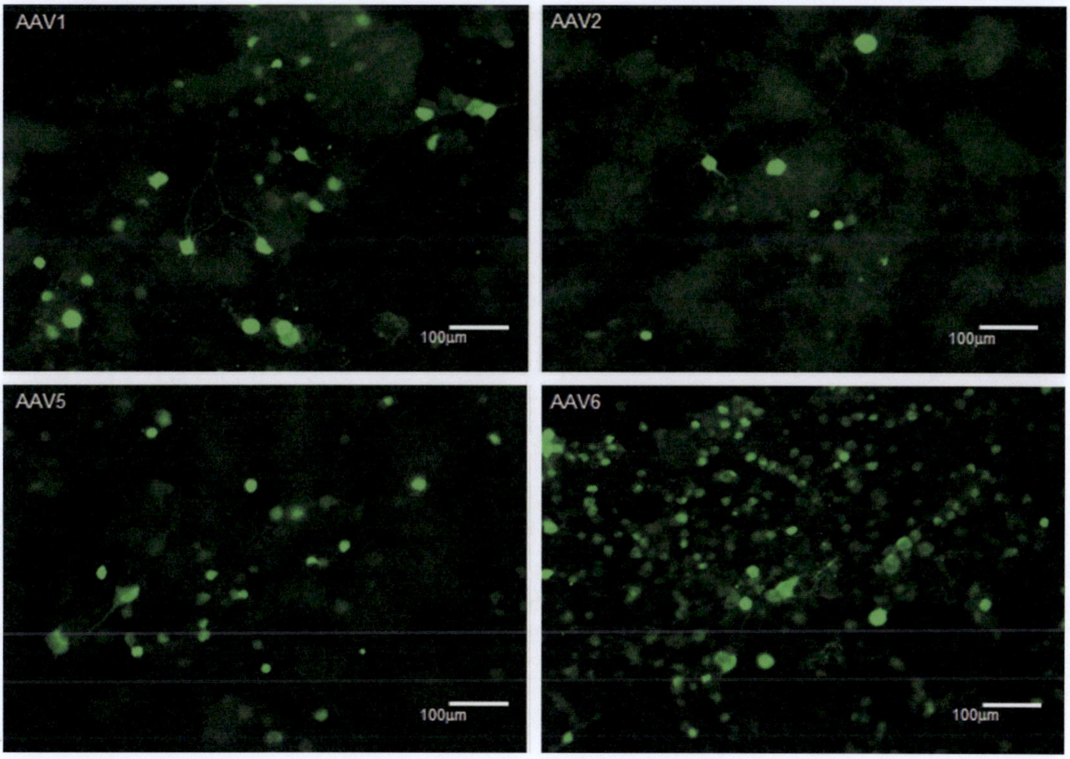

Fig. 2 Expression of GFP using different serotypes of Adeno-associated virus (AAV) in differentiated PC-12 cells. All serotypes express GFP driven by a CMV promoter. Different serotypes of AAV were tested at a concentration of 1×10^{10} gc/ml at an exposure time of 1 h, followed by a 5-day expression period. Cultures were 90 % confluent at the time of image acquisition. AAV6 had the highest transduction rate, AAV2 had the lowest transduction rate

dicular angles, it is easy to penetrate through the entire CG, and inject virus into the abdominal cavity.

9. The glass micropipette may not visibly inject any liquid into the CG. This is the result of too much backpressure at the end of the pipette. To reduce backpressure, retract the glass pipette from the CG 100 μm or try injecting at a higher pressure. If the pipette is retracted, wait 2 min for the CG to reseal around the pipette.

10. Time points earlier than 4 weeks for the viral expression were tested and no staining was discerned in animals that recovered for 2–3 weeks after exposure to AAV (data not shown).

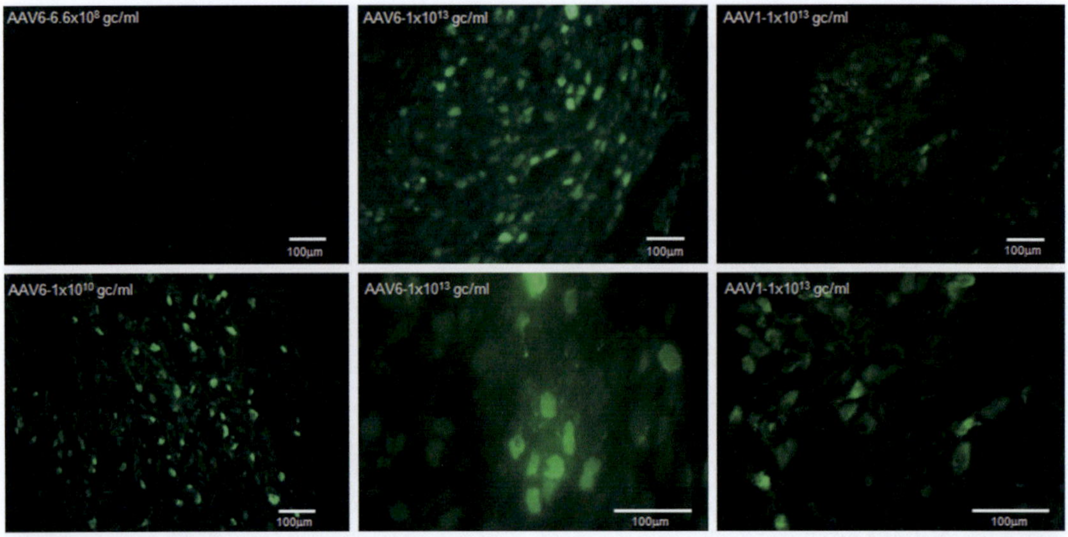

Fig. 3 Expression of GFP in CG 4 weeks after direct injection of Adeno-associated virus (AAV). All serotypes express GFP driven by a CMV promoter. There was transduction of both AAV1 and AAV6 at concentrations of 1×10^{10} and 1×10^{13} gc/ml. GFP expression was present in the cell bodies and processes. AAV6 demonstrated a higher transduction rate when compared to AAV1

Acknowledgement

This work was supported by NIH grant P01HL070687.

References

1. Kaplitt MG, Leone P, Samulski RJ, Xiao X, Pfaff DW (1994) Long-term gene expression and phenotypic correction using adeno-associated virus vectors in the mammalian brain. Nat Genet 1–7
2. Blacklow NR, Hoggan MD, Kapikian AZ, Austin JB, Rowe WP (1968) Epidemiology of adenovirus-associated virus infection in a nursery population. Am J Epidemiol 88(3):368–378
3. Peel AL, Klein RL (2000) Adeno-associated virus vectors: activity and applications in the CNS. J Neurosci Methods 98(2):95–104
4. Ruitenberg MJ, Eggers R, Boer GJ, Verhaagen J (2002) Adeno-associated viral vectors as agents for gene delivery: application in disorders and trauma of the central nervous system. Methods 28(2):182–194
5. Chamberlin NL, Du Du B, de Lacalle S, Saper CB (1998) Recombinant adeno-associated virus vector: use for transgene expression and anterograde tract tracing in the CNS. Brain Res 793(1-2):169–175
6. Glatzel M, Flechsig E, Navarro B, Klein MA, Paterna JC, Büeler H et al (2000) Adenoviral and adeno-associated viral transfer of genes to the peripheral nervous system. Proc Natl Acad Sci U S A 97(1):442–447
7. Mason MR, Ehlert EM, Eggers R, Pool CW, Hermening S, Huseinovic A et al (2010) Comparison of AAV serotypes for gene delivery to dorsal root ganglion neurons. Mol Ther 18(4):715–724
8. Macrae IM, Furness JB, Costa M (1986) Distribution of subgroups of noradrenaline neurons in the coeliac ganglion of the guinea-pig. Cell Tissue Res 244(1):1–8
9. Quinson N, Robbins HL, Clark MJ, Furness JB (2001) Locations and innervation of cell

bodies of sympathetic neurons projecting to the gastrointestinal tract in the rat. Arch Histol Cytol 64(3):281–294

10. Browning KN, Zheng Z, Kreulen DL, Travagli RA (1999) Two populations of sympathetic neurons project selectively to mesenteric artery or vein. Am J Physiol Heart Circ Physiol [Bethesda, Md.]. American Physiological Society; 45(4):H1263–H1272
11. Richardson RJ, Grkovic I, Allen AM, Anderson CR (2006) Separate neurochemical classes of sympathetic postganglionic neurons project to the left ventricle of the rat heart. Cell Tissue Res 324(1):9–16
12. Keast JR, McLachlan EM, Meckler RL (1993) Relation between electrophysiological class and neuropeptide content of guinea pig sympathetic prevertebral neurons. Journal of Neurophysiology Am Physiological Soc; 69(2):384–394
13. Fink GD. (2009). Arthur C. Corcoran Memorial Lecture. Sympathetic activity, vascular capacitance, and long-term regulation of arterial pressure. Hypertension,, 307–12. PMCID: PMC2685147
14. Kandlikar SS, Fink GD (2011) Splanchnic sympathetic nerves in the development of mild DOCA-salt hypertension. AJP: Heart and Circulatory Physiology 301(5):H1965–H1973
15. Greene LA, Tischler AS. (1976). Establishment of a noradrenergic clonal line of rat adrenal pheochromocytoma cells which respond to nerve growth factor. Proceedings of the National Academy of Sciences of the United States of America, 73(7):2424–2428. PMCID: PMC430592

Chapter 21

Convection Enhanced Delivery of Recombinant Adeno-associated Virus into the Mouse Brain

Kevin R. Nash and Marcia N. Gordon

Abstract

Recombinant adeno-associated virus (rAAV) has become an extremely useful tool for the study of gene over expression or knockdown in the central nervous system of experimental animals. One disadvantage of intracranial injections of rAAV vectors into the brain parenchyma has been restricted distribution to relatively small volumes of the brain. Convection enhanced delivery (CED) is a method for delivery of clinically relevant amounts of therapeutic agents to large areas of the brain in a direct intracranial injection procedure. CED uses bulk flow to increase the hydrostatic pressure and thus improve volume distribution. The CED method has shown robust gene transfer and increased distribution within the CNS and can be successfully used for different serotypes of rAAV for increased transduction of the mouse CNS. This chapter details the surgical injection of rAAV by CED into a mouse brain.

Key words Adeno-associated virus, rAAV, Convection-enhanced delivery, CED, Intracranial injection, Step cannula, CNS

1 Introduction

Recombinant adeno-associated virus (rAAV) vectors have recently emerged as one of the most promising gene delivery approaches for the central nervous system (CNS) in a large range of animal species including humans. There are a number of advantageous characteristics including their lack of pathogenicity, low immunogenicity, persistence of the transgene as an episome, removal of all viral genes, and long-term gene expression. Furthermore, there are now a number of different rAAV serotypes available that confer different cellular tropisms [1–3].

Unfortunately, one disadvantage of intracranial injections of rAAV vectors into the brain parenchyma has been restricted distribution to relatively small volumes of the brain. Convection enhanced delivery (CED) is a method for delivery of clinically relevant amounts of therapeutic agents to large areas of the brain in a

Fredric P. Manfredsson (ed.), *Gene Therapy for Neurological Disorders: Methods and Protocols*, Methods in Molecular Biology, vol. 1382, DOI 10.1007/978-1-4939-3271-9_21, © Springer Science+Business Media New York 2016

direct intracranial injection procedure. CED was developed in the early 1990s as a method to deliver drugs that were not blood brain barrier permeable [4]. CED uses bulk flow to increase the hydrostatic pressure and thus improve volume distribution (Fig. 1). An optimal flow rate up to 5 μL/min is generally recommended [5] as flow rates in excess of 10 μL/min have been shown to cause tissue damage in rats [6]. Factors that affect the size of the volume of distribution include the hydrophobicity, charge, viscosity, and affinity to brain tissue of the deliverable molecules [7]. Research into the development and improvement of CED delivery devices is still under current investigation by several researchers.

CED was originally investigated predominantly for treatment of diseases such as glioblastoma because it offers several advantages, including of delivery of high concentration of molecules, robust distribution, targeted treatment, and lower system effects [5]. Since distribution is affected by hydrostatic pressure rather than diffusion, the molecular weight of the molecule is not usually the limiting factor controlling distribution. This supported the investigation of viral delivery into the CNS with the CED method.

CED was initially examined in gene therapy studies as a way to increase the distribution of rAAV2 vectors in the brain. Bankiewicz et al. revealed that CED can significantly increase gene transfer and distribution of rAAV expressing AADC in the striatum of MPTP-treated monkeys [8]. Similar results have been replicated by Cunningham et al. in the rat brain with rAAV2 expressing thymidine kinase (TK) [9]. The CED method showed robust gene

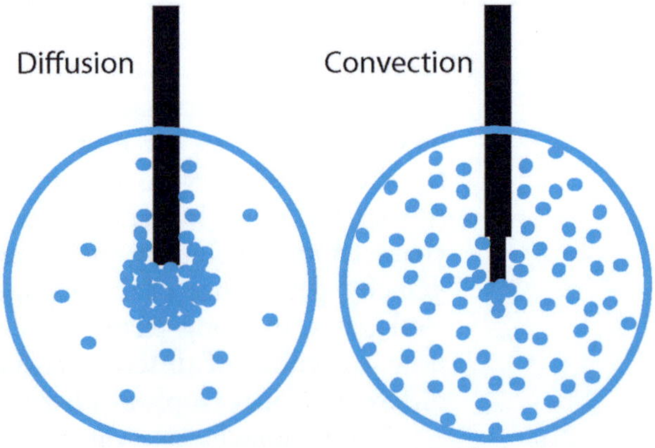

Fig. 1 Diffusion-based delivery results in a limited area of distribution and reflux along the needle track. Increased flow rate with CED results in a pressurized extracellular bulk flow that allows for a more homogeneous distribution of molecules/particles and increased distances from the cannula tip. Reflux is also reduced, especially with the step cannula design resulting in increased volume of distribution

transfer and increased distribution within the putamen, with immunoreactive cells found outside the striatum, including the globus pallidus, subthalamic nucleus, thalamus, and substantia nigra [9, 10]. Carty et al. examined the use of CED in mouse models [11]. They demonstrate that CED can be successfully used for different serotypes of rAAV for increased transduction of the mouse CNS.

Most recently a new method of real time convection delivery (RCD) has been established in order to make the method safer and more efficacious. This currently utilizes magnetic resonance imaging (MRI) [12–14]. Viral samples can be co-administered with an MRI tracer such as gadoteridol or gadopentetic acid (Gd-DTPA) to monitor optimal placement of the injection. This also has the advantages of identifying if any reflux is occurring (especially at higher flow rates), or if the injected sample is leaking into the ventricles and to allow for corrections of flow rate or needle placement if necessary. In nonhuman primates, it has been shown that there is a direct correlation between gadoteridol distribution and AAV2 transduction [15].

Back flow failure using needles and catheters is a significant concern with injections into the CNS and this is one of the theoretical mechanistic limitations of the CED method as well. To limit this problem, Krauze et al. developed a step cannula design [6]. This can effectively reduce reflux by placing a silica tubing within the needle creating a horizontal step that reduces the back flow of fluid. The optimization of more efficient cannula designs coupled with the encouraging results from studies showing enhanced gene transfer and distribution emphasize the therapeutic potential of the CED method in helping overcome some of the mechanical disadvantages of gene delivery for gene therapy [6]. However, further refinements are required before CED becomes a standard clinical practice. CED is, however, an excellent method to achieve significant transduction of virus in research animals. This chapter continues to describe in detail the method used for CED delivery into a rodent brain.

2 Materials

2.1 Step Design Cannula

1. Fused silica tubing (Polymicro Technologies, Pheonix, AZ, USA).
2. Hamilton syringe with 27-gauge stainless steel blunt removable needle (RN) (Hamilton, Reno, NV, USA).
3. Gel super glue.

2.2 Surgery

1. Adult mouse.
2. Preemptive analgesic.

3. Isoflurane (Abbott Animal Health, Abbott Park, IL, USA).
4. Sterile/germicidal surgical swabs.
5. Iodine solution.
6. Stereotaxic apparatus.
7. Sterile adhesive plastic drape material (e.g., Bioclusive by Johnson & Johnson, or equivalent).
8. #10 or #11 scalpel.
9. Mosquito hemostats.
10. Dremel drill with dental drill bit.

3 Methods

3.1 Step Cannula Assembly

1. The step design cannula can be prepared as follows: fused silica tubing is partially inserted into a Hamilton 27-gauge stainless steel blunt removable needle (RN).
2. A small amount of super glue is placed onto the silica tubing and the tubing is pushed further into the needle. Excess glue is dabbed off the needle. The glue is usually allowed 24 h to cure. In our experience, the gel type super glue seems to work more easily than the liquid type.
3. After the glue has set, the end of the silica tubing is cut with a straight razor blade leaving 1–2 mm of tubing protruding from the end of the needle (Fig. 2; *see* **Note 1**).

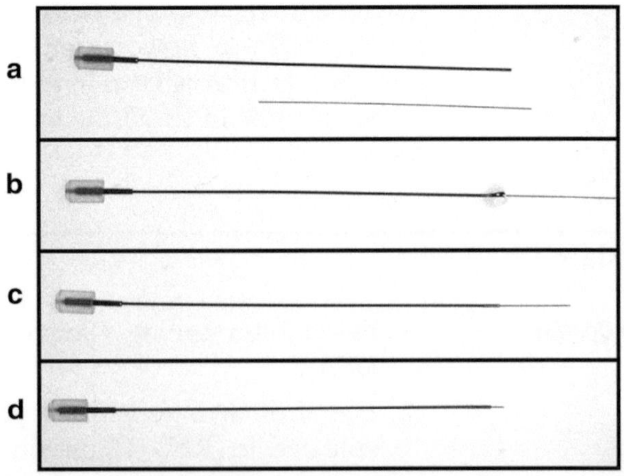

Fig. 2 Production of cannula designed needles. (**a**) Hamilton 27 gauge needle plus silica tubing are depicted. (**b**) Silica tubing is inserted in needle and gel super glue is applied. (**c**) Tubing is inserted slightly further to apply glue to inside of needle and excess glue is removed. (**d**) Silica tubing is cut to size, 1–2 mm extending from needle

3.2 Surgery

Our standard protocol is described, but minor variations in the surgical procedure to match local Institutional Animal Care and Use Committee requirements are easily incorporated. Figure 3 illustrates a typical stereotaxic set up for a mouse surgery (for rats the nose piece can be easily exchanged).

In our experience maximal expression of transgene is achieved within 3–4 weeks post-injection. Transgene expression has been reported to continue for at least 1.5 years in rodents (most likely the life span of the rodent) and as much as 6 years in primates [16, 17]. The spread with serotypes such as rAAV9 into the hippocampus can target the majority of the entire hippocampus (Fig. 5). It should be noted that the level of expression and spread is dependent on the titer of the virus and in our experience also the transgene being expressed.

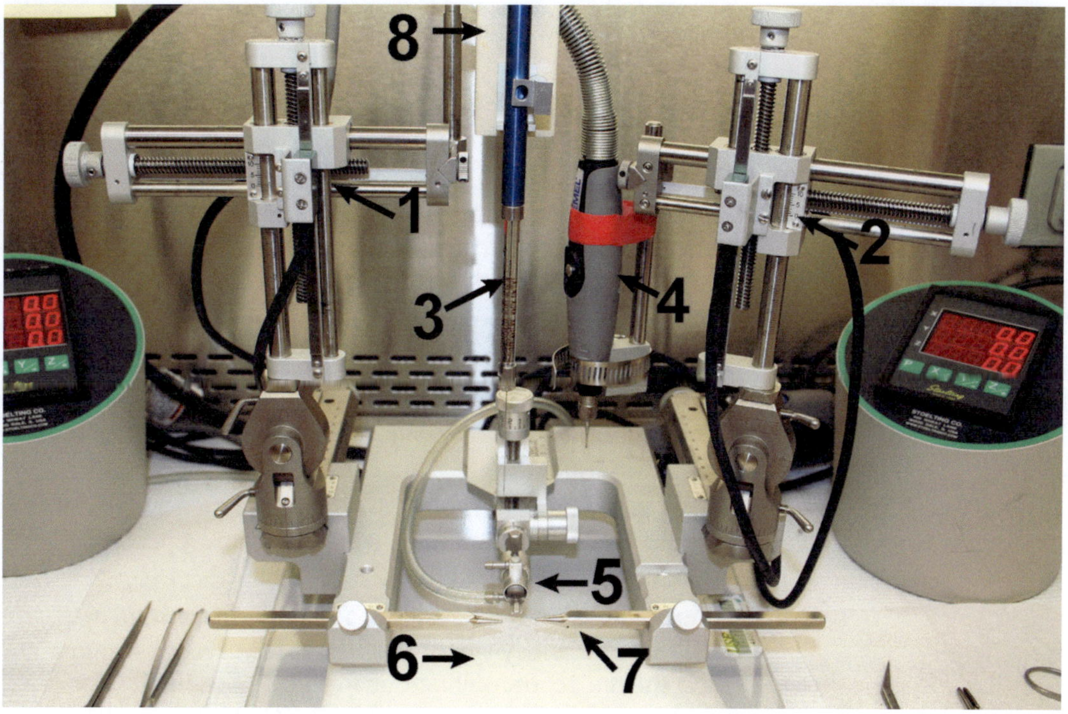

Fig. 3 A typical surgical stereotaxic setup. (*1*) Stereotaxic left arm, (*2*) Stereotaxic right arm, (*3*) Syringe with CED needle, (*4*) Dremel drill, (*5*) mouse nose piece for positioning mouse head and delivery of O_2 and anesthesia, (*6*) warming pad to maintain mouse body temperature, (*7*) ear bars for positioning mouse head, (*8*) syringe pump for automated delivery. The *oval* control units on each side of the stereotaxic frame provide automated movement of the arms to the desired, preset coordinates, and are part of a digital system attached to the stereotaxic arms for accurate needle placement

1. The mouse is weighed, injected with preemptive analgesic, and placed in the anesthesia induction chamber connected to a calibrated vaporizer delivering isoflurane anesthetic.
2. Anesthesia induction is usually performed with an initial vaporizer flow rate of 3–5 % with oxygenation at 0.5 L/min.
3. Once anesthesia is evident, the mouse is removed from the induction chamber. The hair on the top of the head is shaved or plucked from the eyes rostrally to the nape of the neck caudally and to the ears laterally to prevent contaminating the operative site.
4. The skin is cleaned three times by scrubbing with germicidal surgical scrub, wiping from the center of the surgical field outward. The surgical site is painted with a dilute, tamed iodine solution.
5. The mouse is then moved to the surgery area, placed on an isothermal pad, and mounted in the stereotaxic apparatus.
6. Anesthesia is conducted by surgical tubing into the surgical arena and delivered to the mouse using a rodent-specific nose cone apparatus (Fig. 3, arrow 5) and an anesthesia flow rate of 2 %.
7. A sterile, adhesive plastic drape material is used to isolate the surgical field.
8. Figure 4 illustrates the surgical steps described below. A single incision will be made on the top of the skull using a #10 or #11 scalpel, or scissors.
9. The skin will be reflected, and held open using two mosquito hemostats clamped to the subcutaneous material. The area is cleaned and dried with sterile surgical swabs.
10. The location of bregma is ascertained and serves as fiduciary to calculate positions of injection location (*see* **Notes 2** and **3**).
11. Burr holes are drilled in the skull using a dremel drill and a dental bit under stereotaxic guidance to the region of interest (*see* **Notes 4–6**).
12. Using adult mice weighing 30–40 g, our typical coordinates for the right hippocampus would be anteroposterior (AP) −2.7, Lateral +2.7 (*see* **Note 7**).
13. After establishing drill holes, a Hamilton syringe with the CED needle and virus of interest is lowered to bregma and then moved to the defined anatomical coordinates for injection (e.g., right hippocampus AP −2.7, Lat +2.7, dorsal-ventral (DV) −3.0) (*see* **Notes 8–10**).
14. We routinely dispense 2 μL of virus at a flow rate of 2.5 μL/min. The needle remains in position for 1 min after injection before being withdrawn. This allows for at least 1 min for complete dispersion of hydrostatic pressure and the virus, further limiting back flow up the needle track.

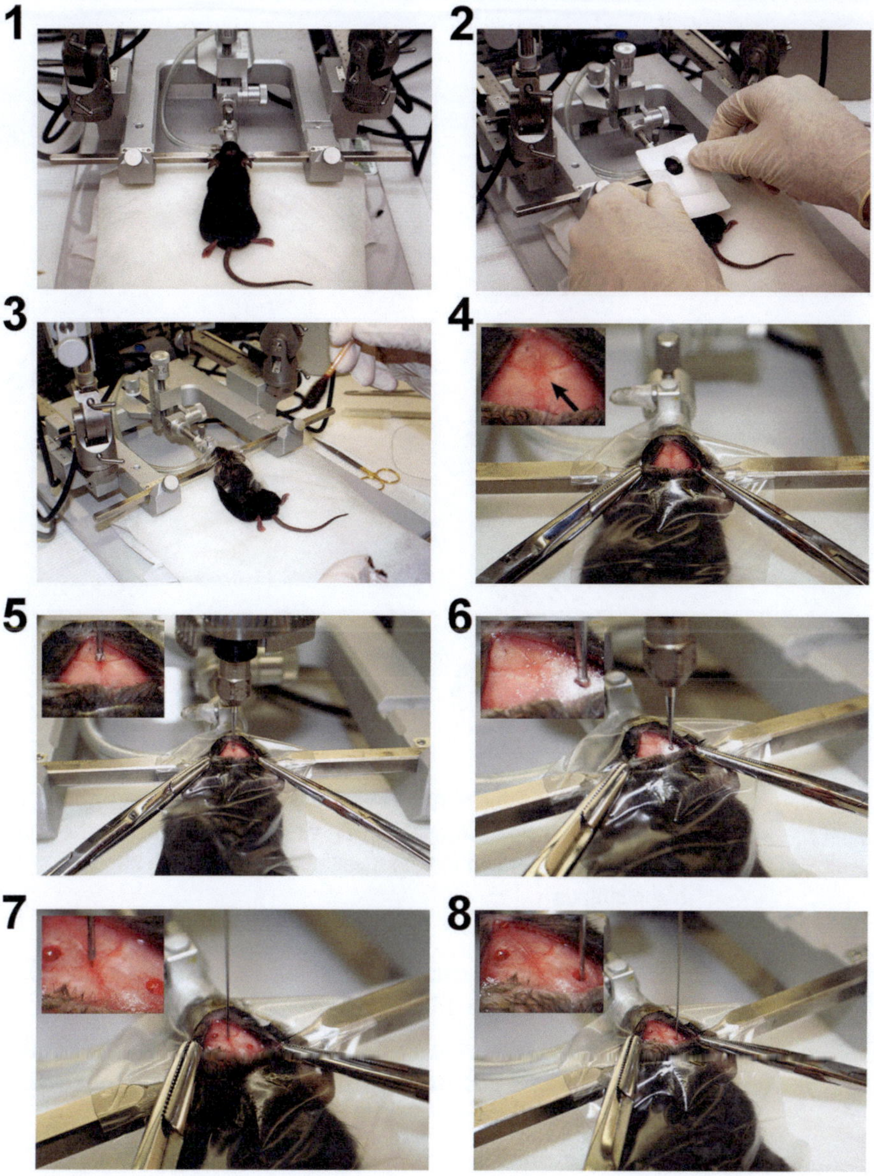

Fig. 4 Surgical steps. (*1*) Positioning of mouse into stereotaxic frame. (*2*) Bioclusive film is cut and measured to fit surgical field of the mouse's head. (*3*) Surgical area is cleaned and iodine is applied to top of the head. (*4*) Incision is made with a scalpel, skull is cleaned with sterile surgical swab, and bregma located (as indicated by *arrow* on *inset*). (*5*) Dremel drill bit is positioned on bregma. (*6*) Coordinates are located with the right stereotaxic arm and a hole is drilled with the dremel (targeted to right hippocampus in this figure). (*7*) Needle tip is placed carefully at bregma. (*8*) Needle is moved to the coordinates for injection and the pump is initiated. Note that a second hole in the left cortex is shown in *inset*. Multiple holes can be drilled so that injections may be performed one right after another with the same syringe

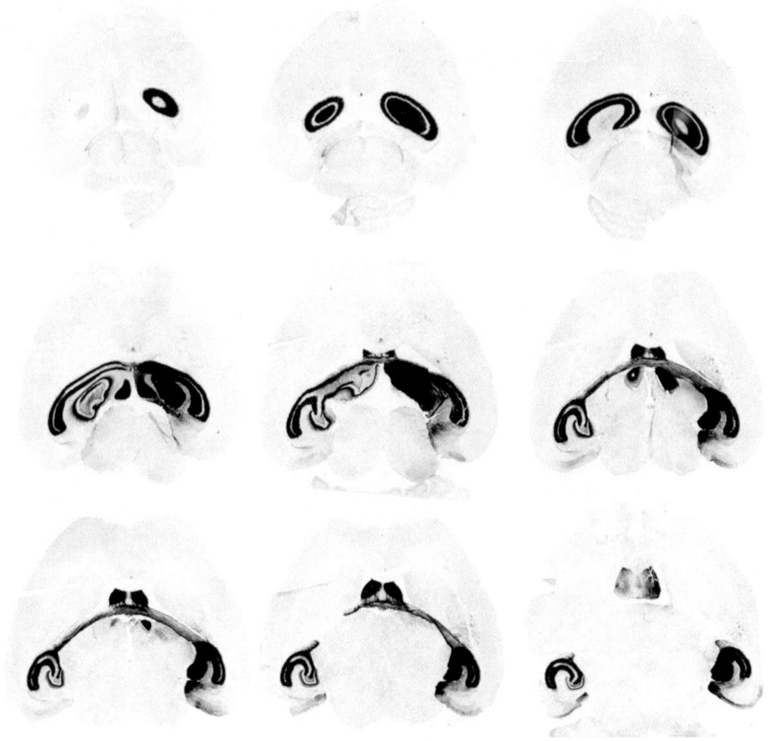

Fig. 5 Immunohistochemical staining for GFP expression from a CED-injected AAVserotype 9 virus expressing GFP. Viral injection was into the right hippocampus. CED delivery can increase volume distribution to almost the entire mouse hippocampus

15. Burr holes in the skull may be sealed with a thin disk of bone wax (*see* **Note 11**).
16. The surgical incision is closed with nylon (Ethilon® or an equivalent product) sutures or with sterile (autoclaved) surgical wound clips.
17. The mouse is placed into its home cage on an isothermal pad and monitored until recovery (typically within 30 min).
18. An example of GFP expression from a CED injection of a rAAV serotype 9 virus ($1x10^{13}$ vg/mL) into the right mouse hippocampus is shown in Fig. 5. Expression is seen throughout the entire hippocampus and some expression in other areas such as the entorhinal cortex.

4 Notes

1. It should be noted that the silica tubing could become broken with repeated uses. When this occurs, the needle may be recycled by placing the tip into acetone for a period of time until

the glue is dissolved, followed by removal of the silica tubing with a needle cleaning wire (supplied with the needles when purchased). New silica tubing can be inserted into the needle, glued and trimmed as described in Subheading 3.1 above.

2. Exact stereotaxic coordinates from bregma are determined by the desired area of injection. Initial stereotaxic coordinates may be estimated using a mouse atlas designed for this purpose (see for example Paxinos and Watson). For injections into large areas such as lateral ventricle, stereotaxic coordinates may vary somewhat and still adequately target the desired area. For smaller structures, such as substantia nigra, reproducible measurement of bregma and exact coordinates are required for targeting.
3. It is sometimes useful to perform pilot injections of 1–5 % India ink or food dye, followed by gross sectioning of the brain with a razor blade to verify placement.
4. If multiple injection sites are to be performed, these holes are drilled in succession.
5. If stereotaxic equipment does not have a separate arm for the dremel attachment, a handheld dremel may be used.
6. The point at which to drill will be determined by the needle point set to the appropriate coordinates from bregma. It is recommended that a small divot be drilled first and the needle position checked with the divot before completely drilling the hole. This allows adjustment of the drill position if it is slightly off the desired injection position.
7. An example list of coordinates that are routinely used in our laboratory for adult mice and rats are indicated in Table 1. Coordinates may have to be adjusted for smaller mice.

Table 1
Stereotaxic coordinates for rAAV viral injections to specified brain structures routinely used in this laboratory

	Lateral (*X*)	Anterior-posterior (*Y*)	Dorsal-ventral (*Z*)
Mouse			
Hippocampus	±2.7	−2.7	−3.0
Anterior cortex	±2.0	+2.0	−2.0
Substantia nigra	±1.4	−2.8	−4.6
Lateral ventricle	±1.0	−0.4	−2.4
Rat			
Substantia nigra	±2.0	−5.2	−8.2

Coordinates are from bregma

8. We typically use one placement to test expression of each vector preparation, to monitor transduced cells and area, or to establish proof of concept for specific vectors.
9. Bilateral targeting may be used to control expression throughout the target of interest.
10. To effect whole brain targeting, we have used four placements, bilaterally into anterior cerebral cortex and hippocampus using four burr holes.
11. In mice, this is less than satisfactory because the skull is so thin that there is no true cavity to fill.

References

1. Davidson BL, Stein CS, Heth JA, Martins I, Kotin RM, Derksen TA, Zabner J, Ghodsi A, Chiorini JA (2000) Recombinant adeno-associated virus type 2, 4, and 5 vectors: transduction of variant cell types and regions in the mammalian central nervous system. Proc Natl Acad Sci U S A 97(7):3428–3432. doi:10.1073/pnas.050581197
2. Cearley CN, Vandenberghe LH, Parente MK, Carnish ER, Wilson JM, Wolfe JH (2008) Expanded repertoire of AAV vector serotypes mediate unique patterns of transduction in mouse brain. Mol Ther 16(10):1710–1718. doi:10.1038/mt.2008.166
3. Zincarelli C, Soltys S, Rengo G, Rabinowitz JE (2008) Analysis of AAV serotypes 1-9 mediated gene expression and tropism in mice after systemic injection. Mol Ther 16(6):1073–1080. doi:10.1038/mt.2008.76
4. Bobo RH, Laske DW, Akbasak A, Morrison PF, Dedrick RL, Oldfield EH (1994) Convection-enhanced delivery of macromolecules in the brain. Proc Natl Acad Sci U S A 91(6):2076–2080
5. Saito R, Tominaga T (2012) Convection-enhanced delivery: from mechanisms to clinical drug delivery for diseases of the central nervous system. Neurol Med Chir 52(8): 531–538
6. Krauze MT, Saito R, Noble C, Tamas M, Bringas J, Park JW, Berger MS, Bankiewicz K (2005) Reflux-free cannula for convection-enhanced high-speed delivery of therapeutic agents. J Neurosurg 103(5):923–929. doi:10.3171/jns.2005.103.5.0923
7. Saito R, Krauze MT, Noble CO, Tamas M, Drummond DC, Kirpotin DB, Berger MS, Park JW, Bankiewicz KS (2006) Tissue affinity of the infusate affects the distribution volume during convection-enhanced delivery into rodent brains: implications for local drug delivery. J Neurosci Methods 154(1-2):225–232. doi:10.1016/j.jneumeth.2005.12.027
8. Bankiewicz KS, Eberling JL, Kohutnicka M, Jagust W, Pivirotto P, Bringas J, Cunningham J, Budinger TF, Harvey-White J (2000) Convection-enhanced delivery of AAV vector in parkinsonian monkeys; in vivo detection of gene expression and restoration of dopaminergic function using pro-drug approach. Exp Neurol 164(1):2–14. doi:10.1006/exnr.2000.7408
9. Cunningham J, Oiwa Y, Nagy D, Podsakoff G, Colosi P, Bankiewicz KS (2000) Distribution of AAV-TK following intracranial convection-enhanced delivery into rats. Cell Transplant 9(5):585–594
10. Hadaczek P, Kohutnicka M, Krauze MT, Bringas J, Pivirotto P, Cunningham J, Bankiewicz K (2006) Convection-enhanced delivery of adeno-associated virus type 2 (AAV2) into the striatum and transport of AAV2 within monkey brain. Hum Gene Ther 17(3):291–302. doi:10.1089/hum.2006.17.291
11. Carty N, Lee D, Dickey C, Ceballos-Diaz C, Jansen-West K, Golde TE, Gordon MN, Morgan D, Nash K (2010) Convection-enhanced delivery and systemic mannitol increase gene product distribution of AAV vectors 5, 8, and 9 and increase gene product in the adult mouse brain. J Neurosci Methods 194(1):144–153. doi:10.1016/j.jneumeth.2010.10.010
12. Mardor Y, Rahav O, Zauberman Y, Lidar Z, Ocherashvilli A, Daniels D, Roth Y, Maier S, Orenstein A, Ram Z (2005) Convection-enhanced drug delivery: increased efficacy and magnetic resonance image monitoring. Cancer Res 65(15):6858–6863
13. Salegio E, Samaranch L, Kells A, Forsayeth J, Bankiewicz K (2012) Guided delivery of

adeno-associated viral vectors into the primate brain. Adv Drug Deliv Rev 64(7):598–604

14. San Sebastian W, Richardson RM, Kells AP, Lamarre C, Bringas J, Pivirotto P, Salegio EA, Dearmond SJ, Forsayeth J, Bankiewicz KS (2012) Safety and tolerability of magnetic resonance imaging-guided convection-enhanced delivery of AAV2-hAADC with a novel delivery platform in nonhuman primate striatum. Hum Gene Ther 23(2):210–217. doi:10.1089/hum.2011.162
15. Su X, Kells AP, Salegio EAA, Richardson RM, Hadaczek P, Beyer J, Bringas J, Pivirotto P, Forsayeth J, Bankiewicz KS (2009) Real-time MR imaging with gadoteridol predicts distribution of transgenes after convection-enhanced delivery of AAV2 vectors. Mol Ther 18(8):1490–1495. doi:10.1038/mt.2010.114
16. Ideno J, Mizukami H, Honda K, Okada T, Hanazono Y, Kume A, Saito T, Ishibashi S, Ozawa K (2003) Persistent phenotypic correction of central diabetes insipidus using adeno-associated virus vector expressing arginine-vasopressin in Brattleboro rats. Mol Ther 8(6): 895–902
17. Rivera VM, Gao GP, Grant RL, Schnell MA, Zoltick PW, Rozamus LW, Clackson T, Wilson JM (2005) Long-term pharmacologically regulated expression of erythropoietin in primates following AAV-mediated gene transfer. Blood 105(4):1424–1430.doi:10.1182/blood-2004-06-2501

Chapter 22

Nonviral Gene Therapy of the Nervous System: Electroporation

Xue-Feng Ding and Ming Fan

Abstract

Electroporation has been widely used to efficiently transfer foreign genes into the mammalian central nervous system (CNS), and thus plays an important role in gene therapeutic studies on some brain disorders. A lot of work concerning electroporation is focused on gene transfer into rodent brains. This technique involves an injection of nucleic acids into the brain ventricle or specific area and then applying appropriate electrical field to the injected area. Here, we briefly introduced the advantages and the basic procedures of gene transfer into the rodent brain using electroporation. Better understanding of electroporation in rodent brain may further facilitate gene therapeutic studies on brain disorders.

Key words Electroporation, Rodent brain, Gene therapy, Braindisorder

1 Introduction

Electroporation has become one of the most popular methods to transfer foreign nucleic acids into mammal tissue, and almost all main organs have been used as the electroporation target, including brain, spinal cord, retina, lung, liver, kidney, muscle, skin, testis, and more [1–10], which greatly facilitates gene gain–loss function research [11, 12]. Here, we will focus on gene transfer into the rodent brain using electroporation. Electric pulse condition of electroporation and injection system varies with the age of operated mouse/rat. The younger the operated brain, finer injection needles and lower voltage is necessary. However, all electroporation in the rodent brain shares the same principle, in which direct current is used to produce a temporary pore on the cell membrane and simultaneously deliver the nucleic acids with negative charge to the tissue close to the anode [12–14]. Thus, theoretically any molecule with negative charge can be transferred into the target tissue in vivo, including DNA, RNA, and other negative charged molecules. Compared with other popular gene transfer methods such as gene targeting techniques and viral systems,

Fredric P. Manfredsson (ed.), *Gene Therapy for Neurological Disorders: Methods and Protocols*, Methods in Molecular Biology, vol. 1382, DOI 10.1007/978-1-4939-3271-9_22,

Table 1
Comparison of in vivo electroporation and other gene alternating methods

	Gene knockout	Transgenic mice	Viral system	In vivo electroporation
Toxicity	No	No	Potential	No
Immunogenicity	No	No	Potential	No
Time specificity	No	No	Yes	Yes
Site specificity	No	No	No	Yes
Cost ($)	>20k	>5k	>2k	<0.2k
Procedure	Very complicated	Very complicated	Complicated	Simple
Integrity	Yes	Yes	Yes	No
Capacity		Unlimited	<10 kb	>10 kb
Multiple genes transfer		Very difficult	Difficult	Easy
Expression regulation	Remove	Upregulation	Upregulation/downregulation	Upregulation/downregulation

Compared with viral system and transgenic/gene knock out techniques, in vivo electroporation is a more time saving method with lowest cost, higher efficiency, bigger capacity, less risk of genomic DNA integration, less toxicity and immunogenicity, better spatiotemporally specificity, and simpler procedure

electroporation is more time and labor saving and exhibits higher efficiency and capacity, less risk of genomic DNA integration, less immunogenicity and toxicity, better spatiotemporally specificity, and is a simpler procedure (as shown in Table 1). Due to these advantages, electroporation may potentially be the most valuable method to perform gene therapy studies on brain disorders, including Alzheimer's disease [15, 16], apoptotic disease [17], psychiatric disorders [18], and malignant brain tumors [19]. A better understanding of this technique will not only further increase transfection efficiency and safety, but also pave the way for gene therapy studies based on electroporation.

2 Materials

Materials and protocol may vary with different experimental purposes. Here we utilize an enhanced green fluorescent protein (EGFP) plasmid as an example to introduce the basic procedure of transferring a foreign gene into the embryonic brain, neonatal brain, and adult brain using electroporation.

2.1 Plasmid, Microinjection System, and Mice

1. Microinjection glass needle for DNA injection in the in utero electroporation and neonatal electroporation: prepared using a P-97/IVF micropipette puller (Sutter Instruments, Novato, CA, USA) or equivalent. Conditions: pressure, 500; heat, 800; pull, 30; velocity, 40; time, 1 s.
2. A modified metal injection needle is used for DNA injection into the adult rodent brain.
3. ECM830 Electroporator with tweezer electrode (diameter: 7 and 10 mm) (Harvard Apparatus, Holliston, MA, USA) or equivalent.
4. Purified pEGFP-N1 plasmid (>0.5 μg/μl) carrying GFP gene (CMV promoter) mixed with 100 μg/ml Fast Green (Amresco, Solon, OH, USA) (*see* **Note 1**).
5. Pregnant ICR mice at embryonic stage 16 days (E16) used for the in utero electroporation.
6. Newborn pups no older than 12 h used in the neonatal electroporation.
7. All experiments must be performed in accordance with the guidelines of the institutional animal care and use committee [20–24].
8. Nembutal.
9. Hamilton syringe with a 30-gauge beveled needle (Hamilton Company, Reno, NV, USA).
10. Microinjection pump.
11. Stereotaxic apparatus.
12. Isolated pulse stimulator (A-M System, Carlsborg, WA, USA).
13. Surgical materials: Surgical scissors, gauze, sterile saline, 70 % ethanol, sutures.

3 Methods

3.1 In Utero Electroporation

1. As shown in Fig. 1, anesthetize a timed-pregnant mouse with an intraperitoneal injection of 1 % Nembutal solution (60 mg/kg body weight).
2. Place the operating board in a dissecting tray and place a stack of paper towels under the operating board to absorb spilled saline and prevent slipping of the board.
3. Fix the mouse with its back on the operating board.
4. Clean the abdomen with 70 % ethanol.
5. Open the abdomen along the midline abdominal with scissors. 2–3 cm long incision is enough to operate.

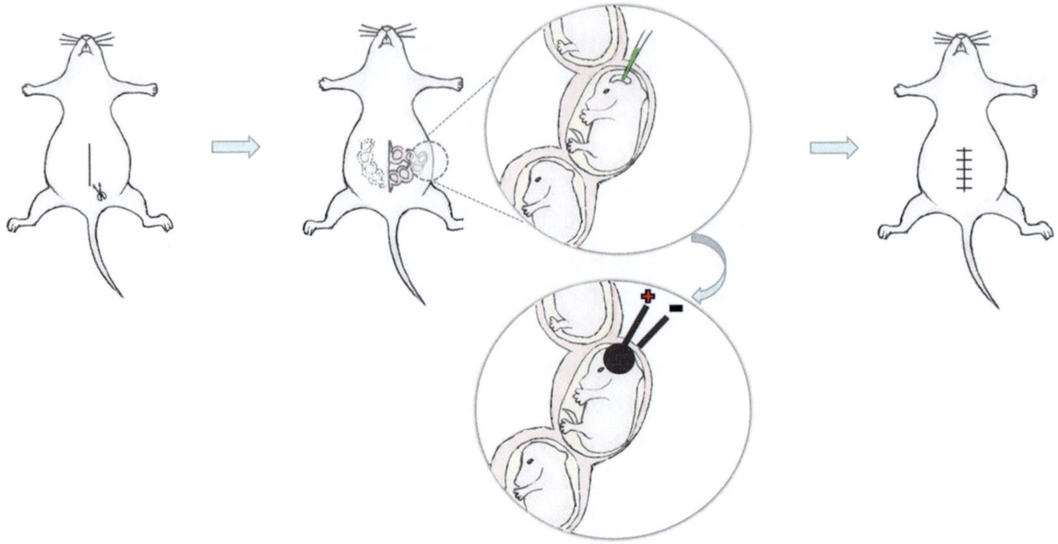

Fig. 1 The schematic diagram of in utero electroporation. (1) The abdominal cavity was opened and the uterus was exposed. (2) The DNA/RNA solution containing fast green was injected into the lateral ventricle. (3) Optimal electric pulses were delivered to the injected embryonic brain using tweezer electrode, and the anode was placed at the injected side. (4) The uterus was returned to the abdominal cavity, and the skin and muscle was sutured

6. Cover the abdomen with a piece of folded gauze, which has an incision with the same size as abdominal incision.
7. Soak the covered gauze and abdominal incision with 37 °C pre-warmed normal saline (*see* **Note 2**).
8. Pull the uterus out of the abdominal cavity, and next inject 1 μl DNA solution into one of the lateral ventricles of the embryonic brain using a mouth controlled injection system (*see* **Notes 3** and **4**).
9. Using the tweezer electrode, with the anode on the injected side, deliver five electric pulses at 50 V with 50 ms duration and 1 s interval to the injected embryonic head through the uterus wall.
10. Place the uterus gently back into the abdominal cavity with hands after operating all the embryonic brain, and then fill the abdominal cavity with 37 °C pre-warmed normal saline.
11. Suture the muscle and skin respectively, and then place the pregnant mouse on the 37 °C heat pad until awake.
12. Sacrifice the mouse mother at a specific time point and dissect the electroporated brain. Keep the electroporated brain in normal saline until finishing the fluorescence observation.

3.2 Neonatal Electroporation

1. As shown in Fig. 2, keep the mouse mother out of the cage until finishing operating on all the pups (*see* **Note 5**). Newborn pups (12 h after birth) were anesthetized with diethyl ether and fixed by one hand.

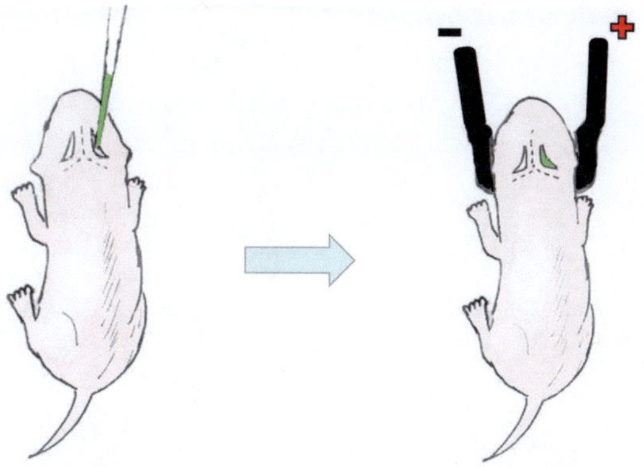

Fig. 2 The schematic diagram of neonatal electroporation. (1) The DNA/RNA solution mixed with fast green was injected into the lateral ventricle. (2) The Optimal electric pulses were applied to the injected head using tweezer electrode with the anode placed at the injected side

Table 2
Neonatal in vivo electroporation

Authors	Animal	Age	Voltage	Electrode	Surgical intervention	Surviving rate	GEP+ rate
Boutin C et al. (2008)	Mouse	P0-4	100 V	Tweezer	Yes	83 %	73 %
Chesler AT et al. (2008)	Rat	P0-4	150 V	Tweezer	Yes	95 %	73 %
Oudin MJ et al. (2011)	Mouse	P3	99.9 V	Tweezer	Not mentioned	Not mentioned	Not mentioned
Pathania M et al. (2012)	Mouse	P0-2	135 V	Tweezer	Not mentioned	Not mentioned	Not mentioned
Breunig JJ et al. (2012)	Mouse	P2	135 V	Tweezer	Not mentioned	Not mentioned	Not mentioned
Ding XF et al. (2012)	Mouse	P0	90 V	Tweezer	No	94 %	100 %

Different neonatal in vivo electroporation conditions described by different researchers

2. Inject 1 μl of plasmid solution (>0.5 μg/μl in normal saline containing 100 μg/ml Fast Green) into the lateral ventricle using a mouth-controlled pipette system (*see* **Note 6**). To facilitate the placement of the injection, the injection should be performed under a cold light source.
3. Immediately after injection, pups were electroporated with the tweezer type electrodes. Deliver five pulses at 90 V with 50 ms duration and 1 s interval to the injected head (Table 2).

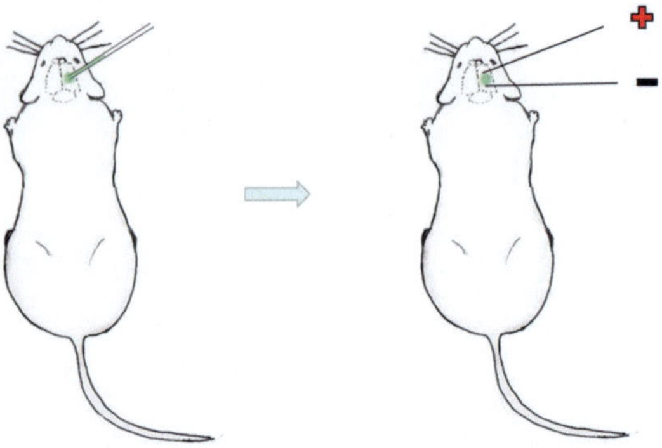

Fig. 3a The schematic diagram of adult electroporation. (**a**) Adult electroporation using needle like electrode. The nucleic acid solution was injected into the substantial area of adult brain, and the electric pulses were delivered to the target area using needle like electrode. (**b**) Adult electroporation using tweezer electrode. The electric pulses were delivered to the injected brain in a noninvasive manner, after the nucleic acid solution was injected into the lateral ventricle

4. Place the pups on a heat pad until they recovered full mobility (*see* **Note 7**) and subsequently return them to their mothers.
5. Sacrifice the successfully injected neonates at a specific time point and observe the fluorescence with fluorescent microscopy.

3.3 Adult Electroporation

1. As shown in Fig. 3a, fix the anesthetized rats in a stereotaxic apparatus (*see* **Note 8**).
2. Microinject the pEGFP-N1 vector into the pACC at the following coordinates: 2.2- and 3.2-mm anterior to bregma, 0.5-mm lateral from the midline, and at a depth of 3.0 mm from the skull surface, using a 30-gauge beveled needle connected to the Hamilton syringe mounted on a microinjection pump (*see* **Note 9**).
3. The needle was left in place for 10 min before being withdrawn slowly.

   After microinjection, a pair of naked silver electrodes (60 gauge) was placed 4.7-mm anterior and 0.5-mm posterior to bregma, 0.5-lateral from the midline, and at a depth of 3.5 mm from the skull surface.
4. Deliver Square-wave electric pulses using an isolated pulse stimulator.

4 Notes

1. Fast Green is used to indicate the successful injection and GFP is used to indicate the successful expression of our foreign gene.
2. It is be better to keep the uterus wet using warmed normal saline (37 °C) during the surgical operation, or the uterus will be filled blood and the embryonic brain will not be observed clearly, which may cause the failure of injection.
3. Check the tip of the glass needle before injecting the DNA solution into the embryonic brain, too fine of a needle will increase the difficult by which to push the DNA with the mouth controlled pipette. However, if the tip is not fine enough, it may also cause failure of injection. Researchers can check the tip by expelling water.
4. Proper strength should be used to fix the embryos in place. Improper strength may cause the death of embryos or injection failure. Thus enough practice should be performed before performing the formal experiment.
5. Changes in smell may cause the mother to eat the operated pups. Thus, keep the mouse mother out of the cage and collect the bedding polluted by the mouse mother's droppings and urine as much as possible. Cover all the operated pups at least 5 min using the above collected sawdust before returning the mother to her pups.
6. To ensure that the exact volume is injected in to the brain, use a transferpettor to transfer exact volume of DNA onto a tube wall, and then pull up the full amount of DNA with the mouth controlled pipette. Make sure that the mouth controlled pipette parallels with tubal wall.
7. It is very important to keep the operated pups on the heat pad until they awaken which will greatly increase the survival rate.
8. Exact fixation of the animal in the stereotaxic apparatus is pivotal in injecting the DNA into the exact brain area.
9. If the researchers want to transfer the foreign gene into the ventricular zone, the tweezer electrode may be better than needle-like electrode (Fig. 3b). In that case, shaving the hair of the mouse/rat will facilitate to observe the injection site and increase the conductivity.

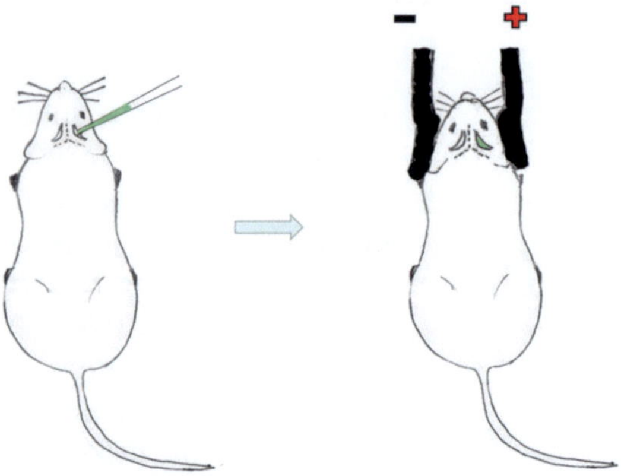

Fig.3b The schematic diagram of adult electroporation. The head skin was opened and the skull above lateral ventricle was polished thin enough to facilitate DNA injection. The tweezer electrode was used to deliver the foreign DNA into the ventricular zone of adult mouse brain

Acknowledgement

This work was supported by the Natural Science Foundation of China (31200822), the grants of Beijing Natural Science Foundation (5122033), and the National Basic Research Programs of China (grant number 2012CB518200).

References

1. Saito T, Nakatsuji N (2001) Efficient gene transfer into the embryonic mouse brain using in vivo electroporation. Dev Biol 240:237–246
2. Lai HC, Klisch TJ, Roberts R et al (2011) In vivo neuronal subtype-specific targets of Atoh1 (Math1) in dorsal spinal cord. J Neurosci 31:10859–10871
3. Dreau GL, Campmany LG, Rabadan MA et al (2012) Canonical BMP7 activity is required for the generation of discrete neuronal populations in the dorsal spinal cord. Development 139:259–268
4. Petros TJ, Shrestha BR, Mason C (2009) Specificity and sufficiency of EphB1 in driving the ipsilateral retinal projection. J Neurosci 29:3463–3474
5. Zhou R, Norton JE, Zhang N et al (2007) Electroporation-mediated transfer of plasmids to the lung results in reduced TLR9 signaling and inflammation. Gene Ther 14:775–780
6. Lagor WR, Heller R, Groh EDD et al (2007) Functional analysis of the hepatic HMG-CoA reductase promoter by in vivo electroporation. Exp Biol Med 232:353–361
7. Franquesa M, Alperovich G, Fresneda IH et al (2005) Direct electrotransfer of hHGF gene into kidney ameliorates ischemic acute renal failure. Gene Ther 12:1551–1558
8. Bertrand A, Muller VN, Hentzen D et al (2003) Muscle electrotransfer as a tool for studying muscle fiber-specific and nerve-dependent activity of promoters. Am J Physiol Cell Physiol 285:1071–1081
9. Markelc B, Bellard E, Sersa G et al (2012) In vivo molecular imaging and histological analysis of changes induced by electric pulses used for plasmid DNA electrotransfer to the skin: a study in a dorsal window chamber in mice. J Membr Biol 245:545–554
10. Yomgogida K (2008) Mammalian testis: a target of in vivo electroporation. Dev Growth Differ 50:513–515
11. Quiroga IN, Chittajallu R, Gallo V et al (2007) Long-term, selective gene expression in developing and adult hippocampal pyramidal neu-

rons using focal in utero electroporation. J Neurosci 27:5007–5011
12. De Vry J, Martinez PM, Losen M et al (2010) In vivo electroporation of the central nervous system: a non-viral approach. Prog Neurobiol 92:227–244
13. Bigey P, Bureau MF, Scherman D (2002) In vivo plasmid DNA electrotransfer. Curr Opin Biotechnol 13:443–447
14. Bloquel C, Fabre E, Bureau MF et al (2004) Plasmid DNA electrotransfer for intracellular and secreted proteins expression: new methodological developments and applications. J Gene Med 6:11–23
15. Erviti LA, Seow YQ, Yin HF et al (2011) Delivery of siRNA to the mouse brain by systemic injection of targeted exosomes. Nat Biotechnol 29:341–345
16. Davtyan H, Ghochikyan A, Movsesyan N (2012) Delivery of a DNA vaccine for Alzheimer's disease by electroporation versus gene gun generates potent and similar immune responses. Neurodegener Dis 10:261–264
17. Miao HH, Yu LY, Hui GZ et al (2005) Antiapoptotic effect both in vivo and in vitro of A20 gene when transfected into rat hippocampal neurons. Acta Pharmacol Sin 26: 33–38
18. Taniguchi Y, Pearse TY, Sawa A et al (2012) In utero electroporation as a tool for genetic manipulation in vivo in order to study psychiatric disorders: from genes to circuits and behaviors. Neuroscientist 18:169–179
19. Tsurushima H, Yuan X, Dillehay LE et al (2007) Radioresponsive tumor necrosis factor-related apoptosis inducing ligand (TRAIL) gene therapy for malignant brain tumors. Cancer Gene Ther 14:706–716
20. Fan J, Wu X, Cao Z et al (2009) Up-regulation of anterior cingulate cortex NR2B receptors contributes to visceral pain responses in rats. Gastroenterology 136:1732–1740
21. Ding XF, Hu ZY, Zhao YQ et al (2014) Specific expression of foreign gene in postnatal mouse glial cells using electroporation. CNS Neurosci Ther 20:98–100
22. Ding XF, Zhao YQ, Fan WH et al (2013) Gene transfer into the both sides of postnatal cerebral cortex using tweezer electrode. CNS Neurosci Ther 19:197–198
23. Ding XF, Zhao YQ, Hu ZY et al (2012) Efficient gene transfer into neonatal mouse brain using electroporation. Neurochem Res 37:1392–1398
24. Saito T (2006) In vivo electroporation in the embryonic mouse central nervous system. Nat Protoc 1:1552–1558

Chapter 23

Non-Viral, Lipid-Mediated DNA and mRNA Gene Therapy of the Central Nervous System (CNS): Chemical-Based Transfection

James G. Hecker

Abstract

Appropriate gene delivery systems are essential for successful gene therapy in clinical medicine. Cationic lipid-mediated delivery is an alternative to viral vector-mediated gene delivery. Lipid-mediated delivery of DNA or mRNA is usually more rapid than viral-mediated delivery, offers a larger payload, and has a nearly zero risk of incorporation. Lipid-mediated delivery of DNA or RNA is therefore preferable to viral DNA delivery in those clinical applications that do not require long-term expression for chronic conditions. Delivery of RNA may be preferable to non-viral DNA delivery in some clinical applications, because transit across the nuclear membrane is not necessary and onset of expression with RNA is therefore even faster than with DNA, although both are faster than most viral vectors. Here, we describe techniques for cationic lipid-mediated delivery of nucleic acids encoding reporter genes in a variety of cell lines. We describe optimized formulations and transfection procedures that we previously assessed by bioluminescence and flow cytometry. RNA transfection demonstrates increased efficiency relative to DNA transfection in non-dividing cells. Delivery of mRNA results in onset of expression within 1 h after transfection and a peak in expression 5–7 h after transfection. Duration of expression in eukaryotic cells after mRNA transcript delivery depends on multiple factors, including transcript stability, protein turnover, and cell type. Delivery of DNA results in onset of expression within 5 h after transfection, a peak in expression 24–48 h after transfection, and a return to baseline that can be as long as several weeks after transfection. In vitro results are consistent with our in vivo delivery results, techniques for which are described as well. RNA delivery is suitable for short-term transient gene expression due to its rapid onset, short duration of expression and greater efficiency, particularly in non-dividing cells, while the longer duration and the higher mean levels of expression per cell that are ultimately obtained following DNA delivery confirm a continuing role for DNA gene delivery in clinical applications that require longer term transient gene expression.

Key words Non-viral, Lipid-mediated, Gene delivery, Transfection, RNA, DNA, siRNA, Primary neurons, Post-mitotic, Molecular therapy, Transient, CHO, NIH3T3

1 Introduction

Gene therapy has the potential to significantly advance clinical medicine, but the risks and duration of gene delivery should be closely matched to the proposed clinical application [1].

Fredric P. Manfredsson (ed.), *Gene Therapy for Neurological Disorders: Methods and Protocols*, Methods in Molecular Biology, vol. 1382, DOI 10.1007/978-1-4939-3271-9_23,

Long-term expression after gene therapy is useful for diseases that require chronic levels of protein expression, such as inherited enzyme deficiencies, or cancer, and for these diseases viral vectors may offer advantages. For clinical applications in which only short-term gene expression is required or warranted, the delivery of nucleic acids (DNA or RNA, including mRNA, siRNA, shRNA, and microRNAs) by means of non-viral lipids provides a far more favorable risk/benefit ratio. Lipid-mediated transfection also offers other advantages over viral vectors, most notably safety, low immunogenicity, ease of preparation, and the ability to deliver payloads of nearly unlimited size [2, 3].

Cationic lipid-mediated gene transfer is particularly suited for transient gene expression, both in basic research and in selected clinical applications. Cationic lipids are commonly comprised of a polar headgroup and non-polar symmetric or dissymmetric carbon-based tail, and may also include Nuclear Localization Signals (NLS), antibodies, polymers, or other targeting moieties. Negatively charged nucleic acids condense and self-assemble into heterogeneous complexes when mixed with cationic lipids [4]. The structure and size of these complexes affect transfection efficacy and vary with temperature, concentration, charge ratio, buffer, time, and lipid composition. These lipid/nucleic acid complexes protect nucleic acids from degradation in the extracellular environment [5]. Numerous laboratories [6, 7], including our own [8], have investigated the limiting parameters of cationic lipid-mediated transfection with the goal of improving transfection efficiency [9]. Four general barriers to lipid-mediated DNA transfection include: (1) transport of the nucleic acid/lipid complex in the extracellular environment to target tissue(s); (2) association and uptake of the nucleic acid/lipid complex by the target cell [10]; (3) intracellular DNA or RNA release from the nucleic acid/lipid complex [11]; and (4) translocation of DNA to the nucleus [12], which is not required after RNA delivery. The primary barrier to DNA transfections in post-mitotic cells is assumed to be translocation of DNA to the nucleus [13]. While cationic lipid-mediated transfections work well with many types of cells [2, 3, 6], transfection of primary cell lines remained a problem [9, 14]. This transfection difficulty was generally attributed to markedly reduced or absent mitotic activity in primary cells, which are almost exclusively post-mitotic [14, 15]. In proliferating cells, nuclear translocation is mainly passive, and occurs during mitosis as the nuclear membrane breaks down [10, 16, 17]. Some nuclear translocation does still occur in non-proliferating cells, probably the result of passive movement through the nuclear pore complex (NPC) [18, 19].

Methods for avoiding the necessity of nuclear translocation of DNA have been reported, for example delivery of T7 promoter DNA plasmid systems to T7 polymerase-expressing cells [20], but these systems have limited clinical applications. Lipid-mediated RNA and DNA delivery to proliferating cells [21] as well as intramuscular injection of naked RNA and DNA have also been previously described [22]. Using RNA instead of DNA eliminates the necessity for nuclear translocation and thus has the potential to greatly improve transfection of post-mitotic cells.

We compared numerous lipids, cationic and otherwise, in a wide variety of in vitro and in vivo experiments, transfection of primary neuronal cells, delivery to proliferation-inhibited dividing cells [15], and in vivo imaging of reporter delivery and expression. Our studies show that mRNA and DNA are rapidly and robustly expressed, and offer safety advantages over viral vectors [15, 23]. RNA is 2–5 times more efficient at cell transfection than DNA based on the percentage of cells transfected. DNA delivery results in a higher level of expression under most cases, although with a longer time to expression, and a longer duration of expression.

This protocol is focused on DNA and mRNA delivery into primary neurons and in vivo. However, it is quite practical at first to work out the main parameters of liposomal formulation on fast growing, easily available cell types. That is why we include the experiments on CHO cells. The protocol includes the use of GFP and luciferase reporter transcripts. Both can serve as suitable reporter constructs and controls. Although the measurement of GFP expression by flow cytometry can be preferable for cell line transfection, flow analysis of neuronal cells is more difficult, and luciferase reporter is more suitable for analysis of neuronal transfection. In trouble shooting mRNA delivery, it is advisable to have DNA transfection controls in parallel.

We recommend the commercially available lipid TransFast™ Transfection Reagent (Promega, Madison, WI, USA) for RNA and DNA delivery. In our long experience Transfast is the closest in efficiency and low cytotoxicity to our novel lipid MLRI, which is available by special order but not commercially (*see* **Note 1**) TransFast™ Transfection Reagent is comprised of the synthetic cationic lipid (+)-*N*,*N*{bis (2-hydroxyehyl)}-*N*-methyl-*N*-{2,3-di(tetradecanolyloxy)propyl} ammonium iodide and the neutral lipid, DOPE (Promega). Transfast is supplied by Promega as a lipid film in glass vials under argon gas that must be rehydrated with DDW (nuclease- and RNase-free). Liposome reagents specifically designed for transfection applications often incorporate synthetic cationic lipids, such as the neutral lipid DOPE or cholesterol, to improve efficacy. DOPE has been demonstrated to enhance the gene transfer ability of certain cationic lipids [24]. We optimized transfection formulations of GFP encoding RNA and DNA vectors using flow cytometry, by varying charge ratio, formulation time, concentration, and temperature, to obtain the maximum percentage of GFP-expressing cells. Ideally, each cell line of interest should have optimal formulations confirmed before proceeding to in vivo experiments.

2 Materials

2.1 Cells

1. CHO cells (ATCC, Rockville, MD, USA).
2. NIH3T3 cells (ATCC, Rockville, MD, USA).
3. Primary neuronal cells dissected from the cortex of day 17 Sprague–Dawley rat fetuses.

2.2 Vectors (See Note 2)

1. β-Globin luciferase DNA vector (βglucβg). This vector (gift of Jon Wolf, MD, University of Wisconsin) [21, 22], which contains β-globin stabilizing elements, was used for time course experiments in primary neuronal cells. β-Globin is one of the most stable naturally occurring mRNAs found in nature, and the stabilizing elements from β-globin mRNA have been used in in vitro mRNA transcripts to design long-lived mRNA. To produce the β-globin stabilized transcripts, cloning procedures were carried out essentially as described in Molecular Cloning: A Laboratory Manual [25]. Xenopus laevis β-globin sequences were derived from the plasmid pSP64 T [26], with the 5′ β-globin sequences obtained as the HindIII/Bgl II fragment and the 3′ β-globin sequences released as the Bgl II/EcoRI fragment. These 3′ sequences include a terminal polynucleotide tract of A23C30. T7 RNA polymerase transcription templates, as well as various mRNAs produced from them, are outlined in [21]. We did not see significant differences in stability between the various mRNAs we produced and analyzed.
2. Luciferase DNA Vector (pNDlux.2). The pND luciferase expression vector (gift of Gary Rhodes, PhD, University of California-Davis) contains the CMV immediate early promoter (HCMV IE1) and CMV IE1 intron, a polylinker cloning site, and the RNA terminator/polyadenylation site from bovine growth hormone (BGH). These elements are contained in a pUC19 replicon. cDNA encoding the luciferase gene from the firefly Photinus pyralis (pGL2+, Promega, Madison, WI) was cloned into expression vector pND to give the plasmid pNDlux.2.
3. GFP DNA Vector (pNDeGFP). Full-length enhanced GFP (eGFP, a variant of jellyfish Aequorea victoria GFP) cDNA (pEGFP, Clontech, Palo Alto, CA) was subcloned into the pND vector described above to give the DNA vector, pNDeGFP (also a gift of Gary Rhodes).

2.3 Vectors for mRNA Transcripts (See Note 3)

1. β-Globin Vector for GFP mRNA Transcript (*see* **Note 4**): Briefly, eGFP was subcloned into the β-globin vector using Nco1 and Xba1 restriction enzymes to form β-globin-enhanced Green Fluorescent Protein (β-globin eGFP) vector.

β-Globin eGFP was linearized with restriction enzyme Dra I after elimination of RNases by proteinase K treatment and phenol/chloroform extraction. After precipitation in ethanol mRNA transcripts encoding the reporter gene sequence for *firefly* luciferase were transcribed in vitro using both mMessage mMachine transcription kit and the Ultra mMessage mMachine transcription kit (Ambion, Austin, TX). The mMessage mMachine and Ultra mMessage Machine kits are identical in content except for the mRNA cap structure included. mMessage mMachine kits contain the "standard" mRNA cap while the Ultra kit contains the Anti Reverse Cap Analog (ARCA).

2. β-Globin Vector for Luciferase mRNA Transcript (*see* **Note 4**): Luciferase-expressing mRNA was also produced using the pT7βglucβg plasmid (gift from Jon Wolff, M.D., University of Wisconsin). Procedures similar to **item 1** above were followed.
3. pT7OmegaGFPA Vector for GFP mRNA Transcript: The plasmid pT7OmegaGFPA_{50} (gift from Robert Malone, University of California-Davis) was also used as a template for GFP expression. The GFP-expressing mRNA contains the Omega 5′ UTR and 3′ poly-adenylated tail from Tobacco Mosaic Virus (TMV). 5′ and 3′ UTR from TMV have also been shown to improve intracellular mRNA stability and lead to translational enhancements in eukaryotic as well as prokaryotic cells, much like the β-globin UTR described above [27–29]. Transfections in our laboratory comparing time course of expression between luciferase vectors containing TMV UTR and β-globin UTR vectors revealed nearly identical kinetics and levels of expression (data not shown).

2.4 Equipment

1. 5 % CO_2 incubator at 37 °C.
2. Low speed centrifuge.
3. Dual channel FACScan or multi-channel Caliber flow cytometer (Becton Dickinson, San Jose, CA, USA) equipped with a single 488 nm argon laser (GFP fluorescence measured with a 530 nm band pass filter and Tri-Color fluorescence measured with a 675 band pass filter).
4. CellQuest (Becton Dickinson, East Rutherford, NJ, USA) or equivalent flow cytometry analysis software.
5. Monolight 2010 (Becton Dickinson, East Rutherford, NJ, USA) or equivalent.
6. Stereotaxic small animal surgery frame (Stoelting, Wood Dale, IL, USA).
7. Syringe infusion pump (Stoelting, Wood Dale, IL, USA).

2.5 Supplies

1. 75 cm^2 cell culture flasks.
2. 60 mm Petri dishes.
3. 15 ml sterile tube.
4. Pasteur pipette tubes.
5. 40 μm nylon nets.
6. Intraventricular canula.
7. Microscope slides (Columbia Diagnostics, Inc., Springfield, VA, USA) poly-L-lysine-coated and air-dried for a minimum of 2 h.

2.6 Media

1. HAMS F12 media (Life Technologies, Gaithersburg, MD, USA).
2. DMEM media.
3. OptiMem media.
4. 10 % bovine calf serum.
5. PBS.
6. Ice-cold sterile PBS.
7. Neural Basal Medium (NBM), containing 0.5 mM L-glutamine, 1× B27, 50 ng/ml neural growth factor, supplemented with 0.25 % trypsin-EDTA and 20 μl DNase I.

2.7 Templates for RNA Synthesis

1. eGFP and luciferase mRNA transcripts as controls for optimization (β-globin eGFP is shown in Subheading 3) (*see* **Note 5**).
2. mRNA or short RNAs (siRNA, hRNA, microRNAs) of your desired sequences using standard techniques (in vitro transcription and purification using RNAse-free laboratory techniques, as in Red Book or other standard lab reference) (*see* **Note 6**).

2.8 Nucleic Acid Delivery

1. Restriction enzymes (we used Nco1, Nde1, Dra I, and Xba1).
2. Proteinase K.
3. Phenol/chloroform.
4. Ethanol.
5. T7 mMessage mMachine transcription kit (Ambion, Austin, TX, USA).
6. Enhanced Luciferase assay Kit (BD Bioscience, East Rutherford, NJ, USA).
7. TransFast™ Transfection Reagent (Promega, Madison, WI, USA).
8. Neutral lipid DOPE (Promega, Madison, WI, USA) or cholesterol.
9. eGFP or luciferase pDNA control vectors (Clontech, Mountain View, CA, USA).

10. Trypsin.
11. Dulbecco's phosphate buffered saline with Ca^{2+} (DPBS).
12. Annexin V conjugated with biotin (CalTag, Burlingame, CA, USA) in DPBS.
13. Streptavidin conjugated with a Tri-Color fluorophore (CalTag, Burlingame, CA, USA) in DPBS.
14. Lysis buffer.
15. CSF (cerebral spinal fluid).

2.9 Tissue Preparation for Reporter Protein Localization After In Vivo DNA or mRNA Delivery

1. Iced saline for perfusion.
2. Iced 4 % paraformaldehyde in 0.1 M, pH 7.4 sodium phosphate-buffered saline (PBS).
3. Paraformaldehyde fixative at 4 °C.
4. PBS containing 20 % glycerol at 4 °C.

2.10 Diaminobenzidine Immunohistochemistry

1. Diaminobenzidine (DAB).
2. 24-well plates for free-floating sections.
3. ExtrAvidin® peroxidase system (Sigma, St. Louis, MO, USA).
4. Poly-L-lysine-coated microscope slides for slide-mounted sections (Columbia Diagnostics, Inc., Springfield, VA, USA).
5. 0.1 % Hydrogen peroxide (H_2O_2).
6. 4 % paraformaldehyde (described above).
7. Blocking buffer: 0.3 % Triton X-100, 3 % bovine serum albumin (BSA), 10 % normal goat serum (NGS), Ca^{2+} and Mg^{2+}-free PBS.
8. Biotin-conjugated secondary Ab (biotinylated goat anti-mouse; Vector Labs, Burlingame, CA, USA), tertiary horseradish peroxidase-conjugated probe (ExtrAvidin® peroxidase system).
9. 50 mM Tris–HCl at pH 7.6.
10. 0.5 mg/ml 3,39-DAB with 0.03 % H_2O_2 as the peroxidase substrate.
11. Mouse monoclonal neuron-specific nuclear protein anti-NeuN (1:50; EMD Millipore, Billerica, MA, USA).
12. Rabbit polyclonal antiluciferase antibody.

2.11 Documentation of Results

1. Nikon 600 microscope with camera mount (Nikon Instruments, Melville, NY, USA) or equivalent.
2. Photoshop and Polaroid SprintScan slide scanner for film negatives or slides.
3. Fuji Pictrography 3000 (Fuji Photo Film, Elmsford, NY, USA), or equivalent.

4. Nikon Eclipse TS100 Inverted microscope (Nikon Instruments, Melville, NY, USA) or equivalent.
5. High-resolution digital camera (SPOT; Diagnostic Instruments, Sterling Heights, MI, USA) or equivalent.

3 Methods

3.1 CHO Cells

1. Culture CHO cells in 75 cm^2 cell culture flasks with HAMS F12 media at 37 ° C in a 5 % CO_2 environment (*see* **Note 7**).
2. Split cells 48 h prior to transfection.
3. Plate at 60 % confluence.

3.2 Primary Neuronal Cells

1. Remove fetal rat brains and place in ice-cold sterile PBS (*see* **Note 8**).
2. Dissect cortex and clear meninges.
3. Transfer cortical sections to a 60 mm Petri dish containing 4 ml PBS.
4. Mince into pieces of approximately 1 mm^3 in size.
5. Transfer into a 15 ml sterile tube.
6. Add 4 ml Neural Basal Medium and shake for 30 min at 37 °C.
7. Extrude the resultant suspension through a Pasteur pipette tube to eliminate residual clumps of brain tissue.
8. Filter through 40 μm nylon net to yield single cell suspensions.
9. Spin cells and resuspend in Neural Basal Medium.
10. Seed 1×10^6 cells per ml onto 24-well tissue culture dishes which have been pre-coated with poly-L-lysine.
11. Culture cells in 5 % CO_2 incubator at 37 °C.
12. Maintain in culture for 2 weeks prior to transfection to allow the development of the phenotype of mature human cortical neurons.
13. Replace one-half of the medium every 3–4 days with medium containing fresh neural growth factor (*see* **Note 9**).

3.3 Formulation of DNA or mRNA with Lipid to form Lipoplexes and CHO Cell Transfections (See Notes 10 and 11)

Use both GFP and Luc transcripts and plasmid DNAs for optimization of lipoplex formulations.

1. Day 1, plate cells.
2. Calculate and prepare enough mRNA (or DNA) and Transfast for as many wells as your experiment calls for. For in vitro experiments in 24-well plates, 1 μg of nucleic acid (mRNA or plasmid DNA) is added to OptiMem for each well of the 24-well plate (*see* **Note 12**).

3. Rehydrate Transfast lipid in nuclease- and RNAse-free DDW water and store at −4 °C overnight.
4. Day 2, dilute mRNA in 20 μl OptiMem solution.
5. Thaw Transfast.
6. Add nuclease- and RNAse-free nucleic acids of sequence of interest to OptiMem for a final volume of 200 μl per well of 24-well plate.
7. Vortex.
8. Add lipid to produce a 3:1 lipid:nucleic acid charge ratio in 200 μl per well of 24-well plate (*see* **Note 12**).
9. Add thawed Transfast to NA (mRNA or DNA) and vortex for 15 s.
10. Incubate for 45–60 min at 37 °C (*see* **Note 13**).
11. Aspirate off growth medium from CHO cells.
12. Transfect cells in each well of 24-well plates with 1 μg of mRNA (or DNA) (*see* **Note 12**).
13. Add 200 ml of transfection solution to each well of a 24-well plate (*see* **Note 14**).
14. Incubate in 37 °C CO_2 incubator for 60 min.
15. Add growth medium to cells but do not aspirate off transfection solution.
16. Return to incubator for appropriate time (5–8 h for mRNA, 24–48 h for DNA).
17. Perform appropriate assay, i.e. flow cytometry or Luminometer for GFP or luciferase (*see* **Note 15**).
18. The final solution, in 200 μl, is vortexed and incubated for 1 h at room temperature (RT) prior to aspiration of growth media from the cells and application of transfection formulation to cells.
19. After 1 h of incubation supplement cells with 1 ml of growth media.
20. Harvest cells for flow cytometry analysis 7 h after RNA transfections (24 h after GFP DNA transfections).

3.4 Cell Toxicity Analysis (Biotin-Conjugated Annexin V)

1. Trypsinize CHO and NIH3T3 cells and wash twice in Dulbecco's phosphate buffered saline with Ca^{2+} (DPBS).
2. Resuspend and incubate for 30 min in annexin V conjugated with biotin in DPBS.
3. Wash again with DPBS, resuspend cells for 30 min in streptavidin conjugated with a Tri-Color fluorophore in DPBS.
4. Analyzed cells using a dual-channel FACScan with a single 488 nm argon laser.

5. Measure GFP fluorescence with a 530 nm band pass filter.
6. Measure Tri-Color fluorescence with a 675 band pass filter.
7. Conduct all experiments at least three times using three identical wells of a 24-well plate each time (*see* **Note 16**).

3.5 Flow Cytometry Analysis (GFP)

1. Trypsinize CHO, NIH3T3 and primary neuronal cells and wash twice in Dulbecco's phosphate buffered saline with Ca^{2+} (DPBS).
2. Resuspend in DPBS.
3. Measure GFP fluorescence using a dual-channel FACScan or multi-channel Caliber flow cytometer equipped with a single 488 nm argon laser.
4. Measure GFP fluorescence with a 530 nm band pass filter. Collect at least 10,000 events per sample.
5. Analyze with CellQuest software or equivalent.
6. Cells can be harvested and analyzed at the time point corresponding to the maximum percentage of cells expressing GFP, 7 h for mRNA transfection and 24 h for DNA transfection.
7. Conduct all experiments at least three times for CHO and NIH3T3 cells, and twice for primary neuronal cells (*see* **Note 16**).

3.6 Luciferase Expression Analysis After mRNA Transfection

1. Lyse CHO cells in 200 μl of lysis buffer.
2. Analyze 20 μl of lysate by luciferase assay using an Enhanced Luciferase assay Kit.
3. Measure Quantitative luminescence using a Monolight 2010 (*see* **Note 17**).

3.7 In Vivo Luciferase-Expressing DNA or mRNA Vector Delivery to Rat Brain (See Note 18)

1. Formulate lipoplexes for in vivo experiments with only mRNA or DNA and lipid (no OptiMem), in order to minimize injected volume.
2. Anesthetize animal subjects adequately (250–300 g Sprague–Dawley rats in this example) after obtaining an approved animal care protocol (*see* **Note 19**).
3. Add lipid to mRNA or DNA to make 3:1 lipid:nucleic acid charge ratio, as above. Incubate mixture at 37 °C for 30 min prior to delivery.
4. Mount animals in stereotaxic small animal surgery frame.
5. Begin infusion approximately 30 min after incubation of complexes at 37 °C.
6. Using sterile techniques, deliver previously optimized formulations at a dose of 50 μg/kg nucleic acid.
7. For 250 gram rats, use coordinates of 0.9–1.0 mm posterior and 1.5 mm lateral of midline relative to bregma, at a depth of approximately 3–3.5 mm.

8. Aspirate of CSF to verify intraventricular canula placement.
9. Infuse transfection formulation over 40 min using a syringe infusion pump.
10. Monitor animals closely for signs of discomfort, toxicity, or neurologic injury, although we rarely observe this.

3.8 Tissue Preparation for Luciferase Protein Localization After DNA or mRNA Delivery

1. Seven hours after mRNA vector delivery, deeply anesthetize the animal subject.
2. Perfuse through the ascending aorta with iced saline until blood flowing from the transected inferior vena cava is clear.
3. Follow with iced 4 % paraformaldehyde in 0.1 M, pH 7.4 sodium phosphate-buffered saline (PBS).
4. Dissect out brain and, post fix in paraformaldehyde fixative overnight at 4 °C.
5. Incubate in PBS containing 20 % glycerol at 4 °C.
6. Block brains and cryosection in the coronal plane following standard techniques, beginning approximately 6–7 mm anterior relative to bregma. Ten series of 30-μm serial sections can be collected from each rat brain.

3.9 Diaminobenzidine Immunohistochemistry (See Note 20)

1. Pretreat with 0.1 % H_2O_2 for 15 min before washing in modified PBS to eliminate endogenous peroxidase activity and staining artifact.
2. Fix sections in 4 % paraformaldehyde.
3. Incubate sections in blocking buffer (0.3 % Triton X-100, 3 % bovine serum albumin (BSA)), 10 % normal goat serum (NGS), Ca^{2+} and Mg^{2+}-free PBS for 2 h at room temperature.
4. Dilute primary antibodies in blocking buffer.
5. Incubate sections with antibody solution at 4 °C overnight.
6. Wash sections and incubate with the biotin-conjugated secondary antibody, which targets the primary antibody host species, for 1 h at room temperature.
7. Wash sections again and incubate with the tertiary horseradish peroxidase-conjugated probe for 1 h at room temperature.
8. Wash again and incubate in 50 mM Tris–HCl at pH 7.6 for 5 min at room temperature.
9. Incubate with 0.5 mg/ml 3,39-DAB with 0.03 % H_2O_2 as the peroxidase substrate.
10. Use the following antibodies and dilutions: Primary antibodies: mouse monoclonal neuron-specific nuclear protein anti-NeuN (MAB377, 1:50); rabbit polyclonal antiluciferase antibody (CR2029R, 1:50) (*see* **Notes 21** and **22**).

11. Use a microscope with camera mount for photographic documentation of results.
12. Scan film negatives or slides into Photoshop using a slide scanner and Photoshop plug-in at a resolution of 2700 dpi.
13. Print photographs using Photoshop on a Fuji Pictrography 3000 at 320 dpi, or equivalent.
14. Later experiments were documented on a Nikon Eclipse TS100 Inverted microscope with a high-resolution digital camera.

4 Notes

1. Lipids matter. The protocols described here have been optimized and confirmed over many years and compared against numerous commercially available lipids. We have found [9] that the length of the carbon tails and the symmetry or lack thereof of the carbon tails greatly affects how lipoplexes are taken up and possible also processed by cells. We believe [9] that the asymmetry and size of the linker and hydrocarbon tails of the lipids determines membrane fusogenicity, and furthermore that cell uptake depends not on specific ubiquitous receptors (as it was observed for some viral vectors), but on cell membrane fusion, a slower process. After individual optimization and simultaneous comparison of several commercially available cationic, neutral, dendrimer, and other lipids, we found the novel MLRI cationic lipid [23, 30, 31] to consistently perform superior to any other lipid, as evaluated by cell toxicity and efficacy of transfection. MLRI is an asymmetric C(14)-C(12) cationic lipid variant of the highly active cationic lipid DMRI [24]. MLRI and similar cationic lipids were described previously by Balasubramaniam et al. [32], Bennett et al. [33, 34], and Felgner et al. [24]. Some commercially available and popular lipids are very efficient at cell killing, if one measures toxicity with flow cytometry. MLRI has extremely low toxicity in addition to the high transfection efficacy. MLRI is not easily commercially available, but MLRI can be synthesized by any competent organic chemist, or a specialty lipids manufacturer such as Avanti polar lipids (Alabaster, Alabama) can manufacture it as a specialty order.
2. Our goal in initial experiments was to demonstrate delivery and expression using several carefully designed vectors. DNA vectors encoding reporter gene sequences for β-galactosidase (β-gal) and firefly luciferase (*P. luciferalis*) were first developed and tested. β-gal had the problem of endogenous β-gal in the CNS in particular. From these we selected the optimized luciferase DNA vector pNDlux.2 for further experiments.

Next, sequences encoding for Green Fluorescent Protein (GFP) were subcloned into pNDlux.2, replacing the luciferase sequence. GFP vectors are therefore identical to the pND luciferase vector, except for the coding region. Because the coding sequence is small relative to the size of the entire vector, we assume that luciferase and GFP vectors are taken up by similar mechanisms when complexed in identical optimized formulations, and that this assumption holds for any other DNA or mRNA vectors of similar size, independent of NLS sequences or RNA stabilizing or promoter regions (*see* **Note 18**).

Cationic lipids are used to protect DNA from degradation in the extracellular environment [5]. We optimized transfections using novel cationic lipids developed by Nantz et al. [32, 33, 35]. DNA dose and cationic lipid to DNA ratio was systematically varied in a series of in vitro transfections to arrive at our optimized formulation. Carrier RNA was also tested in increasing ratios. The addition of carrier RNA was shown previously to enhance expression after transfection of both mRNA and DNA [36], but we obtained robust results without carrier tRNA. Initial analyses of DNA vectors encoding β-galactosidase (β-gal) and luciferase reporter genes were performed using cell lysis and chemiluminescent assay (Galacto-Light, Tropix, Bedford, MA, or Enhanced Luciferase Assay Kit, Analytical Luminescence Laboratories, Ann Arbor, MI) on a Monolight 2010 Luminometer (Analytical Luminescence). Immunohistochemistry techniques using both fluorescent and enzymatic detection methods were then used to further characterize GFP and luciferase expression in vitro and later in vivo.

3. The pT7βglucβg plasmid, which contains a T7 promoter, was a gift of Jon Wolff, MD, of the University of Wisconsin. This plasmid is described in detail elsewhere [23]. The T7 bacteriophage promoter allows in vitro transcription of capped mRNA encoding each desired reporter enzyme. The region encoding the luciferase reporter gene is flanked by Untranslated Regions (UTR) from *Xenopus* laevis β globin [22, 37]. β-Globin UTR [26] contain stabilizing sequences and lack other destabilizing sequences of secondary and tertiary structure. The UTR from β-globin thus create a very stable mRNA transcript, and these 5′–3′ UTR sequences confer increased stability on coding sequences that are inserted within the 5′–3′ UTR [38]. These sequences have been used to provide additional stability for a variety of mRNA coding sequences in vitro and in vivo [39, 40].

4. We have used a variety of mRNA transcripts, and three different plasmids encoding eGFP and luciferase mRNAs used by us are described here. All three mRNAs gave very similar time courses and levels of expression and are described here.

5. GFP and luciferase reporter transcripts are used for different experiments. Times for maximal luciferase expression can be determined first in luciferase vector lipoplex transfections. These results can then be compared and confirmed using GFP vectors by flow cytometry to determine the maximum percentage of cells expressing GFP, which could be different than the time of maximal luciferase expression in cell lysate, depending on the protein and time course of degradation. Due to the difficulty of isolating and preparing primary neuronal cells, and of more difficult flow analysis of neurons, it may only be possible to perform a limited number of flow cytometry measurements with neuronal cells.
6. Prepare mRNA or short RNAs (siRNA, hRNA, microRNAs) of your desired sequences using standard in vitro transcription and purification techniques.
7. Standard cell culture methods are used for cell lines of interest. We present standard methods here for one popular rapidly proliferating cell line, CHOs, and for preparation of a primary neuronal cell culture. NIH3T3 cells, another popular, rapidly growing cell line (ATCC, Rockville, MD) are cultured in 75 cm^2 cell culture flasks with DMEM media (Life Technologies) containing 10 % bovine calf serum at 37 ° C in a 5 % CO_2 environment. Split cells 48 h prior to transfection and plated at 60 % confluence.
8. Primary neuronal cells are dissected from the cortex of day 17 Sprague–Dawley rat fetuses (as described in [41]).
9. Cultures of cortical neuron prepared in this fashion usually contain less than 10 % glial cells [41].
10. See also the Promega Transfection protocol [42].
11. Optimization of Luc transcript transfection of neuronal cells similarly to CHO procedure.
12. Although 1 μg each of DNA and RNA are not equal numbers of nucleic acid copies, lipid nucleic acid complexes are formulated based on an equal lipid to nucleic acid charge ratio. Transfast is singly charged per molecule, but not all lipids have a single charge.
13. Promega recommends only 10–15 min at RT.
14. Smaller wells can also be used with a proportional decrease in mass of nucleic acid per well, and a plate reader used for luminescent or other fluorescent analysis technique. Test mRNA in vitro with assay (for our fluorescent or luminescent reporter vectors, we always test with in vitro transfection and Luminometer measurements). We found that in vitro expression of luciferase had to be greater than a threshold of luminescence per microgram in vitro or in vivo expression was not sufficiently robust for detection.

15. We found 6–7 h to be the peak using either luciferase cell lysis analysis or GFP flow cytometry results. However, for the DNA vectors, we found that the GFP peak in expression as a percentage of cells expressing GFP occurred slightly (24–36 h) earlier than the peak in luciferase expression (36–48 h). Our luciferase assay requires cell lysis and is a measure of total luciferase protein. Signal from the action of luciferase on luciferin will depend on the concentration and activity of the enzyme present. The half-life of GFP in mammalian cells has been reported to be as long as 26 h [43]. The half-life of luciferase in mammalian cells is reported to be from 90 min to 4 h [44–46]. Two reviews of the kinetics and modeling of luciferase concentration and activity can be found in [47] and [45]. However we were most interested in the fraction of cells that were transfected rather than the peak in protein expression, which would include intracellular accumulation, and we thus compared flow cytometry results 24 h after DNA delivery and 7 h after RNA delivery based on the peaks in percentage of cells expressing GFP by flow cytometry measurement. We confirmed these peaks in expression with in vivo imaging of both DNA [30] and RNA (data not yet published) luciferase vectors after CNS delivery.
16. Cells in culture vary day to day, even when they are from identical cell source, identical passage number, and identical cell density. For this reason, it is best to design experiments so that all cells and plates for a single experiment are plated and grown at the same time.
17. 20 μl of lysate was always sufficient for analysis, but if the signal is either too low or too high the lysate volumes can be adjusted to bring the Luminometer signal into a linear range.
18. Widespread distribution, uptake and expression is possible after non-viral, cationic lipid-mediated gene delivery of DNA and mRNA vectors by infusing an optimized formulation of luciferase-encoding (or other nucleic acid sequence of interest) DNA or mRNA transcript into the lateral ventricle or into the cisterna magna of rat, mouse, or monkey brain. Perform direct injections using standard techniques that we previously reported [23, 31]. We have found that for the cationic lipids that we recommend the in vitro optimizations can be used with only minor further modification in in vivo applications as well. This is not always the case for every lipid. Kariko and colleagues have also demonstrated optimized in vitro and in vivo expression after in vitro mRNA transcription and delivery. Although Kariko et al. have not demonstrated the widespread in vivo expression that Hecker and colleagues have reported [31, 48–51], including with mRNA [23, 36, 52], Kariko et al. have achieved remarkable protein expression levels after optimization of both

transcript and formulation. They have also investigated vector modifications that both increase immunogenicity (for vaccine applications) or decrease immunogenicity (for clinical use) [53–56], and for some applications the Kariko enhancements may be critical for optimal protein expression after (m)RNA delivery. We previously reported cationic lipid formulations that afford RNA protection from degradation in human CSF for 4–6 h, whereas non-complexed RNA is immediately degraded [23]. Combining mRNA vectors with delivery to the CSF offers rapid expression in the CNS and avoids the problem of vascular barriers and viral vector safety issues. These same lipid-mediated lipoplex delivery techniques can be used effectively for siRNA delivery. We verified knockdown of GFP and luciferase expression in vitro, using Transfast and MLRI cationic lipids and similar charge ratios.

19. In vivo delivery to the rat CSF requires approximately 15 min after needle localization into the lateral ventricle or cisterna magna, resulting in a total effective incubation time of 45–60 min. Anesthetize animals ahead of time or simultaneously so that they are ready or very nearly ready for infusion prior to formulation of lipoplexes.
20. Diaminobenzidine (DAB) and secondary fluorescent immunohistochemistry protocols were optimized for expression using multiple DNA and mRNA vectors [23, 31]. These optimizations were conducted with no primary and no secondary controls on slide-mounted sections or in 24-well plates. To ensure identical, simultaneous processing free-floating sections in 24-well plates were stained using the ExtrAvidin® peroxidase system. For comparison, experiments were also processed using sections mounted on poly-L-lysine-coated microscope slides.
21. The NeuN antibody was used for comparisons of the number of neurons that can be identified in each section.
22. Autofluorescence is a problem with in vivo imaging of luciferase due in part to bioluminescent bacteria on skin, fur, and gut of the rodents.

References

1. Flotte TR (2007) Gene therapy: the first two decades and the current state-of-the-art. J Cell Physiol 213:301–305
2. Li SD, Huang L (2006) Gene therapy progress and prospects: non-viral gene therapy by systemic delivery. Gene Ther 13:1313–1319
3. Gao X, Kim KS, Liu D (2007) Nonviral gene delivery: what we know and what is next. AAPS J 9:E92–E104
4. Felgner PL, Tsai YJ, Sukhu L, Wheeler CJ, Manthorpe M, Marshall J, Cheng SH (1995) Improved cationic lipid formulations for in vivo gene therapy. Ann N Y Acad Sci 772: 1126–1139
5. Luo D, Saltzman WM (2000) Synthetic DNA delivery systems. Nat Biotechnol 18:33–37
6. Felgner PL, Gadek TR, Holm M, Roman R, Chan HW, Wenz M, Northrop JP, Ringold

GM, Danielsen M (1987) Lipofection: a highly efficient, lipid-mediated DNA-transfection procedure. Proc Natl Acad Sci U S A 84: 7413–7417

7. Byk G, Scherman D (2000) Genetic chemistry: tools for gene therapy coming from unexpected directions. Drug Dev Res 50:566–572
8. Niedzinski EJ, Fujii SK, Lizarzaburu ME, Hecker JG, Nantz MH (2002) A versatile linker for non-toxic polyamine-mediated DNA transfection. Mol Ther 6:279–286
9. Nantz MH, Dicus CW, Hilliard B, Yellayi S, Scarfo KA, Zou S, Hecker JG (2010) Unsymmetrical hydrophobic domains improve in vivo transfection efficiency. Mol Pharm 7:786–794
10. Bally MB, Harvie P, Wong FM, Kong S, Wasan EK, Reimer DL (1999) Biological barriers to cellular delivery of lipid-based DNA carriers. Adv Drug Deliv Rev 38:291–315
11. Girao da Cruz MT, Simoes S, Pires PPC, Nir S, Pedrosa de Lima MC (2001) Kinetic analysis of the initial steps involved in lipoplex-cell interactions: effect of various factors that influence transfection activity. Biochim Biophys Acta 1510:136–151
12. Mortimer I, Tam P, MacLachlan I, Graham RW, Saravolac EG, Joshi PB (1999) Cationic lipid-mediated transfection of cells in culture requires mitotic activity. Gene Ther 6:403–411
13. Zabner J, Fasbender AJ, Moninger T, Poellinger KA, Welsh MJ (1995) Cellular and molecular barriers to gene transfer by a cationic lipid. J Biol Chem 270:18997–19007
14. Wangerek LA, Dahl HH, Senden TJ, Carlin JB, Jans DA, Dunstan DE, Ioannou PA, Williamson R, Forrest SM (2001) Atomic force microscopy imaging of DNA-cationic liposome complexes optimized for gene transfection into neuronal cells. J Gene Med 3:72–81
15. Zou S, Scarfo K, Nantz MH, Hecker JG (2010) Lipid-mediated delivery of RNA is more efficient that DNA in non-dividing cells. Int J Pharm 389:232–243
16. Wilke M, Fortunati E, van den Broek M, Hoogeveen AT, Scholte BJ (1996) Efficacy of a peptide- based gene delivery system depends on mitotic activity. Gene Ther 3:1133–1142
17. Nicolau C, Sene C (1982) Liposome-mediated DNA transfer in eukaryotic cells. Dependence of the transfer efficiency upon the type of liposomes used and the host cell cycle stage. Biochim Biophys Acta 721:185–190
18. Mattaj IW, Englmeier L (1998) Nucleocytoplasmic transport: the soluble phase. Annu Rev Biochem 67:265–306
19. Wilson GL, Dean BS, Wang G, Dean DA (1995) Nuclear import of plasmid DNA in digitonin-permeabilized cells requires both cytoplasmic factors and specific DNA sequences. J Biol Chem 270:22025–22032
20. Brisson B, Tseng W-C, Almonte C, Watkins S, Huang L (1999) Subcellular trafficking of the cytoplasmic expression system. Hum Gene Ther 10:2601–2613
21. Malone RW, Felgner PL, Verma IM (1989) Cationic liposome-mediated RNA transfection. Proc Natl Acad Sci U S A 86:6077–6081
22. Wolff JA, Malone RW, Williams P, Chong W, Acsadi G, Jani A, Felgner PL (1990) Direct gene transfer into mouse muscle in vivo. Science 247:1465–1468
23. Anderson DM, Hall LL, Ayyalapu AR, Irion VR, Nantz MH, Hecker JG (2003) Stability of mRNA/cationic lipid lipoplexes in human and rat cerebrospinal fluid: methods and evidence for nonviral mRNA gene delivery to the central nervous system. Hum Gene Ther 14:191–202
24. Felgner JH, Kumar R, Sridhar CN, Wheeler CJ, Tsai YJ, Border R, Ramsey P, Martin M, Felgner PL (1994) Enhanced gene delivery and mechanism studies with a novel series of cationic lipid formulations. J Biol Chem 269:2550–2561
25. Sambrook D, Fritsch EF, Maniatis T (1989) Molecular cloning: a laboratory manual, 1st edn. Cold Spring Harbor Laboratory Press, Cold Spring Harbor, NY
26. Krieg PA, Melton DA (1984) Functional messenger RNAs are produced by SP6 in vitro transcription of cloned cDNAs. Nucleic Acids Res 12:7057–7070
27. Gallie DR, Walbot V (1992) Identification of the motifs within the tobacco mosaic virus 5′-leader responsible for enhancing translation. Nucleic Acids Res 20:4631–4638
28. Gallie DR (1991) The cap and poly(A) tail function synergistically to regulate mRNA translational efficiency. Genes Dev 5:2108–2116
29. Tanguay RL, Gallie DR (1996) Isolation and characterization of the 102-kilodalton RNA-binding protein that binds to the 5′ and 3′ translational enhancers of Tobacco Mosaic Virus RNA. J Biol Chem 271:14316–14322
30. Hauck ES, Zou S, Scarfo KA, Nantz MH, Hecker JG (2008) Whole animal in vivo imaging after transient, non-viral lipid-mediated gene transfer to the rat central nervous system. Mol Ther 16:1857–1864
31. Hecker JG, Hall LL, Irion VR (2001) Nonviral gene delivery to the lateral ventricles in rat brain: initial evidence for widespread distribution

and expression in the central nervous system. Mol Ther 3:375–384
32. Balasubramaniam RP, Bennett MJ, Aberle AM, Malone JG, Nantz MH, Malone RW (1996) Structural and functional analysis of cationic transfection lipids: the hydrophobic domain. Gene Ther 3:163–172
33. Bennett MJ, Aberle AM, Balasubramaniam RP, Malone JG, Nantz MH, Malone RW (1996) Considerations for the design of improved cationic amphiphile-based transfection reagents. J Liposome Res 6:545–565
34. Bennett MJ, Nantz MH, Balasubramaniam RP, Gruenert DC, Malone RW (1995) Cholesterol enhances cationic liposome-mediated DNA transfection on human respiratory epithelial cells. Biosci Rep 15:47–53
35. Bennett MJ, Malone RW, Nantz MH (1995) A flexible approach to synthetic lipid ammonium salts for polynucleotide transfection. Tetrahedron Lett 36:2207–2210
36. Hecker JG, Irion VR (1998) Advances in self-limited gene expression of protective intracellular proteins in vivo in rat brain using mRNA/cationic lipid complexes. Anesth Analg 86:S346
37. Malone RW (1989) mRNA transfection of cultured eukaryotic cells and embryos using cationic liposomes. Focus 11:61–66
38. Strong TV, Hampton TA, Louro I, Bilbao G, Conry RM, Curiel DT (1997) Incorporation of B-globin untranslated regions into a Sindbis virus vector for augmentation of heterologous mRNA expression. Gene Ther 4:624–627
39. Ross J (1988) Messenger RNA turnover in eukaryotic cells. Mol Biol Med 5:1–14
40. Ross J (1989) The turnover of messenger RNA. Scientific American April, 48–55
41. McKinney JS, Willoughby KA, Liang S, Ellis EF (1996) Stretch-induced injury of cultured neuronal, glial, and endothelial cells. Stroke 27:934–940
42. Promega Corporation (2012) Transfast transfection reagent. Promega Corporation website, catalog number E2431
43. Corish P, Tyler-Smith C (1999) Attenuation of green fluorescent protein half-life in mammalian cells. Protein Eng 12:1035–1040
44. University of Arizona (2008) Southwest animal imaging resource
45. Ignowski JM, Schaffer DV (2004) Kinetic analysis and modeling of firefly luciferase as a quantitative reporter gene in live mammalian cells. Biotechnol Bioeng 86:827–834
46. Allen MS, Wilgus JR, Chewning CS, Sayler GS, Simpson ML (2006) A destabilized bacterial luciferase for dynamic gene expression studies. Syst Synth Biol 1:3–9
47. Bartlett DW, Davis ME (2006) Insights into the kinetics of siRNA-mediated gene silencing from live-cell and live-animal bioluminescent imaging. Nucleic Acids Res 34:322–333
48. Hecker JG, Langer DJ, Marshall BE, Barnathan ES, Kariko K (1995) Feasibility studies of mRNA transfection for intermediate duration, perioperative modulation of CNS responses accessible via the cerebrospinal fluid. Anesthesia & Analgesia AUA Abstracts 9999:1–2
49. Hecker JG, Irion VR, Malone RW (1997) Self -limited gene expression in-vitro in neuronal cell cultures and in-vivo in rat brain using mRNA /cationic lipid complexes. Anesthesia Analgesia 84:S360
50. Hecker JG, Irion VR (1998) Advances in transient expression of neuroprotective intracellular proteins *in-vivo* in rat brain. AUA, 45th Mtg, 7–9 May 1998, 103
51. Hecker JG, Nantz MH (2008) novel methods and models for rapid, widespread delivery of genetic materials to the CNS using non-viral, cationic lipid-mediated vectors. PCT/US2006/048093
52. Hecker JG, Irion VR (1998) Expression of protective intracellular proteins in-vivo in rat brain using mRNA/cationic lipid complexes. American Society Gene Therapy, 1st Mtg, 28–31 May, 154A.
53. Kariko K, Muramatsu H, Ludwig J, Weissman D (2012) Generating the optimal mRNA for therapy: HPLC purification eliminates immune activation and improves translation of nucleoside-modified, protein-encoding mRNA. Nucleic Acids Res 39:e142
54. Kariko K, Muramatsu H, Welsh FA, Ludwig J, Kato H, Akira S, Weissman D (2008) Incorporation of pseudouridine into mRNA yields superior nonimmunogenic vector with increased translational capacity and biological stability. Mol Ther 16:1833–1840
55. Kariko K, Megyeri K, Xiao Q, Barnathan ES (1994) Lipofectin-aided delivery of ribozyme targeted to human urokinase receptor mRNA. FEBS Lett 352:41–44
56. Kariko K, Keller JM, Harris VA, Langer DJ, Welsh FA (2001) In vivo protein expression from mRNA delivered into adult rat brain. J Neurosci Methods 105:77–86

Chapter 24

Ex Vivo Gene Therapy Using Human Mesenchymal Stem Cells to Deliver Growth Factors in the Skeletal Muscle of a Familial ALS Rat Model

Masatoshi Suzuki and Clive N. Svendsen

Abstract

Therapeutic protein and molecule delivery to target sites by transplanted human stem cells holds great promise for ex vivo gene therapy. Our group has demonstrated the therapeutic benefits of ex vivo gene therapy targeting the skeletal muscles in a transgenic rat model of familial amyotrophic lateral sclerosis (ALS). We used human mesenchymal stem cells (hMSCs) and genetically modified them to release neuroprotective growth factors such as glial cell line-derived neurotrophic factor (GDNF) and vascular endothelial growth factor (VEGF). Intramuscular growth factor delivery via hMSCs can enhance neuromuscular innervation and motor neuron survival in a rat model of ALS ($SOD1^{G93A}$ transgenic rats). Here, we describe the protocol of ex vivo delivery of growth factors via lentiviral vector-mediated genetic modification of hMSCs and hMSC transplantation into the skeletal muscle of a familial ALS rat model.

Key words Ex vivo gene therapy, Human mesenchymal stem cells, Growth factors, Lentivirus, Cell transplantation, Glial cell line-derived neurotrophic factor (GDNF), Vascular endothelial growth factor (VEGF), Amyotrophic lateral sclerosis (ALS)

1 Introduction

Amyotrophic lateral sclerosis (ALS) causes motor neuron degeneration, muscular atrophy, and ultimately death by respiratory failure [1, 2]. Because ALS is a devastating disease with no effective treatments and no known cures, novel treatments are greatly needed. ALS is an ideal candidate for novel gene and cell therapy approaches as it is both incurable and terminal [3].

ALS research has focused on motor neuron cell death, however degeneration is also observed in skeletal muscle, particularly at the neuromuscular connection [4]. Glial cell line-derived neurotrophic factor (GDNF) and vascular endothelial growth factor (VEGF) promote survival of motor neurons and their neuromuscular junctions in neuromuscular disorders such as ALS [4].

Fredric P. Manfredsson (ed.), *Gene Therapy for Neurological Disorders: Methods and Protocols*, Methods in Molecular Biology, vol. 1382, DOI 10.1007/978-1-4939-3271-9_24,

Growth factors can be delivered through direct injections or viral vectors to infect endogenous cells. Alternatively, cells genetically engineered to deliver growth factor can be transplanted into focal regions of damage or disease. This ex vivo gene therapy may be advantageous as healthy cells directly and locally provide the growth factor versus pathological host cells.

Human mesenchymal stem cells (hMSCs), found in bone marrow and other mesenchymal tissues, are easy to harvest and can be expanded in vitro to clinically relevant numbers while retaining their normal karyotype and differentiation capacity [5–7]. These cells have been shown to have a significant effect on disease progression in a number of animal models of human disease including heart damage, stroke, Parkinson's disease and ALS [8, 9]. While the mechanism of this protective effect remains elusive, it may involve growth factor release or increased angiogenesis [8, 9].

Our group has demonstrated the therapeutic benefits of ex vivo gene therapy using stem cell-based growth/trophic factor delivery targeting the skeletal muscle to prevent degeneration of motor neurons and associated neuromuscular junctions during ALS [10, 11]. Specifically, hMSCs genetically engineered to stably express a combination of GDNF and/or VEGF can survive transplantation to the muscle, synthesize and release growth factors, and slow disease progression in familial ALS model rats (*see* Fig. 1) [11].

The protocols in this chapter provide detailed information to expand and genetically modify hMSCs to release growth factors by lentiviral infection. Detailed protocols for hMSC transplantation into the rat skeletal muscle are also described.

2 Materials

2.1 Human Mesenchymal Stem Cell (hMSC) Culture

1. hMSCs: hMSCs are established using previously described methods (*see* **Note 1**) [12–14]. The hMSCs are derived from neonatal bone marrow aspirates from healthy donors after informed consent (*see* **Note 2**) and are transduced with a retroviral vector containing enhanced green fluorescence protein (GFP) using second passage hMSCs [12]. Aliquot hMSCs in cryovials and store in liquid nitrogen.
2. hMSC culture medium: Dulbecco's Modified Eagle Medium (DMEM high glucose, GlutaMAX™; Life Technologies, Carlsbad, CA, USA) supplemented with 20 % heat-inactivated fetal bovine serum (FBS; Hyclone, Logan, UT, USA), 50 U/ml penicillin, and 50 mg/ml streptomycin (*see* **Note 3**).
3. 0.05 % trypsin-EDTA (Life Technologies, Carlsbad, CA, USA).
4. Dulbecco's Phosphate-Buffered Saline (DPBS).

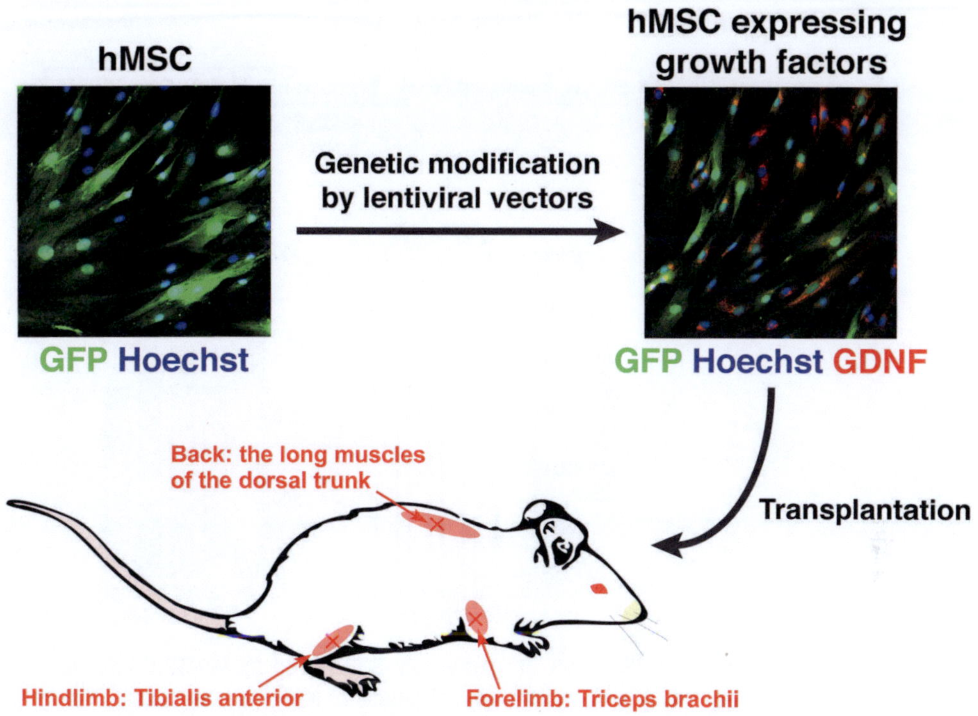

Fig. 1 Ex vivo delivery of growth factors targeting skeletal muscles for ALS. A GFP-expressing hMSC line was genetically modified by lentiviral infection to stably release growth factors such as glial cell line-derived neurotrophic factor [10, 11]. These cells can be used for ex vivo gene therapy to deliver growth factors following intramuscular transplantation in a familial ALS rat model

5. 24-well culture plates.
6. T25-, T75-, or T175-culture flasks.
7. 15- or 50-ml conical tubes.
8. 0.6- or 1.5-ml microcentrifuge tubes.
9. 0.4 % trypan blue solution (Sigma-Aldrich, St. Louis, MO, USA).
10. Hem☺ocytometer.

2.2 Genetic Modification of hMSCs by Lentiviral Vectors

1. Lentivirus: Use a lentiviral construct for constitutive expression of human GDNF or VEGF-165 (VEGF) under the control of the mouse phosphoglycerate kinase 1 (pgk-1) promoter (*see* **Note 4** and Fig. 2a) [10, 11, 15, 16]. Obtain high-titer stocks by ultracentrifugation, and suspend the lentiviral particles in 1 % bovine serum albumin (BSA) in phosphate buffered saline (PBS). Determine particle content of viral batches using p24 antigen ELISA (PerkinElmer Life Sciences, Waltham, MA, USA).

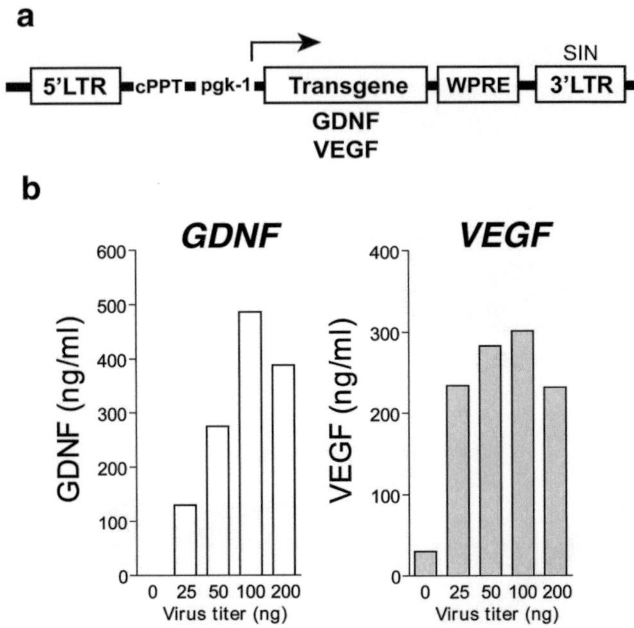

Fig. 2 hMSCs can be genetically engineered by lentiviral infection to stably express growth factors. (**a**) Schematic of lentiviral construct. *LTR* long terminal repeat, *pgk-1* mouse phosphoglycerate kinase 1 promotor, *WPRE* post transcriptional regulatory element of woodchuck hepatitis virus, *cPPT* central polypurine tract, *SIN* self-inactivating. (**b**) After infection with serial dilutions of viruses, ELISA can detect GDNF and VEGF proteins in the conditioned medium [10, 11]

2.3 hMSC Transplantation in the Rat Muscle

1. Sprague-Dawley rats or familial ALS model rats (Taconic, Hudson, NY, USA). This ALS model rat is from a transgenic rat strain overexpressing a G93A mutant form of cytosolic Cu^{2+}/Zn^{2+} superoxide dismutase 1 gene ($SOD1^{G93A}$ transgenic) (*see* **Note 5**) [17–19].
2. Transplant medium: DMEM high glucose (GlutaMAX™; Life technologies, Carlsbad, CA, USA), 0.5 % BSA (Sigma-Aldrich, St. Louis, MO, USA), filtered (*see* **Note 6**).
3. Bupivacaine hydrochloride (Sensorcaine®-MPF 0.75 %; AstraZeneca, Miami, FL, USA) (*see* **Note 7**).
4. Cyclosporine solution: Mix 3 ml of SandImmune (100 mg/ml; Novartis, Basel, Switzerland) and 27 ml of extra virgin olive oil. Final concentration is 10 mg/ml. Filter the mixture using a 0.45 μm bottle top filter. Transfer it in a glass vial, cap with rubber stopper, and cover the bottle with aluminum foil to shield light (*see* **Note 8**).
5. 1 ml disposable syringes.
6. A 100 μl Hamilton syringe (Reno, NV, USA).
7. 30- or 33-gauge needles.

3 Methods

Primary human tissue, hMSC cultures, and any medium removed from these cultures are hazardous waste and should be contained and discarded in appropriate biohazard containers.

All culture procedures should be completed in a laminar-flow culture hood unless indicated otherwise. Cultures should be maintained in a humidified incubator at 37 °C and 5 % CO_2. Media should be warmed in a 37 °C water bath prior to use.

All animal procedures must be carried out in accordance with the guidelines for an Institutional Animal Care and Use Committee (IACUC) and must conform to National Institutes of Health standards of animal care and use.

3.1 Preparation of hMSC Culture

3.1.1 Thawing of hMSCs

1. Thaw the cryovial quickly in a 37 °C water bath.
2. Sterilize the cryovial with isopropanal.
3. Transfer contents to a 15-ml conical tube and slowly add 10 ml of the hMSC culture medium.
4. Centrifuge at 168 × *g* for 5 min at room temperature.
5. Remove the supernatant, re-suspend in 10 ml of the hMSC culture medium, and plate in a T25 culture flask at approximately 30–40 % confluence.
6. Check the cells daily using a microscope. Passage cells at approximately 80–90 % confluency.

3.1.2 Passaging of hMSCs

1. Remove supernatant from the flask.
2. Rinse once with 5 ml of DPBS. Add the PBS gently to not disturb the cell monolayer.
3. Add 3 ml of 0.05 % trypsin-EDTA and incubate in a CO_2 incubator for 3–4 min.
4. Tap the flask vigorously and check under the microscope. Make sure that all cells are properly detached.
5. Add 3 ml of pre-warmed hMSC culture medium.
6. Pipette up and down, rinsing the bottom of the flask. Make sure that all cells are collected and do not remain adhered to the flask.
7. Transfer the cells to a 15 ml conical tube.
8. Centrifuge at 168 × *g* for 5 min at room temperature.
9. Remove the supernatant, re-suspend in 15 ml of the hMSC culture medium, and plate in a new T75 culture flask.
10. Passage hMSC lines approximately once per week (*see* **Note 9**). If starting from a 90 % confluent T75 flask passage in 1:3 ratio, prepare 3T75 flasks with 12 ml of hMSC culture medium each. Re-suspend the cells in 9 ml of the hMSC culture medium and add 3 ml of the cell suspension to each T75.

3.2 Lentiviral Infection of hMSCs

Lentiviral infection can be used to stably integrate a gene of interest. Our previous work has shown that both rat and human neural progenitor cells can be efficiently infected with lentiviral constructs [15, 16]. We used this experience with neural progenitor cells to prepare hMSC expressing GDNF and/or VEGF [10, 11]. The same viral construct has been used for constitutive expression of growth factors under the control of the mouse pgk-1 promoter (*see* **Note 10**).

1. Prepare a single cell suspension of hMSCs in hMSC culture medium as described above (*see* Subheading 3.1).
2. Count cells in a 10 μl aliquot using a hemocytometer and assess viability using 0.4 % trypan blue (*see* **Note 11**).
3. Resuspend the cells in hMSC culture medium as 25 cells/μl in a 15 or 50 ml conical tube (depending on the number of the wells to be used).
4. Add 400 μl of the cell suspension in a 24-well plate and incubate in a CO_2 incubator.
5. After 24 h, remove the medium and add 300 μl of the culture medium containing various virus titers (0–200 ng/p24/per well) to each well (*see* **Note 12**).
6. After incubating for 24–72 h in a CO_2 incubator, collect the conditioned medium and store it in a −20 °C freezer. Rinse cells once with 300 μl DPBS and then add 300 μl per well of 0.05 % trypsin-EDTA. Split the cells as described above (*see* Subheading 3.1) and scale up the culture as necessary for further studies.
7. For preparing a hMSC line stably expressing both GDNF and VEGF [11], subsequently infect the GDNF-expressing hMSC line with lentivirus encoding VEGF, similarly as described above [11].
8. Confirm protein expression of the growth factors by ELISA using conditioned medium (*see* **Note 13** and Fig. 2b) and immunocytochemistry using plated cells (*see* **Note 14** and Fig. 3) [10, 11].

3.3 hMSC Transplantation into the Rodent Muscle

3.3.1 hMSC Preparation for Intramuscular Injection

1. Calculate the number of cells necessary for the transplant experiments (*see* **Note 15**).
2. Expand hMSCs expressing growth factors to the appropriate scale. Trypsinize and suspend cells in 200 μl of hMSC transplantation medium.
3. Count cells in a 10 μl aliquot using a hemocytometer and assess viability using 0.4 % trypan blue (*see* **Note 11**).
4. Adjust cell concentration to 5000 cells/μl using hMSC transplantation medium. Keep the cell suspension in a 0.6 ml- or 1.6 ml-centrifuge tube on ice until ready for intramuscular transplantation.

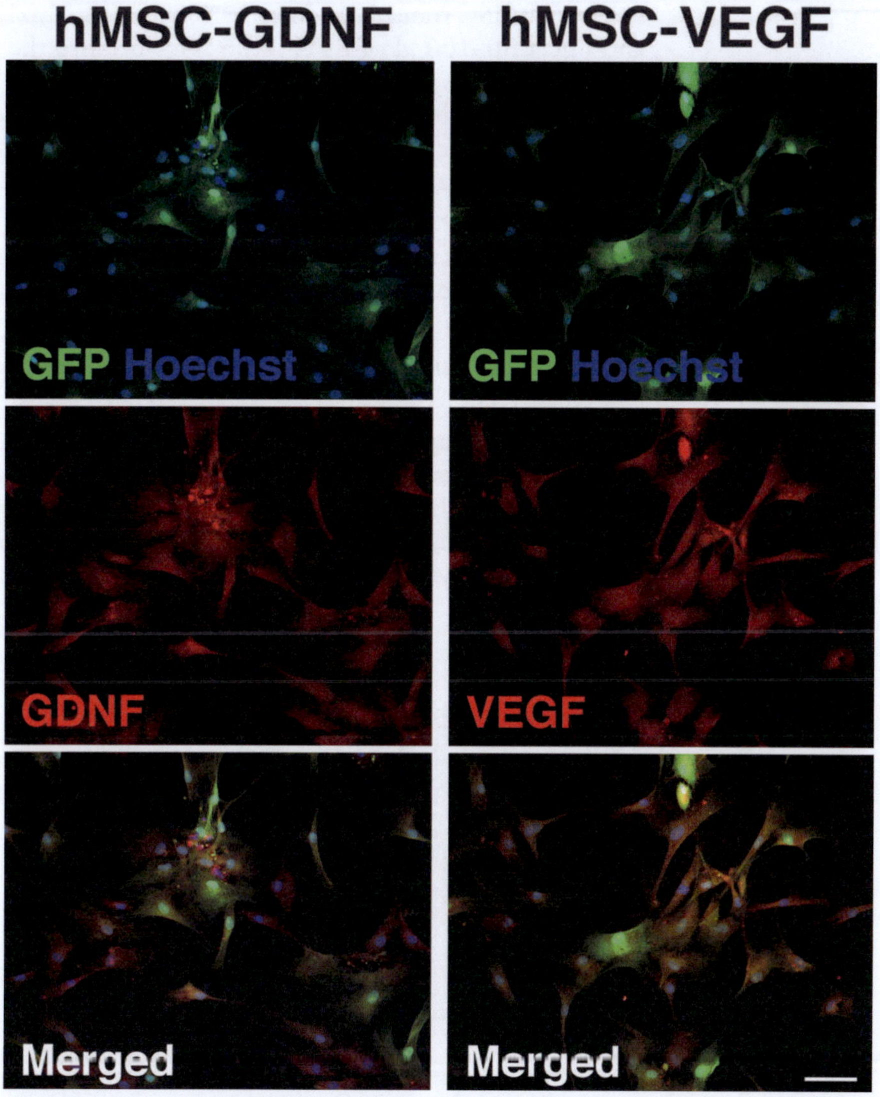

Fig. 3 Immunocytochemistry of the plated hMSCs shows growth factor production. Genetically engineered hMSCs can be plated for immunocytochemistry analysis that confirms robust expression of GDNF or VEGF. Scale bar: 100 μm [11]

3.3.2 Intramuscular hMSC Transplantation

1. At 1–2 days prior to the hMSC transplantation, administer cyclosporine (10 mg/kg/day) intraperitoneally daily for immunosuppression to adult Sprague-Dawley or pre-symptomatic familial ALS model rats (90–110 days of age; $SOD1^{G93A}$ transgenic rat) (*see* **Note 16**).
2. One day prior to hMSC transplantation, induce focal injury in the muscle to promote cell transplant survival (*see* **Note 17**). For this, the local anesthetic bupivacaine hydrochloride (BVC; *see* **Note 7**) should be injected in any muscle that will later

receive a hMSC transplant. Anesthetize the rats with isoflurane and inject 50–100 μl of BVC (0.35 mg/ml) into the rat muscle. Depending on the experiment, inject BVC and the cells bilaterally or unilaterally into the tibialis anterior (TA), forelimb triceps brachii, and/or the long muscles of the dorsal trunk muscles (*see* Fig. 1). A 30-gauge needle connected to a 1 ml disposable syringe is used for injections.

3. After 24 h, inject culture-expanded hMSC (150,000 cells in 30 μl) into the same muscles using a 33-gauge needle connected to a 100 μl Hamilton syringe (*see* **Note 18**).
4. Repeat the injection of hMSCs twice more at 1-week intervals (*see* **Note 19**).
5. After completing the experiments, dissect the hMSC-transplanted muscles and prepare frozen sections (*see* **Note 20**). Determine hMSC survival (*see* Fig. 4a) and growth factor expression (*see* Fig. 4b) in the grafted muscles using immunohistological analyses (*see* **Note 21**).

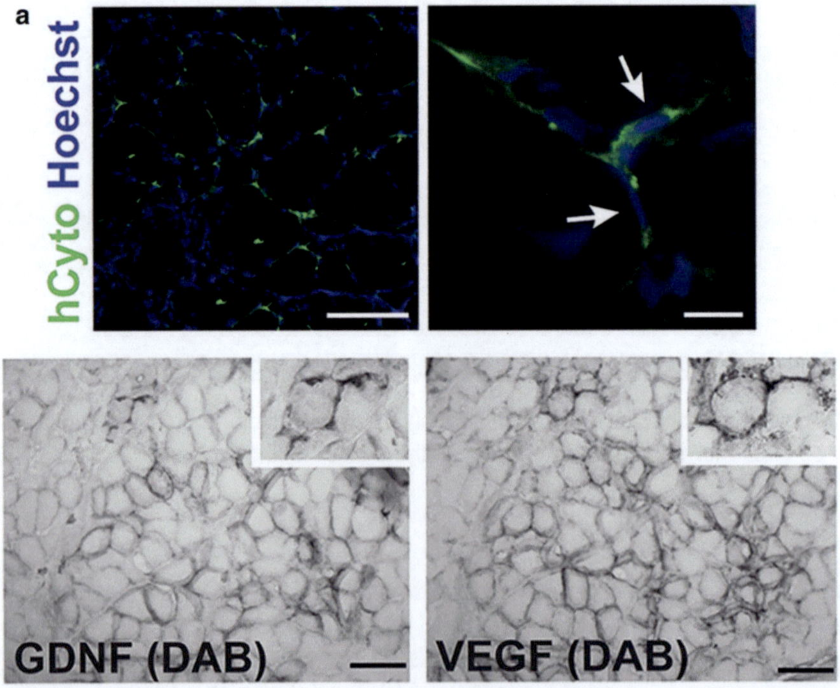

Fig. 4 Transplanted hMSCs survive and release growth factor in rat skeletal muscle. (**a**) Immunohistochemistry with a human cytoplasmic specific (hCyto) antibody confirms hMSCs survival and integration in the muscle. (**b**) Immunohistochemistry with GDNF and VEGF antibodies confirms growth factor release from hMSCs into the transplanted muscle. *DAB* diaminobenzidine. Scare bars: 100 μm in **a**; 20 μm in **b** [11]

4 Notes

1. There are many published papers using a variety of methods for preparing hMSCs from different tissue resources and expanding these cells in culture. The protocol described here is based on our own experience with our hMSC lines. We expect users to optimize these fundamental protocols using the extensive literature on this topic.
2. Single-cell suspensions of neonatal bone marrow were prepared by collecting the bone marrow cells in the humeri and femurs using a syringe and 22-gauge needle. Unselected nucleated cells were plated in 100-mm dishes at a density of 10^5 cell/ml and cultured in the hMSC culture medium. After 3 days, non-adherent cells were removed and the fresh medium was added. When colony size exceeded >500 cells, single colonies were trypsinized and collected using clonal cylinders. The mesenchymal nature of adherent nucleated cells was confirmed by immunophenotyping and multilineage differentiation [12–14].
3. The sterilized medium can be stored at 4 °C for a few weeks.
4. Post-translational *cis*-acting regulatory element of the woodchuck hepatitis virus (WHV) is included and has been shown to significantly enhance transgene expression [10, 11, 15, 16].
5. Breeder $SOD1^{G93A}$ transgenic rats were obtained from Taconic (Hudson, NY, USA). The colonies were developed by crossing male founders with female Sprague-Dawley rats. Heterozygous $SOD1^{G93A}$ progeny were identified with polymerase chain reaction of tail DNA with primers specific for human SOD1 [18]. Rats were maintained in a room with controlled illumination (lights on 0600–1800 h) and temperature (23 ± 1 °C), and given free access to laboratory chow and tap water.
6. Although the transplantation medium is best made up fresh, the sterilized medium can be stored at 4 °C for a few weeks.
7. Different concentrations of BVC solution (0.75 or 0.50 %) are available commercially. They are acceptable to use for this experiment with an appropriate dose.
8. Handle the solutions in a biosafety cabinet. After hood surfaces have been sterilized, place a Wypall paper or bench top paper on your working area, because the oil is difficult to remove from the stainless steel surface of the biosafety cabinet.
9. Depending on the culture scale, different sized flasks can be used for passaging. For T25-, T75-, and T175- culture flasks, 7.5 ml, 15 ml, and 35 ml of the hMSC culture medium should be used, respectively.
10. All viruses should be carefully handled in the appropriate biosafety level environment. Lentivirus can be neutralized with

70 % ethanol, but please check your institutional environmental safety standards for clean-up and disposal.

11. Add 10 μl of cell suspension to 90 μl of the medium and mix well. Add 50 μl of this 10× dilution to 50 μl of 0.4 % trypan blue solution and apply to the hemocytometer with 10 μl of the mixed solution. The final sample dilution is 20×. Count four or five squares on the hemocytometer and determine the average number of live cells. Multiply this value by the dilution factor and then by 10,000 to derive cells/ml. Calculate the total number of cells available.
12. An appropriate titer of the lentivirus would be influenced by differences in viral preparations. In our studies, infecting with 100 ng/p24/well of lentivirus is commonly optimal for protein expression (*see* Fig. 2).
13. Use ELISA kits for human GDNF or human VEGF (R&D Systems, Minneapolis, MN, USA).
14. After plating the hMSCs on coverslips, perform immunocytochemistry as described previously [10, 11]. Our latest immunochytochemical study indicated that in these hMSCs approximately 95-98 % of total cells expressed growth factors [11].
15. The optimal number of cells to transplant will vary depending on the experimental design. We commonly transplant 150,000 cells in 30 μl.
16. Our recent data from the latest $SOD1^{G93A}$ transgenic rat colony indicated that the median to reach disease onset and endpoint were 146 and 184 days, respectively, but these time scales can vary significantly between colonies [20].
17. According to our pilot experiments, inducing focal muscular injury before hMSC transplantation promotes hMSC survival [10]. We use intramuscular injection of the local anesthetic BVC that, similarly to cardiotoxin, has been broadly used to induce a partial muscular lesion [21, 22].
18. Distribute the BVC or hMSCs widely in the injected muscle by directing the needle along the longitudinal direction of the muscle, injecting the solution slowly (for 20–30 s), and moving the needle slightly back and forth in the muscle. Hold the needle in the muscle for 5–10 s after completing the injection to prevent a backflow of the injected fluid and then slowly withdraw. To confirm that the needle for injection is surely placed in the muscle (not in a subcutaneous space), making a small skin cut to expose the muscle is helpful. In this case, the skin opening needs to be sutured.
19. For ex vivo studies using the familial ALS rat model, we inject hMSCs three times at 1-week intervals to improve the likelihood of cell survival. However, hMSC survival can also be confirmed after a single injection. The timing and number of cell injections should be optimized depending on the experimental design.

20. The muscles are dissected, flash-frozen in super-cooled isopentane, sectioned at 20 μm using a cryostat, and placed on glass slides for staining. These sections can be stored at −80 °C.
21. A general protocol for immunohistochemistry is followed to determine hMSC survival and growth factor expression in the transplanted muscle [10, 11]. Briefly, the sections are fixed with ice-cold methanol for 10 min. Use primary antibodies to label human specific cytoplasmic protein (hCyto; mouse monoclonal, 1:200; StemCells, Inc., Cambridge, UK), GFP (mouse monoclonal, 1:100, Life Technologies, Carlsbad, CA, USA), human GDNF, and human VEGF (goat polyclonal, 1:250; R&D Systems, Minneapolis, MN, USA). Use appropriate secondary antibodies conjugated to fluorophores such as Alexa Fluor 488 or Cy3 (donkey anti-mouse IgG, 1:1,000; Jackson ImmunoResearch Laboratories, West Grove, PA, USA). Biotinylated secondary (anti-goat IgG, 1:200; Jackson ImmunoResearch Laboratories) can be used for GDNF and VEGF detection, followed by diaminobenzidine [23] with avidin-biotin and nickel ammonium sulfate enhancement (Vector Laboratories, Burlingame, CA, USA).

Acknowledgments

We gratefully acknowledge Ilaria Bellantuono (The University of Sheffield; UK) for providing the GFP-expressing hMSC lines and Patrick Aebischer (Brain & Mind Institute, Ecole Polytechnique Fédérale de Lausanne, Switzerland) for producing lentiviral vectors. We thank Christina Lewis (University of Wisconsin, Madison, WI) for the help in preparing this manuscript and Soshana Svendsen (Cedars-Sinai Medical Center, Los Angeles, CA) for critical review and editing. This work was supported by grants from the ALS Association, NIH (R21NS06104 and R01NS091540 to M.S) and the University of Wisconsin Foundation.

References

1. Boillee S, Vande VC, Cleveland DW (2006) ALS: a disease of motor neurons and their non-neuronal neighbors. Neuron 52:39–59
2. Kiernan MC, Vucic S, Cheah BC, Turner MR, Eisen A, Hardiman O, Burrell JR, Zoing MC (2011) Amyotrophic lateral sclerosis. Lancet 377:942–955
3. Suzuki M, Svendsen CN (2008) Combining growth factor and stem cell therapy for amyotrophic lateral sclerosis. Trends Neurosci 31:192–198
4. Krakora D, Macrander C, Suzuki M (2012) Neuromuscular junction protection for the potential treatment of amyotrophic lateral sclerosis. Neurol Res Int 2012:379657
5. Prockop DJ (1997) Marrow stromal cells as stem cells for nonhematopoietic tissues. Science 276:71–74
6. Pittenger MF, Mackay AM, Beck SC, Jaiswal RK, Douglas R, Mosca JD, Moorman MA, Simonetti DW, Craig S, Marshak DR (1999) Multilineage potential of adult human mesenchymal stem cells. Science 284:143–147
7. Deans RJ, Moseley AB (2000) Mesenchymal stem cells: biology and potential clinical uses. Exp Hematol 28:875–884

8. Picinich SC, Mishra PJ, Mishra PJ, Glod J, Banerjee D (2007) The therapeutic potential of mesenchymal stem cells. Cell- & tissue-based therapy. Expert Opin Biol Ther 7:965–973
9. Lewis C, Suzuki M (2014) Therapeutic applications of mesenchymal stem cells for amyotrophic lateral sclerosis. Stem Cell Res Ther 5:32
10. Suzuki M, McHugh J, Tork C, Shelley B, Hayes A, Bellantuono I, Aebischer P, Svendsen CN (2008) Direct muscle delivery of GDNF with human mesenchymal stem cells improves motor neuron survival and function in a rat model of familial ALS. Mol Ther 16:2002–2010
11. Krakora D, Mulcrone P, Meyer M, Lewis C, Bernau K, Gowing G, Zimprich C, Aebischer P, Svendsen CN, Suzuki M (2013) Synergistic effects of GDNF and VEGF on lifespan and disease progression in a familial ALS rat model. Mol Ther 21:1602–1610
12. Campagnoli C, Roberts IA, Kumar S, Bennett PR, Bellantuono I, Fisk NM (2001) Identification of mesenchymal stem/progenitor cells in human first-trimester fetal blood, liver, and bone marrow. Blood 98: 2396–2402
13. Baxter MA, Wynn RF, Jowitt SN, Wraith JE, Fairbairn LJ, Bellantuono I (2004) Study of telomere length reveals rapid aging of human marrow stromal cells following in vitro expansion. Stem Cells 22:675–682
14. Wynn RF, Hart CA, Corradi-Perini C, O'Neill L, Evans CA, Wraith JE, Fairbairn LJ, Bellantuono I (2004) A small proportion of mesenchymal stem cells strongly expresses functionally active CXCR4 receptor capable of promoting migration to bone marrow. Blood 104:2643–2645
15. Suzuki M, McHugh J, Tork C, Shelley B, Klein SM, Aebischer P, Svendsen CN (2007) GDNF secreting human neural progenitor cells protect dying motor neurons, but not their projection to muscle, in a rat model of familial ALS. PLoS One 2:e689
16. Capowski EE, Schneider BL, Ebert AD, Seehus CR, Szulc J, Zufferey R, Aebischer P, Svendsen CN (2007) Lentiviral vector-mediated genetic modification of human neural progenitor cells for ex vivo gene therapy. J Neurosci Methods 163:338–349
17. Howland DS, Liu J, She Y, Goad B, Maragakis NJ, Kim B, Erickson J, Kulik J, DeVito L, Psaltis G, DeGennaro LJ, Cleveland DW, Rothstein JD (2002) Focal loss of the glutamate transporter EAAT2 in a transgenic rat model of SOD1 mutant-mediated amyotrophic lateral sclerosis (ALS). Proc Natl Acad Sci U S A 99:1604–1609
18. Suzuki M, Tork C, Shelley B, McHugh J, Wallace K, Klein SM, Lindstrom MJ, Svendsen CN (2007) Sexual dimorphism in disease onset and progression of a rat model of ALS. Amyotroph Lateral Scler 8:20–25
19. Hayes-Punzo A, Mulcrone P, Meyer M, McHugh J, Svendsen CN, Suzuki M (2012) Gonadectomy and dehydroepiandrosterone (DHEA) do not modulate disease progression in the G93A mutant SOD1 rat model of amyotrophic lateral sclerosis. Amyotroph Lateral Scler 13:311–314
20. Suzuki M, Klein S, Wetzel EA, Meyer M, McHugh J, Tork C, Hayes A, Svendsen CN (2010) Acute glial activation by stab injuries does not lead to overt damage or motor neuron degeneration in the G93A mutant SOD1 rat model of amyotrophic lateral sclerosis. Exp Neurol 221:346–352
21. Steer JH, Mastaglia FL, Papadimitriou JM, Van BI (1986) Bupivacaine-induced muscle injury. The role of extracellular calcium 1. J Neurol Sci 73:205–217
22. Hill M, Wernig A, Goldspink G (2003) Muscle satellite (stem) cell activation during local tissue injury and repair 1. J Anat 203:89–99
23. Yang Y, Hentati A, Deng HX, Dabbagh O, Sasaki T, Hirano M, Hung WY, Ouahchi K, Yan J, Azim AC, Cole N, Gascon G, Yagmour A, Ben-Hamida M, Pericak-Vance M, Hentati F, Siddique T (2001) The gene encoding alsin, a protein with three guanine-nucleotide exchange factor domains, is mutated in a form of recessive amyotrophic lateral sclerosis. Nat Genet 29:160–165

Part VI

Gene Therapy Based Modeling of Neurodegenerative Disorders

Chapter 25

Gene Therapy Models of Alzheimer's Disease and Other Dementias

Benjamin Combs, Andrew Kneynsberg, and Nicholas M. Kanaan

Abstract

Dementias are among the most common neurological disorders, and Alzheimer's disease (AD) is the most common cause of dementia worldwide. AD remains a looming health crisis despite great efforts to learn the mechanisms surrounding the neuron dysfunction and neurodegeneration that accompanies AD primarily in the medial temporal lobe. In addition to AD, a group of diseases known as frontotemporal dementias (FTDs) are degenerative diseases involving atrophy and degeneration in the frontal and temporal lobe regions. Importantly, AD and a number of FTDs are collectively known as tauopathies due to the abundant accumulation of pathological tau inclusions in the brain. The precise role tau plays in disease pathogenesis remains an area of strong research focus. A critical component to effectively study any human disease is the availability of models that recapitulate key features of the disease. Accordingly, a number of animal models are currently being pursued to fill the current gaps in our knowledge of the causes of dementias and to develop effective therapeutics. Recent developments in gene therapy-based approaches, particularly in recombinant adeno-associated viruses (rAAVs), have provided new tools to study AD and other related neurodegenerative disorders. Additionally, gene therapy approaches have emerged as an intriguing possibility for treating these diseases in humans. This chapter explores the current state of rAAV models of AD and other dementias, discuss recent efforts to improve these models, and describe current and future possibilities in the use of rAAVs and other viruses in treatments of disease.

Key words Tau protein, Neurofibrillary tangle, Recombinant adeno-associated virus, Hippocampus, Entorhinal cortex, Animal model

1 Alzheimer's Disease and Other Dementias

Alzheimer's disease (AD) is an age-related neurodegenerative disorder involving progressive degeneration in regions of the brain that are important for memory and cognition (i.e., the temporal lobe and hippocampus). The neuropathological hallmarks of AD include accumulations of the amyloid-β peptide (Aβ) and the tau protein. The Aβ pathology is in the form of extracellular plaques (i.e., senile plaques), while the tau pathology is intracellular inclusions known as neurofibrillary tangles (NFTs), neuropil threads, and neuritic plaques [1, 2]. The presence of Aβ plaques

Fredric P. Manfredsson (ed.), *Gene Therapy for Neurological Disorders: Methods and Protocols*, Methods in Molecular Biology, vol. 1382, DOI 10.1007/978-1-4939-3271-9_25,

and tau NFTs are required for a diagnosis of AD; however, the role each pathology plays in the mechanisms of the disease and what initiates their production remain unknown. The majority of AD cases are sporadic in nature but approximately 5 % are due to a specific genetic polymorphism in the amyloid precursor protein gene (APP) or one of the presenilin genes (PS1 or PS2) that make up part of the APP-cleaving complex [3, 4]. Familial AD typically presents with pathology similar to sporadic AD, but with an earlier onset and faster progression indicating that the mechanisms of disease may be similar but more aggressive in inherited cases.

The Aβ peptide was the focus of much of the AD research over the last several decades because the familial mutations are located in APP and its processing machinery. The peptide is the primary component of the extracellular plaques that were originally identified in AD brains [2]. The peptide is cleaved from APP, an integral membrane protein, by various secretase enzymes that determine the length of the peptide. The Aβ40 form (i.e., 40 amino acids long) is more prevalent but the Aβ42 form (i.e., 42 amino acids long) is more prone to aggregate and linked with toxicity [5, 6]. While amyloid plaque density does not correlate closely to progression of AD [7], recent research identified that the levels of soluble oligomers and the ratio of the Aβ42 to Aβ40 are more closely associated with disease-related toxicity [8, 9]. Currently, the role of Aβ in disease remains unclear, but Aβ may directly or indirectly induce neurodegenerative effects such as synaptic defects through interactions with receptors [10], activation of caspases [11], formation of pores [12], and/or axonal transport dysfunction through kinase dysregulation [13]. Many of the toxic effects associated with Aβ expression in animal and cell culture models require the presence of the tau protein, indicating a link between the toxicity of the peptide and the other major pathology associated with AD [14–16].

Indeed, a great deal of recent research has focused on tau pathology in AD and its role in the disease progression. Tau is a microtubule-associated protein that may be involved in stabilizing and spacing microtubules as well as regulating axonal transport, as well as kinases and phosphatases [17–20]. In the adult human central nervous system (CNS), six different isoforms of tau are created by alternative splicing of the 2nd, 3rd, and 10th exons. The 10th exon contains one of four potential microtubule-binding repeat regions that are necessary for microtubule association and contain the regions that make up the core of tau aggregates. Isoforms that contain all four microtubule-binding regions are referred to as 4R, while isoforms without exon 10 are known as 3R isoforms.

In AD, it is believed that tau dissociates from microtubules, begins to aggregate, and is mislocalized from axons to the somatodendritic compartment [21]. Phosphorylation of tau is increased approximately fourfold and some forms of phosphorylated tau do not bind as well to microtubules and aggregate more readily than unmodified tau [22, 23]. Aβ appears to enhance tau aggregation, but not vice versa, in some mouse models that combine disease-related forms of APP and tau [24]. These studies suggest that Aβ may have an upstream role in the disease by altering tau dysfunction in some way [24]. In fact, tau was initially believed by some to be merely a byproduct of Aβ pathology until the discovery of inherited, early-onset FTDs that were caused by mutations in the tau gene [25]. This provided evidence that pathological changes in tau were sufficient to cause a neurodegenerative disease. Further evidence for a role of tau in neurotoxicity came with data suggesting that some of the toxic effects of Aβ appear to be dependent on the presence of tau [14–16]. Ultimately, the toxic effects of abnormal forms of tau are thought to lead to dysfunction and degeneration of synapses and axons that in turn lead to the loss of memory and cognitive deficits that characterizes AD [26, 27].

The progressive cognitive decline in AD correlates well with a stereotypical spatiotemporal distribution of tau pathology [1, 7]. The first stage is called Braak stage I, which is characterized by the initial appearance of tau pathology in the transentorhinal cortex of the temporal lobe. In Braak stage II, tau pathology expands into the entorhinal cortex (EC) and first appears in the CA1 neurons of the hippocampus (HP). Braak stage III is characterized by tau deposits in the subcortical regions of the thalamus and amygdala as well as throughout the entire EC. In Braak stage IV, large numbers of NFTs and neuropil threads are present in the EC along with an increasing number of NFTs in the CA1 and CA4 neurons and the striatum, amygdala, thalamus, and hypothalamus. Tau pathology expands throughout the HP in Braak stage V in conjunction with growing numbers of tau aggregates in the isocortex and subcortical regions. At Braak stage VI, large amounts of tau pathology is found in all regions of the neocortex and isocortex, subiculum, subcortical nuclei, thalamus, hypothalamus, and substantia nigra (SN). As the tau pathology progresses, specific populations of neurons begin to die, worsening the cognitive dysfunction and often leaving behind "ghost tangles" (starting around Stage III and increasing in number through Stage VI), which are the remains of tangles that become extracellular once a neuron dies [28].

NFTs were traditionally thought to be the toxic forms of tau but a number of studies suggest pre-tangle aggregates, such as oligomers, may be the toxic species. Neurons appear to survive for decades containing NFTs [29] and recent work showed that neurons containing NFT-like inclusions remain functionally integrated in the brain of a transgenic tau mouse [30]. Additionally, treatment of cells (i.e., cell lines and primary neurons) with preformed oligomers leads to toxicity in a number of model systems, while treatment with filaments does not [31–33]. Tau pathology appears in degenerating axons prior to the appearance of NFTs in the somatodendritic compartment [34] and the presence of NFTs are not harmful to neurons in the absence of continued tau expression in a transgenic mouse model [35].

In addition to AD, there are a number of less prevalent diseases that are characterized by pathological tau aggregations in the absence of amyloid pathology (reviewed in Ref. [36]). Progressive supranuclear palsy (PSP) is a disease presenting vertical gaze palsy and atypical parkinsonism due to degeneration in regions of the basal ganglia, brainstem, and cerebellum [37, 38]. Pick's disease (PiD) is characterized by dementia and aphasia along with neuropathologic rounded tau aggregates, known as Pick bodies, in the cortex and limbic lobe [39]. Corticobasal degeneration (CBD), is associated with focal cortical degeneration leading to asymmetrical rigidity and apraxia [40]. Phosphorylated tau accumulates in neuronal processes and astrocytes primarily located in the cortex and basal ganglia. Chronic traumatic encephalopathy (CTE), a sporadic tauopathy believed to be the result of repeated head traumas, is associated with frontotemporal and HP atrophy along with varying forms of tau aggregates and symptoms including progressive dementia and behavioral changes [41]. FTD with Parkinsonism linked to Chromosome 17 (FTDP-17) are tauopathies caused by inherited mutations in the tau gene that lead to early-onset dementia and brain atrophy, typically occurring in the cortex, basal ganglia, and limbic lobe [25]. In FTDP-17, the tau pathology occurs in a variety of brain regions and can vary in type (i.e., NFTs, Pick bodies, etc.), and the clinical symptoms can include dementia, parkinsonism, and psychosis. Expression of tau containing FTDP-17 mutations is widely used to generate model systems for studying the protein's role in disease due to their aggressive phenotypes [42, 43]. Many of the mutations are simple missense mutations that change a single amino acid or silent mutations that alter mRNA splicing, but these changes are sufficient to induce a neurodegenerative disease. Interestingly, the tau pathology present in different tauopathies can vary in their composition of tau isoforms. For example, the tau pathology in AD and CTE contains all six of the tau isoforms [41], PSP and CBD display aggregates made up primarily of 4R tau isoforms, and PiD contains mainly 3R tau isoforms (reviewed in Ref. [44]).

Our understanding of the molecular mechanisms underlying the degeneration that occurs in AD and other tauopathies is unclear at this point. A number of proposed hypotheses exist, such as dysfunctional neurotransmission, oxidative damage, reduced cytoskeletal integrity, misregulation of important enzymes (e.g., kinases, phosphatases, proteases), protein degradation impairment, ion imbalances, axonal dysfunction (e.g., transport impairment) and inflammation (reviewed in Refs. [45–53]). Indeed, these diseases likely are an amalgamation of many dysfunctional processes and as such each of these avenues warrants further investigation. One of the most critical components of advancing the field and potentially developing treatments are the animal models that recapitulate key aspects of these diseases.

2 Transgenic Animal Models of AD and Tauopathies

Historically, the majority of in vivo models of AD and other tauopathies are transgenic mice engineered to express proteins related to the diseases and these models are reviewed elsewhere [54–56]. Many of these models attempt to recapitulate one or both of the major pathological hallmarks of the disease, Aβ plaques and tau NFTs, by overexpressing mutant proteins that are associated with familial forms of AD and FTDs. Some of the first mouse transgenic models used wild-type or single, double, or triple familial AD mutations in the APP gene to generate Aβ plaques. All of these models displayed amyloid plaques but with relatively limited cognitive impairment and no neuronal loss [57–61]. The trend toward models with more mutations continued with the addition of mutated forms of presenilin 1, a protein involved in the proteolytic processing of APP. A number of lines were generated by crossing mutant APP mice with mutant PS1 mice, and these crosses typically resulted in enhanced levels of the Aβ peptide as well as early development of plaques and cognitive deficits [62–65] but only the 5×FAD mouse that harbors 3 APP mutations and 2 PS1 mutations showed overt neurodegeneration [66]. Interestingly, when P301L tau transgene was added to mutant APP and PS1 to create the so-called triple transgenic line, there was progressive synapse loss, Aβ and tau pathology accumulation, as well as overt neurodegeneration [67]. A recent transgenic rat model further demonstrated a link between amyloid and tau pathologies and recapitulated several key aspects of AD. The TgF344-AD line included APP_{sw} and PS1ΔE9 transgenes that resulted in age-dependent cerebral amyloidosis that precedes tau pathology and loss of HP and cortical neurons resulting in cognitive defects, which are effects not seen when these mutations are present in mice [68, 69]. One important distinction between mice and rats is that rats contain the full complement of all six tau CNS isoforms, while adult mice do not [70, 71]. Thus, based on the importance of tau for toxicity in these models, it is becoming more evident the tau protein plays a key role in these diseases.

Much like the APP and PS1 mouse models, a number of transgenic lines were generated using inherited FTD tau mutations in the transgenes [43, 72–74]. The initial transgenic mice that expressed wild-type human tau resulted in hyperphosphorylation of the protein but lacked the more mature NFT-like inclusions associated with tauopathies [75, 76]. One of most widely used tau mutations in animal modeling is P301L tau. The JNPL3 mice expressing P301L tau presented aging-related cognitive and motor defects that were associated with NFT-like inclusions in the amygdala, hypothalamus, midbrain, and septal nuclei, and loss of motor and spinal cord neurons [43]. P301L tau was also used in the rTg4510 transgenic line that allowed temporal regulation of transgene expression through a tetracycline-induction system [35]. Most recently, transgenic mice, with locally restricted expression of tau, were used to demonstrate propagation of tau pathology from the EC, demonstrating that trans-synaptic spread of tau could be a mechanism for the progression of tau pathology [77, 78]. The tau transgenic mice have helped move the field forward in important ways; however, there are a number of caveats that must be considered when using transgenic models.

3 Potential Caveats Associated with Transgenic Models

Transgenic mouse models have provided insight into mechanisms of AD, and related disorders, and continue to be useful tools. However, despite their popularity, there is a number of potential limitations that must be considered when using transgenic models to study diseases. First, the combination of genes used in some models makes it difficult to clearly determine what is driving the observed effects. It is worth noting that humans do not contain any of the combinations of gene mutations and, in some cases, the combinations include mutations from different diseases in the same mouse (e.g., AD and FTDP-17). Often, these models develop pathology in a number of brain regions that are largely unaffected in the human diseases and overt neurodegeneration occurs only when multiple genes are combined (except in a number of tau transgenic lines). The vast majority of transgene promoters used do not provide enough specificity to target expression in the select neuronal populations affected in the diseases, and the proteins are typically overexpressed. This limits the ability of the models to recapitulate the region-specific onset of pathology that characterizes the human diseases. The human diseases are adult-onset, with age as a primary risk factor for many of them, which is a difficult aspect for most transgenic mouse models to recapitulate because the mutated forms of human proteins are present in the germ-line cells and expressed constitutively. Lifelong expression could give rise to compensatory changes in the animals that make it difficult to extend results to sporadic adult-onset diseases. Finally, an important practical limitation of transgenic models is that they can take a long time to generate and often require significant financial input making them unsuitable for rapidly and inexpensively testing a number of transgenes.

Some of the problems associated with transgenic models are unavoidable in all model systems, and it must be acknowledged that models are capable of approximating only some aspects the diseases. However, the use of complementary model systems can help to address many of the limitations related to the transgenic approach. One such system uses viral vector technology to deliver genetic material to cells in a safe, controlled, and reproducible manner. Current gene therapy technology has opened new possibilities in AD and tauopathy research by providing tools to develop novel models of these diseases.

3.1 Viral Vector Delivery Systems

Viral vectors are a common vehicle to deliver genetic material in non-transgenic animal model systems (reviewed in Refs. [79, 80]). Here, we will focus on recombinant adeno-associated viruses (rAAVs), which are popular among the available viral vectors (comparisons of other viral vectors can be found in Chapter 1). These are small, single-stranded DNA viruses that require a helper virus, such as an adenovirus, for infection and are non-replicative. The viruses can transduce several cell types within the CNS including neurons, astrocytes, and oligodendrocytes depending on the specific viral serotype [81]. Briefly, viruses are internalized in a receptor-dependent fashion and after infection the DNA is transported to the nucleus. The genetic information is replicated to form double-stranded DNA and then transcribed to produce the gene of interest, which can last for extended periods of time (*see* Chapters 1 and 10 for more details on rAAV biology and transduction mechanisms). Importantly, the transduction specificity (i.e., cell-type selectivity), ability to inject into specific brain regions and at specific times in lifespan, long-term gene expression, and lack of eliciting a strong immunogenic response make rAAVs ideal for modeling neurodegenerative diseases. In addition to their potential in basic research, they also show promise as gene transfer therapeutics for neurodegenerative diseases in the CNS.

4 Advantages of Viral Vector Systems

Viral delivery systems hold a number of advantageous characteristics that are difficult to achieve with other approaches. Viral vector systems provide exquisite control over the temporal expression of the gene of interest. AD and other tauopathies are all adult-onset and aging remains the primary risk factor of developing AD. Thus, studies that introduce the production of disease-related genes of interest should incorporate this important variable by expressing the genes in adult or elderly animals. Delivery of viral vectors is completely under the control of the researcher, which easily facilitates studies where animals are transduced at any point in the lifespan. For example, injection of rAAV2/5-GFP and rAAV2/5-human wild-type tau (2N4R) into the HP of young adult (6 months) and old aged (20 months) Fischer 344 rats results in efficient neuronal

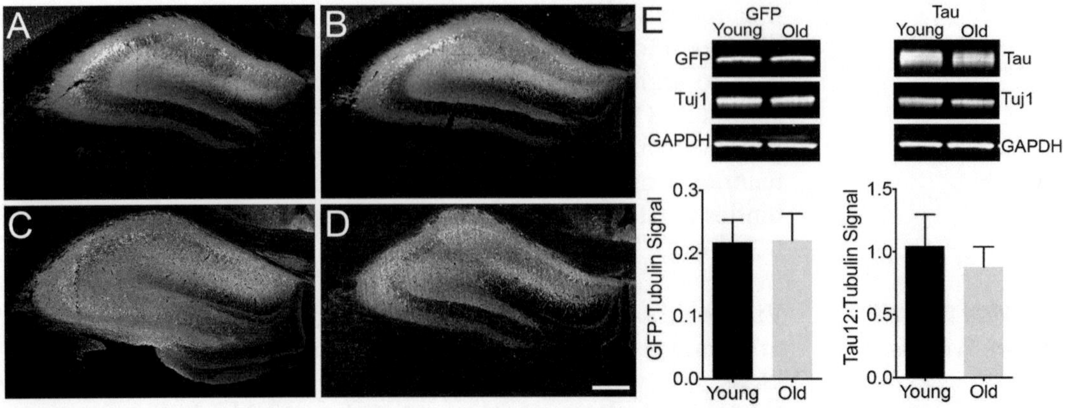

Fig. 1 rAAV2/5 efficiently transduces neurons and produces equal protein expression in the young and aged rat hippocampus (HP). (**a**–**d**) Young adult (**a** and **c**, 6 months, $n = 3$) and old aged (**b** and **d**, 20 months, $n = 3$) Fischer 344 rats were injected with 2 μl (titer 1×10^{13} vg/ml, 0.3 μl/min infusion rate) rAAV-GFP (**a** and **b**) or rAAV-human tau (**c** and **d**) into the dorsal HP (−3.8 mm AP and +2 mm ML from bregma, and −2.6 mm DV from the dura) using pulled glass syringe tips attached to a Hamilton gastight syringe (#7653-01). In all conditions, rAAVs produced widespread transduction of hippocampal neurons. Endogenous GFP signal was used for imaging GFP injected animals, while Tau12 antibody (1:40,000; [144]), a human tau-specific monoclonal antibody, was used to label human tau in rAAV-tau injected animals. Our group has observed efficient transduction of all neuron types in the HP with rAAV2/5 (e.g., CA pyramidal neurons and dentate granule cells). Scale bar = 400 μm. (**e**) Lysates of the dorsal HP were run on a western blot to quantify the amount of rAAV-derived proteins (e.g., GFP and human tau, $n = 3$/group). Blots were probed with GFP antibody (Abcam, ab290, 1:3,000), Tau12 (1:200,000), and two loading controls βIII-tubulin antibody (Tuj1, 1:10,000; [145]) and glyceraldehyde 3-phosphate dehydrogenase (GAPDH, Cell Signaling, 5174, 1:2,000). The signal intensity for each band was quantified using the Licor Image Studio software and signal intensities for GFP or tau are expressed as a ratio to tubulin signal intensities (similar results were obtained when normalized to GAPDH signal). No aging-related differences were observed with either GFP levels (unpaired *t*-test, $p = 0.93$) or human tau levels (unpaired *t*-test, $p = 0.46$). Immunofluorescence and Licor immunoblotting were performed similar to previously published methods [146]

transduction and similar levels of protein expression after 1 month (Fig. 1). Our group recently found that rAAV2/5-GFP transduction in the SN is reduced in aged animals compared to young animals [82], but other studies have shown that rAAV2/9-tau and -GFP transduction is unaffected in the SN [83, 84]. The differences in transduction efficiency with age may reflect the use of different rAAV serotypes. These studies suggest that transduction efficiency in aging animals differs in specific brain regions and with different rAAV serotypes. Virally transduced cells maintain expression of the protein without the addition of other molecules for the remainder of the lifespan. Much like inducible transgenic lines, rAAV-mediated expression can be further regulated if tetracycline regulatory elements are incorporated into the rAAV systems [85].

In addition to great temporal control, viral vectors provide control over the spatial expression of the transduced genes. AD and

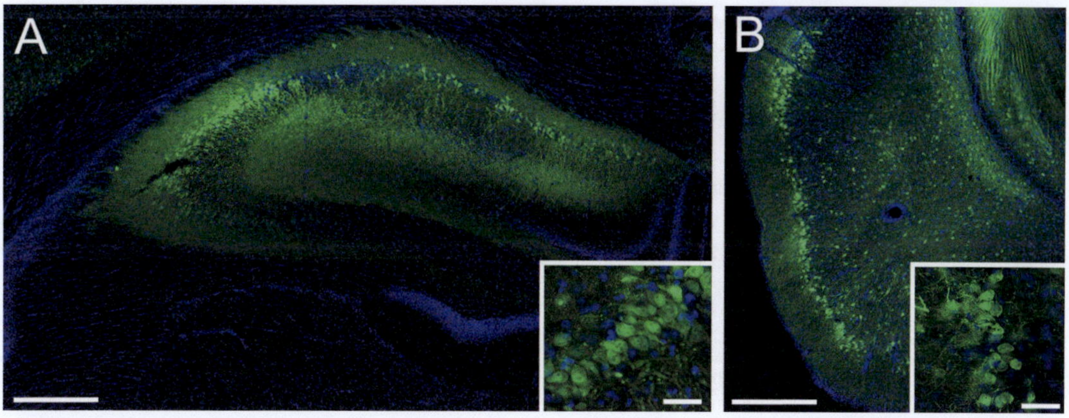

Fig. 2 rAAV2/5 effectively transduces neurons in the hippocampus (HP) and entorhinal cortex (EC) of Fischer 344 rats. (**a**) A representative image of the dorsal HP of a young adult (6 months) rat transduced with rAAV-GFP (as in Fig. 1, scale bar = 400 μm). The *inset* illustrates GFP expressing CA3 pyramidal neurons in the HP (scale bar = 50 μm). (**b**) A representative image of the EC of a young (3 months) rat transduced with rAAV-GFP (titer = 1.9×10^{13}, coordinates: −5.8 mm AP, +3.3 mm ML from bregma, −8.8 mm DV from dura, at 17° angle; scale bar = 400 μm). Note that layer II/III EC neurons are well transduced (*inset*, scale bar = 50 μm); these neurons are particularly susceptible in AD. Endogenous GFP fluorescence was used for imaging transduced cells

other tauopathies are characterized by the degeneration and pathological accumulation of proteins in specific brain regions. For example, the EC and HP are primary affected areas in AD, while other tauopathies involve degeneration in the frontal and temporal cortices, as well as the basal ganglia, brainstem and cerebellum. Viral vectors allow researchers to stereotaxically inject viruses in relatively discrete brain regions of interest. For example, direct injection of rAAVs into the rat HP or EC results in efficient transduction of neurons (Fig. 2). Furthermore, unilateral injections allow the contralateral half of the brain to serve as a control within the same animal, but the contralateral projections of a specific region must be considered. Another level of specificity can be obtained by using cell-specific promoter systems [86, 87], which can also be used in transgenic models [78, 88]. Finally, the serotypes of AAVs exhibit significant differences in cell tropism allowing for increased specificity of transducing different cell types (e.g., neurons, glia, or both) and specific brain regions [89, 90].

Additional advantages of viral vector systems are the relative ease and short timeframe (i.e., a few months) of generating new viral vectors as well as the low cost of production. Typically, the process involves simply cloning the gene of interest into the appropriate rAAV expression plasmid, and then, in the case of many investigators, sending the expression plasmids to a reputable gene therapy core facility for production. Other investigators choose to generate their own viral vectors ([91] and Chapters 7–9).

Thus, viral vector models can be adjusted relatively quickly to test new variants of a particular protein or to examine interactions between proteins. Alternately, viruses can introduce RNA-based inhibitors of gene expression (e.g., shRNA or microRNAs) in a time- and region-specific manner for a relatively simple and selective approach to genetic knockdown [92]. This level of versatility lends itself well to creating numerous novel model systems and easily manipulating the genes/proteins of interest, which is difficult to achieve with other approaches.

5 Potential Caveats with Viral Vector Systems

Every model system has some limitations that are important to acknowledge. rAAVs efficiently transduce cells in the brain, but the level of overexpression can be significant. However, two approaches can help minimize this effect. First, rAAVs can be titered to effectively reduce the copy number per cell, thereby reducing the amount of expression. For example, injection of 2 μl of rAAV-GFP or rAAV-human wild-type tau at a titer of 1×10^{13}, 1×10^{12} or 1×10^{11} viral genomes/ml into the dorsal HP of rats (as above) results in a dose-response of expression levels. The regional HP signal for GFP or human tau levels (using Licor Odyssey densitometry analysis) decreased linearly (GFP – $r^2 = 0.92$, $p < 0.001$; human tau – $r^2 = 0.93$ $p < 0.001$; Fig. 3a, b). Analysis of tau fluorescence levels in individual cells found that the average level of tau expression was approximately ninefold, fourfold or twofold over endogenous tau levels at titers of 1×10^{13}, 1×10^{12} or 1×10^{11}, respectively (Fig. 3c). Second, regulatable promoter systems, such as tetracycline induction systems, provide an effective level of control over transgene expression [85] but leakiness is always a concern with these approaches (as with regulatable transgenic animals). Nonetheless, it is imperative that investigators assess the level of overexpression when using any model system.

Fig. 3 (continued) within individual cells was 1–3 for 1×10^{11}, 2–14 for 1×10^{12}, and 2–17 for 1×10^{13}. Each increase in titer produces a significant increase in signal (one-way ANOVA, $^*p < 0.05$ vs. other groups). Of note, this method can only approximate relative expression levels and does not allow precise quantitative measures of protein levels in each cell. Also, methods that do not allow single-cell analysis (e.g., western blotting or ELISAs) are inappropriate for estimating overexpression levels as the samples contain significant quantities of non-transduced cells/tissue. Finally, the reduction in viral titer reduces expression levels, but also decreases the amount of transduced neurons. Thus, one must balance the level of overexpression with transducing an adequate number of cells

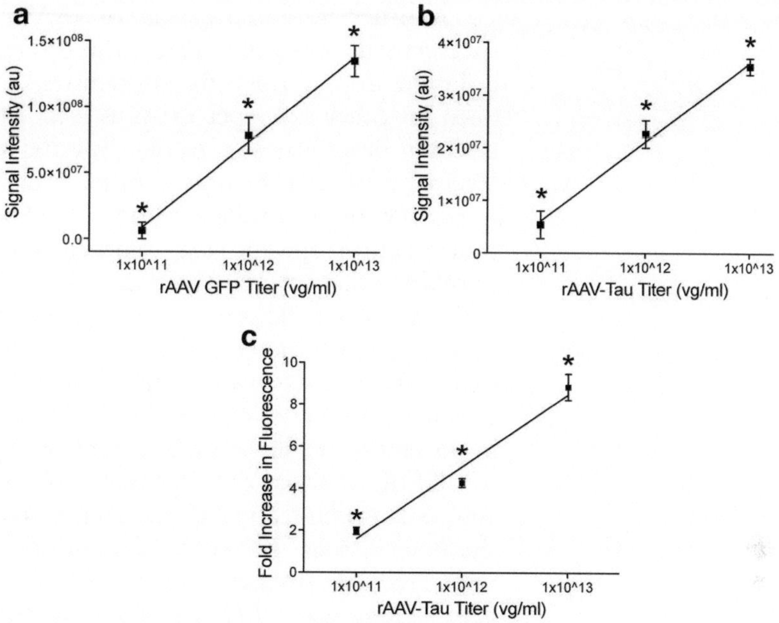

Fig. 3 rAAV2/5 titer-dependent changes in levels of transgene expression and fold-increase in expression are linear in the rat hippocampus (HP). (**a**) Young Fischer 344 rats (3 months, $n = 3$/group) were injected in the dorsal HP with rAAV-GFP at 1×10^{11}, 1×10^{12}, or 1×10^{13} vg/ml (as described in Fig. 1). After 1 month of expression, the tissue was processed for GFP immunofluorescence using GFP antibody (1:28,000) and Licor goat anti-rabbit IRDye 680 (1:500). The fluorescence signal intensity was quantified using Licor Odyssey densitometry analysis in the dorsal hippocampus. Increasing the titer produced a linear increase ($r^2 = 0.92$, $p < 0.001$) in regional GFP signal intensity. Each increase in titer produces a significant increase in signal (one-way ANOVA, $^*p < 0.05$ vs. other groups). (**b**) Using the same parameters, animals ($n = 3$/group) were injected with rAAV-human tau at the three titer doses. Again, a linear increase ($r^2 = 0.93$, $p < 0.001$) in regional human tau (Tau12; 1:40,000, Licor goat anti-mouse 680) signal intensity was observed with increasing titers (using Licor densitometry). Each increase in titer produces a significant increase in signal (one-way ANOVA, $^*p < 0.05$ vs. other groups). (**c**) Sections from rAAV-human tau animals were stained with dual immunofluorescence for Tau5 (a mouse monoclonal pan-tau antibody that labels human and rodent tau equally, 1:50,000; [147]) and a rabbit anti-flag tag antibody (Sigma, F7425, 1:2,000) to obtain estimations of the level of tau overexpression on an individual cell basis. The human tau proteins are flag tagged, which is a small epitope tag (DYKDDDDK), to facilitate distinction of the human tau from endogenous rat tau. Transduced cells were identified with the flag tag staining and levels of Tau5 fluorescence were measured in the cytoplasm of individual neurons. Normal physiological levels of endogenous rat tau were measured in non-transduced cells (i.e., flag tag negative) in the contralateral hemisphere. The data are presented as the fold-increase in fluorescence intensity (i.e., tau protein levels) from non-transduced cells. As with the regional analyses, a linear increase in fluorescence was observed with increasing titers ($r^2 = 0.93$, $p < 0.001$). The range of the fold-increase in expression

Another potential caveat of rAAVs is the lack of absolute specificity in cell-type transduction. Preferential tropism of certain cell types does occur with different rAAV serotypes, but some tropism for other cell types exists as well. For example, rAAV 2/5 is typically described as a neuron-specific serotype, however, a low level of glial transduction is often observed. One mechanism to overcome this is to utilize cell-type specific promoters. Viruses that utilize neuron-specific (e.g., synapsin I) or glial-specific (e.g., glial fibrillary acidic protein) promoters can help address this caveat [87, 93]. Moreover, cell phenotype-specific markers (e.g., tyrosine hydroxylase promoter for monoaminergic neurons) can increase specificity to neuronal subpopulations. As the use of rAAVs becomes more common in the AD and tauopathy field, it will be important to pursue these more refined approaches.

Finally, delivery of rAAV into the brain requires surgical interventions, which lead to disruption of the blood–brain barrier and the brain tissues. To reduce disruption, it is advisable to perform the stereotaxic surgery with the aid of a surgical scope and pulled glass syringe tips. Often, this allows minimal disruption of the dura and in some cases the only disruption is the very fine point of the inserting syringe tip. Details on the surgical procedures used for delivering rAAVs intracerebrally can be found in Chapter 14. Additionally, recent evidence suggests that rAAV 2/9 can effectively enter the brain after a peripheral injection, thereby, circumventing the need for surgery [94].

6 AD and Other Tauopathy Models Using rAAV Tau

The initial rAAV tauopathy models used rAAV-2/2 to express full-length human tau with the P301L mutation in the brains of mice and rats [43, 95]. rAAV-P301L tau was injected into the basal forebrain of adult rats leading to detection of increased levels of the tau protein in the brain for at least 8 months after injection. Within 3–4 weeks hyperphosphorylated tau was found and aggregates resembling NFTs were present although in relatively low numbers. These studies provided the first proof-of-principle that injection of rAAV-P301L tau could result in persistent expression of tau protein and tau aggregation in rodents.

Injection of rAAV-wild-type or triple-mutant APP led to some Aβ plaques but no overt neurodegeneration. In contrast, rAAV-wild-type tau or P301L tau caused significant neurodegeneration of HP pyramidal neurons, but tangles were not present [96]. The tau-induced neurodegeneration was dependent on the microtubule-binding domains, as a truncated version of the protein did not induce toxicity. Interestingly, P301L tau apparently induced cell cycle reentry as indicated by an increase in the presence of cell cycle markers (e.g., cyclins B1 and D2, proliferating cell nuclear antigen,

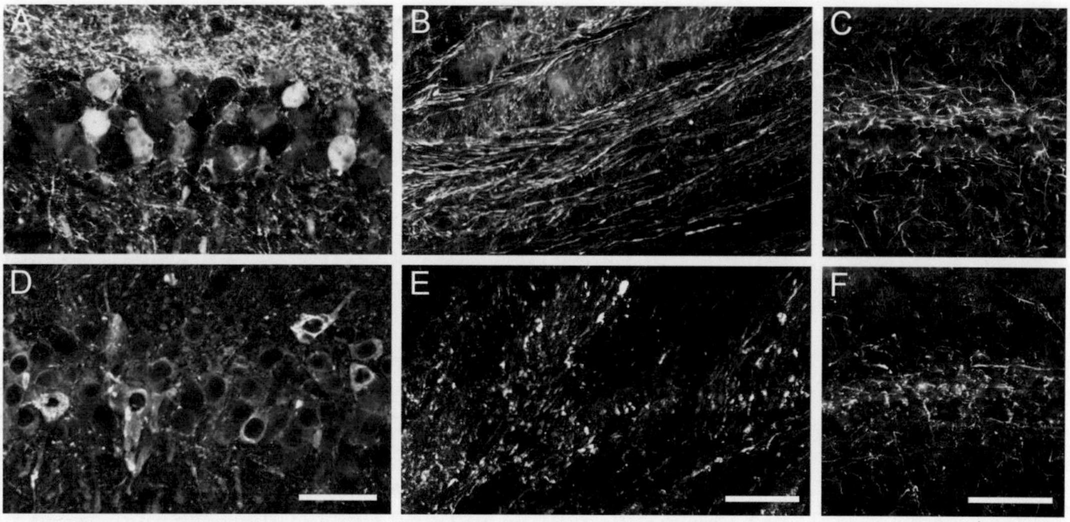

Fig. 4 Expression of human tau using rAAV2/5 causes axonal degeneration in the aged rat hippocampus (HP). Aged Fischer 344 rats (20 months, $n=3$/group) were injected in the dorsal HP (as described in Fig. 1) with either rAAV-GFP (**a**–**c**) or rAAV-human tau (**d**–**f**). (**a**) and (**d**) The CA1 pyramidal neurons were well transduced in both cases (**a**—GFP fluorescence and **d**—Tau12 immunofluorescence). (**b**) and (**e**) Examination of the HP-derived axons in the fimbria revealed that GFP+ axons appeared normal and continuous (**b**), while human tau + axons showed clear signs of degeneration (e.g., fragmentation and dystrophy). (**c**) and (**f**) SMI312, an axon-specific neurofilament monoclonal antibody cocktail (Covance, SMI-312R, 1:10,000) revealed an apparent reduction and fragmentation of axons in the HP (CA1 region depicted) of aged animals expressing human tau (**f**) when compared to those expressing GFP (**e**). Scale bars in **d**, **e,** and **f** = 50 μm. Of note, overt neuron loss in the HP was not assessed in this study

phosphorylated retinoblastoma protein). However, the neurodegenerative effects of injecting rAAV-wild-type tau into the dorsal HP were not replicated in another study using rats [97]. Differences in species, age at rAAV delivery, duration of expression, viral titer and/or neuroinflammatory response may help to explain the discordant findings between these studies.

In a recent experiment, our group used rAAV2/5 to express GFP (used as a control rAAV) or wild-type human tau (longest 4R isoform of 441 amino acids) in the dorsal HP of old Fischer 344 rats (20 months) for 1 month (Fig. 4). Both constructs were well expressed by HP neurons (CA1 pyramidal neurons depicted). Notably, there were clear signs of axonal degeneration, such as fragmentation of tau expressing HP axons in the fimbria, while GFP expressing axons appeared normal. Moreover, the pattern of HP staining with SMI312 antibody, an axon-specific neurofilament antibody cocktail, showed an apparent reduction in axons and the remaining axons appeared dystrophic and fragmented. These data support the usefulness of using rAAV approaches to model early pathological changes such as axonal degeneration and highlight the ability to use aged rats in such studies.

The perforant pathway projects from the EC to the HP and is associated with memory formation [98]. It is also the location of some of the earliest tau pathology in AD and among the regions that undergo significant neurodegeneration [99]. rAAV vectors provide an opportunity to model the early stages of AD and the degeneration associated with the perforant pathway [100]. Expression of rAAV2/9-P301L under the control of a synapsin I promoter produced localized tau expression in the lateral EC and perforant pathway. Multiple forms of phosphorylated and aggregated tau were detected in this region by antibodies and silver staining prior to loss of perforant synapses and neuronal death, seemingly triggered through caspase-mediated apoptosis. HP tau expression was initially present only in the perforant pathway projections but was later observed in the dentate granule neurons as well as in target neurons in CA3. This progressive expression was not replicated in rAAV-eGFP controls and may support trans-synaptic spread of tau [100]. We had similar success in efficiently transducing EC neurons in young (3–4 months) Fischer 344 rats using rAAV-2/5. In this experiment, rAAV-wild-type human tau and a phosphomimetic form of AT8 tau (i.e., pseudophosphorylations at S199, S202, and T205, rAAV-psAT8) was injected into the EC of young animals. AT8 tau is a prominent disease-related phosphoepitope of tau and recombinant psAT8 tau significantly impaired anterograde axonal transport as a monomer in the squid axoplasm, an effect not seen with wild-type tau monomers in this assay [19]. Notably, expression of GFP or wild-type tau for 1 month did not appear to alter perforant pathway axons in young animals (Fig. 5a), but AT8 tau expression induced early signs of axonal degeneration (e.g., spheroids and dystrophic axons) in perforant pathway axons (Fig. 5b). One notable caveat of directly injecting the rodent EC is that the syringe tip must pass through the HP (at an angle). Thus, this approach is not well-suited for studying cell-to-cell transfer or the spread of tau because of the likelihood for injected materials to flow back along the needle tract. Nonetheless, the EC can be targeted and overexpression of disease-related forms of tau (i.e., psAT8) appear to induce axonopathy in young animals. Further investigations that study the long-term and aging-related effects of rAAV tau expression in the EC should be pursued.

Fig. 5 (continued) layer of the dentate gyrus (*arrowheads* in **d–f**), which is a terminal field of the axonal projections from the EC. (**g–i**) Notably, axons within the perforant pathway appeared normal in GFP and human tau expressing animals, but in psAT8 tau expressing animals axons appeared dystrophic (*arrowheads*; i.e., swollen and containing spheroids). Endogenous GFP fluorescence (**a**, **d**, **g**) or Tau12 immunofluorescence (**b**, **c**, **e**, **f**, **h**, **i**) were used to identify rAAV-GFP or rAAV-human tau and rAAV-psAT8, respectively. (**j**) Western blot was used to confirm the expression of rAAV-GFP, rAAV-human tau and rAAV-AT8 tau in the EC (+ samples are the injected side, – samples are the contralateral HP). Immunoblotting antibodies were used as described in Fig. 1, and Tau5 was used at 1:100,000. Overt neurodegeneration was not assessed in this study

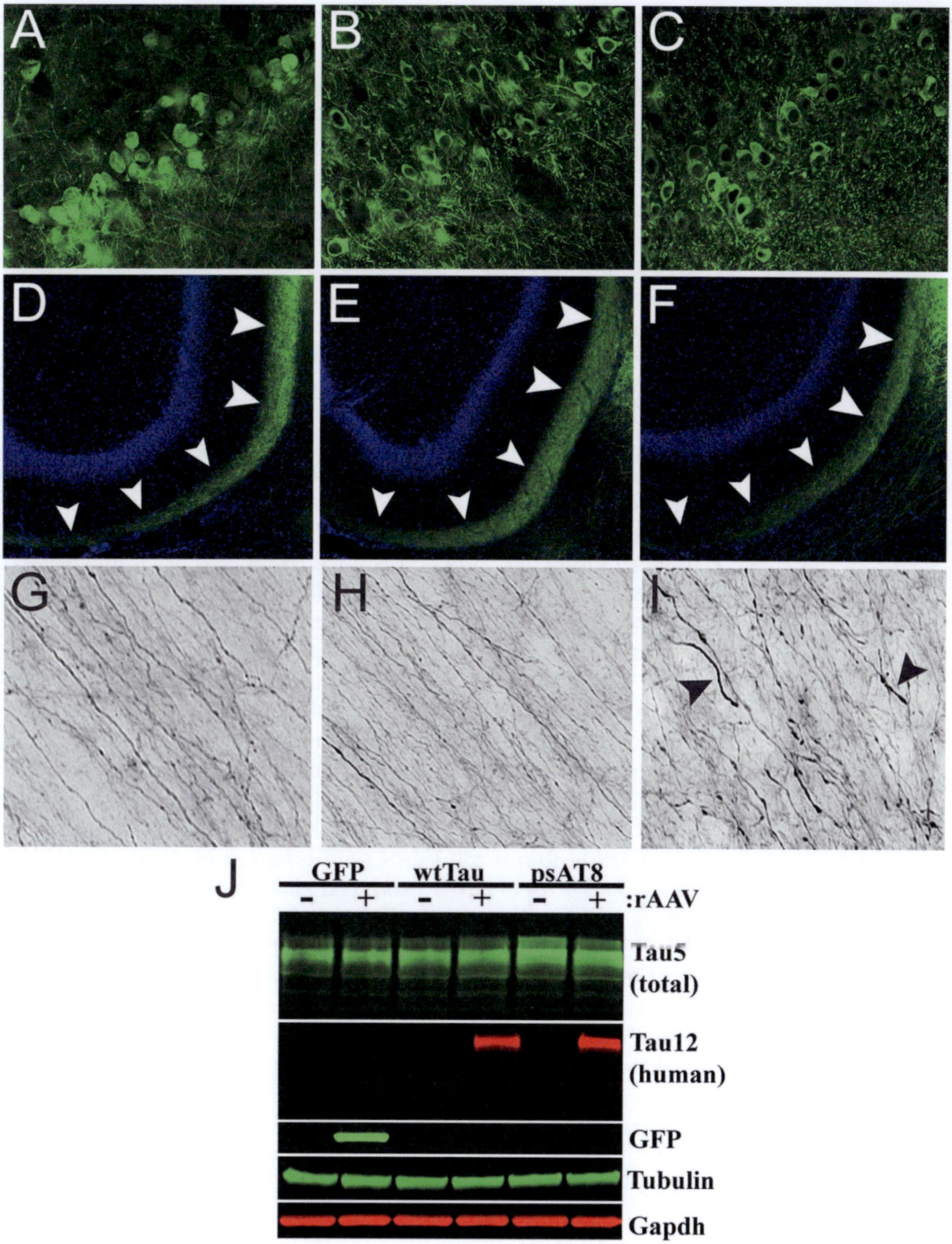

Fig. 5 Expression of a disease-related phosphomimetic form of human tau in the entorhinal cortex (EC) of young rats causes axonal abnormalities in the perforant pathway. Young Fischer 344 rats were injected with rAAV-GFP (**a**, **d**, **g**), rAAV-human tau (**b**, **e**, **h**), or rAAV human tau pseudophosphorylated at the AT8 site (psAT8—S199, S202 and T205) into the EC (**c**, **f**, **i**; as described in Fig. 2; titer = 1.9×10^{13} for all). (**a**–**f**) After 1 month, pyramidal neurons of the EC (**a**–**c**) were transduced and proteins were present in the outer molecular

7 Tau in the Substantia Nigra

The presence of tau pathology and neuronal loss is observed in the SN in many tauopathies, including PSP, CBD, FTDP-17, as well as AD (around Braak Stage VI) and Parkinson's disease [101–104]. rAAV2/2-mediated expression of wild-type or P301L tau in the SN of rat caused early axonal degeneration that was followed by neuronal loss [105]. Accordingly, the animals displayed motor deficits and a directional bias in amphetamine-induced rotational behavior. These behavioral changes occurred in the absence of large numbers of NFT-like inclusions, indicating that tangle-like inclusions may not be required for neuronal dysfunction. In another study, a number of different rAAV serotypes (i.e., rAAV2/2, rAAV2/5, rAAV2/8, rAAV2/9, and rAAV2/10) were tested with P301L tau in the SN [106, 107]. rAAV2/8 P301L induced high tau expression and a loss of approximately 75 % of TH-positive neurons in the SN, while rAAV2/2 and rAAV2/5 induced losses of less than 30 % [106]. rAAV2/9 and rAAV2/10 induced high tau expression and led to the loss of around 90 % of dopaminergic neurons [107]. Interestingly, GFP expression through rAAV2/8 transduction was toxic in the SN at high doses but not in the HP or when expressed through rAAV2/9 [106, 107]. These studies highlight the importance of choosing appropriate control proteins and potential issues with nonspecific toxicity due to dosing effects. Given the losses of 75–90 % of dopaminergic neurons in the SN, this may be of use as a model for some tauopathies that display that particular phenotype, such as PSP, CBD, or FTDP-17 [37, 108, 109].

A study found rAAV2/9 wild-type tau or GFP (at titers of either 2×10^9 or 9×10^9) produced similar levels of expression in young and aged rats in the SN [83, 84]. rAAV-tau expression in the SN caused prolonged microglial activation, neuronal loss and amphetamine-induced rotational behavior in the young and old animals. Interestingly, this effect was enhanced in older rats at lower titers suggesting aged animals are more susceptible to tau toxicity. In fact, expression of wild-type tau with rAAV2/9 induced an upregulation of pro-inflammatory markers and reduced tyrosine hydroxylase mRNA due to loss of dopamine neurons [110], suggesting neuroinflammation might play an important role in tau toxicity in the SN.

8 Mixed Models Using rAAV Tau and Transgenic Mice

In addition to developing novel models, viral transduction can help to improve on existing transgenic models by expressing proteins that introduce or enhance pathology. For example, the PS/APP double transgenic mouse develops amyloid pathology, but only low levels of tau pathology [63]. However, injection of rAAV-P301L tau into the HP of these mice produced moderate tau pathology including early stage tau aggregates and tau-positive neurites surrounding amyloid deposits [95]. Similarly, injection of rAAV2/2-P301L tau into the HP of the triple-mutant APP mouse line, mThy1-hAPP751, resulted in increased phosphorylated tau and tau fibrillar pathology compared to wild-type mice injected with rAAV-P301L tau [58, 111]. Moreover, rAAV-P301L tau induced neuronal loss in the transgenic mice that was not seen in either the wild-type mice or transgenic mice injected with rAAV-GFP, an effect that was attenuated through treatment with a kinase inhibitor [111]. These studies highlight the benefits of combining transgenic mice with rAAV injections to study disease mechanisms and potential therapeutics.

In another series of studies, the TASTPM transgenic mouse line (an APP/PS1 mouse) [112, 113] was injected in the EC with rAAV6 containing wild-type tau, P301S mutant tau, or a modified, enhanced aggregating version of tau named 3PO [114, 115]. The pro-aggregation forms of tau, P301S and 3PO tau, enhanced toxicity around the injection site as well as in CA1 HP neurons and exacerbated microglial activation when compared to wild-type tau. Importantly, the expression of rAAV P301S or 3PO tau resulted in impaired spatial memory. This confirmed previous results demonstrating that transduction of mutant tau into the EC impaired spatial memory of mice [116].

Combining different transgenic lines with rAAV approaches can add novel capability to studying specific processes related to a disease. For example, yellow fluorescent protein transgenic mice were injected with rAAV-P301L tau into the HP, providing a model for tracking axonal degeneration and neuronal loss [117]. In the same study, rAAV-P301L was injected into the HP of transgenic mice harboring the CX3C chemokine receptor 1 gene disrupted with EGFP (produces mice with GFP+ microglia). This study confirmed early degeneration of neuronal processes (i.e., axons and dendrites) prior to overt neuronal degeneration and activation of microglia was detected after the induction of tau expression. These results supported a toxic role for tau prior to the formation of NFT-like aggregates as well as a neurotoxic effect related to microglia.

9 rAAV-Mediated Expression of Other AD- and Dementia-Related Proteins

A number of groups have used rAAVs to model other pathological aspects of AD. For example, rAAVs were used to overexpress Aβ peptides in rodents [118]. Expression of rAAV2/1-BRI-Aβ42, a fusion protein that allows controlled expression and secretion of

Aβ peptides [119, 120], in the HP induced the formation of plaques but this was not seen with rAAV2/1-BRI-Aβ40 expression. Another amyloid-based rAAV study generated GFP-tagged versions of Aβ40 and Aβ42, wild-type C100 (a C-terminal fragment of APP containing the Aβ peptides) and a V717F mutant version of C100 and expressed them in the HP and cerebellum of mice using rAAV2/2 [121]. The Aβ42 and V717F C100 led to greater induction of microgliosis and disruption of the blood–brain barrier compared to the Aβ40 forms but did not induce plaque formation, neurodegeneration or astrocyte activation. The use of rAAV technology to express various forms of Aβ and its precursor proteins will continue to be an important method of studying their role in disease.

Similarly, rAAVs were used to induce expression or alteration of other proteins thought to play a role in the mechanisms of AD. Injection of rAAV1-I_{2NTF} and rAAV1-I_{2CTF}, active cleavage products of an inhibitor of protein phosphatase-2A (PP2A), led to phosphorylation and aggregation of tau and Aβ, synaptic defects, and impaired memory in rats [122]. This is an intriguing demonstration of induction of some of the pathological hallmarks of AD without overexpressing mutant forms of tau or APP. Another novel method used rAAV1/2 to express 5-Lipoxygenase (5-LO) in the brains of transgenic APP (Tg2576) or 3×Tg-AD mice resulting in impaired performance in memory tasks and exacerbated Aβ and tau pathology [123, 124]. 5-LO in an enzyme that oxidizes lipids, such as arachidonic acid, releasing several active compounds into the cell and displays increased levels in the HP of AD patients [125].

Some FTDs are characterized by pathological aggregation of TDP-43, a nucleic acid-binding protein, primarily in the frontotemporal cortex and HP (reviewed in Ref. [126]). Overexpression of rAAV2/9-TDP-43 in the SN of rats caused amphetamine-induced asymmetrical rotational behavior, loss of TH-positive neurons, and the development of cytoplasmic TDP-43 aggregates [127]. Similarly, rAAV2/9-TDP-43 expression in the rat HP caused neuronal loss and some memory impairments [97]. Another study took advantage of the flexible nature of rAAVs to address differential toxicities and pathologies due to expression of mutant TDP-43 in neurons and glia of mice [128]. The synapsin 1 and GFAP promoters limited expression of the M337V TDP-43 to neurons and astrocytes, respectively. Neuronal expression resulted in greater TDP-43 aggregation, neuronal loss, and more severe motor defects in the mice. Astrocytic expression also induced neuronal death but only in older animals and without motor defects. These studies demonstrate that rAAV models are a versatile approach to studying the role of pathological proteins as well as the molecular mechanisms of disease, and may help make important distinctions between different cell-specific pathologies.

10 Lentiviral Models

Other models of AD and related neurodegenerative disorders utilize lentiviral vectors, a retrovirus, to express the desired protein as an alternative to rAAV. These differ from rAAVs because they insert themselves into the host genome and have a capacity for larger genetic material. Some of the AD and tauopathy models developed using lentiviral systems are discussed briefly below.

A number of studies used lentiviruses to express wild-type or mutant forms of tau protein in various brain regions of rodents. Lentivirus-mediated expression of P301S tau in the HP of wild-type or transgenic APP23 mice (harboring 2 Swedish APP mutations) was used to examine interactions between tau and Aβ in adult mice [129]. Based on immunostaining, the APP23 transgenic mice displayed increased tau phosphorylation and aggregation compared to the wild-type mice also injected with lenti-P301S. Another group found that P301L tau expression caused an increase in tau phosphorylation, tau aggregation, and activated microglia when compared to wild-type tau expression in the rat motor cortex rats [130]. Lentivirus expression of wild-type or P301L tau in the rat HP produced a progressive appearance of dystrophic and fragmented neurites prior to the formation of NFT-like inclusions that developed over 8 months [131]. Lentiviral expression of wild-type tau, P301L, and Aβ42 led to microglial activation as well as induction of autophagy as fractalkine signaling was increased [132]. This indicates a possible link between the phosphorylated tau, microglia signaling, and induction of autophagy that could be related to apoptotic death of neurons. Finally, lentiviral Aβ42 led to intracellular Aβ accumulation followed by tau hyperphosphorylation that was ameliorated with clearance of the peptide [133].

Lentiviruses were used to express TDP-43, Aβ42, or both together in the motor cortex of rats [134]. Lenti-TDP-43 induced caspase activity, neuroinflammation, and altered the normal processing of APP, which led to increased Aβ42 production, while lenti-Aβ42 induced some caspase-3 activity and neuroinflammation. Interestingly, the combined expression of TDP-43 and Aβ42 produced pathology similar to TDP-43 alone but overt neuronal loss occurred, indicating a possible synergistic effect between the two proteins. Another study using lentiviruses to express TDP-43 in the motor cortex found altered amino acid metabolism, oxidative stress, and neuronal death [135].

The studies described above demonstrate the versatility and utility of using rAAV and lentiviruses to produce models of neurodegenerative diseases. These tools can be used alone or in conjunction with existing transgenic models to help address potential pitfalls of previous modeling techniques. These valuable techniques will continue to provide insight into neurodegenerative diseases by allowing elucidation of mechanisms and providing platforms to explore new therapeutics.

11 Possible Therapeutic Uses of Viral Vectors

Currently, there are no treatments that effectively slow or prevent AD and other tauopathies. While the precise mechanisms of degeneration are unknown, loss of cholinergic neurons in the basal forebrain and excitotoxicity are believed to contribute to the development of AD [136, 137]. Accordingly, cholinesterase inhibitors (donepezil, rivastigmine, tacrine, and galantamine), which reduce acetylcholine breakdown, are used to treat mild to moderate AD. In addition, a noncompetitive NMDA receptor antagonist (memantine) is prescribed for moderate to severe disease stages [138]. However, neither of these drugs slow or prevent disease progression and only provide temporary symptomatic improvement for some patients [139]. Thus, there is an urgent need to aggressively develop novel therapeutics for treating AD and other tauopathies.

Nerve growth factor (NGF) is an appealing treatment because it potently protects cholinergic neurons and improves the function of basal forebrain neurons [139, 140]. The delivery of NGF to the nucleus basalis of Meynert was approached by direct infusion into the brain, implantation of cell grafts of NGF-secreting fibroblasts, and viral-vector delivery using adeno-associated virus. NGF injection into the lateral cerebral ventricle led to pain syndrome and weight loss with no obvious cognitive improvement in patients [141]. The delivery of NGF via fibroblast provided some therapeutic benefits in nonhuman primates, but the complexity of manufacturing fibroblasts and the diminished protein expression after 18 months makes it less desirable [142]. Using a rAAV vector to deliver NGF to the target region of the forebrain eliminates the hurdles of other NGF delivery methods. A phase I clinical trial (NCT00087789) was conducted using CERE-110 for the treatment of AD patients. The CERE-110 viral vector is an AAV2/2 vector containing full length NGF transgene under the control of the CMV/β-actin promoter and the human polyadenylation signal [143]. There were few adverse effects of the treatment reported, and all were deemed either not related or unlikely related to the rAAV-NGF treatment [139]. The phase I trial confirmed that viral vector therapy was well tolerated with a high level of safety and no systemic toxicity. The success of the phase I trial led to a phase II clinical trial using CERE-110 and results are expected in 2015. As our knowledge of disease mechanisms continues to grow and novel therapies are developed, gene therapy must continue to be pursued as a viable approach for therapeutic interventions in AD and other tauopathies.

Acknowledgements

This work was supported by NIH/NIA grant R01AG044372 (N.M.K.).

References

1. Braak H, Braak E (1991) Neuropathological stageing of Alzheimer-related changes. Acta Neuropathol 82(4):239–259
2. Masters CL, Simms G, Weinman NA, Multhaup G, McDonald BL, Beyreuther K (1985) Amyloid plaque core protein in Alzheimer disease and Down syndrome. Proc Natl Acad Sci U S A 82(12):4245–4249
3. Goate A, Chartier-Harlin MC, Mullan M, Brown J, Crawford F, Fidani L, Giuffra L, Haynes A, Irving N, James L et al (1991) Segregation of a missense mutation in the amyloid precursor protein gene with familial Alzheimer's disease [see comments]. Nature 349(6311):704–706
4. Rogaev EI, Sherrington R, Rogaeva EA, Levesque G, Ikeda M, Liang Y, Chi H, Lin C, Holman K, Tsuda T et al (1995) Familial Alzheimer's disease in kindreds with missense mutations in a gene on chromosome 1 related to the Alzheimer's disease type 3 gene. Nature 376(6543):775–778. doi:10.1038/376775a0
5. Greenfield JP, Tsai J, Gouras GK, Hai B, Thinakaran G, Checler F, Sisodia SS, Greengard P, Xu H (1999) Endoplasmic reticulum and trans-Golgi network generate distinct populations of Alzheimer beta-amyloid peptides. Proc Natl Acad Sci U S A 96(2):742–747
6. Gouras GK, Tsai J, Naslund J, Vincent B, Edgar M, Checler F, Greenfield JP, Haroutunian V, Buxbaum JD, Xu H, Greengard P, Relkin NR (2000) Intraneuronal Abeta42 accumulation in human brain. Am J Pathol 156(1):15–20
7. Arriagada PV, Growdon JH, Hedley-Whyte ET, Hyman BT (1992) Neurofibrillary tangles but not senile plaques parallel duration and severity of Alzheimer's disease. Neurology 42(3 Pt 1):631–639
8. Duering M, Grimm MO, Grimm HS, Schroder J, Hartmann T (2005) Mean age of onset in familial Alzheimer's disease is determined by amyloid beta 42. Neurobiol Aging 26(6):785–788. doi:10.1016/j.neurobiolaging.2004.08.002
9. Deshpande A, Mina E, Glabe C, Busciglio J (2006) Different conformations of amyloid beta induce neurotoxicity by distinct mechanisms in human cortical neurons. J Neurosci 26(22):6011–6018. doi:10.1523/JNEUROSCI.1189-06.2006
10. Shankar GM, Bloodgood BL, Townsend M, Walsh DM, Selkoe DJ, Sabatini BL (2007) Natural oligomers of the Alzheimer amyloid-beta protein induce reversible synapse loss by modulating an NMDA-type glutamate receptor-dependent signaling pathway. J Neurosci 27(11):2866–2875. doi:10.1523/JNEUROSCI.4970-06.2007
11. Marin N, Romero B, Bosch-Morell F, Llansola M, Felipo V, Roma J, Romero FJ (2000) Beta-amyloid-induced activation of caspase-3 in primary cultures of rat neurons. Mech Ageing Dev 119(1–2):63–67
12. Lashuel HA, Hartley D, Petre BM, Walz T, Lansbury PT Jr (2002) Neurodegenerative disease: amyloid pores from pathogenic mutations. Nature 418(6895):291. doi:10.1038/418291a
13. Pigino G, Morfini G, Atagi Y, Deshpande A, Yu C, Jungbauer L, LaDu M, Busciglio J, Brady S (2009) Disruption of fast axonal transport is a pathogenic mechanism for intraneuronal amyloid beta. Proc Natl Acad Sci U S A 106(14):5907–5912. doi:10.1073/pnas.0901229106
14. Rapoport M, Dawson HN, Binder LI, Vitek MP, Ferreira A (2002) Tau is essential to beta-amyloid-induced neurotoxicity. Proc Natl Acad Sci U S A 99(9):6364–6369. doi:10.1073/pnas.092136199, 092136199 [pii]
15. Roberson ED, Scearce-Levie K, Palop JJ, Yan F, Cheng IH, Wu T, Gerstein H, Yu GQ, Mucke L (2007) Reducing endogenous tau ameliorates amyloid beta-induced deficits in an Alzheimer's disease mouse model. Science 316(5825):750–754. doi:10.1126/science.1141736, 316/5825/750 [pii]
16. Ittner LM, Ke YD, Delerue F, Bi M, Gladbach A, van Eersel J, Wolfing H, Chieng BC, Christie MJ, Napier IA, Eckert A, Staufenbiel M, Hardeman E, Gotz J (2010) Dendritic function of tau mediates amyloid-beta toxicity in Alzheimer's disease mouse models. Cell 142(3):387–397. doi:10.1016/j.cell.2010.06.036, S0092-8674(10)00726-9 [pii]

17. Drubin DG, Kirschner MW (1986) Tau protein function in living cells. J Cell Biol 103(6 Pt 2):2739–2746
18. Chen J, Kanai Y, Cowan NJ, Hirokawa N (1992) Projection domains of MAP2 and tau determine spacings between microtubules in dendrites and axons. Nature 360(6405):674–677. doi:10.1038/360674a0
19. Kanaan NM, Morfini GA, LaPointe NE, Pigino GF, Patterson KR, Song Y, Andreadis A, Fu Y, Brady ST, Binder LI (2011) Pathogenic forms of tau inhibit kinesin-dependent axonal transport through a mechanism involving activation of axonal phosphotransferases. J Neurosci 31(27):9858–9868. doi:10.1523/JNEUROSCI.0560-11.2011
20. Lee G, Newman ST, Gard DL, Band H, Panchamoorthy G (1998) Tau interacts with src-family non-receptor tyrosine kinases. J Cell Sci 111(Pt 21):3167–3177
21. Kosik KS, Crandall JE, Mufson EJ, Neve RL (1989) Tau in situ hybridization in normal and Alzheimer brain: localization in the somatodendritic compartment. Ann Neurol 26(3):352–361. doi:10.1002/ana.410260308
22. Grundke-Iqbal I, Iqbal K, Tung YC, Quinlan M, Wisniewski HM, Binder LI (1986) Abnormal phosphorylation of the microtubule-associated protein tau (tau) in Alzheimer cytoskeletal pathology. Proc Natl Acad Sci U S A 83(13):4913–4917
23. Grundke-Iqbal I, Iqbal K, Quinlan M, Tung YC, Zaidi MS, Wisniewski HM (1986) Microtubule-associated protein tau. A component of Alzheimer paired helical filaments. J Biol Chem 261(13):6084–6089
24. Lewis J, Dickson DW, Lin WL, Chisholm L, Corral A, Jones G, Yen SH, Sahara N, Skipper L, Yager D, Eckman C, Hardy J, Hutton M, McGowan E (2001) Enhanced neurofibrillary degeneration in transgenic mice expressing mutant tau and APP. Science 293(5534):1487–1491
25. Hutton M, Lendon CL, Rizzu P, Baker M, Froelich S, Houlden H, Pickering-Brown S, Chakraverty S, Isaacs A, Grover A, Hackett J, Adamson J, Lincoln S, Dickson D, Davies P, Petersen RC, Stevens M, de Graaff E, Wauters E, van Baren J, Hillebrand M, Joosse M, Kwon JM, Nowotny P, Heutink P et al (1998) Association of missense and 5′-splice-site mutations in tau with the inherited dementia FTDP-17. Nature 393(6686):702–705
26. DeKosky ST, Scheff SW (1990) Synapse loss in frontal cortex biopsies in Alzheimer's disease: correlation with cognitive severity. Ann Neurol 27(5):457–464. doi:10.1002/ana.410270502
27. Kowall NW, Kosik KS (1987) Axonal disruption and aberrant localization of tau protein characterize the neuropil pathology of Alzheimer's disease. Ann Neurol 22(5):639–643
28. Bondareff W, Mountjoy CQ, Roth M, Hauser DL (1989) Neurofibrillary degeneration and neuronal loss in Alzheimer's disease. Neurobiol Aging 10(6):709–715
29. Morsch R, Simon W, Coleman PD (1999) Neurons may live for decades with neurofibrillary tangles. J Neuropathol Exp Neurol 58(2):188–197
30. Kuchibhotla KV, Wegmann S, Kopeikina KJ, Hawkes J, Rudinskiy N, Andermann ML, Spires-Jones TL, Bacskai BJ, Hyman BT (2014) Neurofibrillary tangle-bearing neurons are functionally integrated in cortical circuits in vivo. Proc Natl Acad Sci U S A 111(1):510–514. doi:10.1073/pnas.1318807111
31. Lasagna-Reeves CA, Castillo-Carranza DL, Guerrero-Muoz MJ, Jackson GR, Kayed R (2010) Preparation and characterization of neurotoxic tau oligomers. Biochemistry 49(47):10039–10041. doi:10.1021/bi1016233
32. Tian H, Davidowitz E, Lopez P, Emadi S, Moe J, Sierks M (2013) Trimeric tau is toxic to human neuronal cells at low nanomolar concentrations. Int J Cell Biol 2013:260787. doi:10.1155/2013/260787
33. Flach K, Hilbrich I, Schiffmann A, Gartner U, Kruger M, Leonhardt M, Waschipky H, Wick L, Arendt T, Holzer M (2012) Tau oligomers impair artificial membrane integrity and cellular viability. J Biol Chem 287(52):43223–43233. doi:10.1074/jbc.M112.396176
34. Vana L, Kanaan NM, Ugwu IC, Wuu J, Mufson EJ, Binder LI (2011) Progression of tau pathology in cholinergic Basal forebrain neurons in mild cognitive impairment and Alzheimer's disease. Am J Pathol 179(5):2533–2550. doi:10.1016/j.ajpath.2011.07.044
35. Santacruz K, Lewis J, Spires T, Paulson J, Kotilinek L, Ingelsson M, Guimaraes A, DeTure M, Ramsden M, McGowan E, Forster C, Yue M, Orne J, Janus C, Mariash A, Kuskowski M, Hyman B, Hutton M, Ashe KH (2005) Tau suppression in a neurodegenerative mouse model improves memory function. Science 309(5733):476–481
36. Lee VM, Trojanowski JQ (1999) Neurodegenerative tauopathies: human disease and transgenic mouse models. Neuron 24(3):507–510
37. Hauw JJ, Daniel SE, Dickson D, Horoupian DS, Jellinger K, Lantos PL, McKee A, Tabaton M, Litvan I (1994) Preliminary NINDS neuropathologic criteria for Steele-Richardson-Olszewski syndrome (progressive

supranuclear palsy). Neurology 44(11): 2015–2019

38. Cervos-Navarro J, Schumacher K (1994) Neurofibrillary pathology in progressive supranuclear palsy (PSP). J Neural Transm Suppl 42:153–164
39. Munoz-Garcia D, Ludwin SK (1984) Classic and generalized variants of Pick's disease: a clinicopathological, ultrastructural, and immunocytochemical comparative study. Ann Neurol 16(4):467–480
40. Rebeiz JJ, Kolodny EH, Richardson EP Jr (1968) Corticodentatonigral degeneration with neuronal achromasia. Arch Neurol 18(1):20–33
41. McKee AC, Stein TD, Nowinski CJ, Stern RA, Daneshvar DH, Alvarez VE, Lee HS, Hall G, Wojtowicz SM, Baugh CM, Riley DO, Kubilus CA, Cormier KA, Jacobs MA, Martin BR, Abraham CR, Ikezu T, Reichard RR, Wolozin BL, Budson AE, Goldstein LE, Kowall NW, Cantu RC (2012) The spectrum of disease in chronic traumatic encephalopathy. Brain 136(Pt 1):43–64. doi:10.1093/brain/aws307, aws307 [pii]
42. Combs B, Gamblin TC (2012) FTDP-17 tau mutations induce distinct effects on aggregation and microtubule interactions. Biochemistry 51:8597–8607. doi:10.1021/bi3010818
43. Lewis J, McGowan E, Rockwood J, Melrose H, Nacharaju P, Van Slegtenhorst M, Gwinn-Hardy K, Paul Murphy M, Baker M, Yu X, Duff K, Hardy J, Corral A, Lin WL, Yen SH, Dickson DW, Davies P, Hutton M (2000) Neurofibrillary tangles, amyotrophy and progressive motor disturbance in mice expressing mutant (P301L) tau protein. Nat Genet 25(4):402–405
44. Liu F, Gong CX (2008) Tau exon 10 alternative splicing and tauopathies. Mol Neurodegener 3:8
45. Craig LA, Hong NS, McDonald RJ (2011) Revisiting the cholinergic hypothesis in the development of Alzheimer's disease. Neurosci Biobehav Rev 35(6):1397–1409. doi:10.1016/j.neubiorev.2011.03.001
46. Zhao Y, Zhao B (2013) Oxidative stress and the pathogenesis of Alzheimer's disease. Oxid Med Cell Longev 2013:316523. doi:10.1155/2013/316523
47. Bamburg JR, Bloom GS (2009) Cytoskeletal pathologies of Alzheimer disease. Cell Motil Cytoskeleton 66(8):635–649. doi:10.1002/cm.20388
48. Dolan PJ, Johnson GV (2010) The role of tau kinases in Alzheimer's disease. Curr Opin Drug Discov Devel 13(5):595–603
49. Takalo M, Salminen A, Soininen H, Hiltunen M, Haapasalo A (2013) Protein aggregation and degradation mechanisms in neurodegenerative diseases. Am J Neurodegener Dis 2(1):1–14
50. Bush AI, Tanzi RE (2008) Therapeutics for Alzheimer's disease based on the metal hypothesis. Neurotherapeutics 5(3):421–432. doi:10.1016/j.nurt.2008.05.001
51. Kanaan NM, Pigino GF, Brady ST, Lazarov O, Binder LI, Morfini GA (2013) Axonal degeneration in Alzheimer's disease: when signaling abnormalities meet the axonal transport system. Exp Neurol 246:44–53. doi:10.1016/j.expneurol.2012.06.003
52. Wyss-Coray T, Rogers J (2012) Inflammation in Alzheimer disease-a brief review of the basic science and clinical literature. Cold Spring Harb Perspect Med 2(1):a006346. doi:10.1101/cshperspect.a006346
53. Armstrong RA (2013) What causes Alzheimer's disease? Folia Neuropathol 51(3):169–188
54. Gotz J, Ittner LM (2008) Animal models of Alzheimer's disease and frontotemporal dementia. Nat Rev Neurosci 9(7):532–544. doi:10.1038/nrn2420, nrn2420 [pii]
55. Elder GA, Gama Sosa MA, De Gasperi R (2010) Transgenic mouse models of Alzheimer's disease. Mt Sinai J Med 77(1):69–81. doi:10.1002/msj.20159
56. LaFerla FM, Green KN (2012) Animal models of Alzheimer disease. Cold Spring Harb Perspect Med 2(11):a006320. doi:10.1101/cshperspect.a006320
57. Chishti MA, Yang DS, Janus C, Phinney AL, Horne P, Pearson J, Strome R, Zuker N, Loukides J, French J, Turner S, Lozza G, Grilli M, Kunicki S, Morissette C, Paquette J, Gervais F, Bergeron C, Fraser PE, Carlson GA, George-Hyslop PS, Westaway D (2001) Early-onset amyloid deposition and cognitive deficits in transgenic mice expressing a double mutant form of amyloid precursor protein 695. J Biol Chem 276(24):21562–21570. doi:10.1074/jbc.M100710200
58. Rockenstein E, Mallory M, Mante M, Sisk A, Masliaha E (2001) Early formation of mature amyloid-beta protein deposits in a mutant APP transgenic model depends on levels of Abeta(1-42). J Neurosci Res 66(4):573–582
59. Games D, Adams D, Alessandrini R, Barbour R, Berthelette P, Blackwell C, Carr T, Clemens J, Donaldson T, Gillespie F et al (1995) Alzheimer-type neuropathology in transgenic mice overexpressing V717F beta-amyloid precursor protein. Nature 373(6514):523–527. doi:10.1038/373523a0

60. Hsiao K, Chapman P, Nilsen S, Eckman C, Harigaya Y, Younkin S, Yang F, Cole G (1996) Correlative memory deficits, Abeta elevation, and amyloid plaques in transgenic mice. Science 274(5284):99–102

61. Sturchler-Pierrat C, Abramowski D, Duke M, Wiederhold KH, Mistl C, Rothacher S, Ledermann B, Burki K, Frey P, Paganetti PA, Waridel C, Calhoun ME, Jucker M, Probst A, Staufenbiel M, Sommer B (1997) Two amyloid precursor protein transgenic mouse models with Alzheimer disease-like pathology. Proc Natl Acad Sci U S A 94(24): 13287–13292

62. Duff K, Eckman C, Zehr C, Yu X, Prada CM, Perez-tur J, Hutton M, Buee L, Harigaya Y, Yager D, Morgan D, Gordon MN, Holcomb L, Refolo L, Zenk B, Hardy J, Younkin S (1996) Increased amyloid-beta42(43) in brains of mice expressing mutant presenilin 1. Nature 383(6602):710–713. doi:10.1038/383710a0

63. Holcomb L, Gordon MN, McGowan E, Yu X, Benkovic S, Jantzen P, Wright K, Saad I, Mueller R, Morgan D, Sanders S, Zehr C, O'Campo K, Hardy J, Prada CM, Eckman C, Younkin S, Hsiao K, Duff K (1998) Accelerated Alzheimer-type phenotype in transgenic mice carrying both mutant amyloid precursor protein and presenilin 1 transgenes. Nat Med 4(1):97–100

64. Jankowsky JL, Fadale DJ, Anderson J, Xu GM, Gonzales V, Jenkins NA, Copeland NG, Lee MK, Younkin LH, Wagner SL, Younkin SG, Borchelt DR (2004) Mutant presenilins specifically elevate the levels of the 42 residue beta-amyloid peptide in vivo: evidence for augmentation of a 42-specific gamma secretase. Hum Mol Genet 13(2):159–170. doi:10.1093/hmg/ddh019

65. Siman R, Reaume AG, Savage MJ, Trusko S, Lin YG, Scott RW, Flood DG (2000) Presenilin-1 P264L knock-in mutation: differential effects on abeta production, amyloid deposition, and neuronal vulnerability. J Neurosci 20(23):8717–8726

66. Oakley H, Cole SL, Logan S, Maus E, Shao P, Craft J, Guillozet-Bongaarts A, Ohno M, Disterhoft J, Van Eldik L, Berry R, Vassar R (2006) Intraneuronal beta-amyloid aggregates, neurodegeneration, and neuron loss in transgenic mice with five familial Alzheimer's disease mutations: potential factors in amyloid plaque formation. J Neurosci 26(40):10129–10140. doi:10.1523/JNEUROSCI.1202-06.2006

67. Oddo S, Caccamo A, Shepherd JD, Murphy MP, Golde TE, Kayed R, Metherate R, Mattson MP, Akbari Y, LaFerla FM (2003) Triple-transgenic model of Alzheimer's disease with plaques and tangles: intracellular Abeta and synaptic dysfunction. Neuron 39(3):409–421

68. Cohen RM, Rezai-Zadeh K, Weitz TM, Rentsendorj A, Gate D, Spivak I, Bholat Y, Vasilevko V, Glabe CG, Breunig JJ, Rakic P, Davtyan H, Agadjanyan MG, Kepe V, Barrio JR, Bannykh S, Szekely CA, Pechnick RN, Town T (2013) A transgenic Alzheimer rat with plaques, tau pathology, behavioral impairment, oligomeric abeta, and frank neuronal loss. J Neurosci 33(15):6245–6256. doi:10.1523/JNEUROSCI.3672-12.2013

69. Savonenko A, Xu GM, Melnikova T, Morton JL, Gonzales V, Wong MP, Price DL, Tang F, Markowska AL, Borchelt DR (2005) Episodic-like memory deficits in the APPswe/PS1dE9 mouse model of Alzheimer's disease: relationships to beta-amyloid deposition and neurotransmitter abnormalities. Neurobiol Dis 18(3):602–617. doi:10.1016/j.nbd.2004.10.022

70. McMillan P, Korvatska E, Poorkaj P, Evstafjeva Z, Robinson L, Greenup L, Leverenz J, Schellenberg GD, D'Souza I (2008) Tau isoform regulation is region- and cell-specific in mouse brain. J Comp Neurol 511(6):788–803. doi:10.1002/cne.21867

71. Hanes J, Zilka N, Bartkova M, Caletkova M, Dobrota D, Novak M (2009) Rat tau proteome consists of six tau isoforms: implication for animal models of human tauopathies. J Neurochem 108(5):1167–1176. doi:10.1111/j.1471-4159.2009.05869.x

72. Goedert M, Crowther RA, Spillantini MG (1998) Tau mutations cause frontotemporal dementias. Neuron 21(5):955–958

73. Yoshiyama Y, Higuchi M, Zhang B, Huang SM, Iwata N, Saido TC, Maeda J, Suhara T, Trojanowski JQ, Lee VM (2007) Synapse loss and microglial activation precede tangles in a P301S tauopathy mouse model. Neuron 53(3): 337–351. doi:10.1016/j.neuron.2007.01.010, S0896-6273(07)00030-X [pii]

74. Tanemura K, Akagi T, Murayama M, Kikuchi N, Murayama O, Hashikawa T, Yoshiike Y, Park JM, Matsuda K, Nakao S, Sun X, Sato S, Yamaguchi H, Takashima A (2001) Formation of filamentous tau aggregations in transgenic mice expressing V337M human tau. Neurobiol Dis 8(6):1036–1045. doi:10.1006/nbdi.2001.0439, S0969-9961(01)90439-5 [pii]

75. Gotz J, Probst A, Spillantini MG, Schafer T, Jakes R, Burki K, Goedert M (1995) Somatodendritic localization and hyperphos-

phorylation of tau protein in transgenic mice expressing the longest human brain tau isoform. EMBO J 14(7):1304–1313

76. Andorfer C, Acker CM, Kress Y, Hof PR, Duff K, Davies P (2005) Cell-cycle reentry and cell death in transgenic mice expressing nonmutant human tau isoforms. J Neurosci 25(22):5446–5454. doi:10.1523/JNEUROSCI.4637-04.2005, 25/22/5446 [pii]
77. de Calignon A, Polydoro M, Suarez-Calvet M, William C, Adamowicz DH, Kopeikina KJ, Pitstick R, Sahara N, Ashe KH, Carlson GA, Spires-Jones TL, Hyman BT (2012) Propagation of tau pathology in a model of early Alzheimer's disease. Neuron 73(4):685–697. doi:10.1016/j.neuron.2011.11.033, S0896-6273(12)00038-4 [pii]
78. Liu L, Drouet V, Wu JW, Witter MP, Small SA, Clelland C, Duff K (2012) Trans-synaptic spread of tau pathology in vivo. PLoS One 7(2), e31302. doi:10.1371/journal.pone.0031302, PONE-D-11-23353 [pii]
79. Klein RL, Wang DB, King MA (2009) Versatile somatic gene transfer for modeling neurodegenerative diseases. Neurotox Res 16(3):329–342. doi:10.1007/s12640-009-9080-7
80. Low K, Aebischer P (2012) Use of viral vectors to create animal models for Parkinson's disease. Neurobiol Dis 48(2):189–201. doi:10.1016/j.nbd.2011.12.038
81. Terzi D, Zachariou V (2008) Adeno-associated virus-mediated gene delivery approaches for the treatment of CNS disorders. Biotechnol J 3(12):1555–1563. doi:10.1002/biot.200800284
82. Polinski NK, Gombash SE, Manfredsson FP, Lipton JW, Kemp CJ, Cole-Strauss A, Kanaan NM, Steece-Collier K, Kuhn NC, Wohlgenant SL, Sortwell CE (2014) Recombinant adenoassociated virus 2/5-mediated gene transfer is reduced in the aged rat midbrain. Neurobiol Aging 36:1110–1120. doi:10.1016/j.neurobiolaging.2014.07.047
83. Klein RL, Dayton RD, Diaczynsky CG, Wang DB (2010) Pronounced microgliosis and neurodegeneration in aged rats after tau gene transfer. Neurobiol Aging 31(12):2091–2102. doi:10.1016/j.neurobiolaging.2008.12.002
84. Wu K, Meyers CA, Guerra NK, King MA, Meyer EM (2004) The effects of rAAV2-mediated NGF gene delivery in adult and aged rats. Mol Ther 9(2):262–269. doi:10.1016/j.ymthe.2003.11.010
85. Manfredsson FP, Burger C, Rising AC, Zuobi-Hasona K, Sullivan LF, Lewin AS, Huang J, Piercefield E, Muzyczka N, Mandel RJ (2009) Tight Long-term dynamic doxycycline responsive nigrostriatal GDNF using a single rAAV vector. Mol Ther 17(11):1857–1867. doi:10.1038/mt.2009.196
86. Gray SJ, Foti SB, Schwartz JW, Bachaboina L, Taylor-Blake B, Coleman J, Ehlers MD, Zylka MJ, McCown TJ, Samulski RJ (2011) Optimizing promoters for recombinant adeno-associated virus-mediated gene expression in the peripheral and central nervous system using self-complementary vectors. Hum Gene Ther 22(9):1143–1153. doi:10.1089/hum.2010.245
87. von Jonquieres G, Mersmann N, Klugmann CB, Harasta AE, Lutz B, Teahan O, Housley GD, Frohlich D, Kramer-Albers EM, Klugmann M (2013) Glial promoter selectivity following AAV-delivery to the immature brain. PLoS One 8(6), e65646. doi:10.1371/journal.pone.0065646
88. Allen B, Ingram E, Takao M, Smith MJ, Jakes R, Virdee K, Yoshida H, Holzer M, Craxton M, Emson PC, Atzori C, Migheli A, Crowther RA, Ghetti B, Spillantini MG, Goedert M (2002) Abundant tau filaments and nonapoptotic neurodegeneration in transgenic mice expressing human P301S tau protein. J Neurosci 22(21):9340–9351
89. Aschauer DF, Kreuz S, Rumpel S (2013) Analysis of transduction efficiency, tropism and axonal transport of AAV serotypes 1, 2, 5, 6, 8 and 9 in the mouse brain. PLoS One 8(9), e76310. doi:10.1371/journal.pone.0076310
90. Burger C, Gorbatyuk OS, Velardo MJ, Peden CS, Williams P, Zolotukhin S, Reier PJ, Mandel RJ, Muzyczka N (2004) Recombinant AAV viral vectors pseudotyped with viral capsids from serotypes 1, 2, and 5 display differential efficiency and cell tropism after delivery to different regions of the central nervous system. Mol Ther 10(2):302–317. doi:10.1016/j.ymthe.2004.05.024
91. Gao G, Sena-Esteves M (2012) Introducing genes into mammalian cells: viral vectors. In: Green MR, Sambrook J (eds) Molecular cloning: a laboratory manual, vol 2, 4. Cold Spring Harbor Laboratory Press, Cold Spring Harbor, NY, pp 1209–1333
92. Tiscornia G, Singer O, Ikawa M, Verma IM (2003) A general method for gene knockdown in mice by using lentiviral vectors expressing small interfering RNA. Proc Natl Acad Sci U S A 100(4):1844–1848. doi:10.1073/pnas.0437912100
93. Kugler S, Kilic E, Bahr M (2003) Human synapsin 1 gene promoter confers highly neuron-specific long-term transgene expression from an adenoviral vector in the adult rat

brain depending on the transduced area. Gene Ther 10(4):337–347. doi:10.1038/sj.gt.3301905

94. Rahim AA, Wong AM, Hoefer K, Buckley SM, Mattar CN, Cheng SH, Chan JK, Cooper JD, Waddington SN (2011) Intravenous administration of AAV2/9 to the fetal and neonatal mouse leads to differential targeting of CNS cell types and extensive transduction of the nervous system. FASEB J 25(10):3505–3518. doi:10.1096/fj.11-182311
95. Klein RL, Lin WL, Dickson DW, Lewis J, Hutton M, Duff K, Meyer EM, King MA (2004) Rapid neurofibrillary tangle formation after localized gene transfer of mutated tau. Am J Pathol 164(1):347–353. doi:10.1016/S0002-9440(10)63124-0
96. Jaworski T, Dewachter I, Lechat B, Croes S, Termont A, Demedts D, Borghgraef P, Devijver H, Filipkowski RK, Kaczmarek L, Kugler S, Van Leuven F (2009) AAV-tau mediates pyramidal neurodegeneration by cell-cycle re-entry without neurofibrillary tangle formation in wild-type mice. PLoS One 4(10), e7280. doi:10.1371/journal.pone.0007280
97. Dayton RD, Wang DB, Cain CD, Schrott LM, Ramirez JJ, King MA, Klein RL (2012) Frontotemporal lobar degeneration-related proteins induce only subtle memory-related deficits when bilaterally overexpressed in the dorsal hippocampus. Exp Neurol 233(2):807–814. doi:10.1016/j.expneurol.2011.12.002
98. Squire LR, Wixted JT, Clark RE (2007) Recognition memory and the medial temporal lobe: a new perspective. Nat Rev Neurosci 8(11):872–883. doi:10.1038/nrn2154
99. Gomez-Isla T, Price JL, McKeel DW Jr, Morris JC, Growdon JH, Hyman BT (1996) Profound loss of layer II entorhinal cortex neurons occurs in very mild Alzheimer's disease. J Neurosci 16(14):4491–4500
100. Siman R, Lin YG, Malthankar-Phatak G, Dong Y (2013) A rapid gene delivery-based mouse model for early-stage Alzheimer disease-type tauopathy. J Neuropathol Exp Neurol 72(11):1062–1071. doi:10.1097/NEN.0000000000000006
101. Burns JM, Galvin JE, Roe CM, Morris JC, McKeel DW (2005) The pathology of the substantia nigra in Alzheimer disease with extrapyramidal signs. Neurology 64(8):1397–1403. doi:10.1212/01.WNL.0000158423.05224.7F
102. Oyanagi K, Tsuchiya K, Yamazaki M, Ikeda K (2001) Substantia nigra in progressive supranuclear palsy, corticobasal degeneration, and parkinsonism-dementia complex of Guam: specific pathological features. J Neuropathol Exp Neurol 60(4):393–402
103. Spillantini MG, Crowther RA, Kamphorst W, Heutink P, van Swieten JC (1998) Tau pathology in two Dutch families with mutations in the microtubule- binding region of tau. Am J Pathol 153(5):1359–1363
104. Ishizawa T, Mattila P, Davies P, Wang D, Dickson DW (2003) Colocalization of tau and alpha-synuclein epitopes in Lewy bodies. J Neuropathol Exp Neurol 62(4):389–397
105. Klein RL, Dayton RD, Lin WL, Dickson DW (2005) Tau gene transfer, but not alpha-synuclein, induces both progressive dopamine neuron degeneration and rotational behavior in the rat. Neurobiol Dis 20(1):64–73. doi:10.1016/j.nbd.2005.02.001
106. Klein RL, Dayton RD, Leidenheimer NJ, Jansen K, Golde TE, Zweig RM (2006) Efficient neuronal gene transfer with AAV8 leads to neurotoxic levels of tau or green fluorescent proteins. Mol Ther 13(3):517–527. doi:10.1016/j.ymthe.2005.10.008
107. Klein RL, Dayton RD, Tatom JB, Diaczynsky CG, Salvatore MF (2008) Tau expression levels from various adeno-associated virus vector serotypes produce graded neurodegenerative disease states. Eur J Neurosci 27(7):1615–1625. doi:10.1111/j.1460-9568.2008.06161.x
108. Di Maria E, Tabaton M, Vigo T, Abbruzzese G, Bellone E, Donati C, Frasson E, Marchese R, Montagna P, Munoz DG, Pramstaller PP, Zanusso G, Ajmar F, Mandich P (2000) Corticobasal degeneration shares a common genetic background with progressive supranuclear palsy. Ann Neurol 47(3):374–377
109. Wakabayashi K, Oyanagi K, Makifuchi T, Ikuta F, Homma A, Homma Y, Horikawa Y, Tokiguchi S (1994) Corticobasal degeneration: etiopathological significance of the cytoskeletal alterations. Acta Neuropathol 87(6):545–553
110. Wang DB, Dayton RD, Zweig RM, Klein RL (2010) Transcriptome analysis of a tau overexpression model in rats implicates an early pro-inflammatory response. Exp Neurol 224(1):197–206. doi:10.1016/j.expneurol.2010.03.011
111. Ubhi K, Rockenstein E, Doppler E, Mante M, Adame A, Patrick C, Trejo M, Crews L, Paulino A, Moessler H, Masliah E (2009) Neurofibrillary and neurodegenerative pathology in APP-transgenic mice injected with AAV2-mutant TAU: neuroprotective effects of Cerebrolysin. Acta Neuropathol 117(6):699–712. doi:10.1007/s00401-009-0505-4

112. Howlett DR, Richardson JC, Austin A, Parsons AA, Bate ST, Davies DC, Gonzalez MI (2004) Cognitive correlates of Abeta deposition in male and female mice bearing amyloid precursor protein and presenilin-1 mutant transgenes. Brain Res 1017(1-2):130–136. doi:10.1016/j.brainres.2004.05.029
113. Richardson JC, Kendal CE, Anderson R, Priest F, Gower E, Soden P, Gray R, Topps S, Howlett DR, Lavender D, Clarke NJ, Barnes JC, Haworth R, Stewart MG, Rupniak HT (2003) Ultrastructural and behavioural changes precede amyloid deposition in a transgenic model of Alzheimer's disease. Neuroscience 122(1):213–228
114. Dassie E, Andrews MR, Bensadoun JC, Cacquevel M, Schneider BL, Aebischer P, Wouters FS, Richardson JC, Hussain I, Howlett DR, Spillantini MG, Fawcett JW (2013) Focal expression of adeno-associated viral-mutant tau induces widespread impairment in an APP mouse model. Neurobiol Aging 34(5):1355–1368. doi:10.1016/j.neurobiolaging.2012.11.011
115. Iliev AI, Ganesan S, Bunt G, Wouters FS (2006) Removal of pattern-breaking sequences in microtubule binding repeats produces instantaneous tau aggregation and toxicity. J Biol Chem 281(48):37195–37204. doi:10.1074/jbc.M604863200, M604863200 [pii]
116. Ramirez JJ, Poulton WE, Knelson E, Barton C, King MA, Klein RL (2011) Focal expression of mutated tau in entorhinal cortex neurons of rats impairs spatial working memory. Behav Brain Res 216(1):332–340. doi:10.1016/j.bbr.2010.08.013
117. Jaworski T, Lechat B, Demedts D, Gielis L, Devijver H, Borghgraef P, Duimel H, Verheyen F, Kugler S, Van Leuven F (2011) Dendritic degeneration, neurovascular defects, and inflammation precede neuronal loss in a mouse model for tau-mediated neurodegeneration. Am J Pathol 179(4):2001–2015. doi:10.1016/j.ajpath.2011.06.025
118. Lawlor PA, Bland RJ, Das P, Price RW, Holloway V, Smithson L, Dicker BL, During MJ, Young D, Golde TE (2007) Novel rat Alzheimer's disease models based on AAV-mediated gene transfer to selectively increase hippocampal Abeta levels. Mol Neurodegener 2:11. doi:10.1186/1750-1326-2-11
119. Vidal R, Frangione B, Rostagno A, Mead S, Revesz T, Plant G, Ghiso J (1999) A stop-codon mutation in the BRI gene associated with familial British dementia. Nature 399(6738):776–781. doi:10.1038/21637
120. Lewis PA, Piper S, Baker M, Onstead L, Murphy MP, Hardy J, Wang R, McGowan E, Golde TE (2001) Expression of BRI-amyloid beta peptide fusion proteins: a novel method for specific high-level expression of amyloid beta peptides. Biochim Biophys Acta 1537(1):58–62
121. Drummond ES, Muhling J, Martins RN, Wijaya LK, Ehlert EM, Harvey AR (2013) Pathology associated with AAV mediated expression of beta amyloid or C100 in adult mouse hippocampus and cerebellum. PLoS One 8(3), e59166. doi:10.1371/journal.pone.0059166
122. Bolognin S, Blanchard J, Wang X, Basurto-Islas G, Tung YC, Kohlbrenner E, Grundke-Iqbal I, Iqbal K (2012) An experimental rat model of sporadic Alzheimer's disease and rescue of cognitive impairment with a neurotrophic peptide. Acta Neuropathol 123(1):133–151. doi:10.1007/s00401-011-0908-x
123. Chu J, Giannopoulos PF, Ceballos-Diaz C, Golde TE, Pratico D (2012) Adeno-associated virus-mediated brain delivery of 5-lipoxygenase modulates the AD-like phenotype of APP mice. Mol Neurodegener 7(1):1. doi:10.1186/1750-1326-7-1
124. Chu J, Giannopoulos PF, Ceballos-Diaz C, Golde TE, Pratico D (2012) 5-Lipoxygenase gene transfer worsens memory, amyloid, and tau brain pathologies in a mouse model of Alzheimer disease. Ann Neurol 72(3):442–454. doi:10.1002/ana.23642
125. Ikonomovic MD, Abrahamson EE, Uz T, Manev H, Dekosky ST (2008) Increased 5-lipoxygenase immunoreactivity in the hippocampus of patients with Alzheimer's disease. J Histochem Cytochem 56(12):1065–1073. doi:10.1369/jhc.2008.951855
126. Mackenzie IR, Rademakers R, Neumann M (2010) TDP-43 and FUS in amyotrophic lateral sclerosis and frontotemporal dementia. Lancet Neurol 9(10):995–1007. doi:10.1016/S1474-4422(10)70195-2
127. Tatom JB, Wang DB, Dayton RD, Skalli O, Hutton ML, Dickson DW, Klein RL (2009) Mimicking aspects of frontotemporal lobar degeneration and Lou Gehrig's disease in rats via TDP-43 overexpression. Mol Ther 17(4):607–613. doi:10.1038/mt.2009.3
128. Yan S, Wang CE, Wei W, Gaertig MA, Lai L, Li S, Li XJ (2014) TDP-43 causes differential pathology in neuronal versus glial cells in the mouse brain. Hum Mol Genet 23(10):2678–2693. doi:10.1093/hmg/ddt662
129. Osinde M, Clavaguera F, May-Nass R, Tolnay M, Dev KK (2008) Lentivirus Tau (P301S) expression in adult amyloid precursor protein (APP)-transgenic mice leads to tangle formation. Neuropathol Appl Neurobiol 34(5):523–531. doi:10.1111/j.1365-2990.2008.00936.x

130. Khandelwal PJ, Dumanis SB, Herman AM, Rebeck GW, Moussa CE (2012) Wild type and P301L mutant Tau promote neuro-inflammation and alpha-Synuclein accumulation in lentiviral gene delivery models. Mol Cell Neurosci 49(1):44–53. doi:10.1016/j.mcn.2011.09.002
131. Caillierez R, Begard S, Lecolle K, Deramecourt V, Zommer N, Dujardin S, Loyens A, Dufour N, Auregan G, Winderickx J, Hantraye P, Deglon N, Buee L, Colin M (2013) Lentiviral delivery of the human wild-type tau protein mediates a slow and progressive neurodegenerative tau pathology in the rat brain. Mol Ther 21(7):1358–1368. doi:10.1038/mt.2013.66
132. Hebron ML, Algarzae NK, Lonskaya I, Moussa C (2014) Fractalkine signaling and Tau hyper-phosphorylation are associated with autophagic alterations in lentiviral Tau and Abeta1-42 gene transfer models. Exp Neurol 251:127–138. doi:10.1016/j.expneurol.2013.01.009
133. Rebeck GW, Hoe HS, Moussa CE (2010) Beta-amyloid1-42 gene transfer model exhibits intraneuronal amyloid, gliosis, tau phosphorylation, and neuronal loss. J Biol Chem 285(10):7440–7446. doi:10.1074/jbc.M109.083915
134. Herman AM, Khandelwal PJ, Rebeck GW, Moussa CE (2012) Wild type TDP-43 induces neuro-inflammation and alters APP metabolism in lentiviral gene transfer models. Exp Neurol 235(1):297–305. doi:10.1016/j.expneurol.2012.02.011
135. Hebron M, Chen W, Miessau MJ, Lonskaya I, Moussa CE (2014) Parkin reverses TDP-43-induced cell death and failure of amino acid homeostasis. J Neurochem 129(2):350–361. doi:10.1111/jnc.12630
136. Onorato M, Mulvihill P, Connolly J, Galloway P, Whitehouse P, Perry G (1989) Alteration of neuritic cytoarchitecture in Alzheimer disease. Prog Clin Biol Res 317:781–789
137. Greenamyre JT, Young AB (1989) Excitatory amino acids and Alzheimer's disease. Neurobiol Aging 10(5):593–602
138. Bassil N, Grossberg GT (2009) Novel regimens and delivery systems in the pharmacological treatment of Alzheimer's disease. CNS Drugs 23(4):293–307
139. Rafii MS, Baumann TL, Bakay RA, Ostrove JM, Siffert J, Fleisher AS, Herzog CD, Barba D, Pay M, Salmon DP, Chu Y, Kordower JH, Bishop K, Keator D, Potkin S, Bartus RT (2014) A phase1 study of stereotactic gene delivery of AAV2-NGF for Alzheimer's disease. Alzheimer's Dement 10:571–581. doi:10.1016/j.jalz.2013.09.004
140. Fischer W, Wictorin K, Bjorklund A, Williams LR, Varon S, Gage FH (1987) Amelioration of cholinergic neuron atrophy and spatial memory impairment in aged rats by nerve growth factor. Nature 329(6134):65–68. doi:10.1038/329065a0
141. Eriksdotter Jonhagen M, Nordberg A, Amberla K, Backman L, Ebendal T, Meyerson B, Olson L, Seiger SM, Theodorsson E, Viitanen M, Winblad B, Wahlund LO (1998) Intracerebroventricular infusion of nerve growth factor in three patients with Alzheimer's disease. Dement Geriatr Cogn Disord 9(5):246–257
142. Smith DE, Roberts J, Gage FH, Tuszynski MH (1999) Age-associated neuronal atrophy occurs in the primate brain and is reversible by growth factor gene therapy. Proc Natl Acad Sci U S A 96(19):10893–10898
143. Mandel RJ, Gage FH, Clevenger DG, Spratt SK, Snyder RO, Leff SE (1999) Nerve growth factor expressed in the medial septum following in vivo gene delivery using a recombinant adeno-associated viral vector protects cholinergic neurons from fimbria-fornix lesion-induced degeneration. Exp Neurol 155(1):59–64. doi:10.1006/exnr.1998.6961
144. Horowitz PM, Patterson KR, Guillozet-Bongaarts AL, Reynolds MR, Carroll CA, Weintraub ST, Bennett DA, Cryns VL, Berry RW, Binder LI (2004) Early N-terminal changes and caspase-6 cleavage of tau in Alzheimer's disease. J Neurosci 24(36): 7895–7902.doi:10.1523/JNEUROSCI.1988-04.2004
145. Caccamo D, Katsetos CD, Herman MM, Frankfurter A, Collins VP, Rubinstein LJ (1989) Immunohistochemistry of a spontaneous murine ovarian teratoma with neuro-epithelial differentiation. Neuron-associated beta-tubulin as a marker for primitive neuro-epithelium. Lab Invest 60(3):390–398
146. Kanaan NM, Morfini G, Pigino G, LaPointe NE, Andreadis A, Song Y, Leitman E, Binder LI, Brady ST (2012) Phosphorylation in the amino terminus of tau prevents inhibition of anterograde axonal transport. Neurobiol Aging 33(4):826.e815–826.e830. doi:10.1016/j.neurobiolaging.2011.06.006
147. Carmel G, Mager EM, Binder LI, Kuret J (1996) The structural basis of monoclonal antibody Alz50's selectivity for Alzheimer's disease pathology. J Biol Chem 271(51):32789–32795

Chapter 26

Viral Vector-Based Modeling of Neurodegenerative Disorders: Parkinson's Disease

D. Luke Fischer, Sara E. Gombash, Christopher J. Kemp, Fredric P. Manfredsson, Nicole K. Polinski, Megan F. Duffy, and Caryl E. Sortwell

Abstract

Gene therapy methods are increasingly used to model Parkinson's disease (PD) in animals in an effort to test experimental therapeutics within a more relevant context to disease pathophysiology and neuropathology. We have detailed several criteria that are critical or advantageous to accurately modeling PD in a murine model or in a nonhuman primate. Using these criteria, we then evaluate approaches made to model PD using viral vectors to date, including both adeno-associated viruses and lentiviruses. Lastly, we comment on the consideration of aging as a critical factor for modeling PD.

Key words Parkinson's disease, Substantia nigra, Striatum, Lewy body, Tyrosine hydroxylase, Dopamine, α-synuclein, Leucine-rich repeat kinase 2, Parkin, PTEN-induced putative kinase 1, Adeno-associated virus, Lentivirus

1 Introduction to Parkinson's Disease

Parkinson's disease (PD) is the second most prevalent neurodegenerative disorder with approximately "100–300 per 100,000 persons" affected (*see* Chapter 2 of ref. 1). The primary risk factor for PD is aging with PD affecting one-percent of the population over 65 years of age [2]. Akinesia, bradykinesia, rigidity and resting tremor are the most common symptoms. While motor dysfunction is the primary basis for diagnosis, patients may also experience depression, cognitive dysfunction, agnosia and other symptoms at clinical presentation [3]. The progressive degeneration of the dopaminergic cells of the substantia nigra pars compacta (SNc) and their projections to the striatum results in the loss of dopamine to the caudate nucleus and the putamen and the motor symptoms at

Fredric P. Manfredsson (ed.), *Gene Therapy for Neurological Disorders: Methods and Protocols*, Methods in Molecular Biology, vol. 1382, DOI 10.1007/978-1-4939-3271-9_26,

clinical presentation. As a result, current pharmacotherapies (e.g., L-DOPA) attempt to bolster nigrostriatal dopaminergic transmission. However, as disease progression continues, these pharmacotherapies lose symptomatic efficacy and can yield troubling dyskinesias [4], making the development of neuroprotective therapies critical.

While a clinical diagnosis of PD can be made based on symptoms and signs alone, the definitive diagnosis is made *post mortem* by a neuropathologist. The pathological hallmark of PD is the Lewy body, a "proteinaceous neuronal cytoplasmic inclusion" that is often immunoreactive for ubiquitin and most specifically for α-synuclein (α-syn) (*see* Chapter 12 of ref. 1 as well as refs. 5–7). α-syn is a natively unfolded protein of 140 amino acids that binds and bends membranes, though its biological function is not completely understood [8]. (The relationship between α-syn and PD is discussed more below.)

2 Criteria for Evaluating Animal Models of Parkinson's Disease

Our ability to study the mechanisms driving the pathophysiology in PD and to test experimental therapeutics is only as strong as our capacity to accurately model the human condition in a laboratory animal. Neurotoxin-based and transgenic models of PD have proven valuable in advancing our understanding of the consequences of dopamine denervation and the biological function of particular genes within the context of the whole brain environment; however, PD is still without a disease-modifying agent. Through targeted induction of oxidative stress, neurotoxin-based models of PD can be used to produce progressive degeneration of the SNc, but only of the perikarya since the terminals degenerate almost immediately (e.g., 9). Limitations of the neurotoxin-based PD models (e.g., 6-hydroxydopamine, 1-methyl-4-phenyl-1,2,3,6-tetrahydropyridine) include inadequate construct validity, as oxidative stress is only one of many contributors to the disease process, and an absence of the pathological hallmark of PD, Lewy bodies. The overwhelming majority of germ line transgenic models that overexpress normal or mutated forms of genes linked to PD have failed to recapitulate the magnitude of nigrostriatal degeneration observed in the parkinsonian brain. In this chapter, we review viral vector-based models of PD to determine whether this approach more appropriately recapitulates the human condition. In the following sections, we will review the progress that has been made in several different viral vector-based models of PD and evaluate them based on the following criteria that we propose are critical or advantageous to studying the disease and evaluating therapeutics.

2.1 Does the Model Have Construct Validity?

Construct validity requires that the causes and pathophysiological changes that occur in PD patients are comparable to what drives and occurs in the animal model. The causes of PD are for the most part unknown because the majority of cases are idiopathic. A combination of genetic and environmental factors has been posited to increase the risk of developing the disease. Nonetheless, a small percentage of PD cases can be entirely explained by our current understanding of genetics (*see* Chapter 15 of ref. 1, cf. Chapter 16, ib.). With this in mind, viral vector-based animal models of PD have targeted overexpression of normal or mutated forms of some of the causative PD genes to the nigrostriatal system. These include overexpression of normal or mutated forms of α-syn, leucine-rich repeat kinase 2 (LRRK-2) and parkin substrates as well as knockdown of PTEN-induced putative kinase 1 (PINK1). Further, with the concept in mind that PD motor symptoms are primarily driven by loss of nigrostriatal dopamine, some research has explored silencing tyrosine hydroxylase (TH) using shRNA to effectively deplete dopamine in the nigrostriatal system.

2.2 Is the Model Consistently Reproducible?

An important criterion for any model is the level of variability and the ease of reproducibility. We will examine the variability of each model and use the literature to confirm reproducibility both within the same laboratory and between laboratories employing the same model. It should be noted that although viral vector constructs may express the same PD-related gene, these vector constructs may differ in other important aspects such as the promoter used, titer or injection parameters. Therefore, results may vary between specific vector constructs, so some leniency must be afforded when comparing studies. Of note, this lack of standardization has led to some confusion, and perhaps trepidation, by other groups in using certain models. Laboratories employing viral vector-based animal models of PD should take this into account when developing models, comparing their models to those from other groups and communicating their results.

2.3 Is the Model Appropriately Progressive?

Nigrostriatal degeneration occurs over many years both before and after symptom onset and diagnosis. Of great importance, this process proceeds in stages, the very first of which is dysfunction/loss of dopamine (DA) terminals in the caudate nucleus and the putamen (collectively termed the striatum in rodent species). A recent, detailed, postmortem analysis of PD brains confirms that loss of striatal DA innervation is a critical early event that precedes and exceeds loss of DA neuron cell bodies [10]. Therefore, an ideal model of PD should display substantial striatal terminal dysfunction/loss prior to overt nigral DA neuron loss. In accordance with the progressive nature of PD, degeneration of DA neurons of the SNc should occur over months; for the purposes of discussion, we set our minimum at 8 weeks (i.e., degeneration should be complete no sooner than 8 weeks post insult). Finally, it is advantageous for the model to progress to at least 60 % loss of nigral DA neurons in order to mimic late-stage PD, allow for functional evaluations of motor performance and to give the investigator a large window to observe a disease-modifying effect.

2.4 Does the Model Recapitulate the Neuropathological Hallmarks of PD?

Since the definitive diagnosis of PD is made *post mortem* upon confirmation of specific neuropathological hallmarks, a model of PD should also include the relevant neuropathology. Nigral DA neurons should possess Lewy body-like inclusions that are immunoreactive for ubiquitin and α-syn as well as dystrophic (or Lewy) neurites in the striatum. Further, as the nigrostriatal system in the PD brain is associated with neuroinflammation [11–18], markers indicative of reactive microgliosis should be observed. In addition, overt pathology that is not commonly associated with PD should not be observed, thereby demonstrating the model is somewhat specific in its effects.

2.5 Does the Model Result in Quantifiable Parkinsonian Symptoms?

The most important outcome for PD patients is in a therapy's amelioration of symptoms. In this light, PD models should result in quantifiable symptoms analogous to those observed in PD patients to allow for the study of how therapeutic intervention can slow, halt or reverse functional deficits. Therefore, motor symptoms, such as akinesia, bradykinesia, postural instability or tremor, are desirable qualities in a PD model, and their measurement must possess enough sensitivity to observe a therapeutic effect. Furthermore, motor symptoms should ideally present at a time when approximately 50 % of dopaminergic input to the striatum has been lost, thereby mirroring clinical presentation [10, 19].

3 Viral Vector-Based Models of Parkinson's Disease

Using genetic insights into PD pathology, several gene therapy-based--> models have been developed with varying degrees of success at recapitulating the human condition. These models include overexpression of α-syn, leucine-rich repeat kinase 2 (LRRK-2) and parkin substrates as well as knockdown of tyrosine hydroxylase (TH) and PTEN-induced putative kinase 1 (PINK1). Many of these models have been comprehensively reviewed before [20–26], but we will evaluate the currently employed models according to the above criteria in a more prescriptive fashion. A summary is presented in Table 1. Please note, while the following gene therapy-based models have their respective roles in modeling genetic forms of PD and in studying some important cell and molecular biology, we are limiting our discussion to how well they recapitulate the idiopathic form of PD in vivo in the mammalian brain.

Table 1
Overview of Viral Vector-Based Models of Parkinson's Disease

	Construct Validity?	Reproducible?		Progressive nigrostriatal degeneration?				Parkinsonian pathology?				Parkinsonian symptoms?	References		
Model	Construct validity for idiopathic PD	Within group	Across groups	Striatal DA terminal dysfunction/loss (over weeks)	Terminal loss precedes overt cell loss	Nigral DA neuron loss (≥8 weeks)	End stage nigral DA neuron loss >60%	Lewy-body-like inclusions	Dystrophic neurites	No overt, non-PD pathology	Neuro-inflammation	Motor impairment	Rats	Mice	Non-human primate
α-Synuclein overexpression															
Wild type	*Yes*	Yes	Yes	Yes	Yes	Yes	With high expression	Yes	Yes	Yes	Yes	Yes	[34, 35, 37–49, 66]	[50–53]	[54–56, 64]
A30P	*Yes*	Yes	Yes	Yes	Yes	Yes	No	Yes	Yes	Yes	No data	No	[34, 36, 38, 40, 57–59]	[60]	
A53T	*Yes*	Yes	Yes	Yes (on limited data)	No (on limited data)	Yes	No	Yes	Yes	Yes	Yes	Yes	[34, 35, 40, 61–63]	[52]	[54, 55, 64]
E35K, E46K, E57K	*Yes*	Not demonstrated	Not demonstrated	No data	No data	No	No	No data	No data	Yes	No data	No data	[40]		
S129A	*Yes*	Not demonstrated	Yes	Yes	Yes	Yes	Yes	Yes	Yes	No	No data	No data	[37–39, 65]		
LRRK-2 overexpression															
Wild type	*Yes*	Not demonstrated	Yes	No	No	No	No	No	Yes?	No	No	No	[75]	[76]	
G2019S	*Yes*	Not demonstrated	Yes	Yes	No (on limited data)	Yes	No	No	Yes?	No	Yes (on limited data)	No	[75]	[76]	
D1994A	*Yes*	Not demonstrated	Not demonstrated	No	No	No	No	No data	No data	No	No data	No		[76]	
G2019S + D1994A	*Yes*	Not demonstrated	Not demonstrated	No	No	No	No	No data	No data	No	No data	No		[76]	
TH shRNA	*No*	Not demonstrated	Yes	Yes, dysfunction only	No	No	No	No	No	No	No data	Yes	[87]	[86]	

(continued)

Table 1
(continued)

Model	Construct Validity?	Reproducible?		Progressive nigrostriatal degeneration?				Parkinsonian pathology?				Parkinsonian symptoms?	References		
	Construct validity for idiopathic PD	Within group	Across groups	Striatal DA terminal dysfunction/loss (over weeks)	Terminal loss precedes overt cell loss	Nigral DA neuron loss (≥8 weeks)	End stage nigral DA neuron loss >60%	Lewy-body-like inclusions	Dystrophic neurites	No overt, non-PD pathology	Neuro-inflammation	Motor impairment	Rats	Mice	Non-human primate
Parkinsubtrates															
CD-Crel1	*Yes*	Not demonstrated	Not demonstrated	Yes	Yes	Yes	No	No data	No data	No	No	No data	[81]		
Pael-R	*Yes*	Not demonstrated	Yes	Yes	No	Yes	Not quite	No	No	No	No	Yes (stepping test only)	[83]	[82]	
hu p38/JTV	*Yes*	Not demonstrated	Not demonstrated	No data	No data	No	No	No	No	No	No data	No data		[84]	
hu wt synphilin-1	*Yes*	Not demonstrated	Not demonstrated	No data	No data	No	No	No	No data	No	No data	No data		[85]	
hu wt synphilin 1 on A30Pα-syn background	*Yes*	Not demonstrated	Not demonstrated	No data	No data	No	No	Yes	No data	No	No data	No data		[85]	
hu R621C synphilin-1	*Yes*	Not demonstrated	Not demonstrated	No data	No data	No	No	No	No data	No	No data	No data		[85]	
hu R621C synphilin-1 on A30Pα-synback ground	*Yes*	Not demonstrated	Not demonstrated	No data	No data	No	No	Yes	No data	No	No data	No data		[85]	
PINK1shRNA	*Yes*	Not demonstrated	Not demonstrated	No	No	No	No	No	No	No	No data	No data		[90]	

3.1 α-Synuclein Overexpression

Mutations in the gene encoding α-syn, *SNCA*, have been definitively linked to familial PD [27]. Increases in overall α-syn expression by duplications or triplications of *SNCA* and single nucleotide mutations (viz., A30P, A53T and E46K) within *SNCA* are associated with development of PD and an earlier onset of disease symptoms [27–32]. In addition, the primary component of Lewy bodies is aggregated α-syn [33]; hence, targeted overexpression of wild-type or mutated α-syn to the adult, nonhuman nigrostriatal system was the first viral vector-based approach to model PD [34–36]. Since that time, gene transfer of α-syn to the SN has been the most extensively studied with numerous reports in mice, rats and nonhuman primates and using both adeno-associated viruses (AAV) and lentiviral (LV) vector constructs [34–65]. Over time, as technological advances allow for higher vector titers to be achieved and the identification of more efficient promoters, viral vector-mediated overexpression of α-syn has become more consistent and yielded a greater magnitude of effects. Present use of AAV or LV to overexpress α-syn recapitulates several components of PD neuropathology, including: (a) early striatal terminal dysfunction [62, 66], (b) progressive loss of striatal dopaminergic terminals, (c) progressive loss of dopaminergic neurons of the SNc following loss of terminals, (d) Lewy body-like inclusions containing α-syn, (e) dystrophic neurites resembling Lewy neurites [35, 41, 46, 67] and (f) neuroinflammation [44, 51, 62, 68]. In addition, viral vector-mediated α-syn overexpression results in PD-like motor symptoms that correlate with an approximate 50 % loss of striatal DA as observed in PD patients [47, 49].

In our own laboratory, we have characterized the degeneration, pathology and behavioral phenotype induced by intranigral injections of recombinant AAVserotype 2/5 (AAV2/5) in which expression of the human wild-type α-syn transgene is driven by the chicken beta actin/cytomegalovirus (CβA/CMV) enhancer-promoter hybrid that results in efficient gene expression in neurons [49] (*see* Fig. 1). This model results in: (a) transduction of the nigrostriatal system with human wild-type α-syn, (b) 60 % nigral DA neuron loss and 40 % reduction in striatal TH immunoreactivity 8 weeks post injection, (c) marked microgliosis in the SN associated with α-syn overexpression and (d) significant impairment in contralateral forelimb use 8 weeks post injection. Lastly, nigral neuron degeneration correlates with α-syn expression levels and can be adjusted by the investigator through altering the vector titer or construct [49], offering a methodological advantage.

3.2 LRRK-2 Overexpression

LRRK-2, a protein without a definitively known function or substrate, is 2527 amino acids long and located in the cytoplasm (reviewed in ref. 69). Several different mutations in the *LRKK2* gene have been linked to the development of PD in an autosomal dominant inheritance pattern [70–73]. These cases generally include the formation of α-syn immunoreactive Lewy bodies [74].

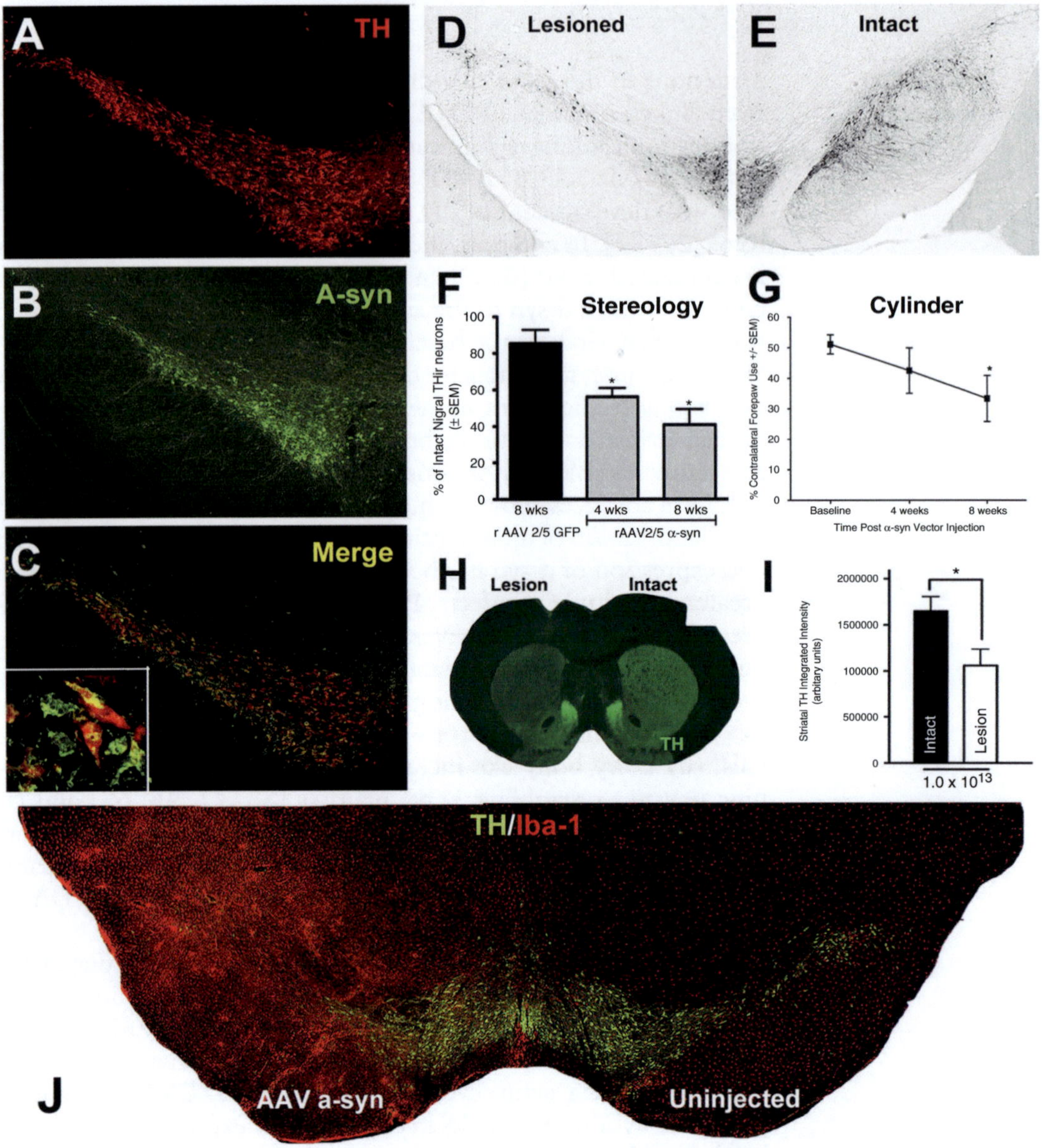

Fig. 1 Overexpression of α-synuclein in the rat nigrostriatal system via AAV2/5 results in nigral degeneration, contralateral forelimb impairment and microgliosis. Intranigral injections of AAV2/5 α-syn (2×2 μl of 1×10^{13} vg/ml, AP −5.3, ML +2.0 mm, DV −7.2 mm; AP −6.0 mm, ML +2.0 mm, DV −7.2 mm) were made to young-adult, male, Sprague Dawley rats as described previously [49]. (**a**–**c**) Coexpression of human, wild-type α-syn (*green*) in TH immunoreactive (THir, *red*) neurons within the SNc at 2 weeks following injections, prior to degeneration. (**d**) Degeneration of THir neurons of the SNc at 8 weeks after AAV2/5 α-syn injections compared to the uninjected, contralateral SN (**e**). (**f**) Stereological assessment revealed that after 4 weeks post α-syn vector injections, there is an approximate 40 % decrease in THir neurons in the SNc that progresses to about 60 % at 8 weeks (*$p < 0.05$ compared to green fluorescent protein (GFP) control). (**g**) Significant deficits in contralateral forelimb use are observed 8 weeks after AAV2/5 α-syn injections (*$p < 0.05$ compared to baseline). (**h**) Partial dopaminergic striatal denervation ipsilateral to α-syn overexpression at 8 weeks visualized using near infrared immunofluorescence and quantified in **i** (*$p < 0.05$ compared to baseline). (**j**) Microgliosis is associated with α-syn overexpression and nigral degeneration. THir SNc neurons (*green*) and ionized calcium binding adaptor molecule 1 (Iba-1, microglia-specific, *red*) immunofluorescence reveals marked microgliosis 8 weeks after AAV2/5 α-syn injections. Adapted from ref. [49]

They also result in the hallmark loss of striatal DA and nigral neuron degeneration found in idiopathic PD. Viral vector-mediated overexpression of LRRK-2 has been used far less extensively (i.e., in only two laboratories) than α-syn-based models [75, 76]. This is primarily due to the prohibitively large size of the LRRK-2 coding sequence, precluding its use in standard vectors such as AAV or lentivirus. Loss of striatal fibers and nigral neurons is of too low a magnitude to model anything more than very early PD. Furthermore, these models produce LRRK-2 complexes that are α-syn deficient and do not resemble Lewy bodies. On the other hand, dystrophic neurites are observed in the striatum with immunoreactivity for the pathological phospho-tau epitope (pSer202/pThr205), but α-syn immunoreactivity is not reported. No data exist on whether motor impairments are present; however, the low magnitude of striatal DA loss suggests motor deficits are unlikely to exist.

3.3 Parkin Substrates

Mutations in the gene for parkin have been linked to an autosomal recessive inheritance pattern of a very early-onset (i.e., juvenile) form of PD [77, 78]. These forms of PD may be better described as parkinsonisms in that they do not show the formation of Lewy bodies, but they do still exhibit loss of nigral DA neurons [79, 80]. Manipulating parkin for modeling PD has proven more difficult for gene therapy approaches since the development of the model requires a complete knockdown of the protein in order to mimic both copies of the gene being mutated in the human condition and to result in pathology. However, since patients with parkin mutations have increased levels of parkin substrates (i.e., parkin is unable to process the increasing supply of substrate), viral vector-mediated overexpression of parkin substrates instead has been used to create a "loss-of-function" paradigm—rather, loss-of-function paradigms, as four parkin substrates have been used with variable results.

In two models overexpressing parkin substrates, CDCrel-1 or Pael-R, a partial loss of striatal DA terminals and concomitant loss of DA is observed, and this loss is progressive on the order of weeks [81–83]. For models overexpressing other parkin substrates, p38/JTV or synphilin, no data on terminal status are available [84, 85]. It is unfortunate that the magnitude of terminal loss matches the loss of nigral DA neurons, suggesting that overt terminal loss does not precede overt loss of perikarya, although this has not been directly examined. Using CDCrel-1 or Pael-R has the advantage over p38/JTV or synphilin in that nigral neuron loss occurs over many weeks proceeding to an eventual loss that corresponds to late-stage disease. Achieving half of the cell loss of these models over a similar time span with p38/JTV or synphilin requires the addition of A30P α-syn expression to create a pro-pathology environment. In these models, A30P α-syn expression lends the advantage of producing some neuropathology, including thioflavin-S positive inclusions. Lastly, the data available do not include behavioral assays to assess motor symptoms, with one exception: Pael-R overexpression will result in contralateral forelimb akinesia in the stepping test, but deficits were not observed in amphetamine- or apomorphine-induced rotations, nor in the cylinder task [83].

3.4 TH Knockdown

The rate-limiting step in DA synthesis requires the enzyme TH. Virally delivered shRNA-mediated knockdown of TH has been used by two laboratories to deplete striatal DA [86, 87]. This approach results in a phenotype that neurochemically resembles PD but not morphologically, so long as the titer is low enough to avoid nonspecific degeneration. While not directly examined, there is no reason to believe that striatal terminals underwent degeneration, nor is there loss of SNc neurons. Other aspects of neuropathology are also absent, and no data exist on the role of neuroinflammation. Lastly, behavioral deficits do exist in this model. It is unfortunate that the vector construct and injection parameters lead to transduced dopaminergic neurons outside the nigrostriatal system, namely those found in the adjacent ventral tegmental area (VTA). As these neurons are actually found to be resilient to degeneration in PD, an impact on their function is not desirable for accurately modeling the disease.

3.5 PINK1 Knockdown

Autosomal recessive loss-of-function mutations in PINK1 have been identified in familial PD [88, 89]. PINK1 plays an important role in mitochondrial homeostasis and is critical for parkin recruitment into mitochondria; therefore, loss-of-function PINK1 mutations lead to deficient mitochondrial homeostasis [89]. Only one laboratory to date has used virally delivered shRNA to knockdown PINK1 [90]. Direct injection of adeno-associated virus expressing PINK1 shRNA into the striatum of mice did not affect nigral dopamine neuron survival during the short interval studied, but it did exacerbate the degeneration induced by MPTP. This study did not report effects of PINK1 knockdown on striatal dopamine levels, dopaminergic terminals or neuroinflammation.

3.6 Consideration of Aging in Vector-Based Models of PD

Aging is known to be the primary risk factor for PD since the vast majority of idiopathic cases occur in patients over the age of 65 [91]. All reports to date in which vector-based modeling of PD has

Fig. 2 (continued) cells are evident in the aged brain (*$p < 0.05$). Values represent mean α-syn cell counts per injected mesencephalon ± SEM for each group. (**f**) Despite transduction of significantly fewer cells and significantly less α-syn expression, aged rats demonstrate equivalent degeneration of the SNc, suggesting that α-syn neurotoxicity is exacerbated in the aged nigrostriatal system. Stereological assessment of THir neurons revealed that young-adult and aged rats injected with 2.2×10^{12} vg/ml AAV2/5 α-syn displayed equivalent loss (≈20 %) of THir SNc neurons 3 months after injections. Young-adult and aged rats injected with 1.0×10^{13} vg/ml displayed equivalent loss (≈35 %) of THir SNc neurons 1 month after injections. Values are expressed as the percent remaining THir SNc neurons as compared to the contralateral hemisphere ± SEM for each group

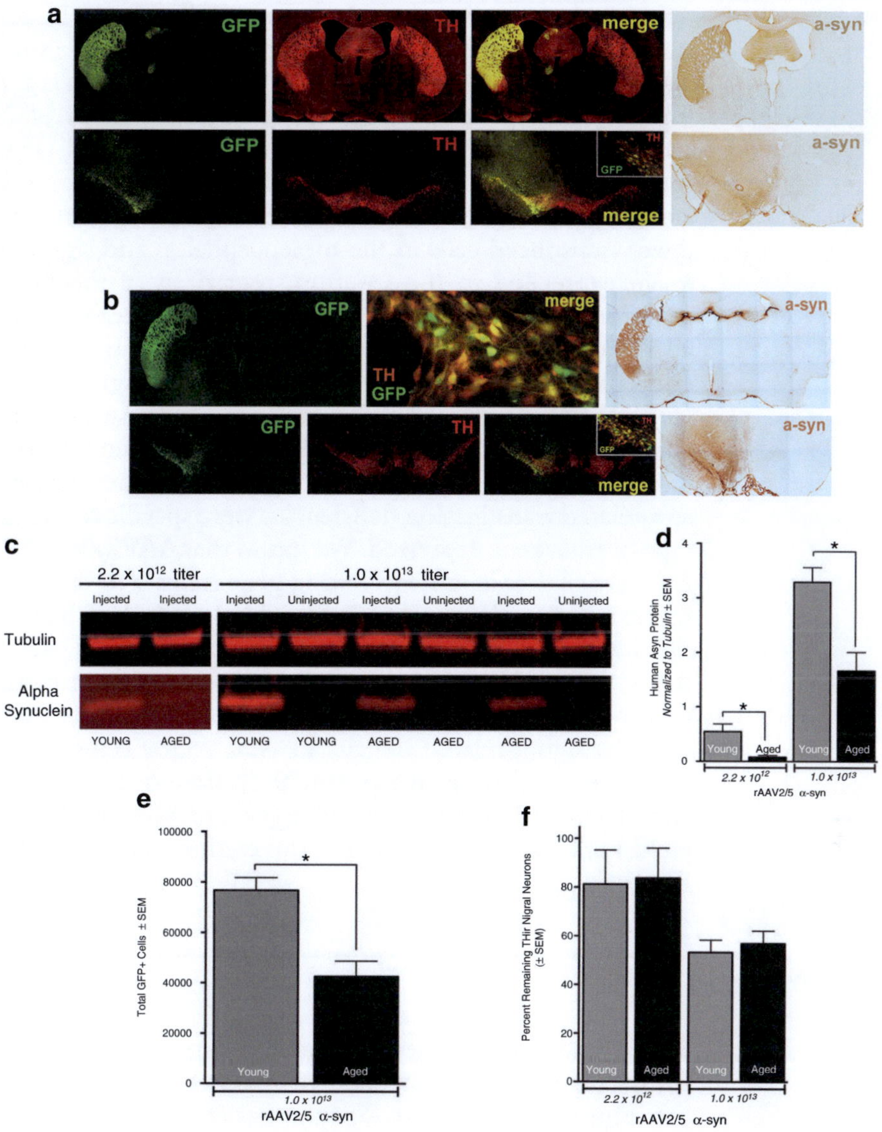

Fig. 2 AAV2/5-mediated α-Synuclein transduction is significantly reduced in the aged nigrostriatal system. Intranigral injections of AAV2/5 α-syn/GFP were made to young-adult and aged (20 months), male, Sprague Dawley rats as described previously [49]. (**a**, **b**) Immunofluorescent and immunohistochemical labeling of transduction of the nigrostriatal system demonstrating GFP (*green*), TH (*red*) and human wild-type α-syn (*brown*) at 1 month following injections into young-adult (**a**) and aged (**b**) rats. Despite the appearance of efficient transduction in both young-adult and aged rats, western blot revealed reduced human α-syn in striatal samples of aged rats compared to young-adult rats. (**c**) Representative western blot of α-syn immunodetection in striatal samples of young-adult and aged rats injected with two different AAV2/5 α-syn titers (2.2×10^{12} vg/ml 3 months after injection or 1.0×10^{13} vg/ml 1 month after injection). The striatum ipsilateral (Injected) and contralateral (Uninjected) to AAV2/5 α-syn injections is depicted. (**d**) Quantification of striatal human α-syn revealed significant deficits in human α-syn expression in aged rats for both vector titers examined ($*p < 0.05$). Values are expressed as the mean optical density scores, normalized to tubulin controls ± SEM for each group. (**e**) Stereological assessment of total α-syn immunoreactive cells in young-adult and aged rats 12 days following AAV2/5 α-syn injections (1.0×10^{13} vg/ml), prior to onset of degeneration. Significantly fewer transduced

occurred have used exclusively young-adult animals. In an effort to examine the impact of nigrostriatal α-syn overexpression in the aged brain environment, we conducted studies using AAV2/5 human wild-type α-syn vector injected into the SN of 3-month-old, young-adult rats and 20-month-old, aged rats (Fig. 2). We found that the transduction efficiency of AAV2/5 α-syn is significantly compromised in the aged brain, resulting in significantly fewer transduced cells in the mesencephalon and significantly less α-syn expression in the striatum, regardless of the titer used or duration of expression. Of interest, despite decreased transduction and lower levels of α-syn expression, the aged brain displayed an equivalent magnitude of α-syn-mediated nigral degeneration (Fig. 2). We continued these experiments using various pseudotypes of AAV or LV expressing the reporter green fluorescent protein (GFP) in young-adult and aged rats to determine whether age-related transduction deficiencies were specific to the AAV serotype or transgene expressed. We found that AAV2/5, AAV2/2 and LV were all deficient in facilitating gene transfer to the aged nigrostriatal system, whereas AAV2/9 yielded equivalent levels of transduction between young-adult and aged rats [92] (unpublished data), confirming previous results [93, 94]. Future studies using vector-based models of PD that include aging as a covariate should consider aging-related deficits in viral vector transduction. With careful control over this potential confound, studies using viral vector-based PD modeling in aged animals will be poised to yield important insights on the interaction of aging and causative PD genes.

4 Conclusions

Gene therapy approaches allow investigators to model PD through the overexpression or knockdown of genes of interest specifically in the nigrostriatal system. Substantial improvements in the last decade have led to more consistent and relevant models of PD than their first generation predecessors. We argue that the most important features of a model for testing experimental therapeutics are: (a) to model progressive striatal terminal dysfunction and DA loss prior to overt nigral neuron degeneration and (b) that this dysfunction and subsequent loss results in a demonstrable behavioral deficit. Based on our criteria, the α-syn overexpression model most closely recapitulates a comprehensive model of idiopathic PD. Even so, several models have come close on several measures. Overexpression of LRRK-2 with the G2019S mutation shows promise on measures of progressive nigrostriatal degeneration and neuropathology, and overexpression of the parkin substrates CD-Crell and Pael-R also show potential, though these require further work to more closely mimic the development of neuropathology and emergence of motor symptoms seen in PD patients.

Since only the α-syn overexpression models have been validated in nonhuman primates, all of the other models discussed deserve consideration for further study in the nonhuman primate, as there may be significant interactions between the model organism and the chosen gene or vector construct that may affect the translational value of the model. After all, the α-syn overexpression model is a result of several iterations by many laboratories around the globe using several model organisms over the last 12 or more years. The other models reviewed in this chapter have received less attention. Lastly, incorporation of aging as a covariate will more closely recapitulate the parkinsonian brain. The fields of PD research and experimental therapeutics stand much to gain in focusing our efforts to refine our use of all of the models reviewed here with respect to our proposed criteria for evaluating a comprehensive model of idiopathic PD.

Acknowledgements

This research was supported by Spectrum Health, Mercy Health Saint Mary's, the Edwin A. Brophy Fund, the Michigan State University (MSU) Neuroscience Graduate Program, the Michael J. Fox Foundation for Parkinson's Research and the Morris K. Udall Center of Excellence for Parkinson's Disease Research at MSU.

References

1. Pahwa R, Lyons KE (2013) Handbook of Parkinson's disease. CRC, Boca Raton, FL, p. xiv, 605 pages
2. Hirtz D et al (2007) How common are the "common" neurologic disorders? Neurology 68(5):326–337
3. Jankovic J (2008) Parkinson's disease: clinical features and diagnosis. J Neurol Neurosurg Psychiatry 79(4):368–376
4. Aminoff MJ (2001) Parkinson's disease. Neurol Clin 19(1):119–128, vi
5. Pollanen MS, Dickson DW, Bergeron C (1993) Pathology and biology of the Lewy body. J Neuropathol Exp Neurol 52(3):183–191
6. Shults CW (2006) Lewy bodies. Proc Natl Acad Sci U S A 103(6):1661–1668
7. Wakabayashi K et al (2013) The Lewy body in Parkinson's disease and related neurodegenerative disorders. Mol Neurobiol 47(2):495–508
8. Bendor JT, Logan TP, Edwards RH (2013) The function of alpha-synuclein. Neuron 79(6):1044–1066
9. Spieles-Engemann AL et al (2010) Stimulation of the rat subthalamic nucleus is neuroprotective following significant nigral dopamine neuron loss. Neurobiol Dis 39(1):105–115
10. Kordower JH et al (2013) Disease duration and the integrity of the nigrostriatal system in Parkinson's disease. Brain 136 (Pt 8): 2419–2431
11. Mogi M et al (1994) Interleukin-1 beta, interleukin-6, epidermal growth factor and transforming growth factor-alpha are elevated in the brain from parkinsonian patients. Neurosci Lett 180(2):147–150
12. Mogi M et al (1994) Tumor necrosis factor-alpha (TNF-alpha) increases both in the brain and in the cerebrospinal fluid from parkinsonian patients. Neurosci Lett 165(1–2): 208–210
13. McGeer PL et al (1988) Reactive microglia are positive for HLA-DR in the substantia nigra of Parkinson's and Alzheimer's disease brains. Neurology 38(8):1285–1291

14. McGeer PL, Itagaki S, McGeer EG (1988) Expression of the histocompatibility glycoprotein HLA-DR in neurological disease. Acta Neuropathol 76(6):550–557
15. Imamura K et al (2003) Distribution of major histocompatibility complex class II-positive microglia and cytokine profile of Parkinson's disease brains. Acta Neuropathol 106(6):518–526
16. Croisier E et al (2005) Microglial inflammation in the parkinsonian substantia nigra: relationship to alpha-synuclein deposition. J Neuroinflammation 2:14
17. Gerhard A et al (2006) In vivo imaging of microglial activation with [11C](R)-PK11195 PET in idiopathic Parkinson's disease. Neurobiol Dis 21(2):404–412
18. Doorn KJ et al (2014) Microglial phenotypes and toll-like receptor 2 in the substantia nigra and hippocampus of incidental Lewy body disease cases and Parkinson inverted question marks disease patients. Acta Neuropathol Commun 2(1):90
19. Cheng HC, Ulane CM, Burke RE (2010) Clinical progression in Parkinson disease and the neurobiology of axons. Ann Neurol 67(6):715–725
20. Kirik D, Bjorklund A (2003) Modeling CNS neurodegeneration by overexpression of disease-causing proteins using viral vectors. Trends Neurosci 26(7):386–392
21. Chesselet MF (2008) In vivo alpha-synuclein overexpression in rodents: a useful model of Parkinson's disease? Exp Neurol 209(1):22–27
22. Schneider B, Zufferey R, Aebischer P (2008) Viral vectors, animal models and new therapies for Parkinson's disease. Parkinsonism Relat Disord 14(Suppl 2):S169–S171
23. Ulusoy A et al (2008) In vivo gene delivery for development of mammalian models for Parkinson's disease. Exp Neurol 209(1):89–100
24. Lindgren HS, Lelos MJ, Dunnett SB (2012) Do alpha-synuclein vector injections provide a better model of Parkinson's disease than the classic 6-hydroxydopamine model? Exp Neurol 237(1):36–42
25. Low K, Aebischer P (2012) Use of viral vectors to create animal models for Parkinson's disease. Neurobiol Dis 48(2):189–201
26. Van der Perren A, Van den Haute C, Baekelandt V (2015) Viral vector-based models of Parkinson's disease. Curr Top Behav Neurosci 22:271–301
27. Polymeropoulos MH et al (1997) Mutation in the alpha-synuclein gene identified in families with Parkinson's disease. Science 276(5321):2045–2047
28. Ibáñez P et al (2004) Causal relation between alpha-synuclein gene duplication and familial Parkinson's disease. Lancet 364(9440):1169–1171
29. Kruger R et al (1998) Ala30Pro mutation in the gene encoding alpha-synuclein in Parkinson's disease. Nat Genet 18(2):106–108
30. Singleton AB et al (2003) alpha-Synuclein locus triplication causes Parkinson's disease. Science 302(5646):841
31. Gasser T, Hardy J, Mizuno Y (2011) Milestones in PD genetics. Mov Disord 26(6):1042–1048
32. Lees AJ, Hardy J, Revesz T (2009) Parkinson's disease. Lancet 373(9680):2055–2066
33. Spillantini MG et al (1997) Alpha-synuclein in Lewy bodies. Nature 388(6645):839–840
34. Lo Bianco C et al (2002) alpha-Synucleinopathy and selective dopaminergic neuron loss in a rat lentiviral-based model of Parkinson's disease. Proc Natl Acad Sci U S A 99(16):10813–10818
35. Kirik D et al (2002) Parkinson-like neurodegeneration induced by targeted overexpression of alpha-synuclein in the nigrostriatal system. J Neurosci 22(7):2780–2791
36. Klein RL et al (2002) Dopaminergic cell loss induced by human A30P alpha-synuclein gene transfer to the rat substantia nigra. Hum Gene Ther 13(5):605–612
37. Gorbatyuk OS et al (2008) The phosphorylation state of Ser-129 in human alpha-synuclein determines neurodegeneration in a rat model of Parkinson disease. Proc Natl Acad Sci U S A 105(2):763–768
38. da Azeredo Silveira S et al (2009) Phosphorylation does not prompt, nor prevent, the formation of alpha-synuclein toxic species in a rat model of Parkinson's disease. Hum Mol Genet 18(5):872–887
39. McFarland NR et al (2009) Alpha-synuclein S129 phosphorylation mutants do not alter nigrostriatal toxicity in a rat model of Parkinson disease. J Neuropathol Exp Neurol 68(5):515–524
40. Winner B et al (2011) In vivo demonstration that alpha-synuclein oligomers are toxic. Proc Natl Acad Sci U S A 108(10):4194–4199
41. Yamada M et al (2004) Overexpression of alpha-synuclein in rat substantia nigra results in loss of dopaminergic neurons, phosphorylation of alpha-synuclein and activation of caspase-9: resemblance to pathogenetic changes in Parkinson's disease. J Neurochem 91(2):451–461
42. Yamada M, Mizuno Y, Mochizuki H (2005) Parkin gene therapy for alpha-synucleinopathy: a rat model of Parkinson's disease. Hum Gene Ther 16(2):262–270

43. Ulusoy A et al (2010) Co-expression of C-terminal truncated alpha-synuclein enhances full-length alpha-synuclein-induced pathology. Eur J Neurosci 32(3):409–422
44. Sanchez-Guajardo V et al (2010) Microglia acquire distinct activation profiles depending on the degree of alpha-synuclein neuropathology in a rAAV based model of Parkinson's disease. PLoS One 5(1):e8784
45. Decressac M et al (2011) GDNF fails to exert neuroprotection in a rat alpha-synuclein model of Parkinson's disease. Brain 134(Pt 8):2302–2311
46. Decressac M et al (2012) Progressive neurodegenerative and behavioural changes induced by AAV-mediated overexpression of alpha-synuclein in midbrain dopamine neurons. Neurobiol Dis 45(3):939–953
47. Decressac M, Mattsson B, Bjorklund A (2012) Comparison of the behavioural and histological characteristics of the 6-OHDA and alpha-synuclein rat models of Parkinson's disease. Exp Neurol 235(1):306–315
48. Mulcahy P et al (2012) Development and characterisation of a novel rat model of Parkinson's disease induced by sequential intranigral administration of AAV-alpha-synuclein and the pesticide, rotenone. Neuroscience 203:170–179
49. Gombash SE et al (2013) Morphological and behavioral impact of AAV2/5-mediated overexpression of human wildtype alpha-synuclein in the rat nigrostriatal system. PLoS One 8(11):e81426
50. St. Martin JL et al (2007) Dopaminergic neuron loss and up-regulation of chaperone protein mRNA induced by targeted over-expression of alpha-synuclein in mouse substantia nigra. J Neurochem 100(6):1449–1457
51. Theodore S et al (2008) Targeted overexpression of human alpha-synuclein triggers microglial activation and an adaptive immune response in a mouse model of Parkinson disease. J Neuropathol Exp Neurol 67(12):1149–1158
52. Oliveras-Salva M et al (2013) rAAV2/7 vector-mediated overexpression of alpha-synuclein in mouse substantia nigra induces protein aggregation and progressive dose-dependent neurodegeneration. Mol Neurodegener 8:44
53. Oliveras-Salva M et al (2014) Alpha-synuclein-induced neurodegeneration is exacerbated in PINK1 knockout mice. Neurobiol Aging 35(11):2625–2636
54. Kirik D et al (2003) Nigrostriatal alpha-synucleinopathy induced by viral vector-mediated overexpression of human alpha-synuclein: a new primate model of Parkinson's disease. Proc Natl Acad Sci U S A 100(5):2884–2889
55. Eslamboli A et al (2007) Long-term consequences of human alpha-synuclein overexpression in the primate ventral midbrain. Brain 130(Pt 3):799–815
56. Yasuda T et al (2007) Neuronal specificity of alpha-synuclein toxicity and effect of Parkin co-expression in primates. Neuroscience 144(2):743–753
57. Lo Bianco C et al (2004) Lentiviral nigral delivery of GDNF does not prevent neurodegeneration in a genetic rat model of Parkinson's disease. Neurobiol Dis 17(2):283–289
58. Lo Bianco C et al (2004) Lentiviral vector delivery of parkin prevents dopaminergic degeneration in an alpha-synuclein rat model of Parkinson's disease. Proc Natl Acad Sci U S A 101(50):17510–17515
59. Lauwers E et al (2007) Non-invasive imaging of neuropathology in a rat model of alpha-synuclein overexpression. Neurobiol Aging 28(2):248–257
60. Lauwers E et al (2003) Neuropathology and neurodegeneration in rodent brain induced by lentiviral vector-mediated overexpression of alpha-synuclein. Brain Pathol 13(3): 364–372
61. Maingay M et al (2006) Ventral tegmental area dopamine neurons are resistant to human mutant alpha-synuclein overexpression. Neurobiol Dis 23(3):522–532
62. Chung CY et al (2009) Dynamic changes in presynaptic and axonal transport proteins combined with striatal neuroinflammation precede dopaminergic neuronal loss in a rat model of AAV alpha-synucleinopathy. J Neurosci 29(11):3365–3373
63. Koprich JB et al (2011) Progressive neurodegeneration or endogenous compensation in an animal model of Parkinson's disease produced by decreasing doses of alpha-synuclein. PLoS One 6(3):e17698
64. Barkholt P et al (2012) Long-term polarization of microglia upon alpha-synuclein overexpression in nonhuman primates. Neuroscience 208:85–96
65. Febbraro F et al (2013) Ser129D mutant alpha-synuclein induces earlier motor dysfunction while S129A results in distinctive pathology in a rat model of Parkinson's disease. Neurobiol Dis 56:47–58
66. Lundblad M et al (2012) Impaired neurotransmission caused by overexpression of alpha-synuclein in nigral dopamine neurons. Proc Natl Acad Sci U S A 109(9):3213–3219

67. Ulusoy A et al (2010) Viral vector-mediated overexpression of alpha-synuclein as a progressive model of Parkinson's disease. Prog Brain Res 184:89–111
68. Sanchez-Guajardo V et al (2013) Neuroimmunological processes in Parkinson's disease and their relation to alpha-synuclein: microglia as the referee between neuronal processes and peripheral immunity. ASN Neuro 5(2):113–139
69. Mata IF et al (2006) LRRK2 in Parkinson's disease: protein domains and functional insights. Trends Neurosci 29(5):286–293
70. Giasson BI et al (2006) Biochemical and pathological characterization of Lrrk2. Ann Neurol 59(2):315–322
71. Paisan-Ruiz C et al (2004) Cloning of the gene containing mutations that cause PARK8-linked Parkinson's disease. Neuron 44(4):595–600
72. Ross OA et al (2006) Lrrk2 and Lewy body disease. Ann Neurol 59(2):388–393
73. Zimprich A et al (2004) The PARK8 locus in autosomal dominant parkinsonism: confirmation of linkage and further delineation of the disease-containing interval. Am J Hum Genet 74(1):11–19
74. Cookson MR, Hardy J, Lewis PA (2008) Genetic neuropathology of Parkinson's disease. Int J Clin Exp Pathol 1(3):217–231
75. Dusonchet J et al (2011) A rat model of progressive nigral neurodegeneration induced by the Parkinson's disease-associated G2019S mutation in LRRK2. J Neurosci 31(3):907–912
76. Lee BD et al (2010) Inhibitors of leucine-rich repeat kinase-2 protect against models of Parkinson's disease. Nat Med 16(9):998–1000
77. Kitada T et al (1998) Mutations in the parkin gene cause autosomal recessive juvenile parkinsonism. Nature 392(6676):605–608
78. Lucking CB et al (2000) Association between early-onset Parkinson's disease and mutations in the parkin gene. N Engl J Med 342(21):1560–1567
79. Shimura H et al (2000) Familial Parkinson disease gene product, parkin, is a ubiquitin-protein ligase. Nat Genet 25(3):302–305
80. Takahashi H et al (1994) Familial juvenile parkinsonism: clinical and pathologic study in a family. Neurology 44(3 Pt 1):437–441
81. Dong Z et al (2003) Dopamine-dependent neurodegeneration in rats induced by viral vector-mediated overexpression of the parkin target protein, CDCrel-1. Proc Natl Acad Sci U S A 100(21):12438–12443
82. Kitao Y et al (2007) Pael receptor induces death of dopaminergic neurons in the substantia nigra via endoplasmic reticulum stress and dopamine toxicity, which is enhanced under condition of parkin inactivation. Hum Mol Genet 16(1):50–60
83. Dusonchet J et al (2009) Targeted overexpression of the parkin substrate Pael-R in the nigrostriatal system of adult rats to model Parkinson's disease. Neurobiol Dis 35(1):32–41
84. Ko HS et al (2005) Accumulation of the authentic parkin substrate aminoacyl-tRNA synthetase cofactor, p38/JTV-1, leads to catecholaminergic cell death. J Neurosci 25(35):7968–7978
85. Krenz A et al (2009) Aggregate formation and toxicity by wild-type and R621C synphilin-1 in the nigrostriatal system of mice using adenoviral vectors. J Neurochem 108(1):139–146
86. Hommel JD et al (2003) Local gene knockdown in the brain using viral-mediated RNA interference. Nat Med 9(12):1539–1544
87. Ulusoy A et al (2009) Dose optimization for long-term rAAV-mediated RNA interference in the nigrostriatal projection neurons. Mol Ther 17(9):1574–1584
88. Houlden H, Singleton AB (2012) The genetics and neuropathology of Parkinson's disease. Acta Neuropathol 124(3):325–338
89. Trempe JF, Fon EA (2013) Structure and function of Parkin, PINK1, and DJ-1, the three musketeers of neuroprotection. Front Neurol 4:38
90. Haque ME et al (2012) Inactivation of Pink1 gene in vivo sensitizes dopamine-producing neurons to 1-methyl-4-phenyl-1,2,3,6-tetrahydropyridine (MPTP) and can be rescued by autosomal recessive Parkinson disease genes, Parkin or DJ-1. J Biol Chem 287(27): 23162–23170
91. Collier TJ, Kanaan NM, Kordower JH (2011) Ageing as a primary risk factor for Parkinson's disease: evidence from studies of non-human primates. Nat Rev Neurosci 12(6):359–366
92. Polinski NK et al (2015) Recombinant adeno-associated virus 2/5-mediated gene transfer is reduced in the aged rat midbrain. Neurobiol Aging 36(2):1110–1120
93. Wu K et al (2004) The effects of rAAV2-mediated NGF gene delivery in adult and aged rats. Mol Ther 9(2):262–269
94. Klein RL et al (2010) Pronounced microgliosis and neurodegeneration in aged rats after tau gene transfer. Neurobiol Aging 31(12): 2091–2102

Chapter 27

Gene Therapy-Based Modeling of Neurodegenerative Disorders: Huntington's Disease

Deborah Young

Abstract

Huntington's disease is a fatal neurodegenerative disease characterized by impairments in motor control, and cognitive and psychiatric disturbances. In this chapter, viral vector-mediated approaches used in modeling the key neuropathological features of the disease including the production of abnormal intracellular protein aggregates, neuronal dysfunction and degeneration and motor impairments in rodents are described.

Key words Viral vector, Huntingtin, Transgenic, Genetic, Neurons, Neurodegenerative disease

1 Introduction

Huntington's disease (HD) is a progressive and devastating neurodegenerative disorder characterized by a triad of symptoms including motor dysfunction, cognitive decline, and psychiatric disturbances. There is currently no effective treatment for HD, a disease that progresses towards death within approximately 20 years of disease onset [1]. HD is inherited in an autosomal dominant manner, with the underlying genetic mutation being an expansion of a CAG repeat sequence in exon 1 of huntingtin (*HTT*). This results in the production of huntingtin protein (HTT) with an expanded polyglutamine (polyQ) tract. Expansion to greater than 36 CAG repeats results in the disease, with increased tract lengths associated with an earlier disease onset. Current hypotheses posit that HD could be caused by the dominant-negative action of mutant HTT resulting in a loss of function of the wild-type protein, but increasing evidence suggests a toxic gain of function of mutant HTT (mHTT) is more likely to be causative of HD pathogenesis. The key neuropathological hallmarks of HD are abnormal intracellular aggregates composed of mHTT and other sequestered proteins including transcription factors,

Fredric P. Manfredsson (ed.), *Gene Therapy for Neurological Disorders: Methods and Protocols*, Methods in Molecular Biology, vol. 1382, DOI 10.1007/978-1-4939-3271-9_27,

neuronal dysfunction and neurodegeneration [2]. The brain regions most profoundly affected are the striatum, where extensive loss of GABAergic projection neurons results in significant atrophy of the striatum. Neuronal loss resulting in the thinning of the cortical layers is also found [3], with the hippocampus, hypothalamus, and substantia nigra also affected [4].

Various animal models of HD have been developed that reproduce specific neuropathological features of the disease. There is no single model that fulfills the criteria of modeling all aspects of the human disease, with each type of model having specific strengths but also limitations [5]. Early studies utilized specific neurotoxins that induce excitotoxicity (glutamate-, kainic acid-, quinolinic acid) [6–9] or impair mitochondrial function (3-nitroprionic acid, malonate [10–12]) to produce the characteristic degeneration of neurons in the striatum in rodents and primates. These models have largely been superseded by genetic approaches that aim to emulate the molecular pathogenic mechanisms underlying HD. Transgenic animal models including *C. elegans* [13, 14], Drosophila [15–17] and mice that express full-length, N-terminal fragments of mHTT or knock-in of mHTT sequences into the endogenous *Htt* locus of the animal have been developed [18–22]. Specific hallmarks of the human disease are reproduced including nuclear and neuritic aggregates that are ubiquitinated, neuronal dysfunction and specific impairments in motor function. Of note, although striatal atrophy and neuronal cell death has been reported in several mouse models [23–25], no striatal neuronal cell death is found in other transgenic rodent models [26, 27].

Viral vector-based modeling of HD offers several advantages over traditional transgenic approaches. Expression cassettes varying in truncated N-terminal fragments and CAG repeat length under the control of regulatable elements or cell or tissue-specific promoters that direct higher levels of transgene expression than can be achieved in transgenic mouse models can be rapidly developed. Expression cassette design coupled with the engineering of viral capsids can be exploited to direct transgene expression to neurons or astrocytes. Viral vectors can also be injected at any developmental age in a broad range of species including mice, rats, and nonhuman primates, making it a very versatile tool for disease modeling purposes. Lentiviral, adenoviral, and adeno-associated viral (AAV) vectors have been exploited as gene transfer agents for HD modeling studies both in an in vitro and in vivo setting as discussed below.

1.1 Using Viral Vectors to Model HD In Vitro

One advantage of many viral vectors is their ability to efficiently transduce primary neuronal cultures, facilitating studies on the molecular mechanisms underlying mHTT neuropathology. Primary striatal neurons infected with lentiviral vectors expressing N-terminal mHTT expressing 82 polyQ (mHTT82Q) under control of the mouse phosphoglycerate kinase 1 (PGK) promoter develop nuclear and neuritic aggregates that subsequently lead to neuronal dysfunction and toxicity by 6–8 weeks [28]. Aggregate neuropathology is also observed following transduction of primary cortical neurons but in contrast, no signs of neuronal dysfunction were observed, consistent with the selective vulnerability of GABAergic medium spiny neurons to mHTT. The rate of formation of HTT inclusions is considerably slower in cells and is also found in neuritic and cytoplasmic compartments expressing full-length mHTT; in comparison truncated mHTT is found predominantly in the form of nuclear aggregates [29]. Furthermore, Senut and colleagues [30] showed that AAV-mediated expression of GFP fused to expanded CAG repeat lengths in HEK293 cells led to the rapid appearance of GFP-positive cytoplasmic and nuclear aggregates 16 h after infection suggesting that expanded polyQ tracts alone can mediate pathogenic effects.

1.2 Viral Vector-Based Rodent Models of HD

Viral vector-based rodent models of HD are a powerful complement to the existing chemical and transgenic mouse and rat models. The most common strategy employed is to inject viral vectors expressing truncated *HTT* transcripts with expanded CAG repeats unilaterally into one striatum, leaving the contralateral side to serve as internal control for subsequent biochemical and molecular analyses. Neuropathological analyses can be compared against control animals that express a matching *HTT* transcript but with normal repeat length (i.e., <36 CAG). Potential new treatments can be initiated either prior or following vector infusion. Most studies have relied on immunohistochemical methods to examine specific aspects of interest in the model including time-dependent changes in mHTT expression levels and intracellular aggregate formation using anti-HTT or antibodies that specifically detect aggregated HTT (e.g., EM48), protein partners that interact with HTT and the effects of mHTT on specific striatal neuronal populations (e.g., using antibodies to NeuN, calbindin-D28K, DARPP32, choline acetyltransferase (ChAT), parvalbumin, Neuropeptide Y (NPY)). Other features such as reactive gliosis can be examined as well as behavioral testing of treated animals to assess any deficits in motor function.

In general, the neuropathological features reported between the different studies are remarkably consistent, with a rapid onset of pathology following vector infusion that progresses to significant neuronal cell loss in the striatum by 5–12 weeks. The earliest abnormalities are the appearance of intranuclear neuronal aggregates and inclusions that sequester ubiquitin and other proteins, progressing to neuronal cell loss and striatal atrophy and motor impairment, with variations in the time course of neuropathology dependent on mHTT expression levels, HTT fragment length, and CAG repeat length. I will now cover some specific variables that can influence the model.

1.2.1 Viral Vectors and Vector Infusion

Lentiviral [31] and AAV vectors including AAV2 [30], chimeric AAV1/2 [32], and mosaic AAV1/8 [33] vectors have been used to deliver mHTT or polyQ transcripts into the mouse and rat striatum. AAV1/2 and AAV1/8 vectors transduce a large volume of the striatum compared to AAV2, and possibly lentiviral vectors [31], leading to a greater volume of distribution of transgene expression and production of larger striatal lesions. For example, we found that 35 % of the injected striatum was devoid of staining for the neuronal marker calbindin-D28K, DARPP-32, and NeuN by 5 weeks [32]. However, differences in transduction and lesion volumes could also be largely due to differences in the promoters used in the AAV1/2 and AAV1/8 studies compared to the lentiviral vector study; the PGK and CMV promoters were used in a lentiviral context, whereas the strong CMV/chicken beta actin hybrid promoter and neuron-specific enolase (NSE) promoters were used in the AAV studies [32, 33]. Indeed we found that the AAV1/2-mediated mHTT expression levels rose to >100-fold that of endogenous rat Hdh expression at 2 weeks, while Drouet et al. [34], found that mHTT expression levels under control of the PGK promoter in lentiviral vectors was 25-fold higher than endogenous HTT. In a subsequent study, Regulier and colleagues used a tetracycline-regulated lentiviral vectors to express the first 853 amino acids of mHTT. Expression levels were 4- to 5-fold higher than that achieved using the PGK promoter and induced an early pathological onset and exacerbated the HD neuropathology [35]. One advantage of the tetracycline-regulated system is that it allows conditional expression of mHTT by systemic administration of doxycycline, enabling exploration of the relationship between mHTT protein expression and disease progression.

A snapshot of mHTT dosage effects on the kinetics of aggregate formation or toxicity can often be captured by analyses of brain sections from a 2 to 5 week time-point, although the specific timing of these appearance of this pathology will vary depending on vector system and mHTT fragment expressed. Typically the highest levels of transgene are detected 1–2 mm from the viral vector injection site, and so the earliest appearance of intranuclear inclusions and neuronal depletion occurs in this region. Surrounding the core region of neuronal depletion, neurons express lower levels of transgene expression presumably reflective of the extent of diffusion of vector from the infusion site. These neurons may show evidence of intracellular aggregates.

In humans with HD, there is preferential loss of GABAergic medium spiny neurons. The demonstration that lentiviral-mediated expression of mHTT leads to selective depletion of DARPP32-positive medium spiny neurons but interneurons in the mHTT-injected rats are largely spared is consistent with that observed in human HD [31]. In contrast, we found that the chimeric AAV1/2 vector used in our study transduced striatal medium spiny neurons but there was increased transduction of cholinergic interneurons compared to other neuronal subtypes. This resulted in toxicity and neuronal death of this interneuron population irrespective of the transgene expressed. In contrast, we found that parvalbumin and NPY-positive interneurons were only susceptible to mHTT [32].

Retrograde and anterograde transport of AAV vectors from the striatal injection site can lead to transgene expression and the appearance of intracellular aggregates in distal basal ganglia regions including the substantia nigra (SN) and globus pallidus [30, 32], brain regions where neurodegeneration and atrophy are also observed in humans with HD [36]. We found axonal transport of a chimeric AAV1/2 vector expressing exon 1 of HTT with 70Q was associated neuronal cell loss and atrophy of the globus pallidus consistent with that observed in human HD. However, we also observed loss of dopaminergic neurons in the SN pars compacta, which is less typical as the SN pars reticulata region is more affected [36]. DeFiglia and colleagues also observed many shrunken and degenerating HTT-labeled neurons in cortical layers 5 and 6 at 2 weeks in mice that received an intra-striatal injection of an AAV1/8 vector expressing a truncated mHTT transcript (400 a.a. with 100Q) [33].

Together these results suggest that viral vector-based modeling can reproduce many of the features of the human disease but artifacts in the disease model might also be introduced by depending on the viral vector type used and thus unexpected results may need to be interpreted cautiously.

More recently, lentiviral vectors optimized for astrocytic targeting by pseudotyping with the Mokola viral envelope and using the astrocyte-specific glial fibrillary protein (GFAP) promoter to drive transgene expression has been used to investigate the effect of expression of mHTT in astrocytes [37]. Reactive gliosis and decreased expression of glutamate transporters leading to altered glutamate uptake and neuronal dysfunction were found, consistent with that found in brain samples from HD disease subjects. This provides an alternative model for examining the contribution of mHTT-mediated astrocyte dysfunction to HD pathogenesis.

1.2.2 Mutant HTT Transcripts and Expression Levels

Various N-terminal truncated *mHTT* transcripts have been expressed using the viral vectors as described above, with comparisons conducted relative to *HTT* fragments matched for size and linked to normal CAG repeat lengths (<30Q). Animals that receive control HTT vectors typically show diffuse neuronal HTT immunostaining in the striatum, and no evidence of intracellular inclusion formation or neuronal toxicity throughout the study.

Senut et al. [30], were the first to demonstrate that cumulative expression of expanded polyQ repeats throughout the life is not required to induce cell death but rather acute overexpression of polyQ is toxic to neurons in vivo. AAV2-mediated expression of expanded polyQ transcripts fused to green fluorescent protein was sufficient to induce the formation of intranuclear aggregates or large inclusion bodies as early as 5 days, with the majority of GFP-expressing neurons showing ubiquitinated aggregates by 12 days, leading to the degeneration of striatal neurons.

Several groups extended this work to characterizing models based on expression of N-terminal truncated mHTT transcripts; de Almeida and colleagues conducted a comprehensive evaluation of the relationship between mHTT expression levels, polyglutamine repeat size (19, 44, 66, or 82 CAG) and protein length (N-terminal fragments consisting of the first 171, 853, or 1520 amino acids of human HTT) [31]. Similarly, studies by deFiglia et al., and Franich et al., use constructs that fall with these ranges (the first 400 amino acids of HTT and 18 or 100Q) [33], (exon 1 of HTT and 20 or 70Q) [32]. In these studies, HD neuropathology at specific time-points ranging from 1 to 12 weeks was examined.

Nuclear HTT aggregates appeared as early as 1 week after viral vector injection, are ubiquitinated by 2 weeks [31] and progressively accumulate over the first 4 weeks. The size of the aggregates are dependent on mHTT expression levels, with lower levels associated with intranuclear and neuritic aggregates and dystrophic neurites, while larger intranuclear inclusions are found with higher transgene expression [33, 31]. Evidence of neuronal loss can occur as early as 2 weeks in the immediate vicinity of the injection site, coinciding with the peak of nuclear inclusion accumulation. A loss of anti-HTT aggregate immunostaining then coincides with extensive neuronal degeneration and reactive gliosis from 5 to 12 weeks after injection and motor impairment [32]. The time course of neuropathology is influenced by the specific transcripts expressed, with shorter mHTT fragments, longer CAG repeats, and higher expression levels resulting in an earlier onset and more severe pathology [31].

Two of these models have been used successfully to demonstrate the efficacy of RNAinterference-based therapies for HD. Therapeutic silencing of mHTT with siRNA or shRNA promoted neuronal survival as well as reduced the numbers of neuropil aggregates and size of inclusions [33, 32], and prevented impairments in motor function as assessed by spontaneous forepaw usage [32].

1.3 Nonhuman Primate Models of HD Using Viral Vector-Mediated Approaches

Viral-vector mediated modeling approaches have been extended to applications to nonhuman primates. In a landmark study, Palfi et al. [38] injected lentiviral vectors expressing mHTT171-19Q or mHTT171-82Q into the dorsolateral putamen of macaque monkeys. Four injection sites were chosen covering the dorsolateral aspect of the commissural and post-commissural putamen, a region known to be involved in dyskinesia in primates with unilateral excitotoxic lesions [39]. Similar to the pathology observed in rats, mHTT171-82Q expression was associated with the formation of neuritic and nuclear ubiquitinated aggregates, loss of staining for the neuronal marker NeuN and astrogliosis at 9 weeks post-vector infusion. By 30 weeks, the size of EM48 HTT inclusions were smaller and atrophy of the putamen was clearly evident. When the vector was infused unilaterally, no changes in spontaneous behavior were observed over the 9 week period but choreiform movements were induced by apomorphine injection. Bilateral infusions of vector led to display of a spectrum of movement deficits beginning at 16 weeks and continued for up to 30 weeks including hand, leg and head dyskinesia, leg dystonia, and even tics. Furthermore, behavioral analysis of these animals showed that their neurological deficits progressed in a manner similar to the progression observed in HD patients.

The first transgenic monkey model of HD has recently been generated by Chan and colleagues [40] using a combination of transgenesis approaches and viral vector technology. Lentiviral vectors expressing exon 1 of the human *HTT* gene with 84 CAG repeats under the control of the human polyubiquitin-C promoter were injected into the perivitelline space of rhesus monkey oocytes. Five live newborns were delivered at full-term, with all transgenic monkeys carrying the transgenic mHTT genes but with variable repeat lengths ranging from 29 to 88 CAG. Three newborns carrying between two and four copies of mHTT developed severe symptoms, with two monkeys died shortly after birth, while the other survived for 1 month. Of the remaining monkeys, each of which caries a single copy of the mutated gene, one has only 29 CAG repeats and so shows no disease symptoms while the other has 83 CAG repeats and has developed low level deficits in movement coordination and involuntary movements such as chorea and dystonia which began 1 week after birth. HTT aggregates or inclusions were found in the striatum and cortex, with increased intensity of EM48 immunostaining correlating with high levels of mHTT with longer repeats. No obvious signs of neurodegeneration were found in striatum of the monkey that died shortly after birth.

The ongoing development and characterisation of these large animal models will be valuable resources for therapeutic screening of new HD therapies. These recapitulate the features of human disease, will be valuable in optimisation of delivery of cell and gene therapies, and help understand relationship between mHTT expression levels, neuropathology, and motor dysfunction.

1.4 Summary

These data show that viral vector-based rodent and primate models of HD can be a powerful complement to existing transgenic models of HD and can play an important role in the screening of new therapies for this devastating disease. Methods for generating a rat model of HD are described below.

2 Materials

1. Viral vectors: We used a chimeric AAV1/2 vector in our study but other AAV serotypes including AAV5, AAV8, AAV9 that transduce large volumes of the rat or mouse striatum would also be suitable as well as the lentiviral vectors as described above and can be obtained through established vector core facilities or suppliers (e.g., GeneDetect.com., Penn Vector Core, University of Pennsylvania). Specific N-terminal mHTT fragments as described above can be expressed; longer fragment sizes will be associated with slower rate of disease progression (also *see* **Note 1**).
2. Animals: We typically use male Wistar or Sprague Dawley rats in the 230–300 g weight range.
3. Stereotaxic frame (David Kopf Instruments, Tujunga, CA, USA).
4. Rotary drill: We use a micro-drill that was purchased from Fine Science Tools Inc. (FST, Foster City, CA, USA). A similar item can be purchased from Stoelting Co. (Stoelting, Wood Dale, IL, USA).
5. Hamilton syringes: 10 μL 701 N point style #2 (Hamilton, Reno, NV, USA).
6. Microinfusion pump: UltraMicroPump with Sys-Micro4 controller (World Precision Instruments, Sarasota, FL, USA). *See* http://www.wpiinc.com for distributors.
7. 0.9 % (v/v) saline: Weigh 9 g of NaCl and add distilled water to a final volume of 1 L.
8. 4 % (w/v) paraformaldehyde (PFA) in 0.1 M phosphate buffer, pH 7.4: Make an 8 % PFA solution by weighing 40 g PFA and dissolve in 500 mL distilled water. Stir solution and heat to 60 °C in a fume hood. Add 1 N NaOH in a dropwise manner using a pipette until the solution clears. Cool and filter to remove any particular matter. Dilute ½ with 0.2 M phosphate buffer, pH 7.4. For 0.2 M phosphate buffer, make a dibasic solution by dissolving 28.39 g Na_2HPO_4 in distilled water and make up to final volume of 1 L. Make a monobasic solution by dissolving 27.6 g $NaH_2PO_4{\cdot}2H_2O$ in water and make up to a final volume of 500 mL. Add the monobasic to the dibasic solution until pH 7.4 is obtained.
9. 10, 20 and 30 % (w/v) sucrose in 1× phosphate-buffered saline (PBS), pH 7.4: Dissolve 100, 200 or 300 g sucrose in 1× PBS to a final volume of 1 L. To make a 10× PBS stock solution, dissolve 80 g NaCl, 2 g KH_2PO_4, 2 g KCl, 11.5 g Na_2HPO_4 in water and make up to a final volume of 1 L and

adjust to pH 6.8 with HCl. Once diluted to 1× PBS, (1/10 with distilled water), this should be at pH 7.4.

10. Spontaneous forelimb use perspex cylinder: made in-house. Dimensions 20 cm diameter × 30 cm height.
11. Automated video-tracking system: Noldus Ethovision XT (Noldus, Leesburg, VA, USA).
12. 2 m long, 20 cm wide runway into an escape box. Made in-house.
13. Rotorod: TSE Rotarod Advanced (TSE Systems Inc., Chesterfield, MO, USA). *See* www.tse-systems.com for distributors.

3 Method

3.1 Stereotaxic Delivery of Vectors into the Rat Brain

This section describes our procedures for infusion of viral vectors into the striatum.

1. Animals are anesthetised according to institutional guidelines and the animal positioned in a stereotaxic frame.
2. Make an incision through the scalp to expose the skull surface and identify bregma and lambda points. Adjust the incisor bar height such that bregma and lambda are at equal dorsoventral height (flat skull).
3. Coordinates for intra-striatal infusion of viral vector into the rat brain are determined with reference to the atlas of Paxinos and Watson [41] and are as follows: anterior–posterior (AP) +0.4 mm, mediolateral (ML) −3.0 mm, dorsoventral (DV) −5.5 mm, bregma = 0.
4. Make a small burr hole in the skull using a high-speed rotary drill at the appropriate AP-ML coordinates and lower the tip of a 10 μL Hamilton syringe with a 25 gauge needle under control of a microinfusion pump controller into the infusion site.
5. AAV vector (up to 3 μL) is infused into the striatum at an infusion rate of 70–100 nL/min (*see* **Note 2**).
6. After completion of the infusion, the syringe is left in place for a further 5 min to allow diffusion of the vector before it is slowly withdrawn over a 5 min period and the scalp sutured.
7. Follow institutional guidelines for application of analgesia and monitoring recovery from surgery as required.
8. At specific time-points following vector infusion, rats or mice are transcardially perfused with 0.9 % (v/v) saline followed by 4 % paraformaldehyde in 0.1 M phosphate buffer, pH 7.4 (4 % PFA).

9. Brains are fixed overnight in 4 % PFA before cryoprotection by immersion in 10 % sucrose in PBS. The following day, remove the 10 % sucrose solution and replace with 20 % sucrose in PBS. The brains will float to the top of the tube. Once the brains have sunk to the bottom of the tube (1–2 days), replace solution with the 30 % sucrose in PBS. The brains are ready for cryosectioning once they have sunk to the bottom.

3.2 Behavioral Testing

Impairments in motor function can be assessed using specific motor function tests. We used the spontaneous exploratory forelimb use test to confirm the motor deficit in our rat HD model [32] but other tests as described below can also be used (*see* **Note 3**).

3.2.1 Spontaneous Exploratory Forelimb Use

Motor impairment following unilateral lesioning of the striatum can be assessed by spontaneous exploratory forelimb use. Rats are placed inside a transparent perspex cylinder (20 cm diameter × 30 cm height) and forelimb use during exploratory activity is recorded by video recording for 5 min. Two mirrors were placed behind the cylinder at an angle to enable recording of forelimb movements when the rat is facing away from the camera. The forelimb used for push-off and landing of vertical movements within the cylinder is counted by reviewing the video footage in slow motion. Net ipsilateral forelimb use values are expressed as a percentage of total forelimb use.

3.2.2 Spontaneous Locomotor Activity in an Open Field

Spontaneous locomotor activity can be assessed by placing rats in a circular enclosure, 1.8 m in diameter divided into segments of equal area. Locomotion under dim lighting condition is video-recorded for 5 min and the number of lines (the boundary between each segment) crossed is manually quantified by the investigator. Monitoring of this type of activity can also be conducted by utilizing automated video-tracking systems.

3.2.3 Footprint Test

The gait of rats can be analyzed to detect deficits in locomotion. Rats are trained to walk along an enclosed 2 m long, 20 cm wide runway into an escape box. On the day of the trial, place the rats' hind- and forepaws in red and blue nontoxic dye, respectively (20 % food coloring in glycerol), before releasing the rat onto a strip of paper on the floor of the runway and allowing the rat to walk into the escape box. The footprint patterns are assessed quantitatively by measurements of forepaw and hindpaw stride length, base width and forepaw/hindpaw footprint overlap. The mean values from each set of six steps are calculated.

3.2.4 Rotarod Performance

Rats' locomotor coordination is assessed on an accelerating rotarod. Rats are placed on the rotarod which accelerates from 4 to 40 rpm over 5 min. Rats are given two training sessions and one trial per day over 3 days when the latency to fall is recorded.

4 Notes

1. Our AAV1/2 HD rat model has a rapid onset of neuropathology with significant cell loss occurring by 2 weeks [32]. We found in subsequent studies that slight differences in virus batches can shift the time course of neuropathology quite dramatically, which can create difficulties when using previous results as a basis for the design of studies aimed at testing new therapies. Similarly, deFiglia et al., [33] reported differences in extent of protection with siRNA against HTT in animals injected with two different batches of AAV vector. To increase the likelihood that the time-course of effects will be consistent between studies, we recommend generating a large batch of viral vector stock, conducting pilot studies in animals to characterize the time course of neuropathology obtained. The same vector stock can then used in subsequent studies.
2. Include appropriate controls such as vectors expressing the same mHTT fragment but with a normal (e.g., <36) CAG repeat length. Performing unilateral infusions also allows comparisons between the injected and uninjected contralateral hemisphere as an internal control.
3. Prior to commencement of the tests, all animals should be prehandled 5 min per day for 3–5 days by the investigator performing the tests to acclimatize them to being handled, preferably in the room where the tests will be performed.

References

1. Martin JB, Gusella JF (1986) Huntington's disease. Pathogenesis and management. N Engl J Med 315(20):1267–1276. doi:10.1056/NEJM198611133152006
2. (1993) A novel gene containing a trinucleotide repeat that is expanded and unstable on Huntington's disease chromosomes. The Huntington's Disease Collaborative Research Group. Cell 72(6):971–983. doi:0092-8674(93)90585-E [pii]
3. Reiner A, Albin RL, Anderson KD, D'Amato CJ, Penney JB, Young AB (1988) Differential loss of striatal projection neurons in Huntington disease. Proc Natl Acad Sci U S A 85(15):5733–5737
4. Vonsattel JP, DiFiglia M (1998) Huntington disease. J Neuropathol Exp Neurol 57(5):369–384
5. Pouladi MA, Morton AJ, Hayden MR (2013) Choosing an animal model for the study of Huntington's disease. Nat Rev Neurosci 14(10):708–721. doi:10.1038/nrn3570
6. McGeer EG, McGeer PL (1976) Duplication of biochemical changes of Huntington's chorea by intrastriatal injections of glutamic and kainic acids. Nature 263(5577):517–519
7. Beal MF, Marshall PE, Burd GD, Landis DM, Martin JB (1985) Excitotoxin lesions do not mimic the alteration of somatostatin in Huntington's disease. Brain Res 361(1-2):135–145
8. Beal MF, Kowall NW, Ellison DW, Mazurek MF, Swartz KJ, Martin JB (1986) Replication of the neurochemical characteristics of Huntington's disease by quinolinic acid. Nature 321(6066):168–171. doi:10.1038/321168a0
9. Roberts RC, Ahn A, Swartz KJ, Beal MF, DiFiglia M (1993) Intrastriatal injections of quinolinic acid or kainic acid: differential patterns of cell survival and the effects of data analysis on outcome. Exp Neurol 124(2):274–282. doi:10.1006/exnr.1993.1197, S0014-4886(83)71197-0 [pii]

10. Beal MF, Brouillet E, Jenkins B, Henshaw R, Rosen B, Hyman BT (1993) Age-dependent striatal excitotoxic lesions produced by the endogenous mitochondrial inhibitor malonate. J Neurochem 61(3):1147–1150
11. Borlongan CV, Koutouzis TK, Freeman TB, Cahill DW, Sanberg PR (1995) Behavioral pathology induced by repeated systemic injections of 3-nitropropionic acid mimics the motoric symptoms of Huntington's disease. Brain Res 697(1-2):254–257, doi:0006-8993(95)00901-2 [pii]
12. Guyot MC, Hantraye P, Dolan R, Palfi S, Maziere M, Brouillet E (1997) Quantifiable bradykinesia, gait abnormalities and Huntington's disease-like striatal lesions in rats chronically treated with 3-nitropropionic acid. Neuroscience 79(1):45–56, doi:S0306452296006021 [pii]
13. Faber PW, Alter JR, MacDonald ME, Hart AC (1999) Polyglutamine-mediated dysfunction and apoptotic death of a Caenorhabditis elegans sensory neuron. Proc Natl Acad Sci U S A 96(1):179–184
14. Parker JA, Connolly JB, Wellington C, Hayden M, Dausset J, Neri C (2001) Expanded polyglutamines in Caenorhabditis elegans cause axonal abnormalities and severe dysfunction of PLM mechanosensory neurons without cell death. Proc Natl Acad Sci U S A 98(23):13318–13323. doi:10.1073/pnas.231476398
15. Jackson GR, Salecker I, Dong X, Yao X, Arnheim N, Faber PW, MacDonald ME, Zipursky SL (1998) Polyglutamine-expanded human huntingtin transgenes induce degeneration of Drosophila photoreceptor neurons. Neuron 21(3):633–642
16. Lee WC, Yoshihara M, Littleton JT (2004) Cytoplasmic aggregates trap polyglutamine-containing proteins and block axonal transport in a Drosophila model of Huntington's disease. Proc Natl Acad Sci U S A 101(9):3224–3229. doi:10.1073/pnas.0400243101
17. Steffan JS, Bodai L, Pallos J, Poelman M, McCampbell A, Apostol BL, Kazantsev A, Schmidt E, Zhu YZ, Greenwald M, Kurokawa R, Housman DE, Jackson GR, Marsh JL, Thompson LM (2001) Histone deacetylase inhibitors arrest polyglutamine-dependent neurodegeneration in Drosophila. Nature 413(6857):739–743. doi:10.1038/35099568
18. Hodgson JG, Agopyan N, Gutekunst CA, Leavitt BR, LePiane F, Singaraja R, Smith DJ, Bissada N, McCutcheon K, Nasir J, Jamot L, Li XJ, Stevens ME, Rosemond E, Roder JC, Phillips AG, Rubin EM, Hersch SM, Hayden MR (1999) A YAC mouse model for Huntington's disease with full-length mutant huntingtin, cytoplasmic toxicity, and selective striatal neurodegeneration. Neuron 23(1):181–192, doi:S0896-6273(00)80764-3 [pii]
19. Lin CH, Tallaksen-Greene S, Chien WM, Cearley JA, Jackson WS, Crouse AB, Ren S, Li XJ, Albin RL, Detloff PJ (2001) Neurological abnormalities in a knock-in mouse model of Huntington's disease. Hum Mol Genet 10(2):137–144
20. Mangiarini L, Sathasivam K, Seller M, Cozens B, Harper A, Hetherington C, Lawton M, Trottier Y, Lehrach H, Davies SW, Bates GP (1996) Exon 1 of the HD gene with an expanded CAG repeat is sufficient to cause a progressive neurological phenotype in transgenic mice. Cell 87(3):493–506, doi:S0092-8674(00)81369-0 [pii]
21. Menalled LB, Sison JD, Wu Y, Olivieri M, Li XJ, Li H, Zeitlin S, Chesselet MF (2002) Early motor dysfunction and striosomal distribution of huntingtin microaggregates in Huntington's disease knock-in mice. J Neurosci 22(18):8266–8276, 22/18/8266 [pii]
22. Shelbourne PF, Killeen N, Hevner RF, Johnston HM, Tecott L, Lewandoski M, Ennis M, Ramirez L, Li Z, Iannicola C, Littman DR, Myers RM (1999) A Huntington's disease CAG expansion at the murine Hdh locus is unstable and associated with behavioural abnormalities in mice. Hum Mol Genet 8(5):763–774, doi:ddc104 [pii]
23. Bayram-Weston Z, Jones L, Dunnett SB, Brooks SP (2012) Light and electron microscopic characterization of the evolution of cellular pathology in the R6/1 Huntington's disease transgenic mice. Brain Res Bull 88(2-3):104–112. doi:10.1016/j.brainresbull.2011.07.009
24. Slow EJ, van Raamsdonk J, Rogers D, Coleman SH, Graham RK, Deng Y, Oh R, Bissada N, Hossain SM, Yang YZ, Li XJ, Simpson EM, Gutekunst CA, Leavitt BR, Hayden MR (2003) Selective striatal neuronal loss in a YAC128 mouse model of Huntington disease. Hum Mol Genet 12(13):1555–1567
25. Stack EC, Kubilus JK, Smith K, Cormier K, Del Signore SJ, Guelin E, Ryu H, Hersch SM, Ferrante RJ (2005) Chronology of behavioral symptoms and neuropathological sequela in R6/2 Huntington's disease transgenic mice. J Comp Neurol 490(4):354–370. doi:10.1002/cne.20680
26. Gray M, Shirasaki DI, Cepeda C, Andre VM, Wilburn B, Lu XH, Tao J, Yamazaki I, Li SH, Sun YE, Li XJ, Levine MS, Yang XW (2008) Full-length human mutant huntingtin with a stable polyglutamine repeat can elicit progressive and selective neuropathogenesis in

BACHD mice. J Neurosci 28(24):6182–6195. doi:10.1523/JNEUROSCI.0857-08.2008

27. Yu-Taeger L, Petrasch-Parwez E, Osmand AP, Redensek A, Metzger S, Clemens LE, Park L, Howland D, Calaminus C, Gu X, Pichler B, Yang XW, Riess O, Nguyen HP (2012) A novel BACHD transgenic rat exhibits characteristic neuropathological features of Huntington disease. J Neurosci 32(44):15426–15438. doi:10.1523/JNEUROSCI.1148-12.2012
28. Zala D, Benchoua A, Brouillet E, Perrin V, Gaillard MC, Zurn AD, Aebischer P, Deglon N (2005) Progressive and selective striatal degeneration in primary neuronal cultures using lentiviral vector coding for a mutant huntingtin fragment. Neurobiol Dis 20(3):785–798. doi:10.1016/j.nbd.2005.05.017
29. Huang B, Schiefer J, Sass C, Kosinski CM, Kochanek S (2008) Inducing huntingtin inclusion formation in primary neuronal cell culture and in vivo by high-capacity adenoviral vectors expressing truncated and full-length huntingtin with polyglutamine expansion. J Gene Med 10(3):269–279. doi:10.1002/jgm.1150
30. Senut MC, Suhr ST, Kaspar B, Gage FH (2000) Intraneuronal aggregate formation and cell death after viral expression of expanded polyglutamine tracts in the adult rat brain. J Neurosci 20(1):219–229
31. de Almeida LP, Ross CA, Zala D, Aebischer P, Deglon N (2002) Lentiviral-mediated delivery of mutant huntingtin in the striatum of rats induces a selective neuropathology modulated by polyglutamine repeat size, huntingtin expression levels, and protein length. J Neurosci 22(9):3473–3483, doi:20026337
32. Franich NR, Fitzsimons HL, Fong DM, Klugmann M, During MJ, Young D (2008) AAV vector-mediated RNAi of mutant huntingtin expression is neuroprotective in a novel genetic rat model of Huntington's disease. Mol Ther 16(5):947–956. doi:10.1038/mt.2008.50
33. DiFiglia M, Sena-Esteves M, Chase K, Sapp E, Pfister E, Sass M, Yoder J, Reeves P, Pandey RK, Rajeev KG, Manoharan M, Sah DW, Zamore PD, Aronin N (2007) Therapeutic silencing of mutant huntingtin with siRNA attenuates striatal and cortical neuropathology and behavioral deficits. Proc Natl Acad Sci U S A 104(43):17204–17209. doi:10.1073/pnas.0708285104
34. Drouet V, Perrin V, Hassig R, Dufour N, Auregan G, Alves S, Bonvento G, Brouillet E, Luthi-Carter R, Hantraye P, Deglon N (2009) Sustained effects of nonallele-specific Huntingtin silencing. Ann Neurol 65(3):276–285. doi:10.1002/ana.21569
35. Regulier E, Trottier Y, Perrin V, Aebischer P, Deglon N (2003) Early and reversible neuropathology induced by tetracycline-regulated lentiviral overexpression of mutant huntingtin in rat striatum. Hum Mol Genet 12(21):2827–2836. doi:10.1093/hmg/ddg305
36. Vonsattel JP, Myers RH, Stevens TJ, Ferrante RJ, Bird ED, Richardson EP Jr (1985) Neuropathological classification of Huntington's disease. J Neuropathol Exp Neurol 44(6):559–577
37. Faideau M, Kim J, Cormier K, Gilmore R, Welch M, Auregan G, Dufour N, Guillermier M, Brouillet E, Hantraye P, Deglon N, Ferrante RJ, Bonvento G (2010) In vivo expression of polyglutamine-expanded huntingtin by mouse striatal astrocytes impairs glutamate transport: a correlation with Huntington's disease subjects. Hum Mol Genet 19(15):3053–3067. doi:10.1093/hmg/ddq212
38. Palfi S, Brouillet E, Jarraya B, Bloch J, Jan C, Shin M, Conde F, Li XJ, Aebischer P, Hantraye P, Deglon N (2007) Expression of mutated huntingtin fragment in the putamen is sufficient to produce abnormal movement in non-human primates. Mol Ther 15(8):1444–1451. doi:10.1038/sj.mt.6300185
39. Burns LH, Pakzaban P, Deacon TW, Brownell AL, Tatter SB, Jenkins BG, Isacson O (1995) Selective putaminal excitotoxic lesions in non-human primates model the movement disorder of Huntington disease. Neuroscience 64(4):1007–1017
40. Yang SH, Cheng PH, Banta H, Piotrowska-Nitsche K, Yang JJ, Cheng EC, Snyder B, Larkin K, Liu J, Orkin J, Fang ZH, Smith Y, Bachevalier J, Zola SM, Li SH, Li XJ, Chan AW (2008) Towards a transgenic model of Huntington's disease in a non-human primate. Nature 453(7197):921–924. doi:10.1038/nature06975
41. Paxinos G, Watson C (1986) The rat brain in stereotaxic coordinates. Academic, San Diego

Part VII

Gene Therapy for the Treatment of Neurological Disorders

Chapter 28

Gene Therapy for the Treatment of Neurological Disorders: Amyotrophic Lateral Sclerosis

Zachary T. McEachin, Anthony Donsante, and Nicholas Boulis

Abstract

Gene therapy is a powerful tool for treating diseases, including neurological disorder such at amyotrophic lateral sclerosis. When delivered to the CNS, gene therapy vectors can provide prosurvival signals to neurons, knock down the expression of toxic proteins, or restore lost function. How to best deliver this type of therapeutic depends on the nature of the disease and the expected function of the transgene. Here we describe a method for parenchymal injection into rodent models, allowing for localized delivery of gene therapy vectors and other therapeutic molecules. This technique has been a robust mechanism for proof-of-principle experiments.

Key words Gene therapy, ALS, Lou Gehrig's, Motor neurons, Spinal cord, Brainstem

1 Introduction

Amyotrophic Lateral Sclerosis (ALS), commonly called Lou Gehrig's Disease, is a progressive and ultimately fatal neurodegenerative disease affecting both the upper motor neurons (UMN) of the cortex and the lower motor neurons (LMN) of the brainstem and spinal cord [1]. Clinically, ALS results in spasticity, fasciculations, muscle atrophy, and paralysis. Death by respiratory failure typically ensues 1–5 years after disease onset due to degeneration of the motor neurons innervating the diaphragm.

ALS is normally subdivided into two forms. Familial ALS (FALS) accounts for approximately 10 % of cases with ~70 % of those resulting from known genetic mutations [2]. To date, mutations in 17 genes have been reported as causal for ALS including superoxide dismutase 1 (*SOD1*), TAR DNA-binding protein (*TARDP*), fused-in-sarcoma (*FUS*), and more recently Chromosome 9 open reading frame 72 (*C9orf72*) [2]. Sporadic ALS (SALS) accounts for the remaining 90 % of ALS. While it is unclear whether these cases are due to genetic or environmental

Fredric P. Manfredsson (ed.), *Gene Therapy for Neurological Disorders: Methods and Protocols*, Methods in Molecular Biology, vol. 1382, DOI 10.1007/978-1-4939-3271-9_28, © Springer Science+Business Media New York 2016

causes, a growing fraction have been found to have mutations in genes associated with FALS.

Unfortunately, there are no known cures or truly effective therapies for ALS. Riluzole, the only FDA approved drug for the treatment of ALS (approved in 1995), shows only modest benefits, extending life by a mere 2–3 months [3]. Given the lack of effective treatments for ALS, many researchers have proposed a number of therapeutic approaches using gene therapy. In this focused review we discuss the current therapeutic strategies, gene therapy vectors, and delivery methods being investigated for the treatment of ALS. In addition, we provide a detailed protocol commonly employed in our laboratory for targeted delivery of a gene therapy vector to the spinal cord.

1.1 Pathophysiology of ALS

Although the endpoint of ALS is clear (loss of innervation of the muscle and neuronal cell death), the sequence of events that leads to disease is not well understood. A number of pathophysiological mechanisms have been proposed, including glutamate-induced excitotoxicity, protein toxicity, mitochondrial dysfunction, aberrant RNA processing, altered ribostasis, and cytoskeletal defects. It should be noted that, although these pathophysiological mechanisms may be present in other cell types, it appears that motor neurons are selectively more vulnerable than other cell types. While these elements are present in both animal models and patients with ALS, the relative contributions of these factors to disease progression are not clear. Since it is likely that more than one of these factors plays an important role in disease progression and that the relative contributions of these factors may vary from patient to patient, successful therapies will probably need to impact multiple aspects of this disease.

1.2 Routes of Administration for Delivery of Gene Vectors (Table 1)

Although ALS affects many cell types, the hallmark of this disease is motor neuron degeneration. Therefore, the potential for any gene therapy strategy to be effective is dependent on successful delivery of the gene vector to the motor neuron population or to a population of non-neuronal cells that directly interacts with motor neurons (e.g., glial cells, skeletal muscle fibers). Various routes of administration are currently employed for gene delivery, including: intraparenchymal [4, 5], intramuscular [6], intrathecal [7], intraventricular [8], and intraneural [9]. Intraparenchymal and intramuscular routes of administration provide a more targeted delivery of the gene vector and are sometimes classified as "segmental" routes of administration. Although they only treat a small region of the cord, these delivery methods can provide very high levels of transduction, which may be desirable in some circumstances, such as providing adequate coverage of motor neurons involved in respiration. Intrathecal and intraventricular delivery are considered to be "diffuse" routes of administration. These routes have the

Table 1
Routes of administration for delivery of gene vectors

Route of administration	Advantages	Disadvantages
Intramuscular (IM)	Noninvasive	Requires retrograde transport of the vector; impractical to treat all skeletal muscles
Intraparenchymal (IP)	Delivery of therapeutic directly to location of interest in the CNS	Invasive; transduction limited to a small area
Intracerebroventricular (ICV)	Diffuse delivery throughout CNS	Invasive
Intrathecal (IT)	Diffuse delivery throughout CNS	Mildly invasive
Intravenous (IV)	Noninvasive; systemic delivery	Vector must cross blood–brain barrier; off-target transduction/effects

advantage that they can transduce the entire length of the spinal cord, although perhaps not to the same degree in any given region as parenchymal delivery [10]. In addition, intrathecal delivery may be easier to translate to the clinic since accessing the thecal sac is a relatively routine procedure. As a result many researchers have begun investigating the efficacy of gene therapy vectors delivered intrathecally [11].

1.3 Gene Therapy Strategies for ALS

Although the mutations associated with ALS affect proteins in seemingly unrelated pathways, they all result in a common endpoint, the degeneration of the upper and lower motor neurons. Therefore, many common gene therapy strategies for ALS have aimed to mitigate the toxicity of known genetic variants by either reducing the expression of the mutant gene or by providing a neuroprotective milieu to increase the survival of functional motor neurons.

1.4 Gene Therapy Strategies for ALS: Neurotrophic Factors

During embryogenesis, neurotrophic factors are responsible for guiding the development of the nervous system. In mature neurons, neurotrophic factors support the survival and promote the growth of specific populations of neurons [12]. Therefore, many neurotrophic factors have been used in gene therapy based strategies for ALS, including brain-derived neurotrophic factor (BDNF), glial cell line-derived neurotrophic factor (GDNF) [13], insulin like-growth factor-1 (IGF-1) [6], and vascular endothelial growth factor (VEGF) [14].

In 2002, Wang et al. demonstrated that AAV2 mediated delivery of GDNF administered intramuscularly into the forelimbs and hindlimbs of an SOD1 mouse allowed for efficient retrograde transport of the GDNF to spinal motor neurons ultimately delaying disease onset and prolonging survival by 16.6 days compared to control ALS mice [13]. In a seminal paper by Kaspar et al., AAV-mediated delivery of IGF-1 via intramuscular administration in a mouse SOD1 model was shown to delay disease onset and prolong life compared to vehicle and AAV-GDNF treated mice. Both the delay of disease onset and increased survival was observed when IGF-1 was administered either at pre-symptomatic (Day 60) or symptomatic (Day 90) time points [6]. These authors also demonstrated that a vesicular stomatitis virus glycoprotein (VSV-G) pseudotyped lentivirus mediated delivery of IGF-1 increased survival, however lentivirus based delivery was not as effective as AAV. Nevertheless others have utilized lentivirus vectors to deliver therapeutic agents intramuscularly. Azzouz et al. showed that a Rabies-G (Rab-G) pseudotyped Equine Infectious anemia virus (EIAV)-based lentiviral vector could be effectively used to deliver the neurotrophic factor VEGF to spinal motor neurons. The authors found that EIAV-LV mediated delivery of VEGF increased survival when administered both prior to or at disease onset [14]. Two similar studies utilized an AAV2-based vector to administer IGF-1 intraparenchymally. Lepore et al. delivered AAV2-IGF-1 directly into the lumbar segment of the spinal cord in an ALS SOD1 mouse model [4] whereas Franz et al. delivered AAV2-IGF-1 directly into the cervical segment of the spinal cord in an ALS SOD1 rat model [15]. Compared to an AAV2-GFP control, a delay in disease onset and an increase in survival were not observed; although both groups report there was a reduction in motor neuron loss [15]. An interesting finding in both studies was that only male rodents presented an increase in either hindlimb (lumbar injection) or foregrip (cervical injection) strength thus indicating that there may be a gender-specific disposition to various gene therapy approaches. In a study by Dodge et al., the authors demonstrated that ICV delivery of AAV4-IGF1 or AAV4-VEGF vectors resulted in broad expression of these genes in many regions of both the brain and spinal cord. Furthermore, delivering AAV4-IGF-1 or AAV4-VEGF intraventricularly provided neuroprotection and resulted in increased motor function and prolonged survival compared to control mice by 12 days and 20 days, respectively [8]. Interestingly, delivery of both neurotrophic factors simultaneously didn't prove to be more efficacious than either IGF-1 or VEGF separately.

1.5 Gene Therapy Strategies for ALS: RNA Interference

RNAinterference (RNAi) is a posttranscriptional regulatory mechanism that specifically targets a messenger RNA for degradation using a short, complimentary RNA [16]. Since many of the mutations associated with ALS result in toxic gain-of-function proteins, RNAi has allowed researchers to silence the disease allele in animal models of this disease. Approximately 20 % of familial ALS cases are having been shown to result from a toxic gain of function mutation in the Cu, Zn superoxide dismutase (SOD1) gene. Many studies have thus investigated the efficacy of delivering noncoding RNA sequences to knockdown mutant SOD1 transcripts in SOD1 animal models of ALS [17, 18].

Using a RNAi approach, Raoul et al. demonstrated that lentiviral-mediated delivery of a shRNA hairpin to the mutant SOD1 transcript resulted in a decrease in the loss of motor neuron pools in a SOD1 mouse model of ALS [5]. Although this decrease was only seen in the areas adjacent to the injection sites, the authors showed that there was a significant increase in the age at onset. Similarly, Ralph et al. sought to mitigate toxicity of the mutant SOD1 protein by using an RNAi approach to knockdown expression of the mutant SOD1 gene. Delivery of a small hairpin RNA (shRNA) via an EIAV-lentivirus vector targeting mutant SOD1 transcripts resulted in a 30 % increase in life expectancy in SOD1 mice [18]. A study conducted by Towne et al. found that intramuscular delivery of an AAV6-shRNA targeting mSOD1 transcripts was able to transduce motor neurons, reduce mutant human SOD1 transcripts, and protect the muscle from SOD1-mediated atrophy; interestingly, however, intramuscular delivery of the AAV6-shRNA was not able to improve motor function or prolong survival in the SOD1 mouse model [19]. A recent study by Wang et al. demonstrated that an AAVserotype, rAAVrh.10, could provide diffuse transduction of the CNS when delivered intrathecally in the lumbar region of an ALS SOD1 mouse model. Furthermore, the authors demonstrated that delivery of a shRNA to the mutant SOD1 gene could extend survival by prolonging disease progression in an SOD1 mouse model [20].

1.6 Conclusion

ALS is a rapidly progressive neurodegenerative disease that ultimately results in paralysis and death. With only one FDA approved therapy that provides modest benefits at best, patients and families have limited options for effective therapeutic treatments. Gene therapy approaches for ALS provide a potentially more effective alternative for therapeutic intervention. It should be noted that many of the gene therapy studies conducted in animal of models of ALS show promising results when gene therapy vectors are delivered at pre-symptomatic or symptomatic time points; however the greatest improvement is seen when animals are treated pre-symptomatically. Advances in the detection of ALS phenotypes, biomarkers of the disease, and family history records will allow for earlier diagnosis of ALS and may provide a unique opportunity to effectively utilize gene therapy approaches.

In this focused mini-review, we have provided an overview of the common strategies, viral vectors, and animal models used in gene therapy research for ALS. Furthermore, we have provided a detailed protocol for delivering gene therapy vectors to the spinal cord. These previous studies discussed here in addition to others have provided proof-of-principle for the use of gene therapy strategies for treating neurodegenerative diseases such as ALS; however improving the safety and efficacy of these gene therapy strategies will ultimately allow for the translation of gene therapy to a clinical setting.

2 Materials

This method demonstrates intraparenchymal delivery of therapeutics into the rodent spinal cord.

All surgical tools are autoclaved prior to the first surgery of the day. For subsequent surgeries, the tools are washed, dried, and then sterilized in a bead sterilizer.

1. Bead sterilizer (Fine Science Tools, Foster City, CA, USA).
2. Handwarmer or regulated heating system.
3. Absorbent pads.
4. Stereotactic apparatus: Stoelting 51650 (Stoelting, Wood Dale, IL, USA) or equivalent.
5. Microinjection unit: Kopf 5000 (Kopf, Turjunga, CA, USA) or equivalent.
6. Sterile cloth, cotton swabs, and gauze.
7. Scale.
8. Oxygen (100 % O_2).
9. Isoflurane and vaporizer.
10. Induction box.
11. Antibiotic eye ointment.
12. Buprenorphine (0.05 mg/kg).
13. Betadine.
14. Ethanol wipes.
15. Surgical drape (General Econopak, Philadelphia, PA, USA).
16. Lidocaine with atropine (Allmedtech, www.allmedtech.com).
17. Surgical tools: scalpel holder, straight Dumont micro-blunted, atraumatic tipped forceps, curved surgical scissors, angled 45 % Dumont micro-blunted, atraumatic tipped forceps, #11 blade, Rongeurs (Fine Science Tools, Foster City, CA, USA).
18. Transverse process clamps (Stoelting, Wood Dale, IL, USA; lumbar injections).
19. Gel foam.

20. Halsey needle holder (Fine Science Tools, Foster City, CA, USA).
21. Friedman-Pearson Rongeurs, Curved, 0.7 mm (Fine Science Tools, Foster City, CA, USA).
22. 10 μl NanoFil syringe with a 34 gauge beveled needle (World Precision Instruments, Sarasota, FL, USA).
23. 4-0 absorbable suture.

3 Methods

The method described below details parenchymal delivery of therapeutics to the rat lumbar spinal cord, the primary target for our lab. However, this procedure can easily be scaled down to mice by adjusting the parameters for the depth of the injection. The method can also be used for injections into the cervical spine with one modification: Due to the depth of the cervical spine, the soft tissue prevents the access of our transverse process clamps. Instead, we forego immobilizing the spine. This modification allows the spinal cord to rise and fall slightly as the animal breaths, but it does not seem to affect our ability to accurately deliver viral vectors.

3.1 Surgical Station Preparation

1. The bead sterilizer is turned on 30 min before the surgery begins.
2. The surgical station is prepared. We use a plastic box to elevate the animal so that the spine will be near the height of the spinal clamps. A hand warmer is placed upon the box and is covered by a disposable absorbent pad. The pad serves to insulate the animal from direct contact with the hand warmer. The hand warmer can be replaced with a regulated heating pad if one is available.
3. A sterile cloth is placed on the bench next to the stereotactic base. The sterilized surgical tools are laid out with their tips placed on sterile gauze. The gauze is a useful visual cue, helping the surgeon keep the ends of the tools sterile during the course of the surgery. This is especially helpful for subsequent surgeries, where only the ends of the tools are sterilized.

3.2 Preparing the Animal for Surgery

1. Prior to being anesthetized, the animal is weighed to determine the dosing of pain medication. The rat is then placed in an induction box connected to an isoflurane vaporizer.
2. Oxygen flow is started (600 cc/min). Anesthesia induction begins at 5 % isoflurane. Every minute, the concentration of isoflurane is dropped by 1 % until 2 % is reached. After 1 min at 2 %, the animal is transferred to from the box to the surgical station.

3. Buprenorphine is administered at a dose of 0.05 mg/kg by subcutaneous injection. Note: Both buprenorphine and isoflurane can suppress respiration. This may become a problem if the surgery is expected to take more than an hour. When planning a large number of spinal injections, we give half of the buprenorphine at the beginning of the surgery and half at the end. This reduces the likelihood that respiratory failure will occur during surgery.
4. To protect the animal's corneas from drying during surgery, a lubricant eye ointment is liberally applied to each eye using a cotton swab.
5. The back of the animal is shaved from the sacral spine up to the mid-thoracic spine and the spine is palpated to locate the 13th thoracic vertebra. The skin is then decontaminated with three rounds of washes (betadine followed by ethanol wipes).
6. The surgeon dons sterile gloves and applies a sterile drape to the back of the animal. The fenestration is centered on the 13th thoracic vertebra.
7. Depth of anesthesia is checked by pinching the toes on all four feet through the sterile drape while watching for a withdrawal response. This is repeated every 5 min during the surgery to ensure the animal is adequately anesthetized.
8. Four 50 μl doses of lidocaine are injected between the skin and muscle to provide additional pain relief and to help control bleeding from the muscle during the operation.

3.3 Exposing the Spinal Cord

1. Using a scalpel, a 5 cm incision is made in the skin along the midline, centered on the 13th thoracic vertebra. The skin adjacent to the incision is released from the underlying fascia and muscle by blunt dissection.
2. The fascia overlying the spine is removed, and incisions are made through the muscle from the T11 to L2, following the grove at the base of the spinous processes.
3. Retractors or scissors can be used stretch the muscle away from the midline and better expose the edges of the mammillary processes.
4. One transverse process clamp is positioned above the groove between the mammillary process and the transverse process of T12 and its screw is tightened to lock into place.
5. Using fine forceps with a 45 degree angle, the spinous process is grasped and the spine gently lifted into position, sliding the clamp into the groove between the mammillary process and the transverse process of T12. While holding the spine in place with one hand, the second clamp is inserted tightly into the groove on the contralateral side and locked into place.

6. Repeat **steps 4** and **5** for L1. The spine from T12 to L1 should now be level and immobile. Applying gentle pressure with a cotton swap from above should not displace it from the transverse process clamps.
7. Using rongeurs, lamina are slowly removed by taking small bites out of the bone. Care should be taken to not apply pressure to the spinal cord itself. The number of lamina that need to be removed will depend on how many injections are desired and how closely they need to be spaced. Our lab routinely performs 10 injections per side, approximately 2 mm apart. To do so, we perform a triple laminectomy, removing the lamina of T12, T13, and L1.

3.4 Injection into the Parenchyma

1. Load the syringe. Due to the use of a 34 gauge needle, some solutions (particularly cells) may be too viscous to simply draw up through the needle. In these cases, we backfill the syringe with the injection material. Next, mount the syringe onto the microinjection unit on the stereotactic platform.
2. To inject into the ventral horn, position the needle 2 mm lateral of midline, lowering the tip of the needle to just rest on the surface of the dura. It may be necessary to move the needle rostrally or caudally to avoid hitting the vasculature. The needle is then lowered 1.3 mm from the surface. It is important that the needle be sharp. Otherwise, it will compress the cord rather than puncture the dura, possibly causing damage to the cord.
3. To better control the injection speed, the plunger is depressed using dial on the microinjection unit. The injection rate we use is approximately 1 μl/min. We generally inject between 0.5 and 2 μl per injection. After completing the injection, the needle is left in place for 1 min and then slowly withdrawn.
4. **Steps 1–3** are repeated for additional injections.

3.5 Closing the Animal and Recovery

1. 4-0 absorbable suture is used to close both the muscle and the skin. Beginning at one end, the muscle adjacent to the spine is drawn together, covering the laminectomy site. The skin is then closed in a similar manner, with stitches approximately 5 mm apart.
2. If applicable, the second half dose of buprenorphine is administered subcutaneously. The wound is cleaned with three alternating treatments of betadine and ethanol wipes.
3. Isoflurane is discontinued and the animal is allowed to breathe oxygen for 1–2 min before being transferred to a clean cage to recover. One half of the cage is placed on a 37 °C heating pad to help maintain the animal's body temperature as it recovers from anesthesia.

4. Once the animal regains consciousness and then again over the next few days, the animal is checked for motor deficits resulting from the surgery. With lumbar injections, this generally presents as loss of function or loss of coordination in one or both hind legs. Since some of our assays involve scoring hind limb function, animals that exhibit surgery-related defects are excluded from our studies.
5. Buprenorphine is administered as needed for 3 days following surgery.

References

1. Rowland LP, Shneider NA (2001) Amyotrophic lateral sclerosis. N Engl J Med 344(22):1688–1700
2. Al-Chalabi A, Hardiman O (2013) The epidemiology of ALS: a conspiracy of genes, environment and time. Nat Rev Neurol 9(11):617–628
3. Bellingham MC (2011) A review of the neural mechanisms of action and clinical efficiency of riluzole in treating amyotrophic lateral sclerosis: what have we learned in the last decade? CNS Neurosci Ther 17(1):4–31
4. Lepore AC et al (2007) Intraparenchymal spinal cord delivery of adeno-associated virus IGF-1 is protective in the SOD1G93A model of ALS. Brain Res 1185:256–265
5. Raoul C et al (2005) Lentiviral-mediated silencing of SOD1 through RNA interference retards disease onset and progression in a mouse model of ALS. Nat Med 11(4):423–428
6. Kaspar BK et al (2003) Retrograde viral delivery of IGF-1 prolongs survival in a mouse ALS model. Science 301(5634):839–842
7. Snyder BR et al (2011) Comparison of adeno-associated viral vector serotypes for spinal cord and motor neuron gene delivery. Hum Gene Ther 22(9):1129–1135
8. Dodge JC et al (2010) AAV4-mediated expression of IGF-1 and VEGF within cellular components of the ventricular system improves survival outcome in familial ALS mice. Mol Ther 18(12):2075–2084
9. Federici T et al (2007) A means for targeting therapeutics to peripheral nervous system neurons with axonal damage. Neurosurgery 60(5):911–918, Discussion 911–918
10. Federici T et al (2012) Robust spinal motor neuron transduction following intrathecal delivery of AAV9 in pigs. Gene Ther 19(8):852–859
11. Federici T, Boulis NM (2012) Gene therapy for amyotrophic lateral sclerosis. Neurobiol Dis 48(2):236–242
12. Ekestern E (2004) Neurotrophic factors and amyotrophic lateral sclerosis. Neurodegener Dis 1(2–3):88–100
13. Wang LJ et al (2002) Neuroprotective effects of glial cell line-derived neurotrophic factor mediated by an adeno-associated virus vector in a transgenic animal model of amyotrophic lateral sclerosis. J Neurosci 22(16): 6920–6928
14. Azzouz M et al (2004) VEGF delivery with retrogradely transported lentivector prolongs survival in a mouse ALS model. Nature 429(6990):413–417
15. Franz CK et al (2009) Intraspinal cord delivery of IGF-I mediated by adeno-associated virus 2 is neuroprotective in a rat model of familial ALS. Neurobiol Dis 33(3):473–481
16. Wilson RC, Doudna JA (2013) Molecular mechanisms of RNA interference. Annu Rev Biophys 42(1):217–239
17. Wu R et al (2009) Nerve injection of viral vectors efficiently transfers transgenes into motor neurons and delivers RNAi therapy against ALS. Antioxid Redox Signal 11(7):1523-U1
18. Ralph GS et al (2005) (S)ilencing mutant SOD1 using RNAi protects against neurodegeneration and extends survival in an ALS model. Nat Med 11(4):429–433
19. Towne C et al (2011) Neuroprotection by gene therapy targeting mutant SOD1 in individual pools of motor neurons does not translate into therapeutic benefit in fALS mice. Mol Ther 19(2):274–283
20. Wang H et al (2014) Widespread spinal cord transduction by intrathecal injection of rAAV delivers efficacious RNAi therapy for amyotrophic lateral sclerosis. Hum Mol Genet 23(3):668–681

Chapter 29

Stereotaxic Surgical Targeting of the Nonhuman Primate Caudate and Putamen: Gene Therapy for Huntington's Disease

Jodi L. McBride and Randall L. Clark

Abstract

Stereotaxic surgery is an invaluable tool to deliver a variety of gene therapy constructs to the nonhuman primate caudate and putamen in preclinical studies for the genetic, neurodegenerative disorder, Huntington's disease (HD). Here we describe in detail how to perform this technique beginning with a pre-surgical magnetic resonance imaging scan to determine surgical coordinates followed by the stereotaxic surgical injection technique. In addition, we include methodology of a full necropsy including brain and peripheral tissue removal and a standard immunohistochemical technique to visualize the injected gene therapy agent.

Key words Huntington's disease, Neurodegeneration, Stereotaxic surgery, Magnetic resonance imaging, Nonhuman primate, Necropsy, Immunohistochemistry

1 Introduction

Huntington's disease (HD) is a genetic, neurological disorder that is characterized by a devastating array of symptoms including a hyperkinetic movement disorder, changes in cognitive capabilities, psychiatric manifestations, and a robust metabolic syndrome [1]. Many of these symptoms are caused by neuronal loss in the caudate and putamen that results from a mutated *HTT* gene containing an abnormally long stretch of the DNA bases cytosine, adenine, and guanine (CAG) [1, 2]. Because of the dramatic cell loss observed in these brain regions, the caudate and putamen have historically been the go-to target structures in both preclinical studies as well as clinical trials investigating potential therapeutic strategies to treat HD.

Stereotaxic surgery is a powerful tool that has been used for decades to accurately inject therapeutic agents into different brain regions in a variety of different animal species. Prior to the

Fredric P. Manfredsson (ed.), *Gene Therapy for Neurological Disorders: Methods and Protocols*, Methods in Molecular Biology, vol. 1382, DOI 10.1007/978-1-4939-3271-9_29, © Springer Science+Business Media New York 2016

identification of the mutant, disease-causing *HTT* gene in 1993 [1], stereotaxic surgery was commonly used to deliver excitotoxins into the nonhuman primate (NHP) caudate and putamen to create a large animal model of HD that replicated some of the key symptoms of the disease [3–6]. More recently, lentiviral vectors (LVs) [7] and adeno-associated viral vectors (AAVs) (ongoing studies in our laboratory) have been used to deliver mutant *HTT* into the caudate and putamen, replicating many of the neuropathological and behavioral changes seen in humans with HD. Preclinical studies have employed stereotaxic surgical targeting of the caudate and putamen to evaluate potential therapeutic strategies in some of these NHP models including fetal tissue transplants and trophic factor administration [8–11]. More recently, stereotaxic surgery has been used to target the naïve NHP putamen to assess the safety of partially reducing HTT expression via AAVs expressing microRNA constructs [12, 13] or siRNAs delivered through a cannula placed into the putamen [14]. With the recent creation of a transgenic HD monkey bearing a fragment of human mutant *HTT* [15], it is anticipated that stereotaxic surgery will be used to evaluate a variety of therapeutics in this NHP model in the near future.

Here, we describe in detail the methodology that our laboratory uses to surgically target the NHP caudate and putamen to deliver both mutant HTT (disease modeling) as well as gene therapeutics (AAV-RNAi) including (1) a pre-surgical magnetic resonance imaging (MRI) scan to determine surgical coordinates, (2) the stereotaxic surgical injection procedure including anesthesia monitoring and pain medication delivery, (3) post-surgical necropsy including brain, blood and peripheral tissue collection, and (4) immunohistochemical processing of brain sections to evaluate the expression and spread of the injected agent.

2 Materials

2.1 Pre-surgical Magnetic Resonance Imaging (MRI)

1. Personal protective equipment including scrubs, a surgical face mask, protective eye wear, nitrile gloves, a hair net, shoe covers and a water-resistant gown.
2. MRI machine (1.5 or 3 T) and associated imaging software.
3. Ketamine HCl.
4. Glycopyrrolate.
5. Isoflurane anesthesia machine on a mobile, wheeled cart or transport board.
6. Shaver.
7. Endotracheal tube (4.0–5.5 mm i.d.) and umbilical tape.
8. Ophthalmic ointment.

9. 22G cephalic vein catheter.
10. Pulse oximeter.
11. Heart rate monitor.
12. Animal heating mechanisms (towels, water blanket, etc.).
13. MRI-compatible nonhuman primate head frame containing ear bars, eye bars, and a palate bar.
14. Mineral oil to fill the ear bars.
15. Allen wrenches of various sizes.
16. Fiducial marker (Vitamin E capsule).
17. Surgical coordinate sheets.
18. MRI surface/head coil.

2.2 Stereotaxic Injection into the Caudate and Putamen

1. PPE including scrubs, hair bonnets, water resistant gown, surgical masks, protective eyewear and gloves (surgical staff) as well as sterile gowns and sterile surgical gloves (surgeons).
2. Anesthesia cart and physiologic monitoring system.
3. Ethylene oxide sterilizing system.
4. Infusate (viral vector, cells, etc.).
5. Bucket of ice or cooling block if infusate must remain cold prior to injection.
6. Primate head frame.
7. Sterile micromanipulator.
8. Sterile Allen wrenches.
9. Sterile T-square (PolySquare™).
10. Sterile Hamilton infusion syringes (10–100 μl or appropriate volume for infusion) (Hamilton, Reno, NV, USA).
11. Sterile Hamilton surgical needles (22 25G, 25 38 mm in length, blunt tip style) (Hamilton, Reno, NV, USA).
12. Sterile stereotaxic infusion pump and controller box.
13. Sterile AP Zeroing bar.
14. Sterile towels and a sterile craniotomy drape.
15. ChloraPrep preoperative skin preparation (CareFusion, San Diego, CA, USA).
16. Hydromorphone (2 mg/ml).
17. Cefazolin (250 mg/ml).
18. Buprenorphine (0.3 mg/ml).
19. Lidocaine (1 %) and epinephrine.
20. Bupivacaine (0.5 %).
21. 1 and 3 cc syringes with 25G needles.

22. Lactated Ringer's solution.
23. Warm forced air system (Bair hugger) and warm water bags (3M, St. Paul, MN, USA).
24. Sterile Parafilm (several 2″ × 2″ squares).
25. Sterile surgical drill and drill bit (2 mm round, carbide burr).
26. Sterile gauze.
27. Sterile surgical cotton swabs.
28. Sterile pipettors and appropriately sized sterile pipette tips.
29. Sterile surgical instruments (Adson forceps, scalpels, scalpel holders, hemostats, scissors, microdissection scissors, rongeurs, needle holders, elevators, Gelpi retractors).
30. Sterile 22G needle.
31. Sterile Sharpie marker.
32. Sterile saline.
33. Sterile 30 ml syringe.
34. Sterile suction tip and suction machine.
35. Sterile Gelfoam (Pfizer, Groton, CT, USA).
36. Sterile 4-0 Monocryl and 3-0 Vicryl suture.
37. Biohazard waste container.
38. Calculator, pencils, and surgical coordinate sheets.

2.3 Necropsy and Tissue Collection

1. PPE: Scrubs, water resistant gown, hair bonnet, nitrile gloves, facemask, protective eye wear.
2. Peristaltic perfusion pump.
3. Polyethylene tubing for perfusion pump (inside diameter: ¼ in.) attached to a 13G cannula.
4. Board and masking tape for securing animal.
5. Scalpel and scalpel blades, Metzenbaum scissors, forceps, assorted sizes of hemostats, heavy duty scissors, bone lever.
6. Bottles of 70 % EtOH and saline.
7. Cutting boards.
8. Weigh boats for organs.
9. 60 cc syringe with 18 gauge needle for drawing terminal serum.
10. Supply of scalpel blades, #22, blunt and sharp.
11. Supply of 4 × 4 gauze.
12. Pruning shears.
13. T-bar.
14. Bone cutting forceps.

15. Pens (sharpie)/pencils for labeling tubes and foil packs.
16. 2 4 ml serum blood tubes.
17. Assorted syringes (1, 3, 5, 20, 30, 60 cc) 7 needles (18–25G, 1″–1.5″).
18. Tourniquet.
19. Sharps containers.
20. Waste disposal bags.
21. Hand saw.
22. Sodium pentobarbital.
23. Ketamine hydrochloride.
24. 0.9 % sterile saline.
25. Paraformaldehyde.
26. Sodium phosphate monobasic monohydrate NaH_2PO_4 (FW137.99).
27. Sodium phosphate dibasic anhydrous Na_2HPO_4 (FW141.96).
28. 4 % paraformaldehyde (40 g paraformaldehyde, 3.2 g monobasic, 10.9 g dibasic, 1 L dH_2O).
29. Ice bucket with wet ice.
30. Dewar with liquid nitrogen.
31. Aluminum foil packs labeled with ID, date and tissue for tissue bank.
32. Microcentrifuge tubes (2.0 ml with screw cap and o-ring/ sterilized).
33. Brain jar on ice filled with sterile saline.
34. Small jars on ice filled with sterile saline.
35. Tissue biopsy cores (2 mm).
36. Rhesus brain matrix.
37. Tissue slicer blades.
38. Sterile petri dishes.

2.4 Cutting and Immunohistochemical Processing of Brain

1. PPE: White lab coat, nitrile gloves, protective eye wear.
2. Sliding microtome and microtome blade.
3. Paint brushes for tissue cutting.
4. Dry ice.
5. Trizma® pre-set crystals (pH 7.4, avg Mw: 151.6) (Sigma-Aldrich, St. Louis, MO, USA).
6. NaCl (FW58.44).
7. Triton® X-100.
8. Sodium phosphate dibasic anhydrous Na_2HPO_4 (FW141.96).

9. Sodium phosphate monobasic monohydrate NaH_2PO_4 (FW137.99).
10. Serum (goat or donkey).
11. Sodium meta-periodate $INaO_4$ (FW213.89).
12. 3,3′-diaminobenzidine (DAB) (FW360.11).
13. Nickel(II) sulfate hexahydrate (FW262.85).
14. VECTASTAIN®*Elite*® ABC Kit (Vector Laboratories, Burlingame, CA, USA).
15. Dilution media (7.46 g Trizma, 8.77 g NaCl, 0.5 ml Triton X-100, 1 l dH_2O).
16. TBS (7.46 g Trizma, 8.77 g NaCl, 1 l dH_2O).
17. PBS (5.47 g Dibasic, 1.60 Monobasic, 9.26 g NaCl, 1 l dH_2O).
18. Cryoprotectant solution (300 g Sucrose, 300 ml ethylene glycol, 0.2 g Sodium Azide, 500 ml PBS).
19. Netted staining dishes (with glass dishes to contain fluid).
20. Orbital shaker.
21. Appropriate primary and secondary antibodies.
22. 6-well tissue culture plates.
23. Large 24-well (4 × 6) compartmented tissue collection box.

3 Methods

3.1 Pre-surgical Magnetic Resonance Imaging (MRI)

1. Sedate the animal while in its home cage with 10–20 mg/kg Ketamine combined with 0.01–0.02 mg/kg Glycopyrrolate (intramuscular—IM) and bring to the preoperative procedure room (food withheld for 12 h prior to MRI and surgery). Shave animal's head from the brow ridge caudally to the foramen magnum and laterally down to the lateral canthus of each eye, place ophthalmic ointment in each eye, shave the left forearm, and establish a 22G cephalic vein catheter for agent administration during surgery.
2. Place an endotracheal tube (for adult rhesus macaques, 4.0–5.5 mm i.d. endotracheal tubes are typically used), secure tube with umbilical tape and start the animal on 1.5 % isoflurane. While waiting to go into MRI, monitor the animal's pulse rate, blood oxygen saturation, and $ETCO_2$ levels.
3. Bring the animal to the MRI facility on a wheeled cart under isoflurane anesthesia.
4. In the MRI procedure room, place the animal's head into the MRI-compatible stereotaxic head frame (Fig. 1a) while still connected to the breathing tube and under anesthesia (*see*

Note 1). First place the animal in the ear bars (filled with contrast dye or mineral oil) such that there is free movement of the animals head up and down (nodding motion), but no movement laterally. Center animal in the head frame both medial-laterally and dorsoventrally. Next, place the palate bar into the animal's mouth, resting against the hard palate, and secure into position such that the animal's head is straight and its eyes are facing forward. Finally, secure the eye bars such that they fit tightly in the ventral grooves of the orbital socket. Ensure that ear bars, palate bar and eye bars are all secured in place.

5. Disconnect the animal's breathing tube from the anesthesia machine and quickly transfer the animal placed in the head frame to the MRI table (still prone and head first) and reconnect to the anesthesia machine located in the MRI unit (*see* **Note 2**).
6. Place a circulating warm water pad and towels under the animal on the MRI table and place extra towels over the animal order to maintain body temperature.
7. Attach peripheral physiological (pulse rate/oxygen saturation, $ETCO_2$, respiration, and/or noninvasive blood pressure) monitors to the animal (digit cuffs) and observe for accuracy and reliability of readings. From the control room, an MRI technician should continue to monitor the animal's respiratory rate, pulse rate, end tidal CO_2 and oxygen saturation (*see* **Note 3**).
8. Tape a fiducial marker to right side of the animal's head to use as a reference on the MR image (*see* **Note 4**).
9. Place the surface/head coil (Fig. 1b) directly above and parallel to the animal's head such that it is centered over the scanning area of interest (caudate and/or putamen, approximately

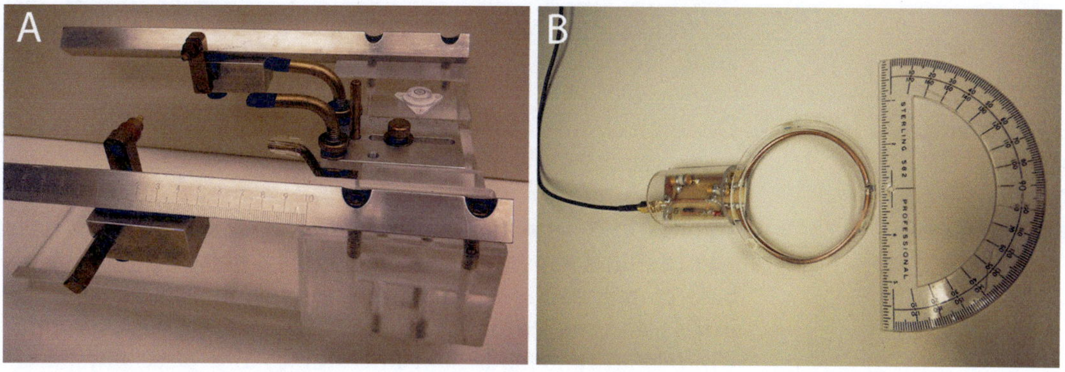

Fig. 1 Pre-surgical MRI to determine surgical coordinates. (**a**) Animals are anesthetized and fitted into an MRI-compatible stereotaxic head frame by first securing the animal into the ear bars, followed by the palate and eye bars. (**b**) The scan is obtained using a surface coil that sits directly above the animal's head

12–25 mm in front of the ear bars) for HD gene therapy studies. The coil should be approximately 0.5–1.0 cm above the surface of the head but not touching the skin (*see* **Note 5**). Center the animal's head on MRI table using the laser crosshair feature on the MRI scanner and then send the animal into the center of the magnet.

10. First collect a quick, single-slice, three-axis localizer scan that gives a view of the animal's head in the three scanner-frame axes (voxel size 1.0 × 1.0 × 1.0). This ensures that the coil is functioning properly and allows for further refinement around the scanning area of interest (1–2 min scan).
11. Next collect a T1-weighted structural scan (MPRAGE) scan to establish the surgical coordinates (voxel size: 0.5 × 0.5 × 0.5, 12 min scan; *see* **Note 6**). Coordinates are determined from coronal scans using the MRI software Distance Tool as follows: (1) Anterior/Posterior (AP) determined by establishing the MR image containing your injection region of interest (i.e., caudate or putamen) and calculating the distance (in millimeters (mm)) in front of or behind the MR image containing the ear bars (ear bar contrast dye is evident on scan), (2) Medial/Lateral (ML) established by determining the distance (mm) from the center of the two hemispheres to the proposed injection site (draw a vertical line through the center of the sagittal sinus vein as a reference), and (3) Dorsal/Ventral (DV) established by determining the distance from the pial surface to the proposed injection site (Fig. 2, red asterisks indicates the sagittal sinus vein and the blue dot indicates the injection site in the rhesus putamen; *see* **Note 7**). Record surgical coordinates on a piece of paper that will go with you into the operating room.
12. When the scan is finished, the animal can be removed from the magnet, disconnected from physiological monitors and anesthesia tubing, placed back onto the mobile anesthesia cart, reconnected to anesthesia on cart, and wheeled into the operating room.

3.2 Stereotaxic Injection into the Caudate and Putamen

1. Disconnect animal from anesthesia and remove from mobile anesthesia cart. Place animal onto the surgical table and connect endotracheal tube to surgical anesthesia unit. Maintain animal on 1–2 % Isoflurane combined with 100 % oxygen administered at a rate of 1–1.5 l/min (*see* **Note 8**). Secure the base of stereotaxic frame to the table with surgical tape (animal's torso and legs must be delicately lifted up to achieve this, while head remains locked into frame) and place physiologic monitoring equipment on the animal.
2. Administer preoperative doses of Hydromorphone HCl (2 mg/ml) (range of doses: <3 kg bodyweight, 0.5 mg, 0.25 ml;

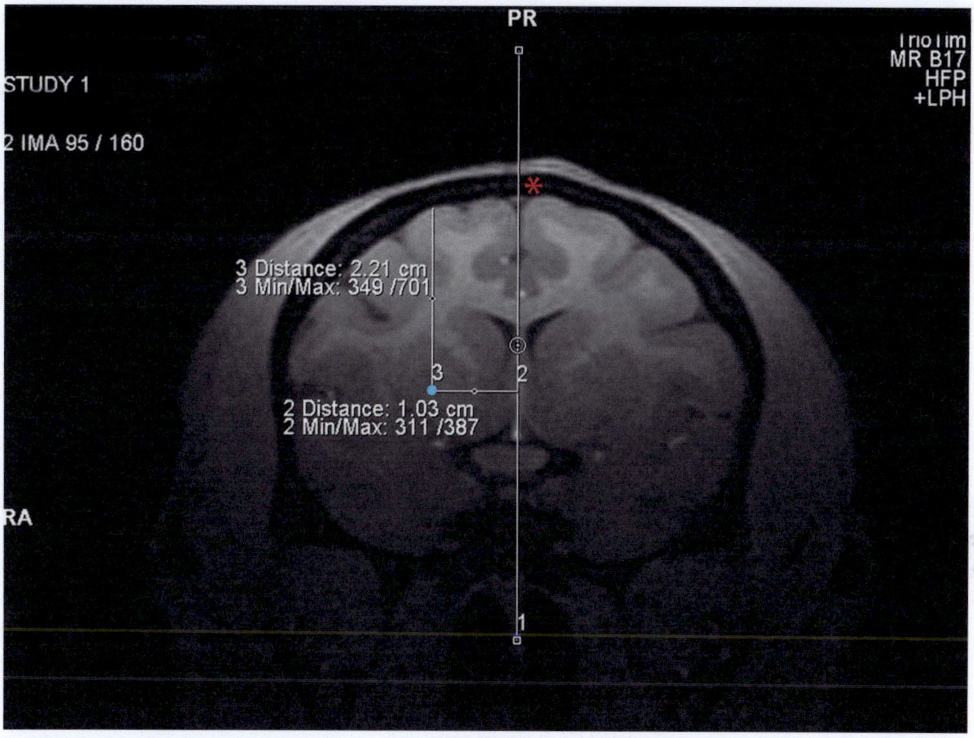

Fig. 2 Determining surgical coordinates from the MRI scan. Surgical coordinates an for injection site into the putamen (shown by a *blue dot*) are created by using the MRI software to establish (1) the distance in mm in front of the ear bars (AP coordinate), (2) the distance in mm lateral to the center of the sagittal sinus vein (ML coordinate, shown by a *red asterisks*) and (3) the distance in mm ventral to the pial surface of the brain (DV coordinate)

3–10 kg bodyweight, 1 mg, 0.5 ml; >10 kg bodyweight, 2 mg, 1 ml) intravenously and 25 mg/kg Cefazolin intravenously as well as every 2 h intraoperatively (for both drugs).

3. Establish intravenous fluids (Lactated Ringers solution) @ 10–20 ml/kg/h. Fluids may be increased based on hydration status of patient or due to blood loss during surgery.
4. Apply a ChlorPrep (a chlorhexidine/alcohol instant solution) to the entire surgical site and allow it to dry. Drape surgical areas with sterile towels.
5. Place a warm forced air system (Bair hugger) around the animal, as well as warm water bags, for the duration of the surgical procedure. Place a full table drape over the animal, towels and Bair hugger to avoid any potential breaks in aseptic technique due to the necessary manipulation of the stereotaxic manipulator.
6. Surgical staff should wear hair bonnets, surgical masks, protective eyewear and gloves prior to entering the operating room.

Surgeons should perform a full pre-surgical sterile scrub with a chlorhexidine or betadine solution. Surgeons should wear sterile gowns and sterile surgical gloves prior to draping the patient and initiating the surgical procedure.

7. Open sterile (ethylene oxide sterilized) surgical packs containing instruments, drill, micromanipulator, AP zeroing bar, surgical pump, needles/syringes and supplies on a flat surface near the operating table.
8. Assemble needles and syringes and wet the syringe barrels by withdrawing and expelling injection vehicle (saline, PBS, buffer, etc.) several times (*see* **Note 9**).
9. Slide the micromanipulator onto the AP zeroing bar and tighten down anywhere.
10. Screw the AP slide attachment onto the micromanipulator and tighten down (use Allen wrench if necessary).
11. Screw the infusion pump onto the AP slide attachment and set/lock AP slide attachment to a set number (example 50 or 0; all AP coordinate measurements from the MRI will be added or subtracted from this number, depending on whether the micromanipulator is placed on the left or right AP bar on the animal's head frame).
12. Place the syringe/needle into the infusion pump and lock into place. Use a T-square (e.g., PolySquare) to make certain that the micromanipulator, pump, syringe and needle are straight and make adjustments if not (Fig. 3).
13. Ensure that all components of the micromanipulator are tightened appropriately (use Allen wrenches).
14. Carefully loosen the micromanipulator from the AP zeroing bar and slide laterally to line up the tip of the needle with the point structure on the AP zeroing plate. Record the number on the AP zeroing bar (not the AP slide attachment). When the micromanipulator is placed at the same number on the animal's head frame AP bar, the needle will be correctly positioned at the ear bars (ear bar zero).
15. Administer 25 mg/kg of Cefazolin intravenously prior to making the initial scalp incision for antibiotic coverage. Inject an intradermal, local block at the proposed incision site consisting of 0.4 ml (up to 0.8 ml depending on size of incision) Bupivacaine (0.5 %) combined with 0.1 ml (up to 0.2 ml depending on size of incision) and Lidocaine (1 %) with epinephrine.
16. Make a linear sagittal incision of approximately 6 cm length across the top of the animal's scalp. Elevate the subcutis and periosteum and retract laterally to expose the skull. Sharply dissect muscle attachments and retract laterally.

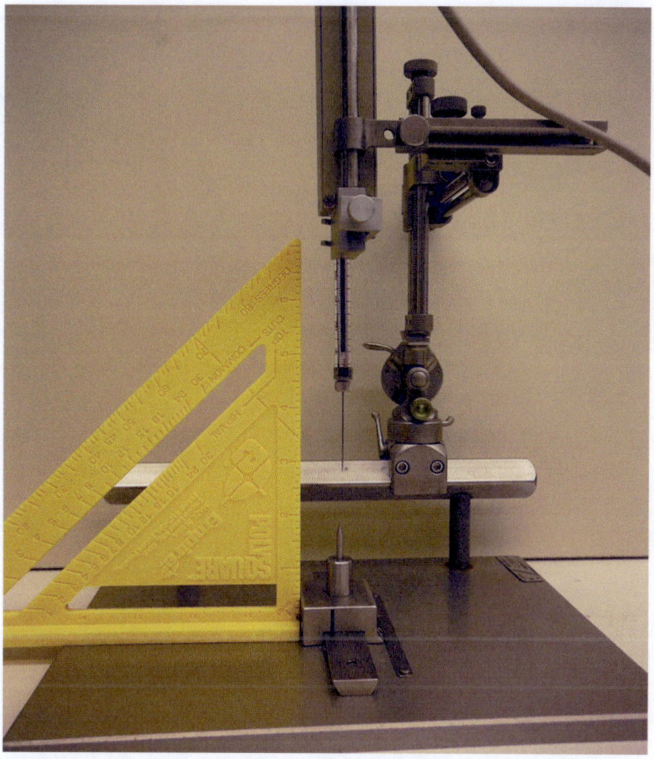

Fig. 3 Zeroing the AP slide bar. Prior to securing the micromanipulator on the animal's head frame in the surgery room, the AP slide bar must be zeroed out by setting the AP slide bar to a predetermined number (example, 50) and lining up the tip of the needle/cannula with the point structure on the AP zeroing plate. When the micromanipulator is placed into the exact position on the AP bar of the head frame, the needle tip will be located at the ear bars. Using a T-square when zeroing the AP slide bar helps to ensure that the needle is straight

17. Drill a 1 cm long line close to the planned injection site (caudate and/or putamen) perpendicular to the sagittal sinus. Extend the drill line down to the level of the dura to expose the dark sagittal sinus vein. The exposed sagittal sinus vein will be used to establish the ML zero point.
18. Bring the micromanipulator (connected to the pump, syringe and needle) from the AP zero bar to the animal's head frame and lock into place on either the left or right AP bar at the number recorded earlier (*see* **Note 10**). Use the AP slide attachment and the ML dials on the micromanipulator to bring the needle to the center of the sagittal sinus vein and record the ML number as the zero point.
19. Calculate the AP and ML coordinates for each injection site (*see* **Note 11**) and use the micromanipulator to bring the needle to those sites and mark the sites on the skull with a pencil or permanent marker (Sharpie).

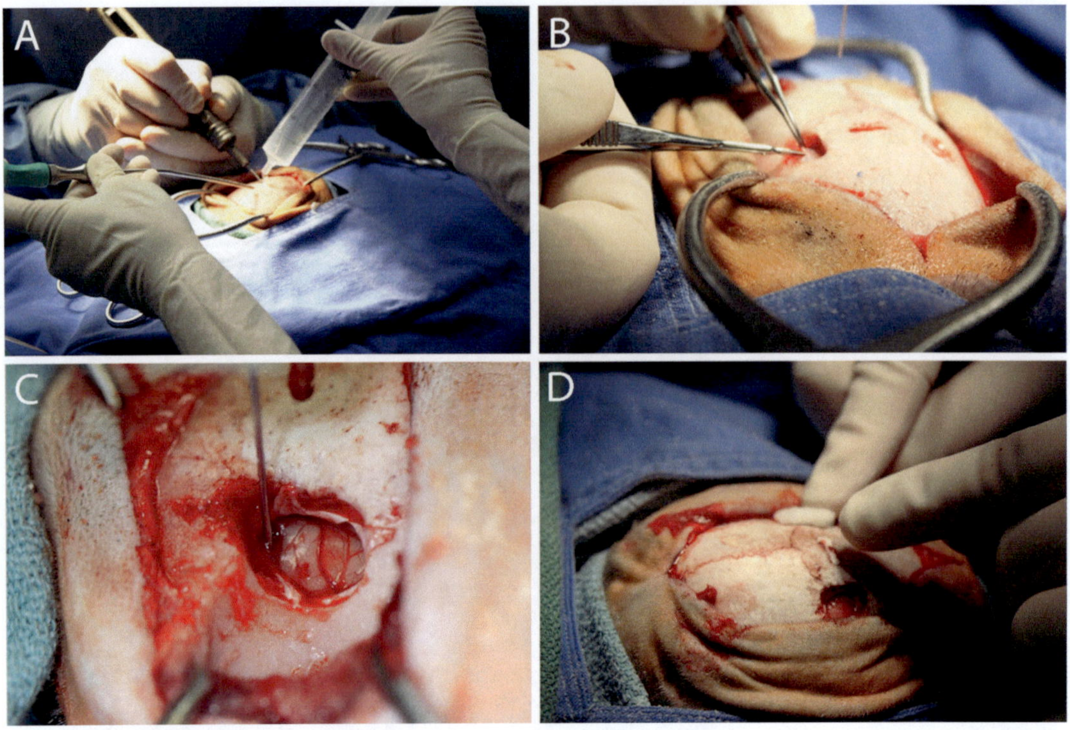

Fig. 4 Stereotaxic injection into the caudate and putamen. (**a**) A craniotomy that encompasses the proposed injection site(s) is established using a high speed drill and a 2 mm carbide burr. The edges are rounded using a rongeur. (**b**) The dura is excised using a micro-scissor and folded back to expose the pial surface of the brain. (**c**) The needle is slowly lowered to the calculated injection site coordinate and the infusate delivered using an infusion pump set to either a constant or ramping injection rate. (**d**) Post-injection, the needle is slowly raised and the craniotomy is filled with Gelfoam prior to suturing

20. Create oval craniotomies over the proposed injection sites using a high speed surgical drill with a 2 mm carbide burr (Fig. 4a). The length and diameter of the craniotomies will vary depending on the number of proposed injection sites in each hemisphere. Use saline and suction to cool the drill bit and clear bone debris from the craniotomy area. Smooth the edges of the craniotomies with rongeurs and incise the dura with a 22 gauge needle followed by microdissection scissors (Fig. 4b).
21. Load the syringe with infusate without removing the syringe from the infusion pump (*see* **Note 12**) and prime the needle by using the infusion pump to infuse until a small amount of liquid is visible at the tip of the needle. Swab the infusate away with gauze or a surgical swab.
22. At each injection site, carefully lower the needle down to the pial surface and record the DV value on the micromanipulator, which will serve as the DV zero value. Calculate the DV coordinate by subtracting the DV value obtained from the MRI from the DV zero value.

23. Slowly lower the needle to the DV coordinate and infuse at the desired rate (Fig. 4c; *see* **Note 13**).
24. After the infusion is complete, allow the needle to rest in place for an additional 3–10 min to allow the infusate to disperse from the needle tip. Slowly raise the needle out of the brain and proceed to the next injection site until all injections have been made.
25. Place Gelfoam sponge into the craniotomy sites (Fig. 4d) and use Cruciate 3-0 Vicryl sutures to bring the muscle back to its original lateral position along the skull. Close the subcutis and appose the skin edges with simple interrupted and intradermal 4-0 Monocryl sutures. Take the animal out of the stereotaxic head frame, discontinue the isoflurane gas anesthesia while continuing oxygen support and then release the animal from the stereotaxic head frame. Allow the animal to recover on the OR table until extubation.
26. Administer the following post-surgical medications: Hydromorphone on the day of surgery every 4 h IM until 2000 h, Buprenorphine for 2 days IM SID at 2000 h and Cefazolin for 3 days IM BID (25 mg/kg, 250 mg/ml) (Table 1).

3.3 Necropsy and Tissue Collection

1. Sedate the animal while in its home cage with 10–20 mg/kg ketamine (IM) and bring to the necropsy room. Secure animal to the board on the necropsy table with masking tape and anesthetize animal with sodium pentobarbital administered IV at 25 mg/kg. Assess depth of anesthesia by loss of palpebral, corneal, pain, and gag reflexes.
2. Incise abdomen with scalpel after adequate anesthesia has been established and collect terminal blood samples from abdominal aorta or caudal vena cava if needed. Place blood in red top serum tubes. Sever aorta to effect exsanguination and euthanasia.

Table 1
Pain medications administered post-surgery

Animal weight	Hydromorphone (2 mg/ml)	Buprenorphine (0.3 mg/ml)
<3 kg	0.5 mg (0.25 ml)	0.15 mg (0.5 ml)
3–10 kg	1 mg (0.5 ml)	0.3 mg (1 ml)
>10 kg	2 mg (1 ml)	0.3 mg (1 ml)

Different doses of Hydromorphone and Buprenorphine are listed according to rhesus macaque weight. Hydromorphone is administered post-surgery every 4 h IM until 2000 h and Buprenorphine (IM) for 2 days post-surgery at 2000 h (once per day)

3. If necessary, collect cerebrospinal fluid (3 cc syringe with 1.5″ 22 gauge sterile needle) and place into sterile Microcentrifuge collection tubes.
4. Open the thoracic cavity and the perfuse brain and spinal cord with 1–2 l of ice-cold, 0.9 % sterile saline via the right carotid artery (*see* **Note 14**).
5. Carefully saw around the entire skull in the axial plane using a hand saw and remove the skull cap using a prying motion with a bone lever. Gently remove the brain from the calvarium and place it in a brain jar (in saline) on ice. It will be necessary to dissect through the cranial nerves and spinal cord prior to removing the brain.
6. Collect relevant samples of tissues/organs (liver, spleen, gastric, kidney, lung, lymph nodes, pancreas, intestine, adrenal, gonads) in microcentrifuge tubes (smaller samples) and foil packs (larger samples for banking). Immediately place the samples in microfuge tubes on dry ice and the samples in foil packs in the Dewar filled with liquid nitrogen (*see* **Note 15**).
7. If of interest, remove spinal cord, divide into three sections (cervical, thoracic, and lumbar), remove meninges, cut a 2 cm long segment from the middle of each section and place each in small jars with sterile saline on ice.
8. Transport the brain and spinal cord on ice back to the laboratory in an enclosed biohazard container. Transport the Dewar full of frozen tissue samples back to laboratory. Transport blood samples in red-top tubes back to the laboratory at room temperature in a biohazard container.
9. Place the brain into a nonhuman primate brain matrix and carefully cut into slabs (2–8 mm thick slabs recommended) using tissue blades (Fig. 5a, showing a rhesus macaque brain matrix). Place each brain slab into a separate, sterile saline-filled petri dish on ice.
10. Use tissue biopsy cores (2 mm) to take samples from regions of interest for future molecular and/or biochemical analyses (Fig. 5b; *see* **Note 16**). Place samples in microcentrifuge tubes and place immediately on dry ice. Store samples at –80°.
11. After collecting brain samples, place slabs of brain tissue in 4 % paraformaldehyde for 48 h for post-fixing. After post-fixing in 4 % paraformaldehyde, place slabs of brain tissue in 30 % sucrose until they have completely sunken to the bottom of the jar.
12. Cut brain slabs using a frozen microtome at a thickness of 40 μm and collect tissues in large 24-well (4 × 6) compartmented tissue collection box filled with cryoprotectant solution.

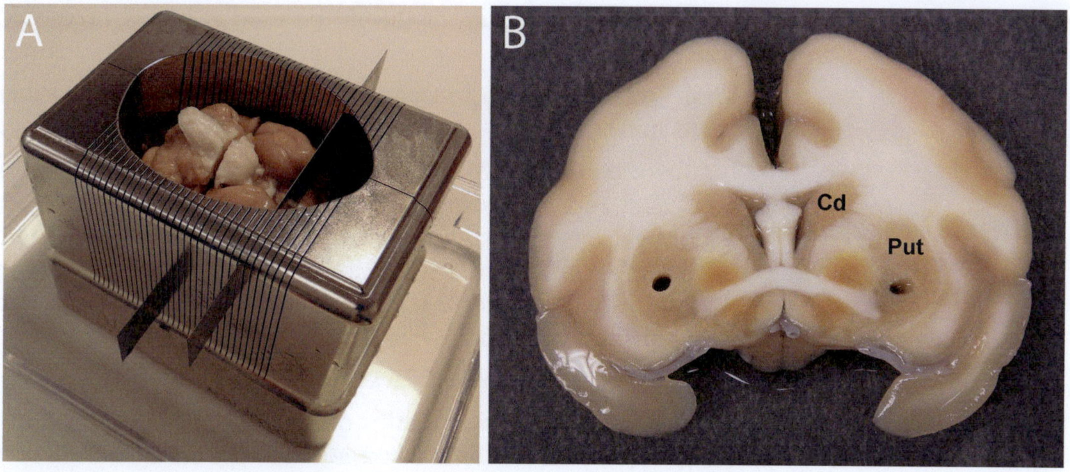

Fig. 5 Brain processing post-necropsy. (**a**) Following removal at necropsy, the brain is placed in to a matrix and sectioned into slabs of varying thickness using tissue slicer blades. (**b**) The slabs are placed into petri dishes containing sterile saline and biopsy punches are taken from key areas to be used for molecular and biochemical studies. Brain slabs are then post-fixed in 4 % paraformaldehyde for 48 h and cryoprotected in 30 % sucrose prior to immunohistochemical staining

3.4 Blood Processing

1. Allow blood to coagulate for 2 h at room temperature.
2. Centrifuge red top serum tubes at $18 \times g$ for 20 min at room temperature (25 °C) using a tabletop centrifuge.
3. Pipet off serum and transfer to microcentrifuge tubes. Store serum samples at −80°.

3.5 Immunohistochemical Processing of Brain to Verify Expression of Gene Therapy Agent

1. Wash tissue in dilution media in netted staining dishes (5×8 min).
2. Block endogenous peroxidase (2.13 g of sodium metaperiodate per 100 ml TBS) for 20 min in netted staining dishes.
3. Wash tissue in dilution media in netted staining dishes (5×8 min).
4. Block tissue in 5 % of the appropriate serum (goat, donkey) in netted staining dishes (5 ml serum per 100 ml dilution media).
5. Incubate tissue in primary antibody solution (primary antibody solution: 3 ml serum and 400 μl Triton-X per 100 ml PBS) in 6-well tissue culture plates (either shaking at room temp overnight or at 4° for 48 h). Each primary antibody should be used at its own concentration.
6. Wash tissue in dilution media in netted staining dishes (5×8 min).
7. Incubate tissue in secondary antibody solution in 6-well tissue culture plates (shaking at room temperature for 1 h). Secondary antibody solution: 3 ml normal serum per 100 ml dilution

media. Each secondary ab should be used at its own required concentration.

8. Wash tissue in dilution media in netted staining dishes (5 × 8 min).
9. Incubate tissue in ABC solution in netted staining dishes for 1 h. (First add 3 ml serum per 100 ml of dilution media. Next add four drops of bottle A and four drops of bottle B to 10 ml of serum/dilution media solution and mix. After waiting 30 min, mix those 10 ml back with the other 90 ml and place tissue into the solution.)
10. Wash tissue in TBS in netted staining dishes (3 × 8 min).
11. DAB reaction to develop color on the tissue: First add 50 mg DAB to 100 ml of TBS. Next add 2.5 g Nickel II Sulfate if nickel intensification needed (makes stain purple versus brown). Immediately prior to tissue incubation, add 20 μl of the 30 % hydrogen peroxide solution. Incubate tissue in DAB solution until the stain develops (depends on the antibody, ranges from ~1 to 10 min) (*see* **Note 17**; Fig. 6).
12. Wash tissue in TBS in netted staining dishes (3 × 8 min).
13. Store tissue at 4° in PBS in netted staining dishes until able to mount onto slides.

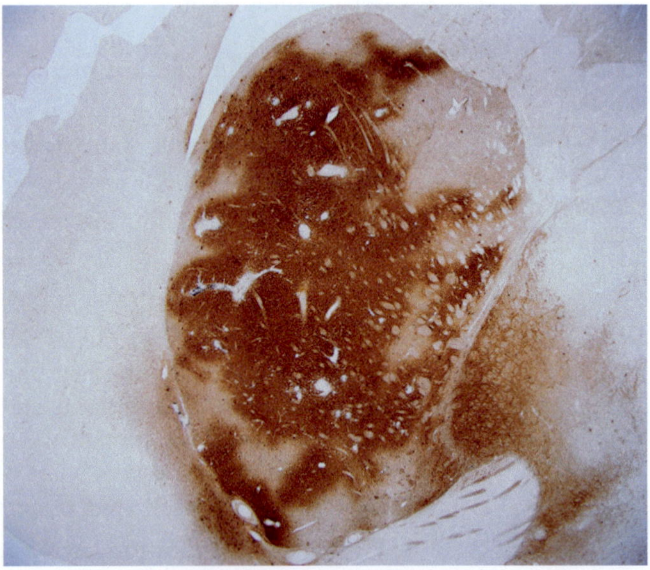

Fig. 6 Immunohistochemical staining of brain sections to identify the injected agent. Anti-eGFP staining demonstrates accurate expression of the transgene following a 50 μl injection of AAV2/1-eGFP (5e12 vg/ml) into the putamen of an adult rhesus macaque

4 Notes

1. The animal is positioned in the frame in the prone/sphinx position, with the stomach facing downward, the head facing forward, the hind legs positioned behind the animal and the forelegs placed in front of the animal. While placing the animal in the frame, it is important to not disturb or move the endotracheal tube. It is best to carefully place the tube laterally and out of the way of the palate bar.
2. Due to the magnetic field in and around the MRI Scanner, a precision vaporizer is located outside of the MRI Scanner room, in the control room. A long (3 m) non-rebreathing tube is utilized which extends from the anesthesia machine, through the wall of the control room, into the MRI room, where it connects to the subject. This setup also allows for adjustments to the oxygen flow rate and anesthetic gas delivery to the subject from the control room. Because of the length of the non-rebreathing tube, high oxygen flow rates are necessary to avoid rebreathing of the exhaled gases in the tube. Typically the oxygen flow rate is set at 1–2 l/min.
3. Approximate normal values: Respiratory rate: 10–20 breaths/min, Heart rate: 110–170 beats/min, O_2Sat (while breathing 100 % O_2): 95–100 %; EtCO_2: 35–45 mmHg.
4. This can be anything that contains an MRI contrast such as a Vitamin E capsule or a small tube filled with contrast dye.
5. Surface/head coils must be affixed to a base that is placed near/next to the animal. In our laboratory, the head coil is attached to a Gorillapod (flexible tripod) that is affixed to a heavy utility bottle filled with water. The tripod affords flexibility when placing the coil directly above the head.
6. Scan details on our Siemen's 3 T MRI unit: TR-2500, TI-1100, TE-3.9, BW-128, MF-2.04
7. Carefully choose pathways to desired injection sites that do not pass through sulci (location of major blood vessels that may rupture) or pass through ventricular walls (may alter needle trajectory to target). The number of injection sites into the caudate and putamen will depend on numerous factors including the diffusion capacity of the infusate (e.g., viruses versus cells), different viral vectors (e.g., lentivirus versus adeno-associated virus), different viral vector serotypes (e.g., AAV2 versus AAV5), different viral vector titers (e.g., high versus low) different cell types (e.g., stem cells versus progenitor cells).
8. The criteria used to assess adequacy of anesthesia and animal intraoperative well-being during the procedure include the

following: body temperature via esophageal temperature probe, heart rate and pulse character (fast or slow) via pulse oximetry and electrocardiography, blood pressure via indirect blood pressure cuff or direct percutaneous arterial line, oxygen saturation via pulse oximetry, respiratory rate and pattern, end tidal CO_2, capillary refill time, absence of palpebral response to touching the medial canthus, jaw tone and the color of mucous membranes at the gums or conjunctiva.

9. When injecting viral vectors, we prime the needle at this point with a separate tube of the virus to coat the barrel of the syringe with viral particles (prevents the vector prep that is injected into the brain from sticking to the barrel). This is done by flushing up and down several times and expelling the used vector into a biohazard container.

10. Ensure that the pump, syringe, and needle are raised as high as possible before removing from the AP zero bar to ensure maximal clearance of the animal's head when being placed onto the head frame. After the manipulator is locked onto the head frame, the manipulator, infusion pump, syringe and needle will remain sterile, while the end of the pump cord will become unsterile as it is attached to pump controller box located on a Mayo stand close to the surgical table. At this point, a non-scrubbed member of the surgical staff can program the infusion pump with the volume and rate of the infusion. The non-scrubbed staff can continue to operate the pump controller box or a sterile clear plastic drape can be placed over the box so that it can be controlled by the surgeon. We find the latter to be the easiest option.

11. AP coordinates: The set point on the AP slide attachment when the needle was zeroed on the zeroing bar (e.g., 50 or 0) added or subtracted from the distance from the ear bar to the injection site established from the MRI. Example: with the AP slide attachment set to 50, the micromanipulator placed on the left AP bar of the head frame and the distance of an injection site in the putamen 20 mm in front of the ear bars on the coronal MR image, the AP for this injection site is $50-20=30$. Example 2: with the AP slide attachment set to 0, the micromanipulator placed on the right AP bar of the head frame and the distance of an injection site in the putamen 20 mm in front of the ear bars on the coronal MR image (the AP for this injection site is $0+20=20$).

    ML coordinates: ML value on the micromanipulator when the needle was zeroed on the animal's sagittal sinus added or subtracted from the distance from the sagittal sinus to the injection site established from the MRI. Example: with the ML zero value at the sinus at 60, the micromanipulator placed on the left AP bar of the head frame and the distance of an injection site in the left putamen 15 mm lateral to the center of the sagittal sinus on the coronal MR image, the ML for this injection site is 60 + 15 = 75. For an injection site in the right putamen the calculation would be 60–15=45

12. Loading the syringe with infusate can be achieved by using a sterile pipetteman and sterile pipette tips to pipette the desired amount of infusate onto a sterile piece of Parafilm, bringing the infusate on the Parafilm over to the needle and using the infusion pump to withdraw the infusate into the syringe. Any extra virus should be placed into a biohazard container.
13. Infusion rates will differ depending on type and expected spread of the infusate. A standard rate of 1–2 μl/min works well for injections into both the caudate and putamen. Spread of infusate can be enhanced with a ramping infusion rate that begins at 0.5 μl/min and increases by 0.5–1.0 μl every 5 min.
14. The pumping rate should be a moderate pulsatile flow versus a steady stream so that blood vessels in the brain are not ruptured. Ruptured blood vessels can lead to background in future immunohistochemical staining processes.
15. Wash instruments (forceps and Metzenbaums) with Tergazyme, 70 % ethanol and rinse in PBS between each tissue to prevent cross-contamination. It is advised to collect peripheral tissues to assess for potential distribution of the infusate out of the brain and into the periphery.
16. Spray biopsy cores and forceps with 70 % ethanol and wiped clean with a Kimwipes between punches of different brain regions.
17. DAB is a carcinogen and should be handled with caution. It is essential that the hydrogen peroxide is not added until immediately prior to placing the tissue in the DAB solution because it will quickly break down into oxygen and water. Figure 6 depicts an example of a coronal section from a rhesus macaque that was injected with 50 μl of AAV2/1-eGFP (titered at 5e12 vg/ml) into the putamen. The MRI, surgery, necropsy, and immunohistochemistry techniques used were those mentioned above and the tissue was stained using an anti-eGFP antibody (Invitrogen, 1:1000).

Acknowledgements

This work was supported by NIH Grant NS069798 (JLM), and Oregon National Primate Research Center (ONPRC) Core Grants RR000163 and RR000163.

References

1. Huntington Disease Collaborative Research Group (1993) A novel gene containing a trinucleotide repeat that is expanded and unstable on Huntington's disease chromosomes. The Huntington's Disease Collaborative Research Group. Cell 72:971–983
2. Vonsattel JP, Myers RH, Stevens TJ, Ferrante RJ, Bird ED, Richardson EP Jr (1985) Neuropathological classification of Huntington's disease. J Neuropathol Exp Neurol 44:559–577
3. Isacson O, Riche D, Hantraye P, Sofroniew MV, Maziere M (1989) A primate model of Huntington's disease: cross-species implantation of striatal precursor cells to the excitotoxically lesioned baboon caudate-putamen. Exp Brain Res 75:213–220
4. Burns LH, Pakzaban P, Deacon TW, Brownell AL, Tatter SB, Jenkins BG, Isacson O (1995) Selective putaminal excitotoxic lesions in nonhuman primates model the movement disorder of Huntington disease. Neuroscience 64:1007–1017
5. Hantraye P, Riche D, Maziere M, Isacson O (1990) A primate model of Huntington's disease: behavioral and anatomical studies of unilateral excitotoxic lesions of the caudate-putamen in the baboon. Exp Neurol 108:91–104
6. Roitberg BZ, Emborg ME, Sramek JG, Palfi S, Kordower JH (2002) Behavioral and morphological comparison of two nonhuman primate models of Huntington's disease. Neurosurgery 50:137–145
7. Palfi S, Brouillet E, Jarraya B, Bloch J, Jan C, Shin M, Conde F, Li XJ, Aebischer P, Hantraye P, Deglon N (2007) Expression of mutated huntingtin fragment in the putamen is sufficient to produce abnormal movement in nonhuman primates. Mol Ther 15:1444–1451
8. Isacson O, Hantraye P, Maziere M, Sofroniew MV, Riche D (1990) Apomorphine-induced dyskinesias after excitotoxic caudate-putamen lesions and the effects of neural transplantation in non-human primates. Prog Brain Res 82:523–533
9. Hantraye P, Riche D, Maziere M, Isacson O (1992) Intrastriatal transplantation of cross-species fetal striatal cells reduces abnormal movements in a primate model of Huntington disease. Proc Natl Acad Sci U S A 89:4187–4191
10. Emerich D, Winn S, Hantraye P, Peschanski M, Chen E, Chu Y, McDermott P, Baetge E, Kordower J (1997) Protective effect of encapsulated cells producing neurotrophic factor CNTF in a monkey model of Huntington's disease. Nature 386:395–399
11. Emerich D, Thanos C, Goddard M, Skinner S, Geany M, Bell W, Bintz B, Schneider P, Chu Y, Babu R, Borlongan C, Boekelheide K, Hall S, Bryand B, Kordower J (2006) Extensive neuroprotection by choroid plexus transplants in excitotoxin lesioned monkeys. Neurobiol Dis 23(2):471–480
12. McBride JL, Pitzer MR, Boudreau RL, Dufour B, Hobbs T, Ojeda SR, Davidson BL (2011) Preclinical safety of RNAi-mediated HTT suppression in the rhesus macaque as a potential therapy for Huntington's disease. Mol Ther 19:2152–2162
13. Grondin R, Kaytor MD, Ai Y, Nelson PT, Thakker DR, Heisel J, Weatherspoon MR, Blum JL, Burright EN, Zhang Z, Kaemmerer WF (2012) Six-month partial suppression of Huntingtin is well tolerated in the adult rhesus striatum. Brain 135:1197–1209
14. Stiles DK, Zhang Z, Ge P, Nelson B, Grondin R, Ai Y, Hardy P, Nelson PT, Guzaev AP, Butt MT, Charisse K, Kosovrasti V, Tchangov L, Meys M, Maier M, Nechev L, Manoharan M, Kaemmerer WF, Gwost D, Stewart GR, Gash DM, Sah DW (2012) Widespread suppression of huntingtin with convection-enhanced delivery of siRNA. Exp Neurol 233:463–471
15. Chan AW, Xu Y, Jiang J, Rahim T, Zhao D, Kocerha J, Chi T, Moran S, Engelhardt H, Larkin K, Neumann A, Cheng H, Li C, Nelson K, Banta H, Zola SM, Villinger F, Yang J, Testa CM, Mao H, Zhang X, Bachevalier J (2014) A two years longitudinal study of a transgenic Huntington disease monkey. BMC Neurosci 15:36

Chapter 30

Gene Therapy for the Treatment of Neurological Disorders: Metabolic Disorders

Dominic J. Gessler and Guangping Gao

Abstract

Metabolic disorders comprise a large group of heterogeneous diseases ranging from very prevalent diseases such as diabetes mellitus to rare genetic disorders like Canavan Disease. Whether either of these diseases is amendable by gene therapy depends to a large degree on the knowledge of their pathomechanism, availability of the therapeutic gene, vector selection, and availability of suitable animal models. In this book chapter, we review three metabolic disorders of the central nervous system (CNS; Canavan Disease, Niemann–Pick disease and Phenylketonuria) to give examples for primary and secondary metabolic disorders of the brain and the attempts that have been made to use adeno-associated virus (AAV) based gene therapy for treatment. Finally, we highlight commonalities and obstacles in the development of gene therapy for metabolic disorders of the CNS exemplified by those three diseases.

Key words Gene therapy, CNS, Metabolic disorders, AAV, Retrovirus, Phenylketonuria, Canavan's disease, Niemann–Pick disease

1 Introduction

Metabolic disorders are a vast group of diseases that can affect every cell, organ, and organism. Generally speaking, each disease that compromises the maintenance of cellular homeostasis could be considered as a metabolic disorder. For example, protein misfolding in Alzheimer disease, serotonin metabolism in major depressive disorders or nucleotide metabolism in Lesch–Nyhan Syndrome involve metabolic pathways. However, such a broad classification might be controversial and of limited use, especially considering that the metabolic contribution to the disease pathomechanism might differ substantially. Thus, we focus in this chapter on three diseases affecting classic metabolic pathways of amino acid and lipid metabolism that exemplify metabolic CNS disorders. We refer to disorders originating in the CNS as primary metabolic disorders and those originating outside of the CNS but displaying a CNS disease phenotype as secondary metabolic disorders of the CNS.

Fredric P. Manfredsson (ed.), *Gene Therapy for Neurological Disorders: Methods and Protocols*, Methods in Molecular Biology, vol. 1382, DOI 10.1007/978-1-4939-3271-9_30, © Springer Science+Business Media New York 2016

This simple differentiation between primary and secondary has substantial implications for the vector selection and the route of vector administration in particular. Primary metabolic disorders of the CNS are amendable to intracranial vector delivery with concomitant CNS transgene restriction by the blood–brain barrier (BBB). In contrast, secondary metabolic disorders of the CNS frequently require expression of therapeutic genes either restricted to a peripheral organ where the mutant gene is expressed or ubiquitously. It has also been shown, however, that muscle can serve as an ectopic biofactory of therapeutic gene products for either functioning as a metabolic sink to reduce the metabolic burden or dissemination of transgene product via the blood circulation (*see* Subheading 3.1).

In addition, metabolic disorders can be subdivided into mono-causative (monogenic) or multifactorial (polygenic) and inherited or acquired. For example, Canavan Disease (CD) is considered to be primary, mono-causative (monogenic) and inherited or acquired. Although patients with CD predominantly display symptoms of the CNS it has not been fully investigated if there is peripheral organ involvement as well. On the other hand, some unpublished animal studies suggest that targeting the CNS only, might be beneficial too. Nevertheless, for didactical purposes, we will consider CD as a primary metabolic disorder of the CNS in this chapter. In contrast, phenylketonuria (PKU) and Niemann–Pick disease (NPD) are secondary monogenic metabolic disorders of the CNS, with mutations in the phenylalanine hydroxylase (PAH) gene restricted to the liver and ubiquitous acidic sphingomyelinase (ASM) deficiency, respectively. Monogenic disorders serve as ideal disease models for gene therapy development, due to one gene replacement strategy and limitations in vector biology. For example, the 5 kilo base (kb) genome size of recombinant adeno-associated virus (rAAV) often constrains space for large transgenes [1, 2].

A different genetic aspect is whether the mutation causes a gain- or loss-of-function. Loss-of-function mutations are amendable to the delivery of the native functional gene, whereas gain-of-function mutations demand reduction of the toxic gain-of-function gene product activity, which raises additional concerns of how the toxic activity can selectively be inactivated and to restore the physiologic function of the native normal gene simultaneously. This may require some dual functional vectors, meaning silencing the mutant gene but supply the normal gene function at the same time [3]. However, to our knowledge no metabolic disease of the CNS in the narrower sense with gain-of-function mutation has been targeted for proof-of-concept gene therapy experiments.

To demonstrate the challenges of gene therapy development for metabolic disorders of the CNS three enzyme deficiencies

(Canavan disease, phenylkentonuria, and Niemann–Pick disease) are reviewed within this chapter that share fundamental preconditions across different disease entities, e.g., availability of transgene cDNA, animal model and vector but might demonstrate features unique to CNS metabolic disorders. One of these features is the intricacy to adjust widespread transgene delivery to peripheral organs and/or the CNS despite the BBB. For example, systemic vector administration exploits the natural ramification of the CNS blood vessels for optimal and wide spread transgene delivery but necessitates means to limit peripheral organ transduction in Canavan disease as opposed to local CNS pathologies (e.g., Huntington disease or CNS tumors) that are accessible by local vector injection. In addition, systemic vector delivery to the CNS is compromised by the first pass effect in the liver and the tight BBB, enhancing peripheral organ transduction as desired in NPD but rendering the neuron and glial cell transduction diminished. This sequestration of systemically delivered vector by peripheral organs entails the administration of higher vector doses to still efficiently transduce the CNS, which in turn may increase the risk for an immune response and the burden of vector manufacturing.

This chapter attempts to address several of these briefly introduced aspects of gene therapy development for metabolic disorders of the CNS and demonstrate how different approaches might provide improvements in these matters.

2 Lipid Metabolism Disorders

2.1 Niemann–Pick Disease (NPD)

2.1.1 Introduction

The first patient with Niemann–Pick disease (type A) was reported by the German pediatrician Albert Niemann in 1914. Niemann–Pick disease (NPD) is a rare autosomal recessive inherited disorder that is characteristic for accumulation of sphingomyelin (SPM), organomegaly and variable central nervous system (CNS) involvement [4]. Of note, Niemann–Pick is an umbrella term that encompasses two different entities, Niemann–Pick disease type A (NPA) and type B (NPB) are classified as acid sphingomyelinase deficiencies whereas Niemann–Pick disease type C (NPC) is characterized by altered trafficking of endocytosed cholesterol [5, 6].

The onset of NPD is quite heterogeneous even within the three subclasses. For example, the occurrence of first symptoms in Niemann–Pick disease type C were reported to fall within a wide range of ages from childhood up to 60 years of age, while Niemann–Pick disease type A manifests in infancy with progressive neurodegenerative decay [7, 8]. Unfortunately, current treatment is limited to symptoms [9–11].

2.1.2 Prevalence

Demographic data for epidemiologic analysis are limited. However, the prevalence of Niemann–Pick A and B together was estimated to be 1:250,000 [12], whereas the prevalence of NPC is about 1:120,000 for live birth in Europe [5]. Some authors report increased incidence and prevalence among descendants of Ashkenazi Jews [11].

2.1.3 Symptoms

Niemann–Pick disease type A is considered to be the neurovisceral form and presents during early ages with hepatosplenomegaly, poor feeding behavior, and loss of motor function and general neurological deterioration eventually resulting in death by the age of 3 years [13]. Low HDL and hypertriglyceridemia may be present [14].

In contrast, NPB is of milder progression with symptom onset between childhood and adulthood; hepatosplenomegaly is one of the most consistent findings. Disease progression is complicated by deteriorating pulmonary function and liver dysfunction. Other symptoms may be growth restriction and impaired eye function [15, 16]. Neurological symptoms are uncommon in this subclass. As in NPA, low HDL and hypertriglyceridemia is often present [14].

Niemann–Pick disease C is both of different genetic origin versus NPD type A and B and might present quite differently. NPC can become symptomatic between infancy and adulthood [17], with cognitive dysfunction, ataxia, dysarthria and loss of language [18]. Although severity of symptoms might differ along with the onset of symptoms, most patients die between the ages of 10 and 25. Of note, even fetal NPC has been described [17].

2.1.4 Genetics

Individuals with subclasses A and B of Niemann–Pick disease share the trait that they are deficient in the enzyme acid sphingomyelinase (ASM). However, they display different phenotypes, which might be explained with no residual sphingomyelinase activity in NPD type A but low remaining enzyme activity in NPD type B [19, 20]. The gene (SMPD1) coding for ASM is localized on human chromosome 11p 15.1–15.4. Its cDNA is about 1895 base pairs (bps) long and comprises six exons, which are transcribed to 629 amino acids [7]. Several transcript variants have been identified.

In contrast, NPD type C is caused by mutations in the NPC gene that was discovered by Carstea et al. in 1997 [21]. Of those identified, 95 % of patients carry a mutation in the NPC-1 and only in about 5 % in the NPC-2 gene [22]. NPC-1 is located on human chromosome 18q11 and NPC-2 on chromosome 14q24.3. Interestingly, NPC-1 (cDNA 3836 bases) is a membrane glycoprotein predominantly located in late endosomes as opposed to NPC-2 (cDNA 455 bases), which is a soluble lysosomal protein [22]. Current hypotheses assume that both genes are involved in the same pathway that processes endocytosed cholesterol [22].

2.1.5 Pathomechanism

Acid sphingomyelinase (ASM) is predominantly found in lysosomes, where it metabolizes the membrane compound sphingomyelin to ceramide (SPM) [23, 24]. ASM follows the trafficking via mannose and mannose-6 phosphate receptors from the rough endoplasmatic reticulum (rRER) to Golgi apparatus and its final destination the lysosomes. Of note, secreted ASM can also be taken up by neighboring cells (cross-correction), which implies that partial cell transduction might be therapeutic [25]; this concept also applies to enzyme replacement therapy (ERT).

The significance of SPM metabolism to ceramide depends on the location of this catabolic step. Different stressors can translocate the hydrolysis of SPM to the cell membrane. Under these conditions, ceramide can be involved in important cell regulatory mechanisms such as autophagy, apoptosis, differentiation and cell growth [26–28]. Thus, NPD type A and B could be considered as an ASM deficiency and also ceramide metabolism disorder. A detailed review of possible molecular pathways involved in the NPD pathomechanism can be found in Smith 2008 and Schuchman 2010 [29, 30].

2.1.6 Animal Models

The first animal models for Niemann–Pick disease type A and B were described in 1980 and 1982 [31, 32]. Although these mice presented with symptoms associated with Niemann–Pick disease, biochemical studies differed from what was found in NPD type A and B patients [33, 34]. This difference between animal and human disease manifestation can frequently be observed, questioning the authenticity of the disease imitation. In fact, Horinouchi et al. demonstrated that these mouse models did not carry any mutation in the ASM gene [35], making these mice inappropriate models for NPA and NPB studies.

2.1.7 NPD Type A Mouse Model

In 1995, Horinochi et al. engineered an acid sphingomyelinase knockout (ASMKO) mouse model by direct gene targeting [36]. This mouse is still the most commonly used model for gene therapy studies. It carries a PGK-neo expression cassette disrupting exon 2 of the ASM gene. Clinically, the ASMKO mouse displays the severe phenotype reflecting NPA. First symptoms can be noticed at 8 weeks of age with ataxia and mild tremor progressing to lethargy, non-responsiveness to stimuli and poor feeding with weight loss at 12–16 weeks of age. Death occurs between 6 and 8 months [36]. Interestingly, mice homozygous for the mutation can still breed, which facilitates the reduction of animal costs and time intensive procedures, e.g., genotyping. Molecular analyses revealed that homozygous mutant mice are deprived of residual ASM activity and accumulate sphingomyelin intracellularly. As with NPD patients, the blood lipid profile might show hypercholesterolemia but with elevated HDL. Macroscopic examination of tissue is significant for decreased size of brain and most peripheral organs; no hepatosplenomegaly was found. The CNS findings were significant for Purkinje cell loss along with cerebellar and midbrain atrophy. In addition, lipid-laden foam cells along with multilamellar inclusions are present in many organs.

Another knockout mouse model designed by Otterbach et al. displays a similar phenotype with first onset of symptoms (tremor and ataxia) at 8–10 weeks of age [37]. The affected mice present with heptosplenomegaly as seen in many patients but not in the ASMKO mouse from Horinochi et al. The disease progresses with severe dyspnea and death by 16 weeks of age, providing a mouse model with more severe phenotype. The pathology revealed an almost complete loss of Purkinje cells and accumulation of sphingomyelin and foam cells. Molecular analysis shows no transcript or enzyme activity. Interestingly, despite the more severe phenotype, which might be beneficial for the evaluation of treatment potency, this mouse model is not commonly used in gene therapy studies.

2.1.8 NPD Type B Mouse Model

In addition, a mouse model for Niemann–Pick disease type B has been engineered that only displays the characteristic visceral phenotype [38]. In view of the reduced capability of many drugs to cross the blood–brain barrier, this animal model seems to be ideal to evaluate treatment options that do not concern the brain, e.g., enzyme replacement therapy.

2.1.9 NPD Type C Mouse Model

Finally, it is noteworthy that different groups developed mouse models for NPD type C with variability in phenotype. However, due to insufficient literature the focus will be on NPD type A and B [39, 40].

2.1.10 Preclinical Gene Therapy Studies

Many efforts have been made to develop potential remedies for Niemann–pick disease with gene therapy being only one of them. Other concepts that have been pursued are: liver transplantation [41, 42], amniotic membrane transplantation [43], bone marrow transplantation [44–46], and enzyme replacement therapy [47]. It would be beyond the scope of this chapter to discuss all these treatment strategies. Instead, the focus will be on the aspects of ex vivo and in vivo gene therapy.

2.1.11 First In Vitro Study

Despite the natural occurrence of Niemann–Pick disease animal models first attempts for gene therapy were not accomplished until 1992. Suchi et al. used retroviral vectors to successfully deliver the full acidic sphingomyelin (ASM) cDNA into fibroblasts from two NPA patients in vitro [48]. The resulting 16-fold increase in ASM activity and decreased sphingomyelin content underlined the success of this first attempt to correct ASM deficiency; yet only in vitro.

2.1.12 Ex Vivo Gene Therapy

Hematopoietic Stem Cell Gene Therapy

Ex vivo gene therapy relies on explantation of patient cells, in vitro transfection with the transgene and re-implantation to the patient, which was first attempted in the ASM knockout (ASMKO) mouse by Miranda et al. Male mice-derived nucleated bone marrow cells were transfected with a retroviral construct carrying the full cDNA of the human ASM (hASM). These cells were administered intravenously into 2-day-old ASMKO neonates after entire body radiation (200 cGy). Five months later, the highest ASM activity was detected in spleen, liver and lung whereas brain, heart and kidney showed insignificant difference to untreated ASM mice. Simultaneously, sphingomyelin content of the same organs was analyzed, showing a positive correlation between ASM activity and sphingomyelin reduction [49]; similar improvements were found on tissue sections. Interestingly, although there was no significant difference in ASM activity of brain in untreated versus treated mice, pathology of the cerebellum showed substantial improvement in treated mice [49]. While, this study using ex vivo gene therapy for the treatment of NPD type A/B could not rescue the phenotype and survival to the level of wild-type mice, overall, this study demonstrated the potential of hematopoietic stem cell gene therapy (HSCGT) for the treatment of NPD type A/B. However, given the limited success to improve the ASM activity in the brain, HSCGT might be more promising for NPD type B at this point.

2.1.13 Mesenchymal Stem Cell Gene Therapy

The same group demonstrated that bone marrow-derived mesenchymal stem cell (MSC) transplantation to the CNS could significantly improve the NPA and NPB phenotype. Again, cells (MSC) were transfected using a retroviral expression system carrying the human ASM cDNA. This time, however, transduced cells were injected directly into hippocampus and cerebellum of 3-week-old ASM knockout mice. Strikingly, the treatment could extend the life expectancy, improve motor function and reduce Purkinje cell loss of the cerebellum [50].

2.1.14 Combined Hematopoietic and Mesenchymal Stem Cell Gene Therapy

Later, this group combined systemic HSCGT and intracranial MSC transplantation for the treatment of the ASM knockout mouse. Three-day-old neonates were infused with hematopoietic stem cells after 400 cGy whole body radiation and the same mice were treated intracerebrally with mesenchymal stem cells at 4 weeks of age. In both cases, cells were transduced with the retroviral construct expressing the human ASM cDNA [51]. The results of this study were especially remarkable for almost 70 % of normal ASM activity in the brain of treated mice, which was in contrast to previous studies of this group [48, 51]. Unfortunately, loss of ASM activity was noted after 24 weeks post-treatment due to the development of anti-ASM antibodies [51]. These studies of ex vivo gene therapy show promising results but also bring several issues relating to the retroviral vector, e.g., efficiency, genotoxicity and in vitro cell manipulation (*see* Chapters 1 and 24).

2.1.15 In Vivo Gene Therapy: Systemic Vector Administration

An alternative approach to ex vivo is in vivo gene therapy, which aims to transduce cells that remain in their natural environment. Hereby, it is not necessary to harvest and re-implant cells of a patient and radiate either certain parts or the whole body, which raise additional safety concerns in clinical trials.

Adeno-associated virus (AAV) has become the new favorable vector for many therapeutic in vivo gene therapies, which was also used by Barbon et al. to demonstrate that intravenously delivered rAAV8hASM under the control of a liver specific promoter could correct the NPD type B phenotype in the ASM knockout mouse and performs superior to rAAV1 by about 100-fold increased ASM expression in liver. In addition, sphingomyelin clearance to almost normal levels was found in liver, spleen and lung within 8 weeks versus 12 weeks in rAAV1hASM treated mice; similar improvements were found on tissue pathology in the rAAV8hASM treated group. One drawback of this study was a humeral immune response against the rAAV capsid as well as the ASM enzyme, when ASM was expressed ubiquitously under the control of a CMV promoter. However, consistent with previous findings, liver restricted ASM expression could circumvent anti-ASM antibody immune response significantly [52]. While this study demonstrated the potential of liver restricted transgene expression for NPB it might not achieve therapeutic efficacy in NPD with CNS disease phenotype, due to the incapability of ASM crossing the blood–brain barrier.

2.1.16 In Vivo Gene Therapy: Intracranial Vector Administration

A particular route of administration for gene therapy in CNS pathologies is intracranial injection, which encompasses either intraparenchymal or intraventricular vector delivery. This could provide a powerful alternative to the usage of tissue specific promoters in systemic delivered gene therapeutics for tissue restricted transgene expression.

This was tested by Passini et al. using intracerebral injection of rAAV2-hASM in the ASMKO mouse, specifically the hippocampus [53]. Although rAAV was only injected on one side of the brain, ASM positive cells were also found on the contralateral site. The authors reasoned that this was due to axonal transport of rAAV2-hASM, which was of particular surprise given that axons in NPD type A are severely affected. Of note, 3.6–5.4 % of ASM positive cells were found on the non-injected site. Interestingly, the improvement of the CNS pathology was observed even in areas, were no ASM positive cell were found, e.g., entorhinal cortex. However, several essential regions of the brain were not reached by the local injection, e.g., thalamus [53].

Similarly, Dodge et al. injected rAAVhASM intracerebrally to attempt the rescue of gross Purkinje cell loss and improved functional outcome in ASMKO mice for NPD type A [54]. In addition, this group compared five different rAAV serotypes (rAAV1, 2, 5, 7 and 8) by unilateral injection into the deep cerebellar nucleus (DCN), with rAAV1 demonstrating the widest tissue distribution and serotype 8 the highest hASM expression. In contrast, the previously used rAAV2 serotype showed the most restricted tissue distribution and lowest protein expression, even when injected bilaterally [53, 54]. Despite the difficulties to achieve wide transgene distribution, some regions negative for hASM expression still showed decreased cholesterol accumulation. Comparing unilaterally versus bilaterally injected mice revealed improvement of motor function in both groups but bilaterally injected mice took the lead. Overall, this study revealed that on average rAAV1 was the most efficient serotype among five tested rAAVs after local injection in the ASMKO mouse CNS in terms of tissue distribution, transgene expression, and cell transduction [54].

Importantly, both studies report that intracerebral delivered rAAVhASM promotes clearing of lipid accumulation but global CNS transduction was still limited. Furthermore, intracerebral injections into the hippocampus or DCN are not efficient to recover the pathology of peripheral organs [55].

2.1.17 *In Vivo Gene Therapy: Combined Intravenous and Intracerebral Vector Administration*

To overcome the limitations of intracranial transgene delivery for NPD, one could inject rAAVhASM intravenously and intracerebrally at the same time. Indeed, Passini et al. showed that combined injection of rAAV8hASM intravenously and rAAV2hASM intracerebrally (2 weeks apart) benefits the overall performance of ASMKO mice over intravenous or intracerebral injection alone. Double injected animals did not only perform better in survival, motor and cognitive function in comparison to intravenous or intracerebral injected mice, the microscopic pathology was significantly improved as well [55]. Interestingly, it was found that the brain hASM level in double-injected mice was higher than in intracerebral injected animals alone despite the use of the same dose for the intracerebral injections. It has to be pointed out, that at this time, it had not been demonstrated that rAAV is capable of crossing the blood–brain barrier (BBB). As expected, intravenously injected mice showed lower levels of hASM in brain versus mice treated intracerebrally. Of note, significantly increased levels of anti-hASM antibodies were found in brain homogenates of intracerebrally treated mice but no difference from the control group in the combined injected mice, suggesting the occurrence of immune tolerance against the transgene contributed to the higher intracerebral hASM levels in the combination group [55].

In summary, the rAAV mediated hASM delivery to peripheral organs and/or the CNS has been demonstrated to improve the phenotype in the Niemann–Pick disease type A/B mouse model

(ASMKO mouse). However, in order to improve safety and feasibility of this gene therapeutic approach, the transition to a large animal model would be desirable. This is especially true in view of the difference in body size between mouse and a large animal model, which is crucial for vector production, transgene delivery, and treatment efficacy.

2.1.18 Large Animal Model Gene Therapy for Niemann–Pick Disease

Salegio et al. evaluated magnet resonance imaging (MRI) guided rAAV2hASM delivery by simultaneous unilateral brainstem and bilateral thalamus infusion in nonhuman primates (NHP) [56]. It was found that 72 % of neurons in the brainstem and 68 % in the thalamus were successfully transduced and vector distribution correlated with infusion volume. In addition, neurons distant from the injection site were transduced as well, indicating axonal transport of the vector, transgene, or transgene product. Although transduced neurons were found in remote anatomical regions, it remains to be elucidated to what extent the low transduction rate of these distant cells will contribute to clinical improvements [56].

Another study attempted the translation of the minimal effective dose of intracranial delivered rAAV1hASM, determined in mouse, to nonhuman primates (NHP) [57]. It is worthwhile noticing that this group around Bu et al. used multiple injections into cortex, subcortical structures, and cerebellum. As hypothesized, the dose correlated with sphingomyelin clearance in the ASMKO mouse. Even mice injected with the lowest dose (4×10^9 genome copies per mouse) tested, still benefited from the treatment and survived significantly longer than untreated control mice. The study group with the highest dose (1.2×10^{11} genome copies per mouse) performed best [57]. Of importance, treatment of the CNS only could significantly improve the survival, but ASMKO mice eventually died due to the visceral manifestation of the disease. This implicates that treating peripheral organs or the CNS only is not efficient for curing NPD type A and has to be taken into account for subsequent studies [55]. To translate these findings, healthy Cynomolgus monkeys were injected a total of six times into each hemisphere: motor and occipital cortex, striatum, thalamus, hippocampus, and cerebellum (total vector dose 2.6×10^{12} genome copies per brain). Five weeks post-injection, almost all CNS regions injected showed increased hASM expression levels that were above that of the mouse group injected with the highest dose. Interestingly, the cortex of the cerebellum did not express significantly increased hASM [57]. Overall, this study demonstrates that rAAV mediated hASM delivery to the NHP CNS can provide hASM levels similar to levels found to be sufficient to treat Niemann–Pick disease in the ASMKO mouse.

2.1.19 Immune Response in Large Animal Model Gene Therapy

Among the most formidable complications in gene therapy are transgene toxicity and severe immune response. This is in accordance with several studies that failed to predict and safely translate findings in mouse or rat to large animal models due to immune response or transgene toxicity, which emphasizes the importance of large animal studies before gene therapeutics can be applied to patients.

These aspects have been recently evaluated by Salegio et al., who conducted a safety study using rAAV2 to express hASM in naïve rats and NHPs [58].

Intra-thalamic infused rats showed increased hASM expression and activity in thalamus, cortex, prefrontal cortex, cerebellum/brainstem, and spinal cord. However, hASM activity was only significantly increased in the thalamus. In addition, monkeys were simultaneously infused into both thalami and into the brainstem and evaluated for either 3 or 9 months. Differences in hASM expression were found at 3 and 9 months post-treatment and between the low and high dose groups. Most importantly, rats tolerated the treatment without side effects, which stands in stark contrast to the rAAVhASM infused NHPs. Independent of the dose, severe side effects of neurological deterioration, weight loss, abnormal blood results were noticed. Most severely affected animals were of the high dose group that resulted in early necropsy. Tissue sections revealed cell infiltrates and gliosis along with activated macrophages, microglia, T cells and antigen-presenting cells. Furthermore, the authors could show a correlation between hASM expression and CCL5 upregulation, which might have been one of the key players in this immune response [58].

2.1.20 Challenges and Future Directions

While rAAV is considered to be nonpathogenic, there is increasing awareness in the field of gene therapy, that immunogenicity against the rAAV capsid, the transgene, or the transgene product might be the main challenge in the transition from bench to bedside. The severe side effects in NHPs reported by Salegio et al. could also have been due to transgene toxicity, which might be an artifact of using the human transgene in preclinical animal models. However, further studies are urgently needed to develop strategies allowing for safe gene therapy in humans.

Furthermore, many metabolic diseases affect multiple cell types and organs demanding wide tissue distribution and tropism for treatment success. This also accounts for Niemann–Pick disease type A and C, which manifest in several peripheral organs and the CNS that might benefit from systemic delivery of a blood–brain barrier (BBB) crossing vector. However, serotypes tested in all studies discussed above, are known to have limitations in this regard. For example, rAAV serotype 8 displays outstanding properties in transducing hepatocytes and thus is commonly selected for liver targeting. Other serotypes (AAV1, 2, 5, and 7) have been shown to be weak performers in the transduction of cells of the CNS when administered systemically. Thus, future studies are warranted to evaluate new serotypes with wide tissue tropism, the capability of crossing the BBB and their therapeutic benefit and safety profile, in addition to approaches to circumvent severe immune responses.

3 Amino Acid Metabolism Disorder

3.1 Phenylketonuria (PKU)

3.1.1 Introduction

Phenylketonuria (PKU) is a rare monogenetic disease that is caused by different mutations in the phenylalanine hydroxylase (PAH) gene [59] and was first described by Følling in 1934 [60]. The prevalence for PKU differs between countries; in the USA about 1 in 15,000 newborn have PKU [61], in Europe 1 in 10,000 [62] and in Turkey, due to the high rate of consanguinity, 1 in 4000 [63].

Clinically, untreated PKU manifests with severe mental retardation, developmental impairment, seizures, eczema, reduced hair, skin and iris pigmentation, and psychosocial problems [64, 65]. However, this is rarely seen in developed countries due to extensive screening programs of the newborn and early nutritional intervention that can reduce or even prevent cognitive impairment [65]. This also means that the patient population has shifted to more adult individuals affected by PKU, who might face other challenges than children in terms of disease management and treatment.

3.1.2 Genetics

Phenylketonuria is an autosomal recessive disorder that correlates in severity with the residual activity of the enzyme PAH. The gene coding for this enzyme is located on chromosome 12, comprises 13 exons and is 79 kilobases (kb) long. PAH is predominantly expressed in liver tissue but has also been found in other organs, e.g., kidney, pancreas, and brain [66]. Three isozymes are known in rat and two in human liver; these are all missing in classical PKU [67]. Until now, over 531 mutations (www.pahdb.mcgill.ca) in the phenylalanine hydroxylase gene have been described. Of particular interest for gene therapy is the 1358 base pair long cDNA that was cloned first from rat liver by Robson et al. and from human by Woo et al. in 1982 and 1983 respectively [59, 68] (*see* Chapter 1).

3.1.3 Pathomechanism

Phenylalanine hydroxylase catabolizes the hydroxylation of phenylalanine to tyrosine [69]. This irreversible reaction depends on molecular oxygen and the cofactor tetrahydrobyopterin, which is abundant in the liver [70]. It is synthesized from guanosie triphosphate (GTP) by the enzymes GTP cyclohydrolase I (GTPCH), 6-pyruvolytetrahydrobiopterin synthase (PTPS), and sepiapterin reductase (SR) [71]. Phenylalanine itself functions as an activator of PAH by inducing conformational changes [70].

Based on the residual activity of PAH, two distinct forms of PKU have been described: Classical PKU and atypical PKU. Whereas classical PKU is characteristic for no residual enzyme function at all, the atypical PKU seems to have about 5 % remaining enzyme function [72, 73]. Although the definite pathomechanism for PKU, especially the severe CNS phenotype has not been conclusively elucidated, several hypotheses have been postulated:

3.1.4 Amino Acids Transport Inhibition Hypothesis

First, in the state of hyperphenylalaninemia, phenylalanine impairs the transport of large neutral amino acids (LNAA, i.e., phenylalanine, tyrosine, tryptophan, leucine, isoleucine, and valine) across the blood–brain barrier (BBB) causing reduced levels of CNS LNAAs. This hypothesis is supported by the finding that PKU animal models with pathologically high concentration of serum phenylalanine have reduced CNS levels of several amino acids, especially LNAAs [74]. Furthermore, a study by Pietz et al. using ^{1}H magnetic resonance spectroscopy and phenylalanine (Phe) challenge test demonstrated that patients receiving Phe only, showed increased brain Phe and disturbed brain activity measured by electroencephalogram (EEG). In contrast, patients that received LNAAs along with Phe showed no increase in brain Phe and normal brain activity [75]. Another group could show that LNAA supplementation in the PKU^{ENU} mouse model reduced brain Phe levels [76]. However, in a prospective, double blind, crossover study in PKU patients, no correlation between Phe levels in plasma and brain could be found. Yet, LNAA supplementation reduced Phe plasma levels [77].

3.1.5 Protein Synthesis Impairment Hypothesis

Second, early experiments by Hughes et al. and Binek et al. in a pharmacologically induced PKU animal model suggested that chronically increased phenylalanine levels impair the protein synthesis of the CNS [74, 78, 79]. These findings could be restored upon administration of LNAA. However, the Phe levels remained unchanged, supporting the rational that reduced LNAA rather than elevated Phe levels contribute to the pathomechanism of PKU [74]. This finding was substantiated by Groot et al. in a positron emission tomography study using ^{11}C-tyrosine in 16 PKU patients that showed inverse association between tyrosine influx and Phe plasma levels. Importantly, ^{11}C-Tyrosine influx was positively associated with ^{11}C-tyrosine protein incorporation.

3.1.6 Neurotransmitter Deficiency Hypothesis

Third, hyperphenylalaninemia causes CNS neurotransmitter deficiency. Several studies reported significant decrease in aminergic neurotransmitter, especially serotonin, dopamine and catecholamine, in the CNS and CSF of PKU patients as well as PKU mouse models [80–85]. Lou et al. reported increased Phe levels in plasma and CSF and decreased plasma levels of tyrosine and tryptophan in patients upon phenylalanine-restricted diet cessation. Interestingly, homovallinic acid (HVA) and 5-hydroxyindoleacetic acid (5-HIAA) metabolites of the dopamine and serotonin synthesis pathway were markedly decreased in three out of four patients examined upon non-restricted diet. The authors concluded that the changes in HVA and 5-HIAA might be due to impaired uptake of tyrosine and tryptophan across the BBB or at the neuronal plasma membrane caused by increased Phe levels in plasma and CSF [86]. Data from a PKU mouse model support these findings that increased Phe levels correlate with decreased levels of dopamine, serotonin and catecholamines [81, 83]. Although there is no doubt about the benefit of Phe-restricted diet in PKU treatment, decreased levels of dopamine and serotonin despite early dietary treatment have also been reported [82].

Additional pathomechanisms have been postulated. For example, oxidative stress has gained popularity in the last decade as contributing factor to many disease phenotypes (*see* Subheading 3.2). Kienzle Hagen et al. could demonstrate that increased levels of Phe are associated with reduced glutathione peroxidase and that Phe inhibits catalase activity [87]. A different group suggested the contribution of NADPH oxidase to the PKU disease phenotype [88]. Despite the progress that has been made to explain different factors leading to the PKU phenotype, further research has to be done.

3.1.7 Animal Models

The importance of adequate animal models for the development of gene therapy cannot be overemphasized since it is one of the three mainstays of gene therapy development.

3.1.8 Transiently Induced Animal Models

The first animal models were developed by feeding PAH inhibitors, e.g., *p*-chlorophenylalanine along with phenylalanine itself [89, 90]. This could mimic the PKU phenotype and appeared crucial to decipher the PKU pathomechanism, but was inappropriate for the development of gene therapy based therapeutics because of the missing genetic defect along with potential side effects of the PAH inhibitor [89, 91]. Even if the PAH inhibition could be maintained, cells could still express the PAH gene and provide the functioning enzyme. Any enzyme expressed from an externally introduced expression cassette would have also been inhibited, therefore making the evaluation of this gene therapeutic treatment ineffective.

3.1.9 The PAH^{enu} Mouse

Later, Bode et al. reported an *N*-ethyl-*N-nitrosourea* (ENU) induced mutant mouse displaying symptoms of hyperphenylalaninemia. However, it was shown later that this was caused by a mutation in the GTP-cyclohydrolase gene [92, 93]. It was not until 1990, when McDonald et al. reported the creation of a mouse model (PAH^{enu1}) carrying a mutation in the PAH gene [94, 95]. However, this mouse model failed to show the severe biological characteristics seen in human PKU patients [94, 96]. Interestingly, it was found later that the PAH$^{hph\text{-}5}$ (PAH^{enu1}) mouse carries a C to

T transition at position 364 in exon 3 [97], which results in a missense mutation with the exchange of a hydrophobic amino acid by another hydrophobic amino acid (valine to alanine), which might explain the mild phenotype.

In a subsequent attempt, Shedlovsky et al. using ENU engineered two mouse strains (PAH^{enu2} and PAH^{enu3}) carrying a PAH mutation on the BTBR (Black and Tan BRachyury) background. These mice demonstrated elevated Phe serum levels along with phenylketonuria. Although the PAH mRNA level as well as protein detection on WB in these two PAH mutant strains differed, they displayed the same clinical phenotype, e.g., slow growing, small head size (microcephaly), hypopigmentation and behavioral abnormalities starting at about week 2. This phenotype even deteriorated under oral Phe stress test [96]. McDonald et al. reported a C to T transition at position 835 in exon 7 of the PAH gene in the PAH^{enu2} mutant [97] that results in a phenylalanine to serine substitution. Interestingly, this mutation reflects the most common missense mutation of the PAH gene found in humans.

3.1.10 Preclinical Gene Therapy Studies

In the last two decades several strategies for PAH transgene delivery have been tested due to the availability of animal models, which are still essential aspects of gene therapy development [98–106]. With the cloning of the phenylalanine hydroxylase (PAH) gene, another mainstay towards gene therapy for PKU was established [59, 68, 107–109].

3.1.11 In Vitro Studies

The first in vitro studies for PKU were conducted in 1985 by Ledley et al., who could demonstrate the functional expression of PAH mRNA and protein after NIH3T3 cell transfection with a PAH cDNA carrying expression vector. This was an important step towards virus based gene therapy [107]. Only 1 year later, in 1986, Ledley et al. successfully transduced NIH 3T3 and hepatoma cell lines using a retrovirus deprived of self-replication. Importantly, the authors demonstrated that the PAH cDNA was expressing the PAH mRNA and functional phenylalanine hydroxylase [110]. Of note, only about 10 % of the PAH activity is sufficient to accomplish therapeutic effects [107]. However, the experiments were, until that point, merely performed in vitro. Furthermore, the retroviral vector used was carrying the bacterial neo gene, which might lead to neomycin resistance in transfected cells. This is of no concern in vitro, but a major hindrance for the application in humans.

3.1.12 Ex Vivo Gene Therapy

In ex vivo gene therapy, patient cells are collected, in vitro transduced with the therapeutic gene and subsequently re-implanted.

Lin et al. transduced T lymphocytes from children with PKU with the PAH cDNA using a retroviral vector system. An important aspect of this study was the demonstration that functional

PAH can be expressed in cells different than liver (heterologous) if sufficient amounts of the cofactor tetrahydrobiopterin are supplied. T lymphocytes are permeable for phenylalanine and contain small quantities of tetrahydrobiopterin that acts as an essential cofactor for the PAH catalyzed reaction. After transfection, T lymphocytes were able to catabolize phenylalanine [106]. Although this study was merely a proof-of-concept and was not further pursued, it demonstrates potential alternatives to in vivo gene therapy.

3.1.13 In Vivo Gene Therapy: Adenovirus Mediated Transgene Delivery

One of the first in vivo studies were conducted by Fang et al. who could normalize hyperphenylalaninemia in mice within 1 week after treatment using a adenoviral vectors delivering the PAH (phenylalanine hydroxylase) transgene. Unfortunately, this effect was not persistent, most likely because of a host immune response. This study also did not provide data about phenotypic changes. However, it determined that only 10–20 % of the normal enzyme activity is sufficient to decrease the phenylalanine level to normal [103].

In 1999, another study reported the normalization of blood phenylalanine (Phe) within 24 hrs using a systemically delivered replication defective adenoviruses carrying the PAH cDNA in the PAH^{enu2} mice [105]. The drawback of this study was that the transgene expression only sufficiently decreased blood phenylalanine for 10 days, due to a strong host immune response against the recombinant virus. Even re-administration of adenovirus-PAH failed to decrease the phenylalanine level again. Interestingly, the simultaneous administration of an immunosuppressant could extend the sufficient phenylalanine decrease in serum to 47 days [105].

Although adenovirus mediated transgene delivery could transfect cells efficiently and decrease blood phenylalanine transiently, the major drawback was a strong host immune response. In contrast, recombinant adeno-associated virus (rAAV) has been considered nonpathogenic with low immunogenicity and high efficiency for gene therapy.

3.1.14 In Vivo Gene Therapy: Recombinant Adeno-associated Virus

The first study using rAAV mediated PAH delivery through portal vein injection, was published in 2004 by Mochizuki et al. This group could demonstrate that liver directed rAAV5mPAH delivery improves the PKU phenotype and decreases serum phenylalanine for up to 40 weeks. These encouraging findings were only weakened by the fact that the response rate of female mice was less effective versus male [102]. These results were confirmed by similar findings in the same year by Oh et al., using rAAV serotype 2 to deliver the PAH transgene [101]. Later, it was suggested that the gender dependency might be due to unknown genetic variability in the PAH^{enu2}animal model [111]. Furthermore, Oh et al. identified 12 genes with altered expression profiles in the untreated PAH^{enu2}

mouse that nearly normalized upon rAAV2PAH treatment and plasma phenylalanine reduction [112]. However, the mechanism behind the gene expression profile alteration remains elusive.

Ding et al. reported the next encouraging step towards gene therapy for PKU in 2006. This group compared portal vein and tail vein injection using a rAAV2-PKU-5/8 vector in the PAH^{enu2} mouse. Although long-term reduction of blood Phe levels were found in both groups, portal vein injected mice performed better than tail vein injected mice (42 and 35 weeks respectively) despite fourfold lower dose in the portal vein group [100]. Of note, this study did not report any gender-dependent response to treatment, which might be due to crossing of the PAH^{enu2} allele in the C57BL/6 background before the start of this study [100]. In addition, several surrogate markers for liver damage (e.g., ALT, AST, γGT), inflammation (interleukin-12 and tumor necrosis factor-α) and humoral immune response against PAH were assed and reported without significant changes [99].

Another study used a pseudotyped rAAV2/8 vector in combination with a liver specific promoter to express PAH in a dose of 5×10^{11} viral particles and achieved significant decrease of blood Phe. However, the blood Phe levels were only monitored for 17 weeks post injection (endpoint of study) and could not provide longer-term expression data. In addition, interpretation of a sex-dependent response to treatment was limited due to low animal number ($n = 5$) and only one treated female mouse [99].

These previous studies focused primarily on transgene expression, biodistribution, safety profile and blood phenylalanine levels. However, it is worth noting that the most severe manifestation of PKU is in the CNS. Although this is most likely secondary to high blood Phe levels, the recovery of the severe CNS phenotype is of greatest importance for patients and families.

3.1.15 rAAV Mediated CNS Phenotype Improvement

In 2007, Embury et al. demonstrated that as early as 4 weeks of age, PAH^{enu2} mice develop neurodegenerative changes in the nigrostriatum with decreased immunoreactivity against tyrosine hydroxylase (TH), cytoplasmic vacuolar degeneration and neuronophagia by macrophage infiltration. This pathology could be reversed upon intraportal rAAV-mPAH-WPRE treatment in 10–14 weeks old mice, supporting the hypothesis that the increased Phe serum level is responsible for the underlying pathomechanism [113].

3.1.16 Muscle as a Biofactory

While common approaches target liver as the primary organ of PAH deficiency, hypotheses supporting the metabolic sink theory suggest that ectopic PAH expression might sufficiently reduce the Phe blood level and thus ameliorating the CNS phenotype by decreasing competitive inhibition of amino acid transport across the BBB (*see* Subheading 3.1.3).

Based on this concept, Ding et al. targeted skeletal muscle of PAH^{enu2} mice expressing phenylalanine hydroxylase to reduce blood Phe. However, the hydroxylation of phenylalanine requires the cofactor tetrahydrobiopterin (BH_4), which has been found in sufficiently high levels in liver but not in skeletal muscle. Consequently, PAH in muscle alone did not show therapeutic effect. Conversely, using rAAV1 expressing PAH, GTPCH (GTP cyclohyrolase I, see above) and PTPS (6-pyruvoyltetrahydrobiopterin synthase, see above) via intramuscular injection of 3.5×10^{12} vector genomes sufficiently decreased Phe for at least 35 weeks [98]. The simultaneous expression of GTPCH and PTPS was necessary to guarantee sufficient levels of BH_4.

Rebuffat et al. pursued the same concept targeting muscle to increase systemic PAH expression. In addition, the authors evaluated rAAV1, 2 and 8 for PAH delivery by intramuscular injection of the hindleg. However, the primary goal was not to decrease Phe via muscle restricted PAH expression rather than using muscle injections to deliver rAAVPAH to the liver. Of note, in most animals the transduction rate was highest in liver and hindleg muscle over other organs. Strikingly, strong gender dependent transduction efficiency was observed for all rAAV serotypes used. Nevertheless, the blood Phe levels could be decreased to therapeutic levels (<360 μmol/l) in males for 40–55 weeks and females for 30–35 weeks dependent on the used rAAV serotype. To extent the therapeutic effect in female mice, previously rAAV1PAH injected females were re-dosed with rAAV8PAH. This second administration of a different serotype could decrease blood Phe for additional 12 weeks [2].

3.1.17 Single-Stranded versus Self-Complementary rAAV

Another group compared intraperitoneal (i.p.) injected single stranded (ss) versus self-complementary (sc) rAAV8 of different doses and reported faster treatment response and improved long-term reduction of blood Phe for the scAAV treated group. Again, independent of the used vector, a strong gender dependent response was observed, with female mice demonstrating a weaker treatment response versus male PAH^{enu2} mice [114].

3.1.18 Progress in Clinical Development of PKU Gene Therapeutics

Despite impressive progress in experimental gene therapy treating PKU, the translation into a clinical setting in humans has not been accomplished. According to "clinicaltrials.gov" no clinical trials for gene therapy of PKU have been conducted.

The closest attempts are a few clinical trials using enzyme replacement therapy (ERT) by injecting a phenylalanine ammonia lyase subcutaneously in order to reduce overall phenylalanine. This enzyme is not expressed in humans. However, it can metabolize phenylalanine to ammonia and cinnamic acid.

3.1.19 Challenges and Future Directions

Over the last decade, different groups could demonstrate promising results in murine PKU animal models using adeno-associated virus [98–102]. Hereby, therapeutic effects were generated by either targeting liver or skeletal muscle, raising the question if ectopic transgene expression is therapeutically sufficient.

3.1.20 Ectopic Transgene Expression

Using PKU as a model to answer the question about ectopic transgene expression, it seems that lowering the phenylalanine burden in general, proves to be therapeutically effective. This cannot account for every disease. However, studies support the hypothesis that a metabolic disorder caused by an enzyme deficiency is treatable by lowering the enzyme specific metabolite. Thus, the missing gene can be expressed ectopically as long as it is accessible by its substrate.

In the case of PKU, where the liver is the primary organ of PAH expression and target for gene therapy, some organ specific problems need to be addressed. First, in order to reduce biodistribution of the rAAV/transgene some groups attempted direct portal vein injection of rAAVPAH. This procedure itself is invasive and carries its own risks. Using tissue specific promoters or rAAVs engineering to increase tissue tropism could reduce this risk. Second, it has to be considered, that organ and tissue specific properties might hinder the transgene delivery or persistence. Under physiologic conditions the liver regenerates with an estimated cell turnover of 200–300 days [115]. rAAV genomes are usually episomal, meaning they do not integrate in the genome of the host cell. Consequently, with the replacement of each hepatocyte, the rAAV genomes vanish, leading to the observation of transgene and transgene product loss with eventual therapeutic inefficacy. Of note, the regeneration rate of the liver increases significantly after liver injury. Whether invasive methods of delivery, e.g., intrahepatic vector administration, trigger this liver-injury related regenerative mechanism needs still to be elucidated. Studies on mice, however, demonstrated rAAV genome loss in liver over time, which might be due to increased hepatocyte turn over. Targeting skeletal muscle, an organ with predominantly none dividing cells, might be a promising alternative to accomplish therapeutic long-term transgene expression.

Mochizuki et al. could demonstrate that rAAV based PAH delivery to the liver reduced the phenylalanine levels to normal for about 40 weeks in male mice; other groups reported up to 25–42 weeks of PKU phenotype correction [100–102]. In contrast, injection of rAAVPAH directly into the muscle accomplished long-term expression of >42 weeks. A caveat, however, is that intramuscular administration of rAAVPAH necessitates either the supplementation of tetrahydrobiopterin (BH4) or the simultaneous expression of the BH_4-biosynthetic genes [2].

3.1.21 Large Animal Model

Another point to consider is that despite remarkable progress in preclinical studies for PKU gene therapy, safety studies in large animal models are still missing. The most recent studies in naïve nonhuman primates (NHP) to evaluate rAAV mediated acidic sphingomyelinase (ASM) delivery for the treatment of Niemann–Pick disease, clearly demonstrated how promising results in mice can fail to be translated to large animal models, commonly due to an immune response. In fact, immunogenicity of vector, transgene, or transgene product has become a serious bottleneck in the translation of preclinical studies to applications in humans and warrants further studies in gene therapy development for PKU.

3.2 Canavan Disease

3.2.1 Introduction

Canavan disease is a recessive inherited genetic disorder, predominantly affecting the central nervous system (CNS). It was discovered by van Bogaert and Bertrand in 1949 [116]. Although the disease phenotype had been described before 1949, no connection to Canavan disease was drawn. This devastating disease is caused by mutations in the aspartoacylase gene (ASPA) that results in the accumulation of its only substrate *N*-acetylaspartate (NAA) in the CNS and urine [117, 118]. NAA is the second most abundant amino acid in the mammalian brain. Its true function is still unknown; although several hypotheses have been postulated over the last decades (see below).

Based on onset of symptoms and the severity of disease progression, three subclasses have been postulated: congenital, infantile and juvenile [119, 120]. The most severe form, congenital, presents early after birth with poor feeding and head control, as well as hypotonia and lethargy. Dramatically, affected individuals usually die within days or weeks after birth [119, 121]. Individuals with the infantile form (most common) become symptomatic months after birth showing signs of hypotonia, arrest in motor function development, macrocephaly, and blindness. Later the hypotonia converses to hypertonia with spasticity. Children with the juvenile form are usually asymptomatic until the age of 5 and later. They still develop spasticity and blindness with optic atrophy [119]. As children with Canavan disease grow up the failure to develop language and motor skills similar to age matched children becomes increasingly apparent [121]. This age and phenotype-based classification is not uniformly accepted and an alternative grouping into typical versus mild form has been suggested [120].

Histopathologic hallmarks of Canavan disease are white matter loss, vacuolation and cellular edema [119]. On the ultrastructural level, swollen astrocytic mitochondria are present as well as disruption of the myelin sheath order [119]. On MRI (magnet resonance imaging) T2 sequences, hyperintensity of mainly the white matter is characteristic, not only for leukodystrophy in general, but for Canavan disease as well [122, 123]. This hyperintensity is due to water accumulation within the subcortical areas.

Unfortunately, symptomatic treatment is the only option that has been available, e.g., antiepileptics. Besides gene therapy, recent attempts focusing on the supplementation with triheptanion in mice but could not demonstrate a break through [124].

3.2.2 Pathomechanism

NAA is predominantly synthesized in neuronal mitochondria from L-aspartate and acetate by the enzyme aspartate-*N*-acetyltransferase (Asp-NAT) [125]. After synthesis, NAA is either further processed in a condensation reaction to *N*-acetylaspartylglutamate (NAAG) or transported to oligodendrocytes. While NAAG acts as a neurotransmitter and is released by certain neurons, NAA is broken down by the enzyme aspartoacylase (ASPA). For a long time, it was believed that ASPA is exclusively expressed in oligodendrocytes, where it performs different functions (see below). However, subsequent studies could demonstrate that ASPA is also expressed in peripheral organs, e.g., kidney or intestines [126–129]. Interestingly, no involvement of peripheral organs has been reported, which raises the question of the role of ASPA and NAA outside the CNS. In the case of Canavan disease, mutations in the Aspartoacylase gene result in NAA accumulation in the CNS and urine. Despite intensive research, the physiologic function of NAA or the molecular pathomechanism for CD is still unclear. Several hypotheses, however, have been postulated, four of them being briefly reviewed here (an excellent review on NAA can be found in [132]).

3.2.3 Osmolyte/ Molecular Water Pump Theory

First, it has been suggested that NAA functions as a protective osmolyte or molecular water pump. Since NAA accumulates in the CNS of CD patients, cells would draw water intracellular due to the shift in oncotic pressure and thus cause cell swelling and disruption of the cellular integrity [130–132]. Taylor et al. could demonstrate that hyposmolar stress causes NAA efflux from neurons, suggestion a protective function against neuronal swelling [133]. However, NAA efflux is substantially lower than the efflux of well-characterized "protective osmolytes" [134]. In addition, Baslow et al. proposed that NAA removes metabolic water and thus prevents potentially harmful intracellular water accumulation [130].

3.2.4 NAA-Acetate-Aspartate Concept

The *second* concept proposes that NAA provides acetate and aspartate, with acetate delivering molecules for the lipid synthesis and energy metabolism in myelin forming cells (oligodendrocytes) [135]. This concept is supported by data showing that aspartoacylase expression increases parallel to increased myelination and that acetate is deceased in the ASPA knockout mouse model [135, 136]. More importantly, labeled acetyl-CoA was found to be incorporated into long fatty acids in the brain, demonstrating its contribution to de novo fatty acid synthesis in the CNS [137].

3.2.5 NAA Neurotoxicity Model

Third, NAA exerts a neurotoxic effect. On the one hand, it was demonstrated that NAA injection into the brain did not cause substantial cell death [138]; on the other hand, Akimitsu et al. and others saw seizures in animals injected with NAA intracerebrally [139, 140]. One caveat is that the injected doses varied between these studies. It has also been shown that NAA and NAAG can act on NMDA receptors. However, NAA or NAAG induced toxicity via NMDA receptor activation in the manner of L-glutamate related excitotoxicity could not be confirmed by in vitro experiments [141].

3.2.6 Oxidative Stress Theory

Finally, aspartoacylase deficiency with NAA accumulation might cause oxidative stress overwhelming the organism in maintaining its homeostatic equilibrium. This hypothesis has been favored in recent years [142, 143]. However, meticulous studies were missing until recently Francis et al. could demonstrate that within the first 2–8 weeks in the Nur7 mouse model, metabolites of the energy metabolism (e.g., ATP, acetyl-CoA, and malonyl-CoA) and endogenous antioxidants are decreased suggesting an increased effort of the organism to fight oxidative stress. This was further supported by elevated marker of lipid peroxidation and cortical cell death [144].

Although it is still not clear what comprises the entire mechanism leading to the Canavan disease phenotype, it seems reasonable that it is of multifactorial etiology. It is possible that the mechanistic understanding of gene therapy might depend on the understanding of the pathomechanism. For example, ASPA in the CNS is primarily expressed in oligodendrocytes. Using rAAVhASPA might not predominantly transduce oligodendrocytes but still rescues the phenotype (see below). One explanation could be that the re-establishment of ASPA expression reduces the dramatically increased NAA and thus ameliorates the shift in oncotic pressure, neurotoxicity and metabolic stress. Thus, AspA expression would function as a "metabolic sink", independent of the transduced cells types. This example illustrates the importance of understanding the pathomechanism to plan and anticipate strategies for the therapeutic expression of a gene of interest (*see* also "cross-correction" in Subheading 2.1).

3.2.7 Animal Models

Canavan disease has become a model disorder for the development of gene therapy for the CNS. Thus, it is of no surprise that several animal models have been created. In this context, it has to be emphasized that the availability of an appropriate animal model is one of the essential determinants for testing gene therapy in preclinical settings.

3.2.8 Tremor Rat and CD Knockout Mouse

The first available animal models were the so-called tremor rat and the CD knockout mouse [139, 145]. The tremor rat shows tremor, seizure-like symptoms and NAA accumulation; Histologically, severe leukodystrophy and vacuole formation are characteristic. Importantly, this animal model has a 200 kilo base pair (kbp) deletion in the tm locus on chromosome 11. Although it was demonstrated that this lesion is composed of three genes with ASPA being one of them, it is unclear if the other two affected genes (olfactory receptor gene and vanilloid receptor subtype I) contribute to the tremor rat phenotype [139].

Around the same time, Matalon et al. developed an ASPA knockout mouse model that presents with early onset of motor dysfunction, accumulation of NAA and early death [145]. On the microscopic level, severe vacuolation and white matter loss develops as early as postnatal day 13 (P13) [146]. In addition, the CD knockout (KO) mouse dies at around 28 days of age. However, the overall severity of the phenotype and length of survival seem to depend on the genetic background of the mouse strain [145]. The advantage of this mouse model for gene therapy is the early onset of symptoms and death, which provides relatively quick feedback about the treatment efficacy. The drawback of this model, however, is intrinsic to the utilized engineering strategy. The ASPA gene was interrupted by the neomycin (neo) gene, which is under the control of the PGK promoter with a strong transcription enhancer. This led to the assumption that the PGK promoter might influence adjacent genes and contributes to the severe phenotype seen in the CD KO mouse.

3.2.9 Nur7 Mouse

Another animal model for CD, designed by Kile et al. and characterized by Traka et al., was designed by injecting *N*-ethyl-*N*-*nitrosourea* (ENU) [147, 148]. It only carries a single point mutation in the ASPA gene and expresses a truncated and nonfunctional aspartoacylase protein. This animal model (Nur7) is genetically closest to the actual disease but resembles symptomatically the less common juvenile or mild from of CD.

3.2.10 LacZ Knock-in CD Mouse

In 2011, a group around Klugmann et al. created the so-called LacZ knock-in CD mouse that resembles the juvenile/mild CD phenotype [128]. In this mouse, the ASPA gene is interrupted by a LacZ gene, which is driven by the endogenous ASPA promoter. Subsequently, LacZ is only expressed in cells that naturally express ASPA. Using this mouse model, Mersmann et al. could demonstrate that AspA is expressed in several other organs as well. However, the LacZ knock-in mouse has not been used for the preclinical testing for gene therapy [128].

3.2.11 Deaf14 Mouse

The most recent mouse model was also created by ENU injection: named deaf14. It was found while screening for deafness causing genes [149]. This mouse carries an ASPA mutation and demonstrates with symptoms of CD. Interestingly, it also develops deafness, which has not been reported in Canavan disease before.

Again, the availability of different animal models for the same disease is an important step towards the understanding of pathomechanism and the development of therapeutics. While some animals might be more suitable for answering questions about gene therapy effectiveness or the pathomechanism, e.g., LacZ knock-in mouse, others are in particular useful for the gene therapy development itself, e.g., CD KO or Nur7 mouse. For example, the CD KO mouse with its severe phenotype provides feedback about treatment efficacy within days to weeks. In contrast, The Nur7 mouse shows symptoms much later and demands more extensive study duration. However, the late onset of symptoms in the Nur7 mouse might help to evaluate questions about the "therapeutic window" or the permeability of the blood–brain barrier (BBB), since mice can be treated after they have reached adulthood. In any case, choosing the appropriate animal model is crucial for the study success.

3.2.12 Preclinical Gene Therapy Studies

Over the last 15 years, several groups demonstrated promising results in gene therapy development for Canavan disease. In 1993, a group around Matalon et al. cloned the AspA cDNA [150]. This achievement occurred to be the start of over 20 years of gene therapy development for the CNS.

3.2.13 Intracranial rAAV Administration

Leone et al. was one of the pioneers for gene therapy for Canavan disease using a lipid-entrapped, polycation-condensed delivery system (LPD) in combination with adeno-associated virus (AAV)-based plasmid containing the recombinant aspartoacylase cDNA [151, 152]. This approach was evaluated in healthy rats and primates due to the lack of a CD knockout model at that time. Although this study could not demonstrate disease phenotype rescue, it provided valuable data on transgene expression and safety (no immune response against transgene or LPD was detected) [152]. In addition, two routes of administration, intracerebral and intraventricular, were evaluated to be effective for CNS restricted long-term expression of aspartoacylase. Finally, Leone et al. moved to a proof-of-concept study in two human individuals, that demonstrated biochemical, radiological and clinical changes but no clinically relevant disease rescue [152]. Although this study demonstrated promising local gene expression with decreased NAA levels and improved radiological signs of CD, widespread transgene delivery of the CNS could again not be achieved with this LPD based approach [152]. However, this study by Leone et al. was the first attempt to transfer in vivo results from rodents and primates to two human patients for the treatment of Canavan disease. In a later phase I clinical trial with 14 individuals enrolled, an improved LPD delivery system carrying the hASPA expression cassette was used, which demonstrated with improvements similar to the first two treated human CD patients [152, 153].

After this pioneering work, several studies using first generation rAAVs in animals could demonstrate biochemical, histopathological as well as clinical improvement in the CD knockout mouse.

3.2.14 Preclinical Studies in the CD Knockout Mouse: Intracranial rAAV Administration

Matalon et al. injected a rAAV-ASPA-GFP construct into thalamus and striatum to demonstrate sufficient transgene expression in the brain of the first AspA knockout mouse (CD mouse). rAAV efficiently transduced CNS cells in the CD mouse and provided functional hAspA expression that not only reduced NAA levels but also improved neuroradiological signs of edema as well as pathological hallmarks of leukodystrophy, e.g., decreased vacuolation. Widespread CNS transduction, however, could not be achieved [154].

Following the experiments in mice, McPhee et al. used rAAV2 in the so-called tremor rat model to deliver the human ASPA (hASPA) gene via intracranial injections into the caudate nucleus and thalamus. Again, local ASPA expression could be restored leading to decreased NAA. Treated animals also showed improved motor function [155].

3.2.15 Systemic rAAV Administration

With the discovery that rAAV9 (second generation rAAV) is capable of crossing the blood–brain barrier (BBB), systemic rAAV9hASPA treatment occurred to be a new and promising approach for Canavan Disease gene therapy [156].

In 2013, Ahmed et al. could demonstrate almost complete phenotype rescue in the CD knockout mouse by global CNS transduction using systemic delivered rAAVhAspA. Furthermore, this study reported improvement of several histological and biochemical parameters upon treatment. In detail, Ahmed et al. could demonstrate profound improvement of the CD mouse phenotype by using rAAV9, rh.8 or rh.10 vectors for the hASPA transgene delivery by a single intravenous injection, which showed significantly decreased levels of NAA, recovered CNS pathology on radiological and histological imaging, restored motor function and overall survival comparable to wild-type mice [146]. In addition, successful treatment was age-dependent in the CD mouse model. Although survival and phenotype could be restored in animals treated until 1 week before the expected point of death, CD mice injected shortly after birth performed better on motor function tests and long term survival in comparison to groups treated later in the disease progression. Finally, to reduce untoward peripheral transgene toxicity in intravenously injected mice, miRNA mediated transgene detargeting was successfully applied and showed similar performance to the rAAVhAspA treated control group [146].

3.2.16 Clinical Gene Therapy Trials

Following the first clinical phase I trial for Canavan disease using liposome based intracranial transgene delivery published in 2000 by Leone et al. (*see* also Subheading 3.2.13), the same group initiated a rAAV2 based phase I clinical trial [153]. This trial was the first of its kind using rAAV in the treatment of a CNS disorder. The 10-year follow-up results were published in 2012 [157].

In addition, McPhee and Leone et al. conducted a phase I study to determine safety and immune response in ten human individuals with CD. Using intracranial delivery of rAAV2hASPA, this study was the first attempt to ameliorate a neurodegenerative disorder using recombinant adeno-associated virus (rAAV) [158]. Strikingly, no neutralizing antibodies against rAAV2 in cerebrospinal fluid (CSF) and minimal to mild systemic immune response were detected, which emphasizes the relative safety of the administered dose of 1×10^{12} rAAV2hASPA genomes.

Based on the preceding work, Leone et al. moved to a phase I/II clinical trial with 28 enrolled patients with CD and reported long-term safety of rAAV2hASPA treatment in humans [157]. Furthermore, decreased NAA levels in the CNS, slowed progression of brain atrophy, fewer seizures and overall stabilization of the disease phenotype was seen. Interestingly, rAAV2hASPA treatment seemed to be more efficient in younger patients, which recently has also been shown in the CD knock out mouse [146]. The early clinical trials for Canavan disease demonstrated relative safety using a first generation rAAV vector. However, it will be of great importance for future studies to determine if newer rAAV serotypes as well as optimization of transgene expression can improve treatment efficacy and safety profile [157].

3.2.17 Challenges and Future Directions

Encouraging progress has been made in the field of gene therapy for Canavan disease. However, several questions remain to be answered before new studies in patients with Canavan disease are warranted and ethically justified.

3.2.18 Therapeutic Window

One aspect that necessitates further investigation is the potential age dependent treatment efficacy (therapeutic window). As with many therapeutics, successful treatment depends on the time point of treatment. Ahmed et al. could demonstrate that as late as 20 days postnatal, a single intravenously injected dose of rAAVhASPA can still rescue the CD phenotype in a mouse model that dies untreated after 28 days of age. However, mice receiving early treatment performed superiorly to mice treated at late time points. One reason could be the mature blood–brain barrier (BBB) limiting the number of viral particles entering the CNS in older mice versus neonates, which would have implications for the curability of Canavan disease. Bypassing the BBB by intracranial injections could still rescue the disease phenotype at late time-points but would fail if myelination, oligodendrocyte maturation or other factors were age-dependent. However, it remains to be elucidated how late is too late for systemically delivered rAAVhASPA and if there is a point of no return, i.e., when treatment at late time points fails to be therapeutic.

3.2.19 Dose Scale

Furthermore, early studies could demonstrate safety of intracranial injected rAAV2hASPA but were unable to demonstrate widespread CNS transduction, potentially due to the route of administration or the injected viral particle number, raising the question about the most effective dose with the least side effects for both routes of administration ("dose scale"). Intravascular delivered rAAV generally necessitates higher doses than intracranial delivery, emphasizing safety concerns for gene therapy for CD and thus necessitates studies to address dose response and safety.

3.2.20 Large Animal Model

There is currently no large animal model available for Canavan Disease. However, it has been shown in other disease models that results from mouse or rat studies could fail to be translated to larger animal models due to immune response or transgene toxicity. Thus, a large animal model might be a key element to study and improve safety as well as efficacy for gene therapy for Canavan disease.

4 Obstacles and Prospective

The three metabolic disorders of the CNS outlined above demonstrate the progress that has been made to investigate the influence of animal models, vector selection, route of administration, targeted tissue, and immune response on the gene therapeutic success in animal models. Canavan disease remains the exception among this triad where early gene therapy experiments were moved to human applications. Otherwise, all three diseases face similar obstacles for human applications and the solution in one disease might also benefit the progress in other disorders. While this chapter displays merely a small part of gene therapy for CNS disorders and several basic aspects of immune response, route of administration, vector serotype, organ and cell type targeting and transgene expression control, it is important to highlight some of these aspects in the context of gene therapy for metabolic disorders of the CNS. Regardless of the achievements to improve the specificity of transgene delivery in animal models, predicting the translation of these strategies to human applications is still challenging. Furthermore, recombinant adeno-associated virus (rAAV) clearly has become the vector of choice for several therapeutic applications due to its serotype dependent tropism, non-pathogenicity and low immunogenicity. Nevertheless, tailoring tissue restricted transgene expression to mimic the physiological condition seems to be crucial to minimize undesired side effects, e.g., transgene toxicity and immune response.

4.1 Immune Response

Over the last years, the paradigm shift in neuroimmunology reevaluated the concept of the CNS being an immune privileged organ [159, 160]. This reflects the challenges to circumvent vector, transgene, or transgene product directed immune response, particularly in metabolic disorders of the CNS, where isolated treatment of the CNS is only applicable in a very limited number of diseases and the simultaneous targeting of peripheral organs exposes vector capsid, vector genome, and the transgene product to at least two immune systems. In addition, substantial vector leakage has been reported in intracranial delivered rAAV, which might challenge the reduction of an immune response by CNS restricted vector delivery.

4.2 Humeral Immune Response

The first line defense in systemically administered rAAV are often pre-existing neutralizing antibodies (NAb) of the adaptive immune system in blood and interstitium due to a high percentage of people being immunized against certain AAV serotypes [161–163]. This occurs to be an increasing obstacle for the systemic administration of rAAV to target the CNS and peripheral organs as well [164, 165]. For primary metabolic disorders of the CNS it might be of advantage to inject rAAV directly in the CNS parenchyma or the cerebrospinal fluid (CSF) to evade pre-existing antibodies in serum. In fact, it has been demonstrated that intracranial injection of rAAV in the presence of pre-existing NAbs in serum limit the intra-parenchymal distribution of rAAV delivered transgene but does not prevent cerebral cell transduction of intracranial transgene delivery in general [166]. Furthermore, in a clinical phase I trial for rAAV2 mediated gene therapy in patients with Canavan disease, no antibodies against rAAV were detected in cerebrospinal fluid after intra-parenchymal rAAV2 injection [158]. However, intracranial injection might not be an option for secondary metabolic disorders of the CNS, e.g., PKU or NPD where additional gene transfer to peripheral tissues is required for successful therapy. Thus, the humeral immune response still remains a serious hindrance.

The adaptive humeral immune response can also be directed against the transgene product via the generation of anti-transgene-product antibodies, as part of the second line defense. This has been a serious challenge in ex vivo gene therapy for NPD where anti-ASM antibodies reduced the therapeutic efficacy at around 24 weeks post-treatment. Similarly, intravenous rAAV1hASM (CMV promoter) treated ASMKO mice, resulting in ubiquitous ASM expression, carried an up to 100-fold increased serum anti-ASM titer as early as 4 weeks after treatment versus the control group with liver restricted ASM expression, underlining the hindrance of treatment success by the humeral immune response and the potential of tissue restricted transgene expression to escape an eventual immune response.

4.3 Cell Mediated Immune Response

Another obstacle towards persistent gene therapeutic success is cell mediated immune response of either the innate or the adaptive immune system towards AAV-based gene therapeutics: After the vector had evaded the first line of the innate immune system and eventual pre-existing NAb and finally transduced cells, the branch of the immune system that is specialized in detecting and eliminating infected cells, comes into play. Several immune cells are activated by the viral capsid or the transgene itself: Natural killer cells (innate response), phagocytizing cells (innate and adaptive response), and cytotoxic T cells (adaptive response). For example, it could be demonstrated that Toll-like receptor (TLR) 9 carrying cells (e.g., monocytes) can be activated by the vector genome of rAAV, especially the self-complementary genome [167]. In contrast, TLR2 recognizes the rAAV capsid [168]. Although this has been shown in systemically delivered rAAV targeting liver, it might also account for CNS directed gene therapy, especially since TLR is expressed in the CNS as well [169, 170].

Furthermore, CD8+ cytotoxic T cells of the adaptive immune response have the capability to detect MHCI presented rAAV capsid and thus infected cells [171, 172]. This can become an important factor for sustained transgene expression, especially if asymptomatic patients carry CD8+ cytotoxic memory T cells, even before the actual treatment [171]. Predicting the consequences of this finding seems to be complicated by results reported in mice, where CD8+ cytotoxic cells were unable to significantly destruct rAAV transduced hepatocytes [173]. Of note, a similar immune response in the CNS might be dependent on whether antigen-presenting cells (APC) have been transduced and the nature of transgenes (i.e., foreign or self-antigen) [174]. Consequently, the avoidance of APC transduction might be an option to limit an eventual CD8+ cell mediated immune response. Although preventive strategies have been described to overcome the first line of capsid specific antibody defense, e.g., use of alternative AAV serotypes, plasmaphoresis, or the use of capsid decoy [164], further studies are warranted for the mechanistic understanding and the development of more effective concepts to circumvent immune responses of both humoral and cellular natures toward vector capsids, transgene, and transgene products in rAAV mediated gene therapy and how simultaneous targeting of CNS and peripheral organs will be impacted by the immune system of CNS and periphery.

4.4 Route of Administration and Serotype Selection

It is crucial for successful preclinical gene therapy evaluation to opt for the rAAV serotype that targets therapeutically relevant cells and organs in combination with the route of administration to avoid eventual transgene toxicity or immune response. The most common rAAV serotypes used for NPD, PKU, and CD were rAAV2 and 8 followed by 1, 5, 7, rh.8, 9 and rh.10. While rAAV2 has been extensively evaluated for intracranial injection in preclinical and clinical trials for Canavan disease, it does not cross the BBB sufficiently, thus making rAAV2 inappropriate for systemic administration in CNS directed gene therapy [153–155, 175, 176]. Intravenous rAAV9, rh.8 and rh.10 administration, however, demonstrated significant phenotype rescue in Canavan disease [146]. Although it is difficult to compare studies using different rAAV serotypes along with different routes of administration, the selection of an appropriate rAAV serotype might enable new treatment strategies and improved therapeutic success. Similarly, rAAV8 is considered to be the most selective and efficient hepatocyte transducer, making it the first choice for intravenous liver directed transgene delivery in PKU. Alternatively, it has been shown that intramuscular rAAV1 delivery or skeletal muscle restricted PAH expressing is sufficient to treat PKU in mice as well [2, 98]. Selecting the best vector targeting muscle also has to take into account the route of administration. An excellent review on rAAV mediated muscle targeting can be found by Wang et al. [1].

4.5 Capsid Engineering

Another strategy to improve organ or even cell restricted rAAV-transgene delivery and potentially escape a rAAV capsid directed immune response is changing the serotype dependent capsid structure by capsid engineering. The newly generated rAAV capsids are then selected for cell type, organ or pathology and reduced untoward cell transduction, thus improving targeted cell and organ specificity. This strategy has generated several rAAV mutants with altered tropism and the potential to evade pre-existing anti-capsid antibodies. On the downside, rAAV mutants generated in animals might not display the same tropism in humans impeding their translation to clinical applications [177].

4.6 Transcriptional and Post-transcriptional Control

In addition to transgene delivery restriction by rAAV serotype selection, tissue specific promoter usage controls translation of the transgene in cells and a time dependent manner. However, promoter availability, promoter size and the limited expression cassette capacity of rAAV could become the bottleneck for this strategy. Although the use of a liver specific promoter was successfully tested for PKU in mice, it remains an obstacle for Canavan disease because of limited availability of oligodendrocyte specific promoters that match all requirements, i.e., size, specificity, development-dependent expression [99].

Even further transgene restriction could be achieved by miRNA-mediated detargeting that provides transgene expression control on the mRNA level. This concept exploits tissue and cell type specific miRNA expression profiles that are largely conserved among different species. For example, mir-1 and mir-122 binding sites (BS) were successfully used in a preclinical study for Canavan disease to limit hASPA expression in heart, skeletal muscle, and liver [146].

It is noteworthy that the discrepancy between small and large animal models resulted in unexpected results in preclinical studies of gene therapy for metabolic disorders of the CNS, due to the immune response. As in NPD where nonhuman primates had to be euthanized or in clinical trials for hemophilia in human, the immune response resulted in failure of the treatment, which was not seen in mice for the same treatment [58, 171]. Consequently, the translation from mouse (or rat) to large animal models in preclinical evaluations appears to be crucial for safe translation of gene therapy for metabolic disorders of the CNS to humans in the future.

References

1. Wang D, Zhong L, Nahid MA, Gao G (2014) The potential of adeno-associated viral vectors for gene delivery to muscle tissue. Expert Opin Drug Deliv 11:345–364
2. Rebuffat A, Harding CO, Ding Z, Thony B (2010) Comparison of adeno-associated virus pseudotype 1, 2, and 8 vectors administered by intramuscular injection in the treatment of murine phenylketonuria. Hum Gene Ther 21:463–477
3. Mueller C et al (2012) Sustained miRNA-mediated knockdown of mutant AAT with simultaneous augmentation of wild-type AAT has minimal effect on global liver miRNA profiles. Mol Ther 20:590–600
4. McGovern MM et al (2008) A prospective, cross-sectional survey study of the natural history of Niemann-Pick disease type B. Pediatrics 122:e341–e349
5. Vanier MT (2010) Niemann-Pick disease type C. Orphanet J Rare Dis 5:16
6. Brady RO, Kanfer JN, Mock MB, Fredrickson DS (1966) The metabolism of sphingomyelin. II. Evidence of an enzymatic deficiency in Niemann-Pick disease. Proc Natl Acad Sci U S A 55:366–369
7. Kolodny EH (2000) Niemann-Pick disease. Curr Opin Hematol 7:48–52
8. McGovern MM, Aron A, Brodie SE, Desnick RJ, Wasserstein MP (2006) Natural history of Type A Niemann-Pick disease: possible endpoints for therapeutic trials. Neurology 66:228–232
9. Walterfang M et al (2012) Dysphagia as a risk factor for mortality in Niemann-Pick disease type C: systematic literature review and evidence from studies with miglustat. Orphanet J Rare Dis 7:76
10. NP-C Guidelines Working Group et al (2009) Recommendations on the diagnosis and management of Niemann-Pick disease type C. Mol Genet Metab 98:152–165
11. Schuchman EH (2007) The pathogenesis and treatment of acid sphingomyelinase-deficient Niemann-Pick disease. J Inherit Metab Dis 30:654–663
12. Meikle PJ, Hopwood JJ, Clague AE, Carey WF (1999) Prevalence of lysosomal storage disorders. JAMA 281:249–254
13. Landrieu P, Said G (1984) Peripheral neuropathy in type A Niemann-Pick disease. A morphological study. Acta Neuropathol 63:66–71
14. McGovern MM et al (2004) Lipid abnormalities in children with types A and B Niemann Pick disease. J Pediatr 145:77–81
15. Wasserstein MP et al (2003) Growth restriction in children with type B Niemann-Pick disease. J Pediatr 142:424–428
16. McGovern MM et al (2004) Ocular manifestations of Niemann-Pick disease type B. Ophthalmology 111:1424–1427
17. Spiegel R et al (2009) The clinical spectrum of fetal Niemann-Pick type C. Am J Med Genet A 149A:446–450
18. Fink JK et al (1989) Clinical spectrum of Niemann-Pick disease type C. Neurology 39:1040–1049
19. Graber D, Salvayre R, Levade T (1994) Accurate differentiation of neuronopathic and nonneuronopathic forms of Niemann-Pick disease by evaluation of the effective residual lysosomal sphingomyelinase activity in intact cells. J Neurochem 63:1060–1068
20. Vanier MT et al (1985) Biochemical studies in Niemann-Pick disease. III. In vitro and in vivo assays of sphingomyelin degradation in cultured skin fibroblasts and amniotic fluid cells for the diagnosis of the various forms of the disease. Clin Genet 27:20–32
21. Carstea ED et al (1997) Niemann-Pick C1 disease gene: homology to mediators of cholesterol homeostasis. Science 277:228–231
22. Vanier MT, Millat G (2003) Niemann-Pick disease type C. Clin Genet 64:269–281
23. Dhami R, Schuchman EH (2004) Mannose 6-phosphate receptor-mediated uptake is defective in acid sphingomyelinase-deficient macrophages: implications for Niemann-Pick disease enzyme replacement therapy. J Biol Chem 279:1526–1532

24. Kornfeld S (1987) Trafficking of lysosomal enzymes. FASEB J 1:462–468
25. Sands MS, Davidson BL (2006) Gene therapy for lysosomal storage diseases. Mol Ther 13:839–849
26. Mencarelli C, Martinez-Martinez P (2013) Ceramide function in the brain: when a slight tilt is enough. Cell Mol Life Sci 70:181–203
27. Jiang W, Ogretmen B (2014) Autophagy paradox and ceramide. Biochim Biophys Acta 1841(5):783–792
28. Rego A et al (2012) Modulation of mitochondrial outer membrane permeabilization and apoptosis by ceramide metabolism. PLoS One 7:e48571
29. Schuchman EH (2010) Acid sphingomyelinase, cell membranes and human disease: lessons from Niemann-Pick disease. FEBS Lett 584:1895–1900
30. Smith EL, Schuchman EH (2008) The unexpected role of acid sphingomyelinase in cell death and the pathophysiology of common diseases. FASEB J 22:3419–3431
31. Miyawaki S, Mitsuoka S, Sakiyama T, Kitagawa T (1982) Sphingomyelinosis, a new mutation in the mouse: a model of Niemann-Pick disease in humans. J Hered 73:257–263
32. Pentchev PG et al (1980) A lysosomal storage disorder in mice characterized by a dual deficiency of sphingomyelinase and glucocerebrosidase. Biochim Biophys Acta 619:669–679
33. Pentchev PG et al (1984) A genetic storage disorder in BALB/C mice with a metabolic block in esterification of exogenous cholesterol. J Biol Chem 259:5784–5791
34. Nakashima S et al (1984) A mouse model for Niemann-Pick disease: phospholipid class and fatty acid composition of various tissues. J Lipid Res 25:219–227
35. Horinouchi K, Sakiyama T, Pereira L, Lalley PA, Schuchman EH (1993) Mouse models of Niemann-Pick disease: mutation analysis and chromosomal mapping rule out the type A and B forms. Genomics 18:450–451
36. Horinouchi K et al (1995) Acid sphingomyelinase deficient mice: a model of types A and B Niemann-Pick disease. Nat Genet 10:288–293
37. Otterbach B, Stoffel W (1995) Acid sphingomyelinase-deficient mice mimic the neurovisceral form of human lysosomal storage disease (Niemann-Pick disease). Cell 81:1053–1061
38. Marathe S et al (2000) Creation of a mouse model for non-neurological (type B) Niemann-Pick disease by stable, low level expression of lysosomal sphingomyelinase in the absence of secretory sphingomyelinase: relationship between brain intra-lysosomal enzyme activity and central nervous system function. Hum Mol Genet 9:1967–1976
39. Maue RA et al (2012) A novel mouse model of Niemann-Pick type C disease carrying a D1005G-Npc1 mutation comparable to commonly observed human mutations. Hum Mol Genet 21:730–750
40. Loftus SK et al (1997) Murine model of Niemann-Pick C disease: mutation in a cholesterol homeostasis gene. Science 277:232–235
41. Gartner JC Jr et al (1986) Progression of neurovisceral storage disease with supranuclear ophthalmoplegia following orthotopic liver transplantation. Pediatrics 77:104–106
42. Daloze P et al (1977) Replacement therapy for inherited enzyme deficiency: liver orthotopic transplantation in Niemann-Pick disease type A. Am J Med Genet 1:229–239
43. Scaggiante B et al (1987) Successful therapy of Niemann-Pick disease by implantation of human amniotic membrane. Transplantation 44:59–61
44. Victor S et al (2003) Niemann-Pick disease: sixteen-year follow-up of allogeneic bone marrow transplantation in a type B variant. J Inherit Metab Dis 26:775–785
45. Bayever E et al (1992) Bone marrow transplantation for Niemann-Pick type IA disease. J Inherit Metab Dis 15:919–928
46. Vellodi A, Hobbs JR, O'Donnell NM, Coulter BS, Hugh-Jones K (1987) Treatment of Niemann-Pick disease type B by allogeneic bone marrow transplantation. Br Med J (Clin Res Ed) 295:1375–1376
47. Miranda SR et al (2000) Infusion of recombinant human acid sphingomyelinase into Niemann-Pick disease mice leads to visceral, but not neurological, correction of the pathophysiology. FASEB J 14:1988–1995
48. Suchi M et al (1992) Retroviral-mediated transfer of the human acid sphingomyelinase cDNA: correction of the metabolic defect in cultured Niemann-Pick disease cells. Proc Natl Acad Sci U S A 89:3227–3231
49. Miranda SR, Erlich S, Friedrich VL Jr, Gatt S, Schuchman EH (2000) Hematopoietic stem cell gene therapy leads to marked visceral organ improvements and a delayed onset of neurological abnormalities in the acid sphingomyelinase deficient mouse model of Niemann-Pick disease. Gene Ther 7:1768–1776
50. Jin HK, Carter JE, Huntley GW, Schuchman EH (2002) Intracerebral transplantation of mesenchymal stem cells into acid sphingomyelinase-deficient mice delays the onset of neurological abnormalities and

extends their life span. J Clin Invest 109:1183–1191
51. Jin HK, Schuchman EH (2003) Ex vivo gene therapy using bone marrow-derived cells: combined effects of intracerebral and intravenous transplantation in a mouse model of Niemann-Pick disease. Mol Ther 8:876–885
52. Barbon CM et al (2005) AAV8-mediated hepatic expression of acid sphingomyelinase corrects the metabolic defect in the visceral organs of a mouse model of Niemann-Pick disease. Mol Ther 12:431–440
53. Passini MA et al (2005) AAV vector-mediated correction of brain pathology in a mouse model of Niemann-Pick A disease. Mol Ther 11:754–762
54. Dodge JC et al (2005) Gene transfer of human acid sphingomyelinase corrects neuropathology and motor deficits in a mouse model of Niemann-Pick type A disease. Proc Natl Acad Sci U S A 102:17822–17827
55. Passini MA et al (2007) Combination brain and systemic injections of AAV provide maximal functional and survival benefits in the Niemann-Pick mouse. Proc Natl Acad Sci U S A 104:9505–9510
56. Salegio EA et al (2010) Magnetic resonance imaging-guided delivery of adeno-associated virus type 2 to the primate brain for the treatment of lysosomal storage disorders. Hum Gene Ther 21:1093–1103
57. Bu J et al (2012) Merits of combination cortical, subcortical, and cerebellar injections for the treatment of Niemann-Pick disease type A. Mol Ther 20:1893–1901
58. Salegio EA et al (2012) Safety study of adeno-associated virus serotype 2-mediated human acid sphingomyelinase expression in the nonhuman primate brain. Hum Gene Ther 23:891–902
59. Woo SL, Lidsky AS, Guttler F, Chandra T, Robson KJ (1983) Cloned human phenylalanine hydroxylase gene allows prenatal diagnosis and carrier detection of classical phenylketonuria. Nature 306:151–155
60. Folling I (1994) The discovery of phenylketonuria. Acta Paediatr 407:4–10
61. National Institutes of Health Consensus Development Panel (2001) National Institutes of Health Consensus Development Conference Statement: phenylketonuria: screening and management, October 16–18, 2000. Pediatrics 108:972–982
62. Loeber JG (2007) Neonatal screening in Europe; the situation in 2004. J Inherit Metab Dis 30:430–438
63. Ozalp I et al (2001) Newborn PKU screening in Turkey: at present and organization for future. Turk J Pediatr 43:97–101
64. White DA, Waisbren S, van Spronsen FJ (2010) The psychology and neuropathology of phenylketonuria. Mol Genet Metab 99(Suppl 1):S1–S2
65. Pietz J, Benninger C, Schmidt H, Scheffner D, Bickel H (1988) Long-term development of intelligence (IQ) and EEG in 34 children with phenylketonuria treated early. Eur J Pediatr 147:361–367
66. Lichter-Konecki U, Hipke CM, Konecki DS (1999) Human phenylalanine hydroxylase gene expression in kidney and other nonhepatic tissues. Mol Genet Metab 67:308–316
67. Barranger JA, Geiger PJ, Huzino A, Bessman SP (1972) Isozymes of phenylalanine hydroxylase. Science 175:903–905
68. Robson KJ, Chandra T, MacGillivray RT, Woo SL (1982) Polysome immunoprecipitation of phenylalanine hydroxylase mRNA from rat liver and cloning of its cDNA. Proc Natl Acad Sci U S A 79:4701–4705
69. Udenfriend S, Cooper JR (1952) The enzymatic conversion of phenylalanine to tyrosine. J Biol Chem 194:503–511
70. Li J, Dangott LJ, Fitzpatrick PF (2010) Regulation of phenylalanine hydroxylase: conformational changes upon phenylalanine binding detected by hydrogen/deuterium exchange and mass spectrometry. Biochemistry 49:3327–3335
71. Surtees R, Blau N (2000) The neurochemistry of phenylketonuria. Eur J Pediatr 159(Suppl 2):S109–S113
72. Friedman PA, Kaufman S, Kang ES (1972) Nature of the molecular defect in phenylketonuria and hyperphenylalaninaemia. Nature 240:157–159
73. Justice P, O'Flynn ME, Hsia DY (1967) Phenylalanine-hydroxylase activity in hyperphenylalaninaemia. Lancet 1:928–929
74. Binek-Singer P, Johnson TC (1982) The effects of chronic hyperphenylalaninaemia on mouse brain protein synthesis can be prevented by other amino acids. Biochem J 206:407–414
75. Pietz J et al (1999) Large neutral amino acids block phenylalanine transport into brain tissue in patients with phenylketonuria. J Clin Invest 103:1169–1178
76. Matalon R et al (2003) Future role of large neutral amino acids in transport of phenylalanine into the brain. Pediatrics 112:1570–1574
77. Schindeler S et al (2007) The effects of large neutral amino acid supplements in PKU: an MRS and neuropsychological study. Mol Genet Metab 91:48–54
78. Hughes JV, Johnson TC (1978) Experimentally induced and natural recovery from the effects of phenylalanine on brain protein synthesis. Biochim Biophys Acta 517:473–485

79. Binek PA, Johnson TC, Kelly CJ (1981) Effect of alpha-methylphenylalanine and phenylalanine on brain polyribosomes and protein synthesis. J Neurochem 36:1476–1484
80. Pascucci T et al (2009) 5-Hydroxytryptophan rescues serotonin response to stress in prefrontal cortex of hyperphenylalaninaemic mice. Int J Neuropsychopharmacol 12:1067–1079
81. Puglisi-Allegra S et al (2000) Dramatic brain aminergic deficit in a genetic mouse model of phenylketonuria. Neuroreport 11:1361–1364
82. Burlina AB et al (2000) Measurement of neurotransmitter metabolites in the cerebrospinal fluid of phenylketonuric patients under dietary treatment. J Inherit Metab Dis 23:313–316
83. Pascucci T, Ventura R, Puglisi-Allegra S, Cabib S (2002) Deficits in brain serotonin synthesis in a genetic mouse model of phenylketonuria. Neuroreport 13:2561–2564
84. McKean CM (1972) The effects of high phenylalanine concentrations on serotonin and catecholamine metabolism in the human brain. Brain Res 47:469–476
85. Curtius HC et al (1981) Serotonin and dopamine synthesis in phenylketonuria. Adv Exp Med Biol 133:277–291
86. Lou HC, Guttler F, Lykkelund C, Bruhn P, Niederwieser A (1985) Decreased vigilance and neurotransmitter synthesis after discontinuation of dietary treatment for phenylketonuria in adolescents. Eur J Pediatr 144:17–20
87. Kienzle Hagen ME et al (2002) Experimental hyperphenylalaninemia provokes oxidative stress in rat brain. Biochim Biophys Acta 1586:344–352
88. Lu L et al (2011) Mechanisms regulating superoxide generation in experimental models of phenylketonuria: an essential role of NADPH oxidase. Mol Genet Metab 104:241–248
89. Poncet IB, Berry HK, Butcher RE, Kazmaier KJ (1975) Biochemical effects of induced phenylketonuria in rats. Biol Neonate 26:88–101
90. Dhondt JL, Dautrevaux M, Biserte G, Farriaux JP (1977) A new experimental model of hyperphenylalaninemia in rat. Effect of p-chlorophenylalanine and cotrimoxazole. Biochimie 59:713–717
91. Schalock RL, Brown WJ, Copenhaver JH, Gunter R (1975) Model phenylketonuria (PKU) in the albino rat: behaviroal, biochemical, and neuroanatomical effects. J Comp Physiol Psychol 89:655–666
92. McDonald JD et al (1988) Biochemical defect of the hph-1 mouse mutant is a deficiency in GTP-cyclohydrolase activity. J Neurochem 50:655–657
93. Bode VC, McDonald JD, Guenet JL, Simon D (1988) hph-1: a mouse mutant with hereditary hyperphenylalaninemia induced by ethylnitrosourea mutagenesis. Genetics 118:299–305
94. McDonald JD, Bode VC, Dove WF, Shedlovsky A (1990) The use of N-ethyl-N-nitrosourea to produce mouse models for human phenylketonuria and hyperphenylalaninemia. Prog Clin Biol Res 340C:407–413
95. McDonald JD, Bode VC, Dove WF, Shedlovsky A (1990) Pahhph-5: a mouse mutant deficient in phenylalanine hydroxylase. Proc Natl Acad Sci U S A 87:1965–1967
96. Shedlovsky A, McDonald JD, Symula D, Dove WF (1993) Mouse models of human phenylketonuria. Genetics 134:1205–1210
97. McDonald JD, Charlton CK (1997) Characterization of mutations at the mouse phenylalanine hydroxylase locus. Genomics 39:402–405
98. Ding Z et al (2008) Correction of murine PKU following AAV-mediated intramuscular expression of a complete phenylalanine hydroxylating system. Mol Ther 16:673–681
99. Harding CO et al (2006) Complete correction of hyperphenylalaninemia following liver-directed, recombinant AAV2/8 vector-mediated gene therapy in murine phenylketonuria. Gene Ther 13:457–462
100. Ding Z, Georgiev P, Thony B (2006) Administration-route and gender-independent long-term therapeutic correction of phenylketonuria (PKU) in a mouse model by recombinant adeno-associated virus 8 pseudotyped vector-mediated gene transfer. Gene Ther 13:587–593
101. Oh HJ, Park ES, Kang S, Jo I, Jung SC (2004) Long-term enzymatic and phenotypic correction in the phenylketonuria mouse model by adeno-associated virus vector-mediated gene transfer. Pediatr Res 56:278–284
102. Mochizuki S et al (2004) Long-term correction of hyperphenylalaninemia by AAV-mediated gene transfer leads to behavioral recovery in phenylketonuria mice. Gene Ther 11:1081–1086
103. Fang B et al (1994) Gene therapy for phenylketonuria: phenotypic correction in a genetically deficient mouse model by adenovirus-mediated hepatic gene transfer. Gene Ther 1:247–254
104. Harding CO, Neff M, Jones K, Wild K, Wolff JA (2003) Expression of phenylalanine hydroxylase (PAH) in erythrogenic bone marrow does not correct hyperphenylalanin-

emia in Pah(enu2) mice. J Gene Med 5:984–993
105. Nagasaki Y et al (1999) Reversal of hypopigmentation in phenylketonuria mice by adenovirus-mediated gene transfer. Pediatr Res 45:465–473
106. Lin CM, Tan Y, Lee YM, Chang CC, Hsiao KJ (1997) Expression of human phenylalanine hydroxylase activity in T lymphocytes of classical phenylketonuria children by retroviral-mediated gene transfer. J Inherit Metab Dis 20:742–754
107. Ledley FD, Grenett HE, DiLella AG, Kwok SC, Woo SL (1985) Gene transfer and expression of human phenylalanine hydroxylase. Science 228:77–79
108. Robson KJ et al (1984) Sequence comparison of rat liver phenylalanine hydroxylase and its cDNA clones. Biochemistry 23:5671–5675
109. Kwok SC, Ledley FD, DiLella AG, Robson KJ, Woo SL (1985) Nucleotide sequence of a full-length complementary DNA clone and amino acid sequence of human phenylalanine hydroxylase. Biochemistry 24:556–561
110. Ledley FD, Grenett HE, McGinnis-Shelnutt M, Woo SL (1986) Retroviral-mediated gene transfer of human phenylalanine hydroxylase into NIH 3T3 and hepatoma cells. Proc Natl Acad Sci U S A 83:409–413
111. Ding Z, Harding CO, Thony B (2004) State-of-the-art 2003 on PKU gene therapy. Mol Genet Metab 81:3–8
112. Oh HJ et al (2005) Reversal of gene expression profile in the phenylketonuria mouse model after adeno-associated virus vector-mediated gene therapy. Mol Genet Metab 86(Suppl 1):S124–S132
113. Embury JE et al (2007) PKU is a reversible neurodegenerative process within the nigrostriatum that begins as early as 4 weeks of age in Pah(enu2) mice. Brain Res 1127:136–150
114. Yagi H et al (2011) Complete restoration of phenylalanine oxidation in phenylketonuria mouse by a self-complementary adeno-associated virus vector. J Gene Med 13:114–122
115. Duncan AW, Dorrell C, Grompe M (2009) Stem cells and liver regeneration. Gastroenterology 137:466–481
116. van Bogaert L, Bertrand I (1949) Sur une idiotie familiale avec degerescence sponglieuse de neuraxe (note preliminaire). Acta Neurol Belg 49:572–587
117. Matalon R et al (1988) Aspartoacylase deficiency and N-acetylaspartic aciduria in patients with Canavan disease. Am J Med Genet 29:463–471
118. Matalon R, Michals K, Kaul R (1995) Canavan disease: from spongy degeneration to molecular analysis. J Pediatr 127:511–517
119. Adachi M, Schneck L, Cara J, Volk BW (1973) Spongy degeneration of the central nervous system (van Bogaert and Bertrand type; Canavan's disease). A review. Hum Pathol 4:331–347
120. Traeger EC, Rapin I (1998) The clinical course of Canavan disease. Pediatr Neurol 18:207–212
121. Matalon R, Michals-Matalon K (1998) Molecular basis of Canavan disease. Eur J Paediatr Neurol 2:69–76
122. Sreenivasan P, Purushothaman KK (2013) Radiological clue to diagnosis of Canavan disease. Indian J Pediatr 80(1):75–77
123. Pradhan S, Goyal G (2011) Teaching NeuroImages: honeycomb appearance of the brain in a patient with Canavan disease. Neurology 76:e68
124. Francis JS, Markov V, Leone P (2014) Dietary triheptanoin rescues oligodendrocyte loss, dysmyelination and motor function in the nur7 mouse model of Canavan disease. J Inherit Metab Dis 37(3):369–381
125. Ariyannur PS, Madhavarao CN, Namboodiri AM (2008) N-acetylaspartate synthesis in the brain: mitochondria vs. microsomes. Brain Res 1227:34–41
126. Urenjak J, Williams SR, Gadian DG, Noble M (1992) Specific expression of N-acetylaspartate in neurons, oligodendrocyte-type-2 astrocyte progenitors, and immature oligodendrocytes in vitro. J Neurochem 59:55–61
127. Moffett JR, Namboodiri MA, Cangro CB, Neale JH (1991) Immunohistochemical localization of N-acetylaspartate in rat brain. Neuroreport 2:131–134
128. Mersmann N et al (2011) Aspartoacylase-lacZ knockin mice: an engineered model of Canavan disease. PLoS One 6:e20336
129. Kirmani BF, Jacobowitz DM, Kallarakal AT, Namboodiri MA (2002) Aspartoacylase is restricted primarily to myelin synthesizing cells in the CNS: therapeutic implications for Canavan disease. Brain Res Mol Brain Res 107:176–182
130. Baslow MH (1999) The existence of molecular water pumps in the nervous system: a review of the evidence. Neurochem Int 34:77–90
131. Baslow MH (1999) Molecular water pumps and the aetiology of Canavan disease: a case of the sorcerer's apprentice. J Inherit Metab Dis 22:99–101
132. Moffett JR, Ross B, Arun P, Madhavarao CN, Namboodiri AM (2007) N-Acetylaspartate in the CNS: from neurodiagnostics to neurobiology. Prog Neurobiol 81:89–131
133. Taylor DL et al (1995) Investigation into the role of N-acetylaspartate in cerebral osmoregulation. J Neurochem 65:275–281

134. Davies SE, Gotoh M, Richards DA, Obrenovitch TP (1998) Hypoosmolarity induces an increase of extracellular N-acetylaspartate concentration in the rat striatum. Neurochem Res 23:1021–1025
135. Namboodiri AM et al (2006) Canavan disease and the role of N-acetylaspartate in myelin synthesis. Mol Cell Endocrinol 252:216–223
136. Kirmani BF, Jacobowitz DM, Namboodiri MA (2003) Developmental increase of aspartoacylase in oligodendrocytes parallels CNS myelination. Brain Res Dev Brain Res 140:105–115
137. D'Adamo AF Jr, Gidez LI, Yatsu FM (1968) Acetyl transport mechanisms. Involvement of N-acetyl aspartic acid in de novo fatty acid biosynthesis in the developing rat brain. Exp Brain Res 5:267–273
138. Pliss L et al (2003) Morphology and ultrastructure of rat hippocampal formation after i.c.v. administration of N-acetyl-L-aspartyl-L-glutamate. Neuroscience 122:93–101
139. Kitada K et al (2000) Accumulation of N-acetyl-L-aspartate in the brain of the tremor rat, a mutant exhibiting absence-like seizure and spongiform degeneration in the central nervous system. J Neurochem 74:2512–2519
140. Akimitsu T et al (2000) Epileptic seizures induced by N-acetyl-L-aspartate in rats: in vivo and in vitro studies. Brain Res 861:143–150
141. Kolodziejczyk K, Hamilton NB, Wade A, Karadottir R, Attwell D (2009) The effect of N-acetyl-aspartyl-glutamate and N-acetyl-aspartate on white matter oligodendrocytes. Brain 132:1496–1508
142. Surendran S (2010) Upregulation of N-acetylaspartic acid resulting nitric oxide toxicity induces aspartoacylase mutations and protein interaction to cause pathophysiology seen in Canavan disease. Med Hypotheses 75:533–534
143. Surendran S, Bhatnagar M (2011) Upregulation of N-acetylaspartic acid induces oxidative stress to contribute in disease pathophysiology. Int J Neurosci 121:305–309
144. Francis JS, Strande L, Markov V, Leone P (2012) Aspartoacylase supports oxidative energy metabolism during myelination. J Cereb Blood Flow Metab 32:1725–1736
145. Matalon R et al (2000) Knock-out mouse for Canavan disease: a model for gene transfer to the central nervous system. J Gene Med 2:165–175
146. Ahmed SS et al (2013) A single intravenous rAAV injection as late as P20 achieves efficacious and sustained CNS gene therapy in Canavan mice. Mol Ther 21:2136–2147
147. Traka M et al (2008) Nur7 is a nonsense mutation in the mouse aspartoacylase gene that causes spongy degeneration of the CNS. J Neurosci 28:11537–11549
148. Kile BT et al (2003) Functional genetic analysis of mouse chromosome 11. Nature 425:81–86
149. Carpinelli MR et al (2014) A new mouse model of Canavan leukodystrophy displays hearing impairment due to central nervous system dysmyelination. Dis Model Mech 7(6):649–657
150. Kaul R, Gao GP, Balamurugan K, Matalon R (1993) Cloning of the human aspartoacylase cDNA and a common missense mutation in Canavan disease. Nat Genet 5:118–123
151. Leone P, Janson CG, McPhee SJ, During MJ (1999) Global CNS gene transfer for a childhood neurogenetic enzyme deficiency: Canavan disease. Curr Opin Mol Ther 1:487–492
152. Leone P et al (2000) Aspartoacylase gene transfer to the mammalian central nervous system with therapeutic implications for Canavan disease. Ann Neurol 48:27–38
153. Janson C et al (2002) Clinical protocol. Gene therapy of Canavan disease: AAV-2 vector for neurosurgical delivery of aspartoacylase gene (ASPA) to the human brain. Hum Gene Ther 13:1391–1412
154. Matalon R et al (2003) Adeno-associated virus-mediated aspartoacylase gene transfer to the brain of knockout mouse for canavan disease. Mol Ther 7:580–587
155. McPhee SW et al (2005) Effects of AAV-2-mediated aspartoacylase gene transfer in the tremor rat model of Canavan disease. Brain Res Mol Brain Res 135:112–121
156. Foust KD et al (2009) Intravascular AAV9 preferentially targets neonatal neurons and adult astrocytes. Nat Biotechnol 27:59–65
157. Leone P et al (2012) Long-term follow-up after gene therapy for canavan disease. Sci Transl Med 4:165ra163
158. McPhee SW et al (2006) Immune responses to AAV in a phase I study for Canavan disease. J Gene Med 8:577–588
159. Ransohoff RM, Brown MA (2012) Innate immunity in the central nervous system. J Clin Invest 122:1164–1171
160. Stein-Streilein J, Caspi RR (2014) Immune privilege and the philosophy of immunology. Front Immunol 5:110
161. Calcedo R, Vandenberghe LH, Gao G, Lin J, Wilson JM (2009) Worldwide epidemiology of neutralizing antibodies to adeno-associated viruses. J Infect Dis 199:381–390
162. Mingozzi F et al (2013) Prevalence and pharmacological modulation of humoral immunity to AAV vectors in gene transfer to synovial tissue. Gene Ther 20:417–424
163. Louis Jeune V, Joergensen JA, Hajjar RJ, Weber T (2013) Pre-existing anti-adeno-

associated virus antibodies as a challenge in AAV gene therapy. Hum Gene Ther Methods 24:59–67
164. Mingozzi F et al (2013) Overcoming preexisting humoral immunity to AAV using capsid decoys. Sci Transl Med 5:194ra192
165. Basner-Tschakarjan E, Bijjiga E, Martino AT (2014) Pre-clinical assessment of immune responses to adeno-associated virus (AAV) vectors. Front Immunol 5:28
166. Sanftner LM et al (2004) Striatal delivery of rAAV-hAADC to rats with preexisting immunity to AAV. Mol Ther 9:403–409
167. Martino AT et al (2011) The genome of self-complementary adeno-associated viral vectors increases Toll-like receptor 9-dependent innate immune responses in the liver. Blood 117:6459–6468
168. Hosel M et al (2012) Toll-like receptor 2-mediated innate immune response in human nonparenchymal liver cells toward adeno-associated viral vectors. Hepatology 55:287–297
169. Olson JK, Miller SD (2004) Microglia initiate central nervous system innate and adaptive immune responses through multiple TLRs. J Immunol 173:3916–3924
170. Bsibsi M, Ravid R, Gveric D, van Noort JM (2002) Broad expression of Toll-like receptors in the human central nervous system. J Neuropathol Exp Neurol 61:1013–1021
171. Mingozzi F et al (2007) CD8(+) T-cell responses to adeno-associated virus capsid in humans. Nat Med 13:419–422
172. Sabatino DE et al (2005) Identification of mouse AAV capsid-specific CD8+ T cell epitopes. Mol Ther 12:1023–1033
173. Li H et al (2007) Pre-existing AAV capsid-specific CD8+ T cells are unable to eliminate AAV-transduced hepatocytes. Mol Ther 15:792–800
174. Ciesielska A et al (2013) Cerebral infusion of AAV9 vector-encoding non-self proteins can elicit cell-mediated immune responses. Mol Ther 21:158–166
175. Yang B et al (2014) Global CNS transduction of adult mice by intravenously delivered rAAVrh.8 and rAAVrh.10 and nonhuman primates by rAAVrh.10. Mol Ther 22(7):1299–1309
176. Zhang H et al (2011) Several rAAV vectors efficiently cross the blood-brain barrier and transduce neurons and astrocytes in the neonatal mouse central nervous system. Mol Ther 19:1440–1448
177. Kotterman MA, Schaffer DV (2014) Engineering adeno-associated viruses for clinical gene therapy. Nat Rev Genet 15(7):445–451

Chapter 31

Gene Therapy for the Treatment of Neurological Disorders: Central Nervous System Neoplasms

Neha Kamran, Marianela Candolfi, Gregory J. Baker, Mariela Moreno Ayala, Marta Dzaman, Pedro R. Lowenstein, and Maria G. Castro

Abstract

Glioblastoma multiforme (GBM) is the most common primary brain tumor in adults with a median survival of 16.2–21.2 months post diagnosis (Stupp et al., N Engl J Med 352(10): 987–996, 2005). Because of its location, complete surgical resection is impossible; additionally because GBM is also resistant to chemotherapeutic and radiotherapy approaches, development of novel therapies is urgently needed. In this chapter we describe the development of preclinical animal models and a conditionally cytotoxic and immune-stimulatory gene therapy strategy that successfully causes tumor regression in several rodent GBM models.

Key words GBM models, Immunotherapy, Adenoviral gene therapy, T cell activation assays

1 Introduction

The advantages of implantation GBM models (syngeneic or xenograft) are their efficient tumorigenesis, their reproducible and relatively fast growth rates and the accurate knowledge of tumor location [1–3]. They exhibit many of the histopathological features of human GBM, i.e., infiltration of tumor cells throughout the surrounding brain parenchyma (*see* Figs. 1 and 2), areas of pseudopalisading necrosis, and neovascularization [1, 4, 5]. Hence they have proven to be a valuable tool for the preclinical assessment of novel therapies. Syngeneic GBM models in rodents encompass the implantation of GBM cells that originated in the same mouse or rat breed so that they are immunologically compatible. Since the tumor and the host match immunologically, this tumor model does not require immunodeficient animals, allowing testing of immunotherapeutic strategies. Xenograft tumor formation involves the implantation of human GBM-derived short-term

Fredric P. Manfredsson (ed.), *Gene Therapy for Neurological Disorders: Methods and Protocols*, Methods in Molecular Biology, vol. 1382, DOI 10.1007/978-1-4939-3271-9_31, © Springer Science+Business Media New York 2016

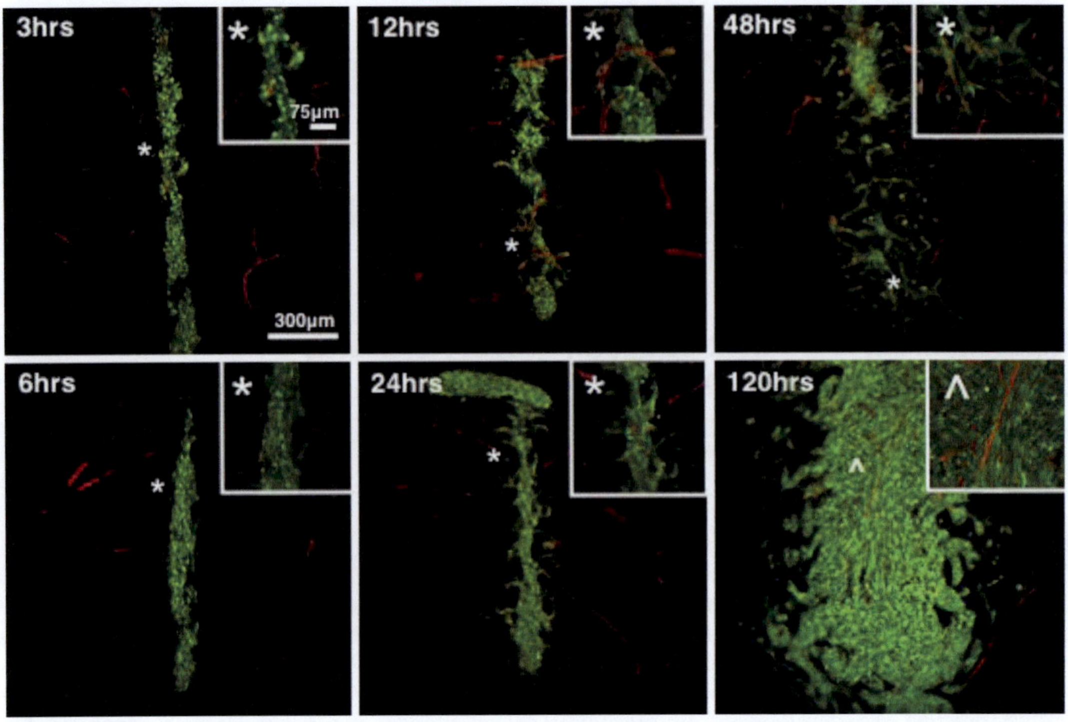

Fig. 1 Time-course analysis of early GL26-Cit glioma growth in the striatum in mice. To visualize perivascular glioma growth GL26-Cit cells were implanted in RA/EGxdelCre mice that express GFP in the brain endothelium. Invasive glioma cells maintain close vascular contact at all time-points analyzed over 120 h as they disseminate throughout the brain. GFP^+ brain microvasculature has been pseudocolored *red*. Corresponding high-magnification micrographs (*insets*) detail perivascular invasion at the tumor border. *White asterisks* (*) relate the image area shown in the high-magnification micrographs with the corresponding area in the low-magnification micrographs. Perivascular tumor invasion begins 24 h post-implantation. *Inset* denoted by the *carrot* (^) in the 120 h micrograph is included to demonstrate the trapping of normal brain microvessels within the growing tumor mass as perivascular invasion is followed by tumor cell proliferation within the perivascular space. Time-points progress from *top* to *bottom* and from *left* to *right*

primary cultures or cell lines in immunosuppressed or immunodeficient mice. For this reason, they are not suitable for the evaluation of immunotherapeutic approaches and will not be part of our focus here.

A number of key physiological processes such as the relative paucity of dendritic cells (DCs) in the brain, lack of lymphatic drainage, production of anti-inflammatory mediators such as TGF-β and nitric oxide (NO) by cells in the central nervous system (CNS), as well as low MHCII expression on infiltrating microglia, protect the brain from an immune-mediated attack but at the same time, also contribute to its immune suppressive environment [6–9]. We have succeeded in developing an adenoviral mediated immunotherapy for brain tumors that is dependent on the expression of two genes: Thymidine kinase (TK), that phosphorylates the

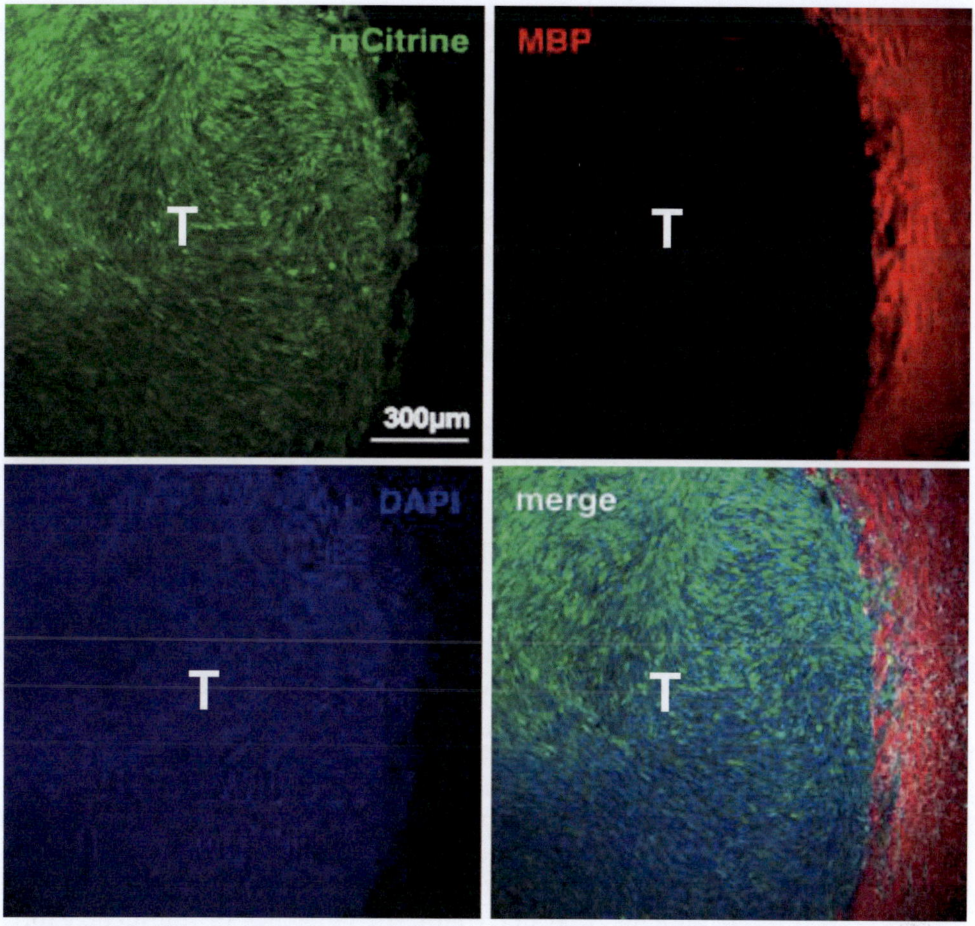

Fig. 2 Late-stage GL26-Cit intracranial glioma tumor. Scanning fluorescence confocal micrograph of syngeneic GL26-Cit mouse glioma (shown in *green*) 21-days post-tumor implantation into the striatum. Glioma cells were genetically modified to express mCitrine fluorescent protein prior to tumor implantation to facilitate direct tumor visualization by fluorescence microscopy. Individual and merged channels are shown. Tumor-bearing brain tissue was sectioned 50 μm thick and immunolabeled with anti-myelin basic protein (MBP) antibodies (shown in *red*) then counterstained with 4′,6-diamidino-2-phenylindole (DAPI) (shown in *blue*). Note that this late-stage tumor has well-defined borders compared to GL26-Cit tumors at earlier stages, which lack definition in their tumor borders and which exhibit extensive perivascular invasion. Also note that large myelinated axonal bundles that become compressed towards the outside of the growing tumor mass. Scale bar corresponds to 300 μm. T denotes tumor area

prodrug Ganciclovir which induces DNA cross-linking followed by cell death and *fms*-like tyrosine kinase-3 ligand (Flt3L), a potent DC growth factor that recruits DCs within the tumor microenvironment [10–12]. Tumor cell death induced by TK causes the release of intracellular proinflammatory molecules and tumor antigens, which are taken up by surveying DCs that subsequently prime the T cells to elicit an antigen specific cytotoxic antitumor immune response [10–12]. Combination therapy using these two adenoviruses induces tumor regression, long-term survival, and immunological memory in several mouse and rat GBM models [10, 12, 13].

Gene therapy vectors can also be used for the treatment of neurodegenerative diseases. In order to be useful for gene therapy of chronic neurological disorders, vector systems should allow long-term transduction of brain cells in the absence of undesirable long-term side effects. Recombinant adenovirus vectors are powerful tools for gene delivery to the CNS [14, 15]. Adenoviruses can be easily purified to high titers and efficiently transduce differentiated cells such as neurons and glial cells. Additionally, transgene expression is restricted to the area of vector administration or areas that project to the injection site [16, 17]. The majority of current adenoviral-mediated gene therapy protocols utilize first-generation vectors, which are recombinant vectors that are non-replicative because of the deletion of the E1 region from the viral genome [18, 19]. While in the absence of prior immune priming to adenovirus, first-generation adenoviral vectors injected into the brain parenchyma can sustain prolonged transgene expression for months, activation of antiviral T cells by peripheral immunization leads to loss of vector-mediated transgene expression [17, 20]. To overcome this instability, "gutless," high-capacity adenoviral (HC-Ad) vectors have been engineered in which all viral encoding genes have been eliminated and replaced with noncoding stuffer DNA sequences. Even in the presence of an anti-adenoviral immune response, expression from HC-Ad vectors remains stable for at least 1 year [21, 22]. Critical parameters that determine the efficiency of transgene expression are (1) dose and volume of adenoviruses administered, (2) vector backbone and the choice of promoter, (3) immune status of the animal, and (4) purity of the adenoviral vector stock [23, 24]. In this unit we provide detailed instructions for the implantation of various syngeneic GBM models in rodents and adenoviral-mediated gene therapy using Ad-TK and Ad-Flt3L. Adenoviral-mediated transgene expression can be readily detected in treated brain tumors by immunocytochemistry. Figure 3 shows representative images from syngeneic rat CNS-1 model, U251 and U87 xenografts models injected intratumorally with Ad adenoviral vectors encoding TK. TK expression was assessed by immunohistochemistry and immunofluorescence. ELISPOT and T cell proliferation are extremely useful assays for testing the quality and quantity of T cell antitumor responses induced by immunotherapeutic approaches; hence the protocols for setting up these assays are also described [10, 12, 13, 25–29].

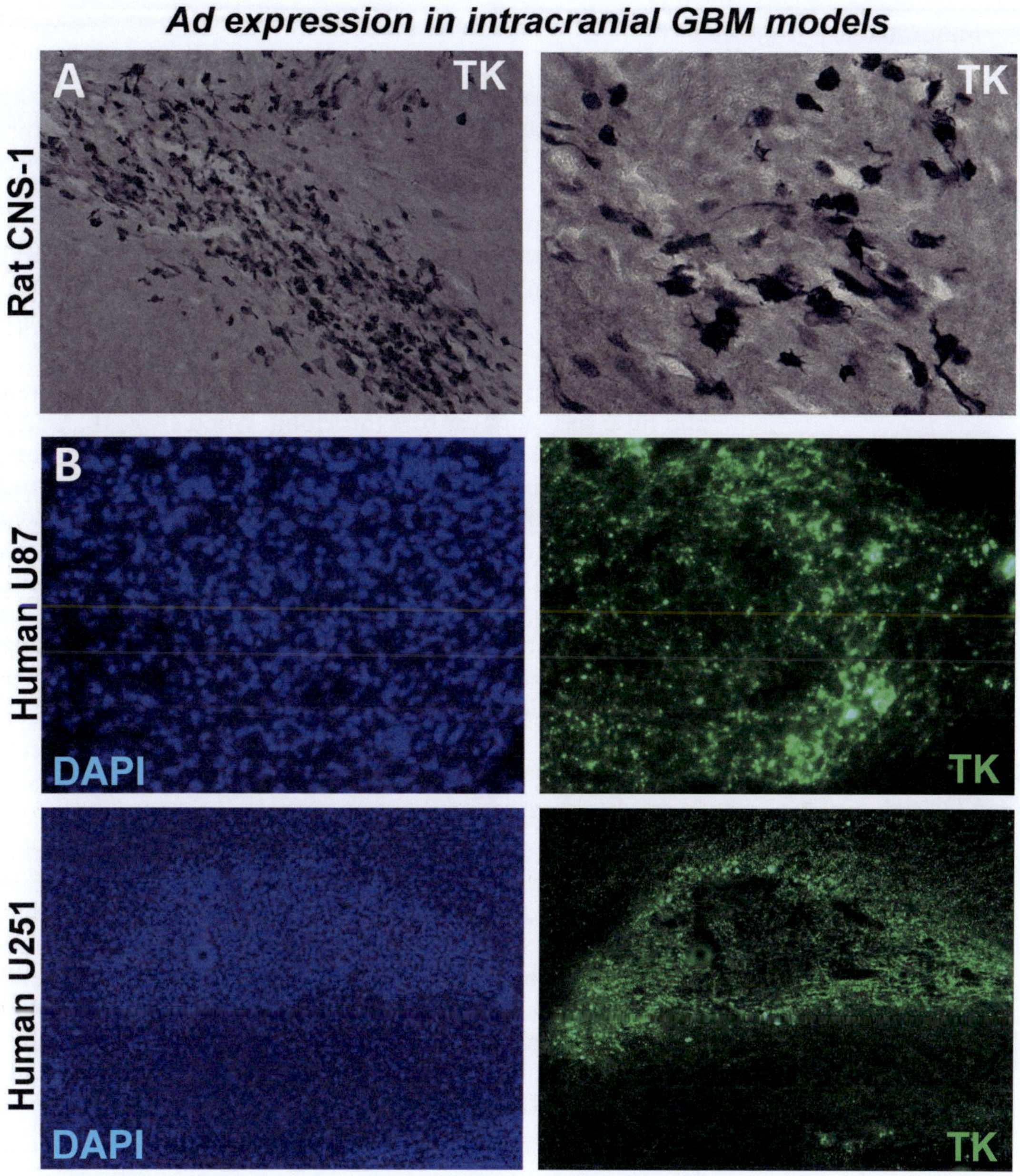

Fig. 3 Adenoviral-mediated transgene delivery into intracranial GBM in rodent models. Microphotographs show TK expression within intracranial GBMs implanted in the rodent brain. Lewis rats bearing intracranial syngeneic CNS-1 GBMs (**a**) and nude mice bearing intracranial human U251 or U87 GBM (**b**) were intratumorally injected with Ad vectors encoding thymidine kinase (TK). TK expression was assessed by immunocytochemistry (**a**) and immunofluorescence (**b**). Nuclei were stained with DAPI (**b**, *left* panels)

2 Materials

2.1 For Intracranial Tumor Implantation

1. Anesthetics/analgesics and other drugs: ketamine (75 mg/kg body weight, Orion Pharma), Dexdomitor (0.25 mg/kg body weight; Orion Pharma, Espoo, Finland), Rimadyl/carprofen (5 mg/kg body weight; Orion Pharma, Espoo, Finland), atipamezole (1 mg/kg body weight; Orion Pharma, Espoo, Finland), buprenorphine (20 μl/g body weight; MWI Veterinary Supply, Boise, ID, USA), povidone–iodine solution, Paralube eye ointment, saline solution (0.9 % sodium chloride).
2. Surgical equipment: Stereotactic frame with adaptor for rats and mice and blunt ear bars (Stoelting, Wood Dale, IL, USA), Zeiss Stemi 1000 Zoom stereomicroscope (Zeiss, Jena, Germany) or equivalent, equipped with 16× eye-pieces and 0.4× auxiliary objective lens and mounted on hinged coupling arm and a heavy foot stand, surgical lamp, electric drill with 0.6 mm drill bit for mice and 1.75 mm drill bit for rats (Stoelting, Wood Dale, IL, USA), surgical shavers, scalpel and blades, skin retractors, ethanol swabs, curved and straight forceps, holding scissors, sharp scissors, 3-0 nylon sutures, 5 μl Hamilton syringe with 26G needle for rats and 33G needle for mice.

2.2 For Adenoviral Gene Therapy

1. Anesthetics/analgesics and other drugs: as above.
2. Surgical equipment: as above.
3. Adenoviral vectors: Detailed methodologies to clone, purify, and quantitate adenoviral vectors for gene therapy are detailed elsewhere [16, 23] (*see* chapter 9). Also included in these references are protocols for the analysis of transgene expression using immunohistochemistry (Fig. 3), enzymatic assays, and flow cytometry. Protocols for the detection of anti-adenoviral immune responses are also described in this chapter. The adenoviral vectors used in this protocol are first generation Ad.hCMV.hsFLT3L + Ad.hCMV.TK.
4. Phosphate buffered saline: Dulbecco's PBS, pH 7.4, without calcium or magnesium (Life Technologies, Carlsbad, CA, USA).
5. Ganciclovir: To prepare 7 mg/ml stock solution, weigh 250 mg of Ganciclovir (Biotang USA, Lexington, MA, USA) and add 20 ml of Milli-Q H_2O to it. Adjust pH to 12 with 1 M NaOH. The solution will clear once the pH reaches 12. Next lower the pH to 11 using 1 N HCl and add Milli-Q H_2O to a final volume of 35.71 ml. Dilute further with saline for injections.

2.3 For Functional Assays Post Gene Therapy Administration

1. Tyrodes solution: 132 mM NaCl, 1.8 mM $CaCl_2{\cdot}H_2O$, 0.32 mM NaH_2PO_4, 5.56 mM glucose, 11.6 mM $NaHCO_3$, and 2.68 mM KCl. To prepare 1 L add 8 g NaCl, 0.264 g $CaCl_2{\cdot}H_2O$, 0.05 g NaH_2PO_4, 1 g glucose, 1 g $NaHCO_3$, and 0.2 g of KCl. Add 100 μl of heparin to prevent blood from clotting while perfusing.
2. ACK buffer: Add 8.29 g NH_4Cl, 1 g $KHCO_3$, 37.2 mg Na_2EDTA to 800 ml of pure H_2O. Adjust pH to 7.4 with 1 N HCl and top up the volume to 1 L with H_2O.
3. Complete media: RPMI/DMEM medium supplemented with 10 % FBS, 50 units/ml penicillin and 50 μg/ml streptomycin.
4. ELISPOT reagents: Mouse IFN-γ ELISPOT Ready-SET-Go kit (eBioscience, San Diego, CA, USA), Immobilon-P 96 well plates, Ethanol (USP grade, mixed 35 % vol/vol in sterile Milli-Q H_2O), ELISPOT Wash Buffer (0.01 % Tween 20 in PBS), Cell stimulation cocktail (eBioscience, San Diego, CA, USA), TMB substrate (Mabtech, Cincinnati, OH, USA), ELISPOT plate reader (AID Autoimmun Diagnostika GmbH, Strassberg, Germany).
5. For T cell proliferation: 5-(and 6)-carboxyfluorescein diacetate succinimidyl ester (CFSE) cell proliferation kit (Life Technologies, Carlsbad, CA, USA), bovine serum albumin, anti-mouse and anti-rat CD3, CD4 and CD8 fluorochrome conjugated antibodies for flow staining, flow staining buffer (2 % FBS in PBS).

3 Methods

3.1 Intracranial GBM Implantation

1. Resuspend glioma cell lines in serum free media at the required concentration (Injectable volume 1–5 μl).
2. The surgical area needs to be clean and organized with all the required instruments, drugs and sterile tools (*see* **Note 1**).
3. Place the animal under anesthesia by intraperitoneal (i.p.) injection of ketamine and Dexdomitor. Once the animal has lost the footpad reflex it is sedated enough to start the surgical procedure. Administer a subcutaneous injection of Carprofen to ensure analgesia during and after surgery (*see* **Note 2**).
4. Shave the fur from the head of the animal.
5. Mount the animal onto the stereotactic frame, immobilize its head in the incisor bar and tighten it gently with the ear bars and the nose clamp (*see* **Note 3**).
6. Clean the incision area thoroughly with alcohol wipes and povidone–iodine solution.

7. Make a ~1 cm (for mice) or ~1.5 cm (for rats) midline incision along the top of the head between the eyes and the ears. Using a mouse or rat skin retractor hold back the skin on both sides of the incision.
8. Direct the light beams onto the exposed skull and focus the microscope on the bregma, the junction of the sagittal and transverse sutures (*see* **Note 4**).
9. Position the Hamilton syringe using the manipulator arms so that the tip of the needle is exactly over bregma. To inject into the striatum of a mouse or rat, move the manipulator arm forward x mm for the anteroposterior coordinate and then y mm lateral away from the bregma. (Refer to Table 1 for the x and y coordinates for the various cell lines.) Watching through the microscope eyepiece, use a bent 26G1/2 needle to itch a small mark in the skull where the needle will penetrate. Lift the needle so that it does not obstruct the area.
10. Drill a small burr hole with a 0.6 mm bit for mice and 1.75 mm bit for rats using wide circular drill motions. The burr hole should be wide to provide a large open area for the insertion of the needle. Drilling into the skull generates considerable heat, so it is advisable to drill in short bursts, while intermittently

Table 1
Cell numbers, implantation coordinates, and survival duration for various GBM models are shown

Cell line	Host	Cell number for implantation	Injection coordinates	Median survival
GL261	C57BL/6	2×10^4	+0.5 mm AP, +2.2 mm ML from bregma and −3.2 mm DV from dura (1 μl)	28–35 days
GL26	C57BL/6	2×10^4	+0.5 mm AP, +2.1 mm ML from bregma and −3.2 mm DV from dura (1 μl)	26–32 days
SMA560	VM/Dk	5×10^3	+0.5 mm AP, +2.1 mm ML from bregma and −3.2 mm DV from dura (1 μl)	30 days
CNS1	Lewis rat	5×10^4	+1.0 mm AP, +3.0 mm ML from bregma and −5.5, −5.0, and −4.5 mm DV from dura (1 μl per site)	13–19 days
9 L	Fisher rat	5×10^5	+1.0 mm AP, +3.2 mm ML from bregma and −5.5, −5.0 and −4.5 mm DV from dura (1 μl per site)	28.1 ±1.3 days
F98	Fisher rat	5×10^4	+1.0 mm AP, +3.2 mm ML from bregma and −6.5, −5.5 and −4.5 mm DV from the dura (1 μl per site)	29 days
RG2	Fisher rat	2×10^4	+1.0 mm AP, +3.0 mm ML from bregma and +6.0 mm DV from dura (3 μl)	20 days

AP anteroposterior, *ML* media-lateral, *DV* dorsoventral

bathing the skull with ice cold saline solution, which can be removed using a cotton swab. Avoid severing any blood vessels during drilling. In the event of bleeding, clean the burr hole to prevent the blood clot from blocking the needle entry and retraction (*see* **Notes 5** and **6**).

11. Load the Hamilton syringe (33G needle for mice, 26G for rats) with the proper dose of cells (*see* **Note 7**). Lower the needle such that it is leveled with the dura. Read the dorsoventral coordinates and lower the needle to 0.5 mm plus the appropriate coordinates, depending on the GBM model. Pull the needle up toward the dura 0.5 mm and wait for 2 min. The extra 0.5 mm provides a pocket for the cells at the time of injection.
12. Administer the injection slowly over the course of 0.5 μl/min. Keep the needle in place for 5 min post injection to allow tumor cells to settle before slowly withdrawing the needle from the brain. Clean the syringe thoroughly by flushing three times with saline (*see* **Note 8**).
13. Flush the skull with sterile saline three times to remove any residual cells from the brain surface and dry the area with a cotton swab.
14. Remove the skin retractor and close the incision using 3-0 nylon suture.
15. Resuscitate the animal by i.p. injection of atipamezole. Administer buprenorphine subcutaneously. Monitor the animal until it fully recovers from anesthesia and return them to their cage.
16. Provide the animals with water soaked chow in a petri dish and monitor for any surgical complications. Remove any remaining sutures at 10–14 days post surgery.

3.2 Adenoviral Gene Therapy

1. Dilute the adenovirus preparation in sterile PBS such that the required number of infectious units can be administered in the appropriate volume (*see* **Note 9**).
2. Anesthetize tumor bearing animals with an i.p. injection of ketamine and Dexdomitor. Ensure that the animal is fully anesthetized by checking for the lack of responses to footpad and tail pinching. Place the anesthetized mice in a stereotactic apparatus.
3. By now the sutures from the surgery for tumor implantation would have fallen off and the old skin incision would have healed. Using a scalpel, make a 1.5 cm midline incision into the skin at the same location as before. Use the scalpel blade to gently separate the healed skin from the underlying tissue.

4. Use skin retractors to hold back the skin on either side of the incision.
5. Remove the fibrous tissue covering the site of the tumor injection by gently scraping with the scalpel blade. Wash the area with cold saline to stop any bleeding and clean with ethanol swab.
6. The old burr hole should now be clearly visible. At this point, it is not required to drill again through the bone to provide access to the needle into the site of tumor implantation. Use a bent 26G needle to remove the scar tissue that forms at the burr hole. Use ethanol swabs to clean the burr hole to remove any clotted blood.
7. Lower the needle into the brain to the dorsoventral coordinate of tumor injection plus 0.5 mm ventrally and wait for 2 min before slowly injecting 1/3rd of the vector suspension. Wait for 1 min for the vector solution to infuse into the tissue. Move 0.5 mm up dorsally and inject another 1/3rd of the vector and wait a further minute. Repeat for one last time to inject the remaining vector suspension. Wait for 5 min after the final administration and then slowly draw the needle out.
8. Starting 24 h post gene therapy, administer 25 mg/kg of Ganciclovir i.p. twice daily for 7 days for mice or 10 days for rats (*see* **Note 10**). Monitor the animals for signs of moribund behavior (hunched posture, lack of grooming, porphyrin staining around the eyes) and euthanize when their health status reaches the criteria established by the institutional animal care guidelines. Animals will be humanely killed by terminal perfusion with oxygenated, heparinized Tyrode's solution under deep anesthesia.

3.3 Analysis of Antigen-Specific T Cells Responses: IFNγ-ELISPOT

1. ELISPOT assays will be carried out using the mouse IFN-γ ELISPOT Ready Set Go! kit.
2. One day prior to the start of the assay, dilute the IFN-γ capture antibody in sterile ELISPOT Coating Buffer, as noted on Certificate of Analysis included with the reagent set. Coat ELISPOT plate (Immobilon-P plates from Millipore) with 100 μl/well of capture antibody solution diluted in the coating buffer included in the kit. Cover the plate and seal with Parafilm. Incubate at 4 °C overnight (*see* **Note 11**).
3. On the day of the assay, decant or aspirate coating antibody from plate and wash plates two times with 200 μl/well sterile ELISPOT coating buffer. Decant.
4. Block plate with 200 μl/well of complete RPMI-1640 at 37 °C for 2 h. Decant or aspirate plate (*see* **Note 12**).

5. Harvest the spleens from the treated animals between 7 and 10 days post gene therapy by making an incision in the abdominal cavity on the left side of the mouse, inferior to the stomach (*see* **Note 13**).
6. Place the excised spleen on a 70 μm cell strainer that is put on a 50 ml conical tube.
7. Mash the spleen gently using the plunger end of a 1 ml syringe. Wash the cells through the strainer using 15–20 ml of complete media. Pellet the cells at 500 × *g* for 5 min at 4 °C. Discard the supernatant.
8. Remove RBCs by ACK lysis as follows: Resuspend the cell pellet in 3 ml of ice cold ACK buffer per spleen. Incubate on ice for 3 min. Dilute out the ACK buffer by adding 10 ml of 1× PBS. Pellet and discard the buffer as in **step 7**. Wash once more with complete media (*see* **Note 14**).
9. Resuspend the splenocytes in complete media to a concentration of 0.5×10^6 cells/100 μl. Add 100 μl of this cell suspension to the ELISPOT plate.
10. Aliquot mitogen (cell stimulation cocktail from eBioscience, 1:500 dilution), antigen (100–200 μg/ml) or controls diluted in complete medium to appropriate wells at 100 μl/well.
11. Incubate the plate for 48 h at 37 °C, in a 5 % CO_2 humidified incubator before discarding the cells (*see* **Note 15**).
12. Wash plate three times with ELISPOT Wash Buffer. Dilute biotinylated detection antibody in Assay Diluent according to instructions on the Certificate of Analysis provided with the kit. Add 100 μl/well to the plate and incubate at 4 °C overnight.
13. Decant antibody solution. Wash four times with ELISPOT Wash Buffer.
14. Dilute Avidin–HRP reagent in Assay Diluent according to instructions on the Certificate of Analysis provided with the reagent set. Add 100 μl/well of Avidin–HRP and incubate at room temperature for 2 h.
15. Decant Avidin–HRP solution. Wash plate three times with ELISPOT Wash Buffer, and then two times with 1× PBS.
16. Add 100 μl/well of TMB substrate and develop at room temperature for 5–10 min; monitor development of spots (*see* **Note 16**).
17. Stop the substrate reaction by washing wells three times with 200 μl/well distilled water.
18. Air-dry the plate. Count spots using a dissecting microscope or automated ELISPOT plate reader.

3.4 Analysis of Antigen-Specific T Cells Responses: T Cell Proliferation Assay

1. Purify splenocytes from tumor-bearing mice treated with intracranial injections of Ad-TK + Ad-Flt3L as above in Subheading 3.3, **step 1**.
2. Wash splenocytes twice with 1× PBS. Resuspend splenocytes in PBS containing 0.1 % BSA at a concentration of 10^7/ml (*see* **Note 17**).
3. Add CFSE dye to a final concentration of 4 μM. Incubate at 37 °C for 10 min with occasional shaking. At the end of the incubation, add the same volume of complete media and incubate on ice for 5 min to quench the staining reaction. Pellet the splenocytes at 1500 rpm for 5 min at 4 °C followed by one wash with complete media.
4. Plate the cells at a density of 1×10^5–5×10^5 splenocytes in 100 μl in a 96 well flat bottom plate. Add appropriate antigens or mitogens in a 100 μl volume to the wells and incubate at 37 °C, 5 % CO_2 humidified incubator for 5 days.
5. At the end of the culture period, pellet the cells as above, wash twice with flow staining buffer.
6. Resuspend the cells in 200 μl of flow staining buffer and add CD3, CD4, and CD8 antibodies at concentrations specified in the respective product sheets. Stain for 30 min at 4 °C in the dark.
7. Pellet as above and wash two times with flow staining buffer and resuspend in 250 μl of flow staining buffer. Acquire data using an appropriate flow cytometer. Figure 4 depicts the workflow for T cell proliferation assay and shows the gating strategy for $CD8^+$ T cells. Histograms show examples of CFSE dye dilution in unstimulated and stimulated $CD8^+$ T cells.

4 Notes

1. Ensure that the Hamilton syringe and the needle are in good working condition prior to commencing the surgery. Only minimal adjustments should be done after positioning the animal.
2. It is imperative to apply Paralube ointment to the eyes prior to surgery to prevent the eyes from drying out.
3. Care should be taken that the positioning of the animal into the stereotactic frame does not cause any respiratory distress. The breathing rate of the animal should be monitored throughout the surgical procedure.
4. To prevent the heat from the lamps causing burn injuries to the animals, maintain the lamps at a sufficient distance from

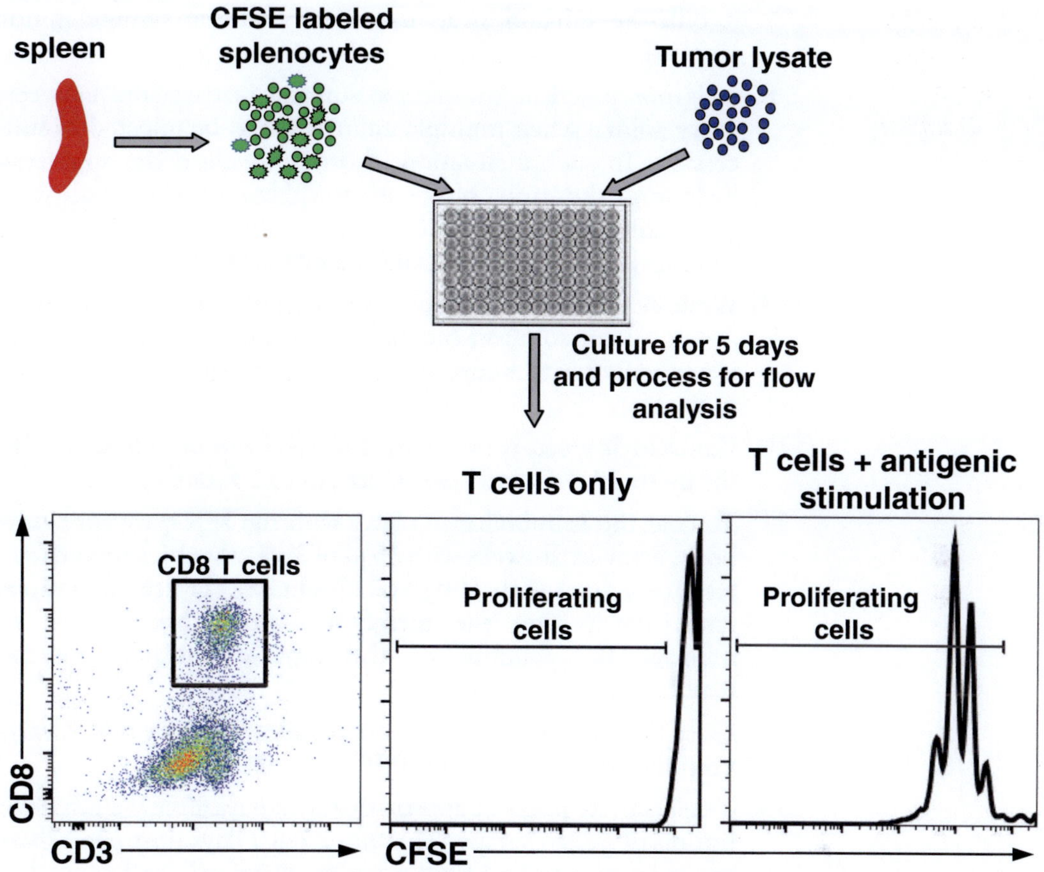

Fig. 4 Workflow for T cell proliferation assay: Purify splenocytes, remove RBCs, label with CFSE dye, and culture with tumor lysate followed by processing for flow cytometric analysis. Dot plots show gating strategy for CD8 T cells. Histograms show representative images of CFSE stains in unstimulated and stimulated CD8 T cells

the animals and only use them when using the microscope. At other times, the beam can be directed away from the animal.

5. Stop drilling when the base of the skull becomes translucent. Perforate the remaining thin layer of skull with a sterile needle and use a pair of curved forceps, carefully remove this layer of bone to expose the dura mater. The duramater (which is whitish, opalescent and elastic) and the brain surface (darker and yellowish in color) should be distinguished carefully to prevent lesioning of the brain surface with the needle.
6. Clear cerebrospinal fluid (CSF) often leaks into the burr hole when the skull is perforated. Soak up this CSF with an ethanol swab to maintain visibility of the injection site.
7. It is recommended to load a small amount of extra volume of cells or adenoviral vectors into the Hamilton syringe to expel

it onto an ethanol swab to ensure that the syringe is not clogged.

8. It is not practical to sterilize surgical instruments between every animal when multiple animals are to be injected in succession. In such a situation, thoroughly clean the apparatus including the drill bits and the Hamilton syringe with enzymatic cleaning solutions such as Endozime followed by rinsing with 70 % ethanol and finally sterile saline.
9. While cell suspension should not be kept on ice for more than 3–4 h, vector solution can be kept on ice for several hours. It is not advisable to freeze thaw the vectors which leads to a loss in titer.
10. Ganciclovir when reconstituted in H_2O can be stored at 4 °C for up to 14 days and −20 °C for up to 2 months.
11. To coat the Immobilon-P plates with the IFN-γ capture antibody, first wet the wells with 35 μl of 35 % ethanol. Immediately rinse two times with 100 μl of PBS before the ethanol evaporates. Pre-wetting the membrane with ethanol helps to improve the binding of the capture antibody to the membrane.
12. Once the plate has been wet, do not allow it to dry out. Always keep the wells wetted with PBS or media.
13. T cell activity peaks at approximately 7 days after the administration of gene therapy. Therefore 7–10 days post gene therapy is an appropriate time point to examine T cell activation induced by gene therapy.
14. Splenocytes can be frozen at this stage to continue the assay at a later time point. To freeze the cells, resuspend $50–70 \times 10^6$ cells in 1 ml of freezing media and store at −80 °C overnight. Then transfer to liquid nitrogen for long-term storage. Splenocytes when frozen carefully can be stored for at least 4–6 weeks before the assay is run.
15. Fill the empty wells with 200 μl of PBS or media during the 48 h incubation at 37 °C. This will prevent unequal evaporation from different areas of the plate and reduce edge effect.
16. Filter the TMB substrate through a 0.45 μm syringe filter prior to adding it to the plate for color development. The spot development using TMB substrate is rapid and should be carefully monitored to prevent merging of individual spots.
17. Proteins in FBS can inhibit CFSE dye uptake by the cells. Therefore wash the cells two times with 1× PBS to remove proteins prior to staining with CFSE.

Acknowledgements

This work was supported by National Institutes of Health/ National Institute of Neurological Disorders & Stroke (NIH/ NINDS) Grants U01-NS052465, U01-NS052465-S1, R01-NS074387, and R01-NS057711 to M.G.C.; NIH/NINDS Grants R01-NS054193, R01-NS061107, R01-NS082311, and R21-NS084275 to P.R.L.; the Department of Neurosurgery, University of Michigan School of Medicine; the Michigan Institute for Clinical and Health Research, NIH 2UL1-TR000433; University of Michigan Cancer Biology Training Grant, NIH/ NCI (National Cancer Institute) T32-CA009676; University of Michigan Training in Clinical and Basic Neuroscience, NIH/ NINDS T32-NS007222; and the University of Michigan Medical Scientist Training Program, NIH/NIGMS (National Institute of General Medicine Sciences) T32-GM007863. M.C. and M.A.M.A were supported by the Consejo Nacional de Ciencia y Tecnologia (CONICET PIP 114-201101-00353) and the Agencia Nacional de Promocion Cientifica y Tecnologica (PICT-2012-0830 and PICT-2013-0310).

References

1. Candolfi M et al (2007) Intracranial glioblastoma models in preclinical neuro-oncology: neuropathological characterization and tumor progression. J Neurooncol 85(2):133–148
2. Castro MG et al (2011) Gene therapy and targeted toxins for glioma. Curr Gene Ther 11(3):155–180
3. King GD et al (2005) Gene therapy and targeted toxins for glioma. Curr Gene Ther 5(6):535–557
4. Radaelli E et al (2009) Immunohistopathological and neuroimaging characterization of murine orthotopic xenograft models of glioblastoma multiforme recapitulating the most salient features of human disease. Histol Histopathol 24(7):879–891
5. Baker GJ et al (2014) Mechanisms of glioma formation: iterative perivascular glioma growth and invasion leads to tumor progression, VEGF-independent vascularization, and resistance to antiangiogenic therapy. Neoplasia 16(7):543–561
6. Assi H et al (2012) Gene therapy for brain tumors: basic developments and clinical implementation. Neurosci Lett 527(2):71–77
7. Finsen B, Owens T (2011) Innate immune responses in central nervous system inflammation. FEBS Lett 585(23):3806–3812
8. Fabry Z et al (2008) Sensing the microenvironment of the central nervous system: immune cells in the central nervous system and their pharmacological manipulation. Curr Opin Pharmacol 8(4):496–507
9. Ransohoff RM, Brown MA (2012) Innate immunity in the central nervous system. J Clin Invest 122(4):1164–1171
10. Ali S et al (2005) Combined immunostimulation and conditional cytotoxic gene therapy provide long-term survival in a large glioma model. Cancer Res 65(16):7194–7204
11. Curtin JF et al (2005) Combining cytotoxic and immune-mediated gene therapy to treat brain tumors. Curr Top Med Chem 5(12):1151–1170
12. Ghulam Muhammad AK et al (2009) Antiglioma immunological memory in response to conditional cytotoxic/immune-stimulatory gene therapy: humoral and cellular immunity lead to tumor regression. Clin Cancer Res 15(19):6113–6127
13. Curtin JF et al (2009) HMGB1 mediates endogenous TLR2 activation and brain tumor regression. PLoS Med 6(1):e10
14. Akli S et al (1993) Transfer of a foreign gene into the brain using adenovirus vectors. Nat Genet 3(3):224–228

15. Le Gal La Salle G et al (1993) An adenovirus vector for gene transfer into neurons and glia in the brain. Science 259(5097):988–990
16. Southgate T et al (2008) Gene transfer into neural cells in vitro using adenoviral vectors. Curr Protoc Neurosci. Chapter 4: p. Unit 4 23
17. Barcia C et al (2006) Immunological thresholds in neurological gene therapy: highly efficient elimination of transduced cells might be related to the specific formation of immunological synapses between T cells and virus-infected brain cells. Neuron Glia Biol 2(4):309–322
18. Ng P, Graham FL (2002) Construction of first-generation adenoviral vectors. Methods Mol Med 69:389–414
19. Bett AJ et al (1994) An efficient and flexible system for construction of adenovirus vectors with insertions or deletions in early regions 1 and 3. Proc Natl Acad Sci U S A 91(19):8802–8806
20. Thomas CE et al (2001) Acute direct adenoviral vector cytotoxicity and chronic, but not acute, inflammatory responses correlate with decreased vector-mediated transgene expression in the brain. Mol Ther 3(1):36–46
21. Kreppel F et al (2002) Long-term transgene expression in the RPE after gene transfer with a high-capacity adenoviral vector. Invest Ophthalmol Vis Sci 43(6):1965–1970
22. King GD et al (2008) High-capacity adenovirus vector-mediated anti-glioma gene therapy in the presence of systemic antiadenovirus immunity. J Virol 82(9):4680–4684
23. Puntel M et al (2010) Gene transfer into rat brain using adenoviral vectors. Curr Protoc Neurosci. Chapter 4: p. Unit 4 24
24. Glover CP et al (2003) Long-term transgene expression can be mediated in the brain by adenoviral vectors when powerful neuron-specific promoters are used. J Gene Med 5(7):554–559
25. Candolfi M et al (2011) B cells are critical to T-cell-mediated antitumor immunity induced by a combined immune-stimulatory/conditionally cytotoxic therapy for glioblastoma. Neoplasia 13(10):947–960
26. Candolfi M et al (2014) Temozolomide does not impair gene therapy-mediated antitumor immunity in syngeneic brain tumor models. Clin Cancer Res 20(6):1555–1565
27. King GD et al (2011) Combined Flt3L/TK gene therapy induces immunological surveillance which mediates an immune response against a surrogate brain tumor neoantigen. Mol Ther 19(10):1793–1801
28. Mineharu Y et al (2014) Blockade of mTOR signaling via rapamycin combined with immunotherapy augments antiglioma cytotoxic and memory T-cell functions. Mol Cancer Ther 13(12):3024–3036
29. Mineharu Y et al (2011) Engineering the brain tumor microenvironment enhances the efficacy of dendritic cell vaccination: implications for clinical trial design. Clin Cancer Res 17(14):4705–4718

Part VIII

Clinical Trials

Chapter 32

AAV2-Neurturin for Parkinson's Disease: What Lessons Have We Learned?

Jeffrey H. Kordower

Abstract

The dream that trophic factors could be effectively delivered and potently forestall and reverse the symptoms of Parkinson's disease (PD) has yet to be realized. Research in this area has been active for 20 years, but after much work, the prospects for utilizing trophic factors in the treatment of PD are currently dim. Millions of dollars have been spent, numerous academic, foundation, and government resources have been invested, and hundreds of patient research volunteers have contributed their time and hope to this effort without a therapeutic breakthrough. As a scientist who has journeyed these events from the beginning and participated in many of the decisions that navigated this field, I consider it important for the movement disorder scientific community to reflect on the evolution of thought and to participate in the dialog over whether the investments were worthwhile.

The most studied group of trophic factor for PD is the glial cell derived family of ligands, of which glial cell derived neurotrophic factor (GDNF) and neurturin are members, and are the best studied. I trace the development of these factors chronologically with commentary on the key decision-making points. Before we collectively invest further, I offer this scientific reflection on the past and offer my own view on the next steps of research in the field of neurotrophins as potential therapeutic agents in PD.

Key words AAV, Clinical trial, Parkinson's disease, Neurturin, GDNF, Nigrostriatal system

1 Introduction

GDNF was first characterized by Lin and coworkers in 1993 [1] as demonstrated to provide potent trophic effects on dopaminergic midbrain neurons in culture. Since that initial observation, GDNF has been found to protect and/or augment dopaminergic nigrostriatal function in normal animals, aged animals, and virtually every toxin animal model in multiple species [2]. These studies lead to open label and blinded clinical trials testing initially the safety and tolerability and then efficacy of intraventricular and then intraputamenal GDNF protein infusions [3–5]. The intraventricular studies were based upon weak preclinical data and were found to not only be nonefficacious but also to have serious side effects. As an aside, given their potency and potentially widespread effects, putting any

Fredric P. Manfredsson (ed.), *Gene Therapy for Neurological Disorders: Methods and Protocols*, Methods in Molecular Biology, vol. 1382, DOI 10.1007/978-1-4939-3271-9_32, © Springer Science+Business Media New York 2016

trophic factor into the ventricular system is probably not a good idea [3, 6]. The rationale for intraputamenal delivery is much stronger. However, this clinical trial studies failed to demonstrate efficacy and was stopped prematurely due to safety concerns although it is likely that most these safety concerns were not very serious [7]. At the time these trials were being planned and ongoing, Ceregene inc. attempted unsuccessfully to license the gene for GDNF from Amgen Inc., the stakeholder for GDNF, for a gene therapy trial testing whether viral vector delivery of GDNF would be safe and efficacious in patients with PD. Since there was no reasonable path to the clinic without a license to operate with GDNF, Ceregene approached Gene Johnson, Jeffrey Milbrandt, and Washington University and successfully licensed another member of the GFL family of ligands, neurturin. While my lab and many other demonstrated strong efficacy in rodent and nonhuman primate models with GDNF [2], there was far less data at this time testing neurturin with regard to distribution by gene delivery, efficacy in these same models, as well as safety and tolerability. Thus it was back to the drawing board and Ceregene's inability to secure a license for GDNF put the field back a number of years.

Ceregene performed numerous experiments and demonstrated that neurturin was equipotent to GDNF in a number of rodent PD models [8, 9]. The distribution of neurturin was less than that of GDNF [9] although differences in the sensitivity between the GDNF and neurturin antibodies likely exaggerated GDNF's superiority in distribution. Both were extremely well tolerated. With regard to efficacy, CERE-120 (the name for gene delivery of neurturin) increased dopamine markers in young rats, aged rats, 6-hydroxydopamine lesioned rats, normal monkeys, aged monkeys, and monkeys intoxicated with MPTP. The preclinical package for both efficacy and safety was extremely strong and Ceregene initiated a Phase I clinical trial in 2004 [10]. This trial consisted of 12 patients with half receiving a low dose (0.3×10^{11} vector genomes (vg)/patient) and half receiving a high dose (5.4×10^{11} vg/patient). The trial lasted 1 year post-treatment. Safety and tolerability was excellent as there were no serious adverse events due to the test article [10]. In the manuscript describing the open label trial, we reported that measures of motor function showed improvement at 1 year; for example, a mean improvement in the off-medication motor subscore of the Unified Parkinson's Disease Rating Scale (UPDRS) of 14 points (SD 8; $p = 0.000121$ [36 % mean increase; $p = 0.000123$]) and a mean increase of 2.3 h (2; 25 % group mean increase; $p = 0.0250$) in on time without troublesome dyskinesia were seen. Interestingly, the expected increased in fluorodopa positron emission tomography was not observed. This was an open label trial that was not powered for efficacy. Open labeled trials, especially for PD, have a high potential for placebo effects and experimenter bias [11] making the efficacy data virtually meaningless. Indeed, the 36 % improvement in UPDRS is clearly within the range of placebo yet, and likely appropriately so, Ceregene continued on. But should we have continued on? At the time we had a signal, but the signal was not very strong. In an efficacy study, when the signal is not strong in an open label analysis it often does not pass muster when submitted to the more rigorous double blind assessment. In addition, perhaps the failure of CERE-120 to enhance fluorodopa uptake should have been a second red flag. If our hypothesis was that CERE-120 would enhance nigrostriatal function, we failed to do so on the one measure, albeit indirect, that we had. However, the fiscal realities of the biotech world demanded that we go forward. As a member of the scientific advisory board for Ceregene, I, as well as all my colleagues, fully supported that decision.

Following the success (safety and tolerability) of the Phase I trial, Ceregene initiated a Phase II efficacy trial. Fifty-eight PD patients were randomized to CERE-120 and placebo in a 2:1 ratio. Multiple injections into the putamen we made bilaterally using higher dose (5.4×10^{11} vg/patient) from the Phase I trial was employed. The primary endpoint was change from baseline to 12 months in the motor subscore of the unified Parkinson's disease rating scale in the practically defined off state. The trial failed to meet this primary endpoint. So was this the end? No because another "signal" of efficacy was observed. When the trial ended 12 months after the last patient received surgery, most patients had been followed in a blinded fashion at 15 and 18 months. An ad hoc statistical analysis suggested that these points were significant at the 0.025 level. Multiple secondary analyses were significantly improved at 18 months relative to 12 months. However, these analyses were not corrected for multiple comparisons and thus do not have the appropriate scientific strength or rigor. However, they were still strong enough to for Ceregene to initiate a third clinical trial.

About the time this trial ended, Ceregene and their scientific advisory board began to question the status of the nigrostriatal system at the time the patients received CERE-120. Inclusion criterion was a good response to levodopa in the opinion of the treating investigator which had to be confirmed by a committee of noted PD experts. In addition, levodopa-induced motor complications could not be satisfactorily controlled with medical therapy in these subjects; at least 2 h per day of off time (poor motor function) according to home diaries; a score of at least 30 on the motor section (part 3) of the Unified Parkinson's disease rating scale (UPDRS) in the off state (range 0–108, with higher scores suggesting more severe disease). We questioned whether there was sufficient neuroanatomical substrate at this time in their disease for neurturin to bind to a cognate receptor, be retrogradely transported to the nigral perikarya, and initiate the expression of a series of survival and phenotype genes that hopefully would improve cardinal symptoms. The literature had a few papers examining nigral neuronal number [12]. However, many of these studies were performed using outdated counting techniques. More importantly, virtually no papers existed examining striatal dopaminergic markers in the striatum. This is critical since striatal dopamine insufficiency, not loss of nigral neurons, is the key pathology underlying the cardinal symptoms of PD. We [13] examined the putamen from 28 patients with PD with disease durations of 1–27 years and compared them to age-matched controls ($n=9$). The findings were striking. By 4 years post-diagnosis there was virtually no visible tyrosine hydroxylase or dopamine transporter-immunoreactive fibers in the putamen in PD. The only remaining fibers were found in the ventral aspect of the putamen or coursing along the external medullary lamina of the external globus pallidus en route to the caudate nucleus. It is impossible to know whether these fibers are lost or just downregulated their dopaminergic phenotype to an undetectable level. Still these data were somewhat disheartening. The degeneration of dopaminergic fibers in the putamen was virtually complete and also virtually completed by 4 years disease duration. This was earlier than most of the patients who were in the trial.

About this time, two patients from the Phase I trial died from events unrelated to the surgery. One died 6 weeks following surgery and one died 3 months following surgery. These unique cases were quite informative. The first observation was that very little (about 15 %) of the putamen expressed gene delivered neurturin. To be fair, this is likely an underestimate for a number of reasons; (1) preclinical studies examine neurturin distribution in perfused animals and do not have to overcome the postmortem interval that occurs in human cases. (2) The neurturin antibody is notoriously poor and requires aggressive antigen retrieval to get any staining. Still 15 % of measurable neurturin-immunoreactivity is low. Even so, we were able to detect an extremely small area of dopaminergic sprouting within the halo of neurturin staining. This suggests that at least a little neurturin can make it back to the nigra and induce regeneration.

Based on the fact that we could induce some sprouting combined with the "clinical signal" seen at 18 months (not correcting for multiple comparisons) from the Phase II trial, another Phase II trial was performed. Changes were made in an attempt to enhance distribution. In this study, the volume of CERE-120 injected into the putamen on each side was increased from 40 to 120 μl. Secondly, an injection was made on each side directly into the nigra. The rationale behind the nigral injections was that since most of the fibers were already lost or dysfunctional, we would try to jumpstart the nigrostriatal system from the cell body; a very tough challenge. This trial failed as well.

So why do such powerful treatments work so well preclinically yet fail in clinical trials. One can easily make the argument that the delivery methods in the neurturin trials were inadequate. However, other factors need to be recognized. Preclinical scientists often design studies to ensure that they work rather than accurately mimic the disease they are studying. There are numerous studies in which gene delivery of a therapeutic occurs *prior* to the nigrostriatal lesion. In 2000, I tested lentiviral delivery of GDNF in aged and parkinsonian monkeys [14]. Robust augmentation of the nigrostriatal system and neuroprotection was observed. However, the gene delivery was only made 7 days after the MPTP lesion. This was due to trying to optimize the success of the study to work and in doing so, we poorly modeled what the nigrostriatal system looks like at the time of surgery (even though that data was not yet available). In this regard, the study was designed to initiate the lesion first, but we still intervened at a time in which there were substantial fibers remaining and, as discussed above, that does not accurately model the patients that received the gene delivery of GFL's. In addition, the choice of model may be critical. As mentioned earlier, GDNF and neurturin work in virtually every PD toxin model. However, PD is a synucleinopathy and GDNF (and presumably neurturin) does not provide protection in the alpha synuclein viral overexpression model due to down regulation of the ret receptor and the transcription factor nurr1 [15, 16].

2 Conclusions

Trophic factors remain as potent molecules that can alter the biology of the systems they target. However, there remain important variables that have yet to be successfully addressed. The potency of a molecule is negated if it is not appropriately distributed to the areas for which it is needed. Preclinical models need to be applied with a better understanding of the diseases for which they are modeling. We also need to better appreciate the stage of disease for which we are intervening. Once there is a better handle on these and other relevant issues, then these potent molecules can get the test they deserve.

References

1. Lin LF, Doherty DH, Lile JD, Bektesh S, Collins F (1993) GDNF: a glial cell line-derived neurotrophic factor for midbrain dopaminergic neurons. Science 260(5111):1130–1132
2. Kordower JH, Bjorklund A (2013) Trophic factor gene therapy for Parkinson's disease. Mov Disord 28(1):96–109
3. Kordower JH, Palfi S, Chen EY, Ma SY, Sendera T, Cochran EJ, Cochran EJ, Mufson EJ, Penn R, Goetz CG, Comella CD (1999) Clinicopathological findings following intraventricular glial-derived neurotrophic factor treatment in a patient with Parkinson's disease. Ann Neurol 46(3):419–424

4. Nutt JG, Burchiel KJ, Comella CL, Jankovic J, Lang AE, Laws ER Jr, Lozano AM, Penn RD, Simpson RK Jr, Stacy M, Wooten GF, ICV GDNF Study Group (2003) Randomized, double-blind trial of glial cell line-derived neurotrophic factor (GDNF) in PD. Neurology 60(1):69–73
5. Lang AE, Gill S, Patel NK, Lozano A, Nutt JG, Penn R, Brooks DJ, Hotton G, Moro E, Heywood P, Brodsky MA, Burchiel K, Kelly P, Dalvi A, Scott B, Stacy M, Turner D, Wooten VG, Elias WJ, Laws ER, Dhawan V, Stoessl AJ, Matcham J, Coffey RJ, Traub M (2006) Randomized controlled trial of intraputamenal glial cell line-derived neurotrophic factor infusion in Parkinson disease. Ann Neurol 59(3):459–466
6. Eriksdotter Jönhagen M, Nordberg A, Amberla K, Bäckman L, Ebendal T, Meyerson B, Olson L, Seiger, Shigeta M, Theodorsson E, Viitanen M, Winblad B, Wahlund LO (1998) Intracerebroventricular infusion of nerve growth factor in three patients with Alzheimer's disease. Dement Geriatr Cogn Disord 9(5):246–257
7. Sherer TB, Fiske BK, Svendsen CN, Lang AE, Langston JW (2006) Crossroads in GDNF therapy for Parkinson's disease. Mov Disord 21(2):136–141
8. Gasmi M, Herzog CD, Brandon EP, Cunningham JJ, Ramirez GA, Ketchum ET, Bartus RT (2007) Striatal delivery of neurturin by CERE-120, an AAV2 vector for the treatment of dopaminergic neuron degeneration in Parkinson's disease. Mol Ther 15(1):62–68
9. Bartus RT, Baumann TL, Brown L, Kruegel BR, Ostrove JM, Herzog CD (2013) Advancing neurotrophic factors as treatments for age-related neurodegenerative diseases: developing and demonstrating "clinical proof-of-concept" for AAV-neurturin (CERE-120) in Parkinson's disease. Neurobiol Aging 34(1):35–61
10. Marks WJ Jr, Ostrem JL, Verhagen L, Starr PA, Larson PS, Bakay RA, Taylor R, Cahn-Weiner DA, Stoessl AJ, Olanow CW, Bartus RT (2008) Safety and tolerability of intraputaminal delivery of CERE-120 (adeno-associated virus serotype 2-neurturin) to patients with idiopathic Parkinson's disease: an open-label, phase I trial. Lancet Neurol 7(5):400–408
11. Goetz CG, Wuu J, McDermott MP, Adler CH, Fahn S, Freed CR, Hauser RA, Olanow WC, Shoulson I, Tandon PK, Parkinson Study Group, Leurgans S (2008) Placebo response in Parkinson's disease: comparisons among 11 trials covering medical and surgical interventions. Mov Disord 23(5):690–699
12. Fearnley JM, Lees AJ (1991) Ageing and Parkinson's disease: substantia nigra regional selectivity. Brain 114(Pt 5):2283–2301
13. Kordower JH, Olanow CW, Dodiya HB, Chu Y, Beach TG, Adler CH, Halliday GM, Bartus RT (2013) Disease duration and the integrity of the nigrostriatal system in Parkinson's disease. Brain 136(Pt 8):2419–2431
14. Kordower JH, Emborg ME, Bloch J, Ma SY, Chu Y, Leventhal L, McBride J, Chen EY, Palfi S, Roitberg BZ, Brown WD, Holden JE, Pyzalski R, Taylor MD, Carvey P, Ling Z, Trono D, Hantraye P, Déglon N, Aebischer P (2000) Neurodegeneration prevented by lentiviral vector delivery of GDNF in primate models of Parkinson's disease. Science 290(5492):767–773
15. Lo Bianco C, Déglon N, Pralong W, Aebischer P (2004) Lentiviral nigral delivery of GDNF does not prevent neurodegeneration in a genetic rat model of Parkinson's disease. Neurobiol Dis 17(2):283–289
16. Decressac M, Ulusoy A, Mattsson B, Georgievska B, Romero-Ramos M, Kirik D, Björklund A (2011) GDNF fails to exert neuroprotection in a rat α-synuclein model of Parkinson's disease. Brain 134(Pt 8):2302–2311

Index

A

B

C

D

E

Fredric P. Manfredsson (ed.), *Gene Therapy for Neurological Disorders: Methods and Protocols*, Methods in Molecular Biology, vol. 1382, DOI 10.1007/978-1-4939-3271-9, © Springer Science+Business Media New York 2016

P

R

S

T

V

Z

MIX
Papier aus verantwortungsvollen Quellen
Paper from responsible sources
FSC® C105338

If you have any concerns about our products,
you can contact us on
ProductSafety@springernature.com

In case Publisher is established outside the EU,
the EU authorized representative is:
Springer Nature Customer Service Center GmbH
Europaplatz 3, 69115 Heidelberg, Germany

Printed by Libri Plureos GmbH
in Hamburg, Germany